Structural Mineralogy

An Introduction

Solid Earth Sciences Library

Volume 7

Structural Mineralogy

An Introduction

by

J. LIMA-DE-FARIA

Centro de Cristalografia e Mineralogia,
Instituto de Investigação Cientifica Tropical, Lisbon, Portugal

Kluwer Academic Publishers

Dordrecht / Boston / London

Library of Congress Cataloging-in-Publication Data

Lima-de-Faria, J. (José), 1925-
Structural mineralogy : an introduction / by J. Lima-de-Faria.
p. cm. -- (Solid earth sciences library ; v. 7)
Includes bibliographical references and indexes.
ISBN 0-7923-2821-3
1. Minerals--Classification. 2. Crystal structure. I. Title.
II. Series: Solid earth sciences library ; SESL 7.
QE388.L56 1994
549'.12--dc20 94-10374

ISBN 0-7923-2821-3

Published by Kluwer Academic Publishers,
P.O. Box 17, 3300 AA Dordrecht, The Netherlands.

Kluwer Academic Publishers incorporates
the publishing programmes of
D. Reidel, Martinus Nijhoff, Dr W Junk and MTP Press.

Sold and distributed in the U.S.A. and Canada
by Kluwer Academic Publishers,
101 Philip Drive, Norwell, MA 02061, U.S.A.

In all other countries, sold and distributed
by Kluwer Academic Publishers Group,
P.O. Box 322, 3300 AH Dordrecht, The Netherlands.

Printed on acid-free paper

Cover illustration: Olivine crystal structure, (Mg, Fe)^{o} $Si^{t}[O_4]^{h}$, *(condensed model representation).*

Printed in the Netherlands

To Natacha and Arnaldo Silvério

In spite of the variety of the mineral crystalline world, the whole 'mineralogical game' just reduces to various modes of filling gaps in uniform close packing with various corresponding patterns.

N.V. Belov [a]

The greatest pleasure in the study of science is understanding.

Leopold Infeld [b]

[a] From 'Structure of Ionic Crystals and Metallic Phases' (1947). Izd. Akad. Nauk SSSR. Moscow.

[b] From 'Albert Einstein. His work and its influences on our world' (1950). Scribner Publishers, New York.

Contents

Foreword

Recently, many fine textbooks in mineralogy have appeared. The great tradition of mineral science continued for over 100 years in *Dana's* systems, textbooks, and manuals, replete with discussion on crystallographic characters, and short statements on the mineral species as then known. The more recent superb Ramdohr/Strunz *Klockmann's Lehrbuch der Mineralogie* is a mine of rich crystal-chemical information, perhaps the finest contemporary pedagogic book of its kind. Within the past three years, a new and ambitious project – the *Handbook of Mineralogy* by several keen mineralogists – receives much assistance from up-to-date computer technology and promises to be a very fine series. The little *Mineral Reference Manual* by Nickel and Nichols is a handy means of quick access to the original literature on species and reference to crystal structure determination. Do we really need another book? For the *Structural Mineralogy* of José Lima-de-Faria, I say indeed we do!

Mineralogy is one of the natural sciences, and about the oldest one at that. It antedated inorganic chemistry, crystallography, and so much geology. Mineralogy is at the scientific root of mining, prospecting, mineral dressing, and a host of other utilities stimulated by our science. Minerals were practically the first crystalline materials to be investigated after the demonstration by von Laue and the Braggs of X-ray diffraction by crystals in 1912. The phenomenal growth of crystallography as a workhorse for structure and crystal analysis, from Patterson's *Faltung* to Hauptmann and Karle's *direct methods* through probabilities of structure semi-variants are in many respects the crowning tiara of 20th century science. After all, without Watson and Crick's unveiling of elusive DNA, where would molecular biology and biochemistry stand today? Dr. Linus Pauling, arguably the greatest thinker in chemistry of the century, made his most notable early achievements in unravelling the crystal structures of minerals. I would include Zachariasen and Belov as well, two other giants in crystal structure analysis. How many contributed in their own way to our science? Ten? Fifty? One hundred? It depends on how you weight the count, but I suspect it is closest to the biggest number just mentioned.

The explosive growth in information (Ger.: *Fach*) on crystal structures began about 1950, when the big computers and automated diffractometers were just emerging. It contributes mightily to our science (Ger.: *Naturwissenschaft*). Unfortunately, technological knowledge seems to come easily now, and *Fach* appears to be outstripping *Naturwissenschaft*, a sorry state of affairs. Most practicing mineralogists today are *Fachidioten* and are hardly *Naturwissenschafters*. The science has become tedious; it no longer dances with nimble feet as Nietzsche instructed us. Mineralogists don't know their minerals. Mineralogists don't go into the field anymore. Crystal structures, the products of blood, sweat, and tears at the diffractometers, computers, and pocket calculators are slipping into a grey gloom, abandoned orphans of science. What do we do with all this information? Is structural knowledge hortative, beyond mere confirmation or determination of a correct formula, or a challenge to an old one?

It seems to me that mineralogy today is far from a dead science, although most mineralogists may fall into that category. These are really exciting times! With so much gruelling and repetetitive data already accumulated, and much structure solving and information collating out of the way, it seems that mineralogy is about to blossom again. This is in the greatest and most ambitious intellectual exercise of them all: the link between crystal structure and mineral paragenesis. Certain more serious thinkers in mineralogy have felt, especially since WW II, that structure types just don't pop up out of nowhere like mushrooms, but that structure type is a kind of Rosetta Stone to a mineral's genesis. The famous kyanite-andalusite-sillimanite trimorphs serve this point well, but isn't it a rather sparse example of such a structure-paragenesis link?! After all, we have about 2000 structure types to explain and understand more fully.

One of the big hindrances in understanding structure is found in structure *representation*. Most structure representations in the literature resist the reader's perception. Several projections should really be shown as structures, like plumbing assemblies, have joints which relate to other structures in different directions. Of course, closest-packing with at least one h (= ABA) in the stacking sequence has just one principal direction. In such a structure, once t^* (the vector normal to the c.-p. layers) is defined, layer-by-layer sequences within the asymmetric unit of translation can be shown. Some of these may be building blocks for other structures. Often, each layer tells a little story by itself, and it may reveal a 'secret' to the reader. If t^* falls along general [*hkl*], then it is the task of the investigator to transform the cell into a new more reducible representation, but one that opens up the window of closest-packing. Fractional coordinates are transformed according to the transpose of the inverse of the adjoint of the transformation matrix, that is $\mathbf{A'} = {}^{T}(\mathrm{Adj}\ \mathbf{A})$, where $\mathbf{A}$ is the transformation matrix from the old cell to the new (pseudo)-orthogonal cell with one axis parallel to t^*. To make it easy, transform the inversion centers according to the same matrix. Projection along t^* layer-by-layer through the asymmetric unit then finishes the job. Surprisingly, this approach has had very few practitioners in the past, thus most structures which are in fact based on principles of that fundamental law of nature, closest-packing, are misrepresented!

José Lima-de-Faria prefers to cast the closest-packed representation as a sphere packing model. I find that rather hard to perceive, as I happen to be a polyhedrist by inclination. Because the closest-packed representation is the real goal of the study, I would even submit a table listing the transformation matrices, the old and new cell, the old and new transformed coordinates, the mean layer separation h, and *difference*, $\Delta\AA$, or the coordinate differences (expressed as 'interatomic' distance) between real and ideal (calculable) perfect closest-packed representation. Such a table is akin to the table of structure factors so familiar to all crystallographers, a certificate of sorts to see whether or not one is cheating or just misinterpreting the information at hand. Such a transformed cell also makes listing of shared polyhedral edges much easier to find, visualize, and enumerate.

Be that as it may, José's book covers the problem of structure description, utility, classification, and representation quite well, and it tries to place some guidelines toward more uniformity in structure descriptions, most of which in the literature are really quite shoddy.

Structural Mineralogy is really not an end in itself. Rather, it is a rallying call to urge further clarification, representation, and systematization of already known structures. Committees cannot delegate such notions as ephemeral and individual as structure description. Much more airing, many more new ways at looking at old crystal structures are required. But José's book we can admire in common, criticize in common, comprehend in common.

Chicago
5 February, 1994

Paul B. Moore

Preface

The classification of animals and plants has not changed much since Linnaeus' proposal (1735), because, at the time, its detailed study was already possible, particularly with the help of the microscope, and consequently a natural classification based on their internal structures could be established. On the contrary, the criterion for the classification of minerals has changed (from practical purposes, to physical properties and to chemical composition) following the development of the mineralogical science. *These changes were always a step further in the direction of the internal structure of minerals.* Only after the first determination of a crystal structure was carried out (Bragg, 1913) was it possible to reach the internal structure of minerals. Since then most of the mineral structures have been studied, and time is now ripe to develop a natural classification by replacing the classical chemical classification by a structural classification of minerals.

Today it appears obvious that the mineral classification had to pass through several stages until it reached the internal structure of minerals. However, *what appears obvious but is against the mental habits of scientists takes normally a long time to become well understood and widely accepted.* New ideas always develop strong reactions against them, and the history of science is full of such examples. It seems that many scientists do not learn much from the history of science – that there exists a kind of cleavage between the latter and their scientific activity – and that they do not use history to enlighten or guide their scientific research. Their negative reactions appear as intrinsic to the process of scientific development; before adopting a new approach, they try to preserve what has shown to be fruitful for so many years.

The author (1965a) proposed the elaboration of a general table ordering the inorganic structures and, with the collaboration of Figueiredo, presented a general table and a structural classification of inorganic structure types (Lima-de-Faria & Figueiredo, 1976). Minerals are an important part of the inorganic domain, and it would seem evident that this classification could be applied to the mineral kingdom. However, it took several years for the author to realize that the general classification of inorganic structures would have important implications on the classification of minerals. It was through the preparation of the 'Historical Atlas of Crystallography' (Lima-de-Faria, Buerger, Glusker, Megaw, Moore, Senechal & Wooster, 1990) that an intimate contact with the history of mineralogy – the mother of crystallography – has clearly shown the changing aspect of the mineral classifications, and the difficulties in replacing them by more appropriate ones. The big fight against the chemical classification of minerals first proposed by Cronstedt in 1758 lasted for more than one hundred years and very much impressed the author. One of the arguments against the chemical approach was the fact that only a few chemical compositions of minerals were known at Cronstedt time, and therefore the chemical criterion could not be applied generally; on the contrary, the physical properties could be determined for all minerals. Only when chemistry reached certain maturity, that is, when chemical formulas were known for most of the minerals, could the chemical approach be applied with success, and it was only with Dana (1850, 1854) and Groth (1874), that the chemical classification gradually became accepted.

The same would have happened with a structural approach to the classification of minerals, had it been proposed at the beginning of crystal structure determination. The strong argument against it would be its lack of generality, the structural criterion being only applicable to a small number of minerals. But now, that so many mineral structures are known, this negative argument does not stand any more. A clear parallelism exists between the replacement of the physical by the chemical classification and the proposed replacement of the chemical by the structural classification. Dana said in his 'System of Mineralogy', third edition (1850),

> [...] chemistry has opened to us a better knowledge of the nature and relation of compounds; and philosophy has thrown new light on the principles of classification. To change is always seeming fickleness. But not to change with the advance of science is worse; it is persistence in error [...]

A similar statement could now be applied to the structural classification. However, with the present knowledge of the structural characteristics of minerals and the development of the history of mineralogy, it may well be that the acceptance of the structural classification of minerals, instead of taking one hundred will take only a few years. Let us hope for the best.

This book is mainly concerned with the understanding of the mineral structures and the structural classification of minerals, and will not treat in detail the other several items of its contents. It is really an introduction to the subject.

Lisbon
February 1994

José Lima-de-Faria

Acknowledgements

I want to express my profound gratitude to the colleagues who encouraged the elaboration of this book, in particular to A. Preisinger, S. Hafner, P.B. Moore, S. Menchetti, S. Ghose and S. Merlino.

I am much indebted to my family for moral and material support. My children, Manuel and Margarida, contributed to the offer of a personal computer, which skillfully used by my wife, Natacha, provided the complete typing of the text. This was really a fundamental help, which enabled me to improve the work step by step.

My friend Arnaldo Silvério made himself available to peruse the manuscript. He checked the English and contributed comments that improved the text intelligibility. I thank him most heartily.

Paul Moore besides having accepted to write the forward has kindly corrected the whole manuscript, which was really a very hard job requiring much attention and patience. I am most indebted to him for such an act of generous friendship.

Many thanks are due to my colleague Maria Ondina Figueiredo who put at my disposal some condensed models, in particular those of the sheet silicates, and who was very closely associated with the early stages of my research work, namely the theory of condensed models and the structural classification of inorganic crystal structures.

The administrative officer Francisco Raposo gave an important contribution to the typing of the systematic part, with patience and great sense of correctness, and the technician Germano Bernardino helped in the elaboration of many condensed models. My friend Rui Paula assisted in the design of the book cover. To all of them my sincere thanks.

Grateful acknowledgement is made of the support and facilities afforded by the *Instituto de Investigação Científica Tropical,* in particular by its President, Professor Joaquim Cruz e Silva, and of the financial support and encouragement provided by the Gulbenkian Foundation during so many years.

I had to ask permission to several authors and publishers for the reproduction of many figures. Among these authors I am particularly grateful to T. Zoltai who kindly put at my disposal his notes on close-packed structures, and to F. Liebau, I. Kostov, H. Strunz, and F.D. Bloss for the kind permission to use some figures of their books. I gratefully acknowledge the following publishers: Plenum Press, Macmillan Publishing Company, John Wiley & Sons, Oliver and Boyd Copyright Protection Agency, Ferdinand Enke Verlag, Springer-Verlag, Academic Press, Gordon and Breach, Oldbourne Press, Keter Publishing House Ltd, McGraw-Hill Publishing Company, International Union of Crystallography, Zeitschrift für Kristallographie, Oosthoek Scheltema & Holkema Publishing Co., Mineralogical Society of America, American Mineralogist, Harper Collins Publishers, Longman Group UK, Mineralogical Association of Canada, Oxford University Press, K. Sutter Parthé, Akademische Verlagsgesellschaft G. & P.K.G., Mineralogical Society (London), Pergamon Press, Cambridge University Press, Saunders College Publishing (Rinehart and Winston), University Chicago Press, Cornell University Press, American Institute of Physics, Consultants Bureau, Société Française de Mineralogie at Cristallographie, The American Society for Metals, Reinhold Publishing Corporation, Chapman & Hall, Nature, University of California Press, Kluwer Academic Publishers, Tschermaks Miner. Petr. Mitteilungen, and United States Atomic Energy Commission.

Finally, thanks are due to the Acquisition Editor of the Kluwer Academic Publishers, Mrs Petra van Steenbergen, for her immediate interest in publishing this book, and her kind and effective support during all the printing process.

José Lima-de-Faria

CHAPTER 1

The structural classification of minerals

1.1. A very brief history of mineralogical classifications

The classification of minerals has changed throughout the ages, the criterion of classification following the development of the mineralogical science. In ancient times, the classification of minerals was mainly based on their practical purposes. According to Theophrastus (372–287 B.C.), and Plinius (77 A.D.), minerals were classified as gemstones, ores, pigments, etc. In the Middle Ages, Geber (Jabir Ibn Hayyân, 721-c.803), proposed a classification based on the external characteristics and some physical properties of minerals such as hardness, fusibility, malleability and fracture; this physical classification was developed later by Avicenna (Ibn Siná, 980–1037), and Agricola (1546). With Werner (1774) the physical classification attained its maturity and was generally adopted at the end of the XVIII century.

Cronstedt (1758) seems to be the first to have outlined a classification whereby the chemical properties were taken first, followed by the physical properties. This chemical classification was referred to the chemical elements and their compounds. Berzelius (1819) improved the chemical classification of minerals by considering chemical radicals as the main factor, instead of the elements. He ordered minerals as chlorides, sulphates, silicates, etc., and not as minerals of zinc, of copper, etc., as had been done before. The main criticism to the chemical classification, at that time, was that it could not be applied to all minerals, because the chemical composition of many of them was not known. For this reason, the chemical classification was strongly opposed during approximately one century, and only with the advance of chemistry could it be adopted. Even Werner (1817), who strongly defended the physical classification, modified the physical classification at the end of his career, and elaborated a physical-plus-chemical classification. The chemical classification was only widely accepted at the middle of the XIX century, due primarily to the effort of James Dwight Dana (1850, 1854) and of Groth (1874). (More details in the history of mineralogical classification may be gained from Povarennykh (1972), pp. 3–26.)

After 1913, when the first structures of minerals were determined, a structural criterion for classification started to be considered. Fedorov (1913) wrote:

> Only very recently has the principle of crystallochemical analysis become capable of leading to a classification in which every mineral has a strictly defined place. Such a classification cannot be called artificial, since its basis is the crystal structure of the mineral, i.e., that which essentially characterizes the nature of each.

Already in 1837, Whewell insisted on the structural aspect:

> We cannot get rid of the fundamental conviction that the elementary composition of bodies, since it fixes their essence, must determine their properties [...] We may begin with the outside, but it is only in order to reach the internal structure.

The structural criterion of classification was first applied to a restricted domain of minerals, viz., the silicates, with great success (Machatschki, 1928; Bragg, 1930; Náray-Szabó, 1930). The structural classification of silicates was a much better systematics than the old chemical subdivision, and was soon widely adopted. This chemical-plus-structural classification has been applied to many other domains of mineralogy, such as fluoaluminates (Pabst, 1950), aluminates (Liebau, 1956; Zoltai, 1960), and phosphates (Liebau, 1966; Corbridge, 1971). More recently, Povarennykh

Table 1. Sketch of the evolution of the basic criteria of mineralogical classification (after Lima-de-Faria, 1983)

Date	Criterion
~340 B.C	Practical uses - THEOPHRASTOS
77 A.D.	Practical uses - PLINIUS
~ 760	Physical properties - GEBER (Jabir Ibn Hayyân)
~ 1020	Physical properties - AVICENNA (Ibn Siná)
1546	Physical properties - AGRICOLA (Georg Bauer)
1750-	
1758	Chemical + physical properties - CRONSTEDT
1774	Physical properties - WERNER
1784	Chemical + physical properties - BERGMAN
1800-	
1801	Chemical + crystallographic properties - HAÜY
1817	Physical + chemical properties - WERNER
1819	Chemical properties - BERZELIUS
1832	Chemical properties - SOKOLOV
1850-	
1854	Chemical properties - DANA
1884	Paragenetic properties - LAPPARENT
1900-	
1928	Class. silicates based on structural prop. - MACHATSCHKI
1930	Develop. class. silicates b. on st. prop.- BRAGG; NÁRAY-SZABÓ
1940	Geochemical properties - UKLONSKII
1950-	
1953	Crystallochemical properties - MACHATSCHKI
1954	Paragenetic + chemical properties - KOSTOV
1966	Chemical + structural properties - POVARENNYKH

—— Well established criterion of classification

---- Intermediate criterion of classification

(1972), in a systematic way, applied this chemical-plus-structural classification to the whole domain of minerals.

A sketch of the evolution of the basic criteria of the mineralogical classification has been already presented (Lima-de-Faria, 1983) (Table 1). It is apparent from this sketch that the classification of minerals has passed through different stages, from practical to physical,to physical-plus-chemical, to chemical and to chemical-plus-structural, *each stage being a step further in the direction of the internal structure.* Although the chemical classification of minerals seemed to be well established, some dissatisfaction emerged during the last decades, and other criteria have been proposed for the classification of minerals, some related to particular domains of application, such as geological (Lapparent, 1884) or geochemical (Uklonskii, 1940; Kostov, 1954); others more theoretical, as the one proposed by Machatschki (1953) which was based on crystal-chemical grounds and the hierarchy of formula complexity.

On the other hand, many structures of minerals have been determined in the last 80 years, and the data thus accumulated call for an appropriate systematization, in order to enable that a better use be made of them. The potentiality of this structural information is certainly enormous.

Moreover, we have now reached the internal structure of minerals and this is what leads to a natural and much better scientific approach to the classification of minerals. It is time to invert the classification criterion and replace the chemical-plus-structural by a structural-plus-chemical classification. This new classification will possibly show many unexpected and interesting relations among mineral structures.

1.2. The structural classification of inorganic structures

When thinking on a structural classification of minerals one should have in mind that it must be based on structural factors, and that these should pertain not only to mineral structures but to inorganic structures in general, because the former are a part of the latter. On the other hand, the structural principles which have been considered on the structural classification of silicates, and which have proved so fruitful, should be preserved in any classification extension to the whole mineral kingdom. Consequently, *the structural classification of minerals must fit in a more general classification of inorganic structures, and also, not to contradict the widely accepted structural classification of silicates.*

One of the first important approaches to a general structural classification of inorganic crystal structures was proposed by Goldschmidt (1929), whose classification was based on the hierarchy of the chemical formula and on the number and arrangement of neighbours around any single atom in the crystal lattice, that is, the coordination of the atoms.

The structural classification of silicates was based on the bond strength distribution. In silicates the Si-O bonds are usually stronger than the bonds between the other chemical elements and oxygen, and therefore, there is a tendency to form complexes of SiO_4 tetrahedra more tightly linked than other parts of the structure, and which correspond to a kind of 'skeleton' of the crystal structure. These complexes of linked SiO_4 tetrahedra may form finite groups, infinite chains, infinite sheets or infinite three-dimensional frameworks. An extension of this bond strength distribution theory, made by Laves in 1930, leads to the general concept of *structural units* which correspond to the assemblages of atoms more tightly linked together in the crystal structure.

Niggli (1945), Garrido and Orlando (1946), Bokii (1954), De Jong (1959) and Wells (1962) also presented general classifications on the same basis, but none of these authors has systematically applied them; theirs were only general proposals.

A provisional classification of crystal structures on the basis of their interatomic bonds was made by Evans (1939); however, he abandoned it (Evans, 1964) due to the difficulties in ascribing, with certainty, the kind of bonds in crystal structures, and to the fact that they are normally not of one kind only, but intermixed.

Bokii (1954) was possibly the first to note a difference between structures such as halite and quartz, calling halite a 'coordination' structure and quartz a framework structure. Although this distinction was a bit ambiguous, it was later clarified by Lima-de-Faria & Figueiredo (1976) in their general classification of inorganic structures.

The bond strength distribution in a crystal structure may be relatively homogeneous, as in many oxides, halogenides, alloys and intermetallic compounds, where the bonds are predominantly nondirectional, and such structures are formed by the close packing of the larger atoms with the smaller atoms occupying the interstices within the packing. What forms the 'skeleton' of this kind of structure, that is, the structural units, are individual atoms, the packing atoms, normally the anions, while cations act as interstitial atoms. For instance in halite, NaCl, the larger atoms, the Cl ions, are the structural units. Speaking of the halite structure in terms of 'coordination' is rather inappropriate because in all structures the atoms are coordinated. What distinguishes halite from the quartz structure is the fact that the former is based on the close packing of its larger atoms, while quartz corresponds to a three dimensional linkage, a framework, with large holes in it.

However, in structures where the directional forces are dominant, the bond strength distribution is normally heterogeneous; the structural units may form finite groups, infinite chains, infinite sheets or infinite frameworks.

According to this classification (Lima-de-Faria & Figueiredo, 1976), there are *five main categories of structural units*: individual atoms, finite groups, infinite chains, infinite sheets and infinite frameworks; the *corresponding main categories of structures* are called close-packed, (at an early stage also called atomic by the author) group, chain, sheet and framework structures, respectively.

In the case of homogeneous bond strength distribution, two limiting situations may occur: either the structure is based on a simple packing of individual atoms, with no-directional bonds (examples are helium, copper and sodium chloride), or it is based on a three-dimensional framework, with directional bonds (examples are diamond and cristobalite).

The bond strength distribution must be combined with the directional character of the bonds in order to avoid certain ambiguities. For instance, the melonite structure, $NiTe_2$, which corresponds to the hexagonal closest packing of Te with Ni atoms filling octahedral voids in alternate layers, could be confused with a sheet structure of $NiTe_2$ layers of octahedra. This ambiguity can be solved if we give to the directional character of the bonds an important role. In fact, in melonite, both the Ni—Te bonds (ionic) and the Te—Te bonds (van der Walls) are non- directional bonds, which explains the formation of the hexagonal closest packing of Te atoms. Therefore, $NiTe_2$ is a close-packed structure. The distribution of Ni in alternate layers, occupying octahedral voids, is just a possible stable distribution of the interstitial atoms.

The structural classification of inorganic crystal structures of Lima-de-Faria & Figueiredo (1976) was systematically applied to 782 structure types, which correspond to approximately five thousand structures, and was presented on a general table (Table 2). These authors have also produced (1978) a chart containing the 270 structural units which form the structure types in this general table (Chart 1). That chart is divided into five columns corresponding to the five categories of structural units. On the first column, concerning the individual atoms, the different atomic layers corresponding to the various close packings are presented; on the second, third and fourth columns, the structural units, corresponding to groups, infinite chains and infinite sheets, respectively, are represented. On the fifth column, because it would be difficult to represent complete frameworks, only parts are shown – the so-called 'connected units' – which, by 'connection', lead to the whole frameworks.

Table 2. Partial representation of the general table of inorganic structure types (after Lima-de-Faria & Figueiredo, 1976) Colours have been added to distinguish the main kinds of inorganic structures, although a certain ambiguity may arise from the fact that certain structure types may represent more than one of these main kinds. [See fold-out in this chapter]

Other general structural classifications of inorganic structures have been proposed recently by Hawthorne and by Hellner.

Hawthorne (1983) has suggested a general structural classification of inorganic structures based on the polymerization of coordination polyhedra (not necessarily of the same type) with higher bond valences, as a consequence of the application of bond-valence theory to inorganic structures. The structures are organized by families based on different arrangements of fundamental building blocks or modules. The possible modules are tightly bounded units within the structure, and correspond to clusters of polyhedra that are the most strongly bonded. Although starting from different roots and with some significant differences, the structural classification of Hawthorne is quite similar to the one developed by Lima-de-Faria & Figueiredo (1976).

A general geometrical-plus-structural classification of inorganic structures was proposed by Hellner (1984, 1986), based on the concept of 'Bauverband' (connection) introduced by Laves (1930) and using a symbolism based on the concept of lattice complex introduced by Niggli (1919), and further developed by Hermann (1935, 1960) and

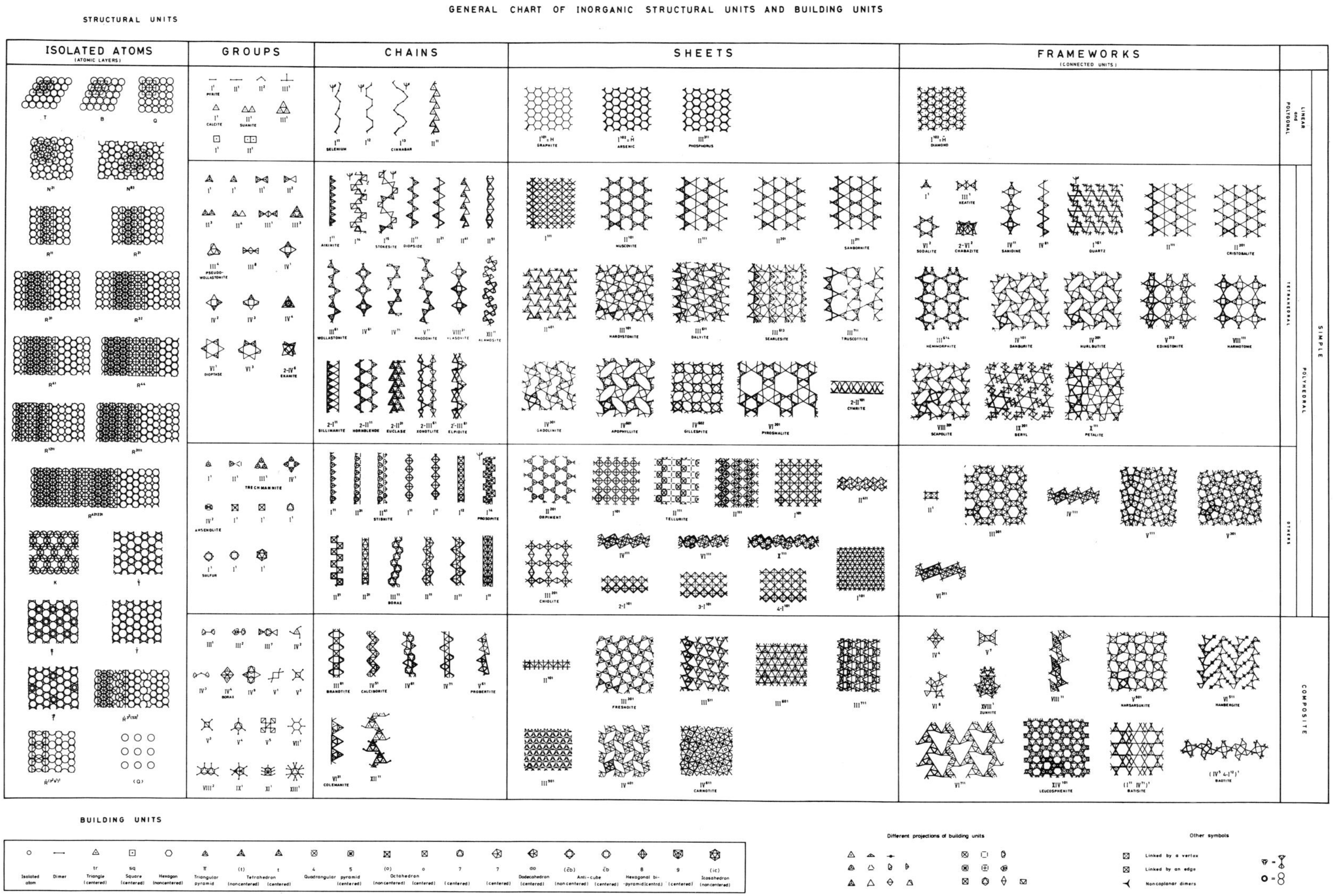

Chart. 1. General chart of inorganic structural units and building units (after Lima-de-Faria & Figueiredo, 1978). Reproduced in a very reduced size just to give an idea of its organization and appearence.

Hellner (1965). The 'Bauverband' may be defined as a three-dimensional periodic arrangement of points in a structure occupied by atoms or polyhedra of atoms, with definite geometric properties; it represents a typical connectivity pattern for a given structure type and, in certain cases, approximates a sphere packing with typical self-coordination and several types of voids for interstitial atoms. The 'Bauverband' describes the essential part or 'skeleton' of the structure types. The symbolism based on the lattice complexes is intended to enable the reconstruction of the structure, and also to show the relationship between different structure types. The *lattice complex* is the infinite array of points derived by applying all the symmetry operations of the space group to a particular point. The lattice complex corresponds, in space groups, to the concept of form in point groups, bearing in mind that a form is generated by the application of the symmetry operations of the point group to a particular face.

Under Hellner's system, the structures are grouped into families, main classes and subclasses. This classification is particularly useful when relating structures with high symmetry, such as cubic and tetragonal, where the symbolism is relatively simple. However, for less symmetrical structures, the symbolism becomes very complex and very difficult to deal with.

1.3. The structural classification of minerals

1.3.1. General considerations

As it has been said before, the structural classification of minerals must fit in a more general structural classification of inorganic structures, and also to conform with the widely accepted structural classification of silicates. The general structural classification of inorganic structures proposed by Lima-de-Faria & Figueiredo (1976), which is a kind of extension of the structural classification of silicates, may be applied to the mineral kingdom, because it will fulfil the two requirements mentioned above.

However the adaptation of the general structural classification of inorganic structures to the mineral kingdom has to take in consideration certain peculiarities of the mineral structures. The general structural classification of inorganic structures was strongly influenced by the work on the systematic derivation of inorganic structures (Lima-de-Faria, 1965a; Lima-de-Faria & Figueiredo, 1969), where structures were derived from the simplest to more complex crystal-chemical formulas. This way of description was good enough for the systematic derivation. However, when elaborating a classification of minerals, one has to group together the closely related mineral structures, that is, those which have a similar arrangement though differing by slight distortions, or by substitution of their atoms. This means that we need to group the mineral structures belonging to the same *structure type* together with their distortion and substitution derivatives, regardless of the complexity of their crystal-chemical formulas. One also has to add to these their defect and insertion (or stuffed) derivatives. Taking these differences into consideration, mineral structures may then be classified according to five main categories: close-packed, group, chain, sheet and framework, corresponding to the structural units: individual atoms, finite groups, infinite chains, infinite sheets, and frameworks (Figure 1). Within these categories they are ordered from the structurally simple to the more complex.

An outline of this structural classification of minerals has been presented (Lima-de-Faria, 1983) and a small section will be shown here (Table 3). Among the mineral structures belonging to the same structure type the simplest and best known, the aristotypes in the designation of Megaw (1973), were chosen to represent the structure type. Within each structure type, the other mineral structures, called its population, are ordered by the chemical sequence generally adopted in the classical chemical classification of minerals. The other related structures, distortion, substitutional and insertion (or stuffed) derivatives, follow later in sequence.

This outline of structural classification (Table 3) has been elaborated mainly on the basis of the data contained in the general table of inorganic structure types (Lima-de-Faria & Figueiredo, 1976) and on the work of Povarennykh (1972).

Another structural classification of minerals has been proposed by Hawthorne (1984, 1985), also as a result of the application of his general classification of inorganic structures to the mineral

Table 3. First part of the outline of a structural classification of minerals (after Lima-de-Faria, 1983). Notice the designation of 'atomic' which has now been replaced by 'close-packed'

A T O M I C

Homogeneous

Layered

Closest-packed

Non-interstitial

$[Cu]^c$ Copper $Fm3m$
$[Pt]^c$ Platinum — $[Ag]^c$ Silver
$[Pd]^c$ Palladium — $[Au]^c$ Gold
$[Ir]^c$ Iridium — $[(Ag,Au)]^c$ Electrum
$[Ni]^c$ Nickel — $[(Ag,Hg)]^c$ Kongsbergite
$[Pb]^c$ Lead

$[Hg]^c$ Mercury $R\bar{3}m$

$[In]^c$ Indium $I4/mmm$

$[AuCu]^c$ Cuproaurite $P4/mmm$
$[FePt]^c$ Ferroplatinum — $[PdHg]^c$ Potarite

$[AuCu_3]^c$ Tricuproaurite $Pm3m$
$[PbPd_3]^c$ Zvyagintsevite

$[(Os,Ir)]^h$ Iridosmine $P6_3/mmc$
$[Zn^{12}]^h$ Zinc $P6_3/mmc$

Interstitial

Octahedral

$Na^o[Cl]^c$ Halite $Fm3m$
$Na^o[F]^c$ Villiaumite — $Ca^o[O]^c$ Lime
$K^o[Cl]^c$ Sylvine — $Cd^o[O]^c$ Monteponite
$Ag^o[(Br,Cl)]^c$ Cerargyrite — $Ca^o[S]^c$ Oldhamite
$Mg^o[O]^c$ Periclase — $Mn^o[S]^c$ Alabandite
$Ni^o[O]^c$ Bunsenite — $Pb^o[S]^c$ Galena
$Fe^o[O]^c$ Wüstite — $Pb^o[Te]^c$ Altaite
$Mn^o[O]^c$ Manganosite — $Ti^o[N]^c$ Osbornite
$Fe^7[Si]^c$ Fersilicite $P2_13$

$Ni^o[As]^h$ Niccolite $P6_3/mmc$
$Co^o[As]^h$ Langisite — $Fe^o[Se]^h$ Achavalite
$Ni^o[Sb]^h$ Breithauptite — $Co^o[Se]^h$ Freboldite
$Pd^o[Sb]^h$ Palstibite — $(Ni^{II},Ni^{III})^o[Se]^h$ Sederholmite
$Pt^o[Sb]^h$ Platstibite
$Fe^o[S]^h$ Troilite — $Ni^o[Te]^h$ Imgreite
$(Fe^{II},Fe^{III})^o[S]^h$ Pyrrhotine — $Pd^o[Te]^h$ Kotulskite
$Co^o[S]^h$ Jaipurite — $Pt^o[Sn]^h$ Niggliite
$Co^o[As]^h$ Modderite $Pbnm$

$Mg^o[Cl_2]^c$ Chloromagnesite $R\bar{3}m$
$Mn^o[Cl_2]^c$ Scacchite — $Fe^o[Cl_2]^c$ Lawrencite

$Ni^o[Te_2]^h$ Melonite $P\bar{3}m1$
$Sn^o[S_2]^h$ Berndtite — $Pt^o[Te_2]^h$ Moncheite
$Ni^o[Se_2]^h$ Nidiselite — $Pd^o[Te_2]^h$ Merenskyite
$Ni^o[SeTe]^h$ Kitkaite

$Ti^o[O_2]^c$ Anatase $I4_1/amd$

$(Fe,Sn)^oTa^o[O_4]^h$ Ixiolite $Pbcn$
$Fe^oW^o[O_4]^h$ Wolframite $P2/c$
$Zn^oW^o[O_4]^h$ Sanmartinite
$Nb_2^oFe^o[O_6]^h$ Columbite $Pbcn$
$Ta_2^oFe^o[O_6]^h$ Tantalite — $Nb_2^oMg^o[O_6]^h$ Magnocolumbite

$Ti^o[O_2]^{ch}$ Brookite $Pbca$

$Ca^o[Cl_2]^h$ Hydrophilite $Pnnm$
$Fe^o[AsS]^h$ Arsenopyrite $P2_1/c$

$Mn^o[O_2]^h$ Ramsdellite $Pbnm$
$V^o[O_2]^h$ Paramontroseite

$Al^o[(OH)_3]^h$ Bayerite $P2_1/a$

$Al_2^o[O_3]^h$ Corundum $R\bar{3}c$
$Fe_2^o[O_3]^h$ Hematite — $V_2^o[O_3]^h$ Karelianite
$Cr_2^o[O_3]^h$ Eskolaite
$Fe^oTi^o[O_3]^h$ Ilmenite $R\bar{3}$
$Mn^oTi^o[O_3]^h$ Pyrophanite — $Mn^o(Fe,Sb)^o[O_3]^h$ Melanostibite

$Bi_2^o[Te_3]^{chh}$ Tellurobismuthite $R\bar{3}m$
$Bi_2^o[Te_2S]^{chh}$ Tetradymite $R\bar{3}m$
$Bi_2^o[Te_2Se]^{chh}$ Kawazulite

$Ti^o[CaO_3]^c$ Perovskite $Pm3m$ (ideal)
$Ti^o[(Na,Ce)O_3]^c$ Loparite — $Nb^o[NaO_3]^c$ Igdloite
$(Ti,Nb)^o[(Na,Ca)O_3]^c$ Latrappite $Pcmn$
$Nb^o[NaO_3]^c$ Lueshite $P222_1$
$Ti^o[PbO_3]^c$ Makedonite $P4/mmm$
$Na^oAl^o[K_2F_6]^c$ Elpasolite $Pa3$

$Fe_3^o[S_4]^{cchh}$ Smythite $R\bar{3}m$

Tetrahedral

$Zn^t[S]^c$ Sphalerite $F\bar{4}3m$
$Cu^t[Cl]^c$ Nantokite — $Hg^t[S]^c$ Metacinnabarite
$Cu^t[I]^c$ Marshite — $Zn^t[Se]^c$ Stilleite
$Ag^t[I]^c$ Miersite — $Hg^t[Te]^c$ Coloradoite
$Cd^t[S]^c$ Hawleyite

$Cu^tFe^t[S_2]^c$ Chalcopyrite $I\bar{4}2d$
$Cu^tGa^t[S_2]^c$ Gallite — $Tl^tFe^t[S_2]^c$ Raguinite
$Cu^tIn^t[S_2]^c$ Roquesite
$Cu_3^tSb^t[S_4]^c$ Famatinite $I\bar{4}2m$
$Cu_2^tFe^tSn^t[S_4]^c$ Stannite $I\bar{4}2m$
$Cu_2^t(Fe,Zn)^tGe^t[S_4]^c$ Briartite — $Cu_2^tZn^tSn^t[S_4]^c$ Kösterite
$Cu_2^tZn^tIn^t[S_4]^c$ Sakuraiite — $Ag_2^tFe^tSn^t[S_4]^c$ Hocartite

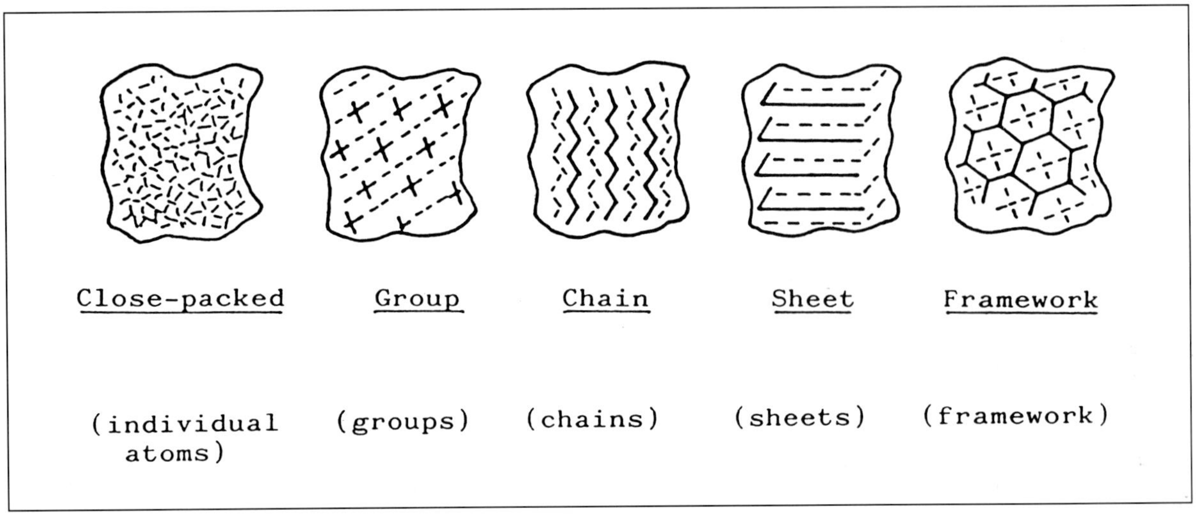

Fig. 1. Very schematic representation of the five main categories of mineral structures according to the bond strength distribution of their structures and to the character of the bonds (directional or non-directional). The corresponding structural units are written in brackets (after Lima-de-Faria, 1986).

kingdom. The Hawthorne classification leads also to five main categories of mineral structures, namely, unconnected polyhedra, finite clusters, infinite chains, infinite sheets and infinite frameworks. The main difference from Lima-de-Faria's classification (1983) hinges on the fact that, for Hawthorne, the structural 'modules', corresponding to the structural units, are only polyhedra, and never individual atoms, as admitted by Lima-de-Faria.

When a new classification replaces another, the former often continues to be valid and useful. The physical classification is still convenient for identification purposes, and the chemical classification for tackling certain paragenetic problems. In order to preserve the advantages of both chemical and structural classifications, a computer data base system with two main entries, the chemical and the structural, has been suggested by Figueiredo (1985).

1.3.2. Implications of the structural classification of minerals in the classification of silicates

When applying the general structural classification of minerals (Lima-de-Faria, 1983) to silicates a few adjustments have to be introduced. A new category of structures is considered, namely, the close-packed structures that correspond to isolated SiO_4 tetrahedra: the island and ring silicates have to be placed together under the designation of group silicates, because both their structural units correspond to finite groups. This way of classifying the silicates is also in agreement with the one proposed by Bragg in 1930: orthosilicates, self-contained groups, silicon-oxygen chains, silicon-oxygen sheets, and three-dimensional silicon-oxygen networks. In Bragg's classification, the orthosilicates (isolated tetrahedra) correspond to silicates based on close packings.

Some silicates may in fact be described in terms of the close packing of oxygen atoms, with silicon and other cations in the interstices. An example is olivine, Mg_2SiO_4, which belongs to the same structure type as Al_2BeO_4 and Mg_2SnO_4. These last two structures are clearly not group structures, but close-packed structures. Megaw (1973) discusses the necessity to distinguish between 'packing structures' and 'linkage structures', placing the olivine in the first category. She describes olivine as based on a hexagonal closest packing of oxygen atoms with certain tetrahedral interstices occupied by silicon, and octahedral sites by magnesium, exactly in the same way as Bragg did in 1929. A similar situation applies to the high pressure form of olivine, Mg_2SiO_4, which corresponds to the spinel structure type Al_2MgO_4; obviously, spinel is not a group structure.

CHAPTER 2

The anatomy of crystal structures

2.1. What is a crystal structure?

A crystal structure is a definite arrangement and linkage of atoms in a periodic orderly array, where the periodicity results from infinite translations in three-dimensions. Other categories of orderly arrays correspond to the so-called quasicrystals, obeying certain non-periodic matching rules. Each individual structure can be described by its symmetry (space group), unit cell parameters, chemical composition, unit cell content, the positions of the atoms within the unit cell, their coordination, and the kind and strength of the bonds between them.

Consequently, a crystal structure is not merely a geometric arrangement of atoms, since the particular characteristics of the chemical elements involved, their linkage and the way interatomic bonds are distributed also play an important role. Unfortunately, we do not yet possess exact information on the characteristics and quantitative data on the strength of the bonds. Only qualitative and semi-quantitative results are available.

2.2. The constitution of atoms

To gain an understanding of the anatomy of crystal structure, it suffices that atoms be regarded as miniature planetary systems of which the central sun is represented by the atomic nucleus, while the role of planets is played by the atomic electrons. Admittedly, this is quite an elementary approach; however, it is well suited to the discussion of a large number of problems in physics. In order that electric neutrality be achieved, the negatively charged electrons have to be balanced by a positively charged nucleus; the latter is assumed to consist of neutrons with no electric charge, as well as of protons, positively charged and in such numbers that they will match all the planetary electrons. The number of either the electrons or protons in the atom of a chemical element is called the atomic number of the element.

The electrons are located around the nucleus in several levels of energy, forming successive shells which are called K,L,M,...,Q, from the inner part to the outside. These electron shells may also be referred to shell quantum numbers 1,2,3,..., respectively. Consequently, to say shell L or shell 2 has the same meaning. The further away an electron shell is from the nucleus the more electrons it may contain. The maximum number of electrons in each shell is $2n^2$, where n is the shell quantum number.

Each shell is divided into sub-shells called orbitals, which are designated by s, p, d and f, beginning with the closest to the nucleus. The orbital belonging to a certain shell is described by the corresponding letter, preceded by the quantum number of the shell. As an example we may consider the 's' orbital of the shell 2 (or L), which is designated by 2s. The orbitals define the probability of finding an electron in a certain region of the atom, along different directions and distances from the nucleus. These probabilities are called 'angular' and 'radial' probabilities, respectively.

In a single shell there can only be one 's' orbital, three 'p' orbitals along three directions x, y and z (Figure 2), five 'd' orbitals, and seven 'f' orbitals. Moreover, an orbital can contain, at most, two electrons, which spin in opposite directions, as required by the Pauli exclusion principle. Consequently, in a K shell (quantum number 1), only a single orbital s and $2 \times 1^2 = 2$ electrons can exist. The L shell (quantum number 2) may have $2 \times 2^2 = 8$ electrons, two in s and $2 \times 3 = 6$ in p orbitals. The M shell may contain the maximum of $2 \times 3^2 = 18$ electrons: 2 in s, 6 in p, and 10 in d. The N

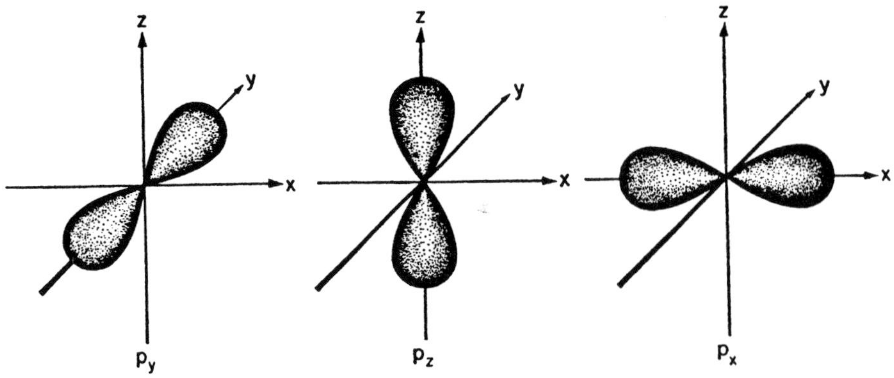

Fig. 2. Angular probability functions for p_x, p_y and p_z electrons (after Spice, 1964)

shell has a maximum capacity of 32 electrons. On Table 4, where superscripts indicate the number of electrons per orbital, the electronic configurations of the first ten chemical elements are presented.

The periodic table of the chemical elements (Table 5) is organized so that those elements in one of the vertical columns headed I, II, III,...,VIII have atoms whose outer shells contain electrons equal in number to the Roman numeral designating the group. The chemical elements of group I with atomic numbers 1,3,11,19,..., have just one electron in their outermost shell, which occupies an s orbital. Similarly, the elements of group II have two s orbital electrons in their outer shell.

The chemical properties of an atom depend on the ability of its outermost or valence shell to lose, gain or share electrons (called valence electrons).

The atoms with orbitals s and p of their outer shell completely filled, that is $s^2 + p^6$, are the most stable because they have no tendency to lose or gain electrons. This is the reason for the existence of inert rare gases, which, with 8 electrons in the outer shell, have the most stable configuration of electrons. The elements of groups V, VI and VII, have a tendency to gain electrons in order to complete their outer shells, thus forming anions. The elements of groups I, II and III tend to lose electrons, and form cations. The chemical elements of group IV, instead of gaining or losing electrons, tend to share electrons with other atoms, in order to achieve the configuration of an inert gas.

Between group II and group III are located the so-called *transition elements* because they may have the same filled orbitals in the outer shell while differing by the number of electrons in the inner shells. For instance, the elements 21 to 30 possess the same

Table 4. The electronic configurations of the first ten elements. A single arrow represents a single electron; a pair of arrows, two electrons of opposite spin occupying a single orbital (after Evans, 1964)

Element	Z	K	L				
		$1s$	$2s$	$2p_x$	$2p_y$	$2p_z$	
H	1	↑					$1s^1$
He	2	↑↓					$1s^2$
Li	3	↑↓	↑				$1s^2, 2s^1$
Be	4	↑↓	↑↓				$1s^2, 2s^2$
B	5	↑↓	↑↓	↑			$1s^2, 2s^2 2p^1$
C	6	↑↓	↑↓	↑	↑		$1s^2, 2s^2 2p^2$
N	7	↑↓	↑↓	↑	↑	↑	$1s^2, 2s^2 2p^3$
O	8	↑↓	↑↓	↑↓	↑	↑	$1s^2, 2s^2 2p^4$
F	9	↑↓	↑↓	↑↓	↑↓	↑	$1s^2, 2s^2 2p^5$
Ne	10	↑↓	↑↓	↑↓	↑↓	↑↓	$1s^2, 2s^2 2p^6$

Table 5. The periodic table of chemical elements (after Bloss, 1971)

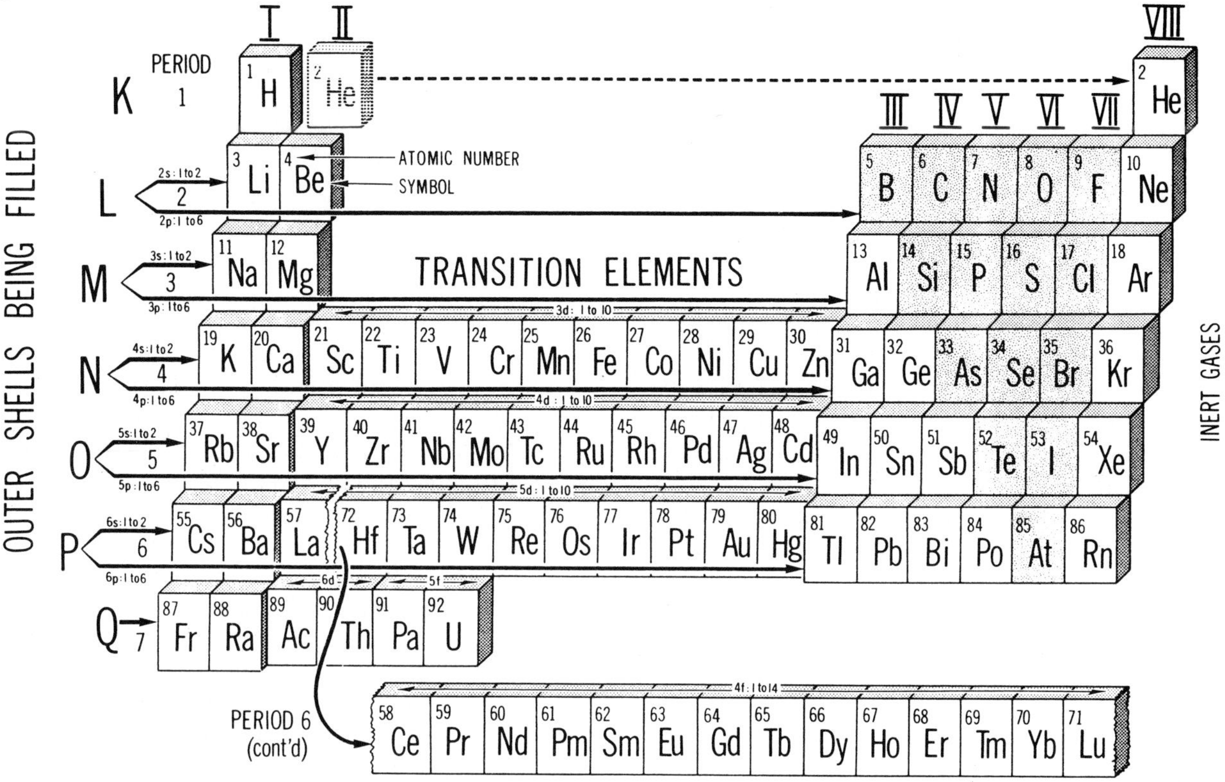

electron configuration of the outer shell, $4s^2$, but they exceed calcium by a certain occupancy of the 3d orbitals of the more interior M shell. Only after this more interior shell is filled to its capacity of ten, as in zinc, is there a resumption of filling of the outer shell p-orbitals (elements 31 to 36).

Cations are smaller than electrically neutral atoms, not only because electrons have been lost, but also because the attractive action of the nucleus becomes more important with respect to the remaining electrons.

The tendency to gain electrons is called 'electronegativity', and it will be larger the nearer its configuration will be to the configuration of the nearest inert gas. The minimum value of the electronegativity is 0.7 for the element francium and the maximum value is 4.0 for fluorine.

According to their electronegativities the elements are divided into 'metals', if their electronegativities are smaller than 1.9, and they tend to lose electrons and form cations; 'non-metals', if their electronegativities are higher than 2.1 and they tend to gain electrons and originate anions; and 'metaloids', if their electronegativities are in the intermediate range between 1.9 and 2.1, and they show properties intermediate between those of metals and of the non-metals.

2.3. The bonds between atoms

The forces of attraction between atoms are known as chemical bonds, of which important types are: 'ionic', 'covalent', 'metallic' and 'van der Waals' bonds. In inorganic compounds, in particular in minerals, the ionic bonds are predominant. In minerals, the metallic bonds appear only in the elements (gold, silver, etc.), but they are very important in the alloys. The van der Waals bonds appear mostly in organic molecular compounds, but also between certain anions in some inorganic compounds.

If the difference of electronegativity between two atoms exceeds 2.0, the one with higher electronegativity will attract and capture electrons from the other atom to its outer shell. As a conse-

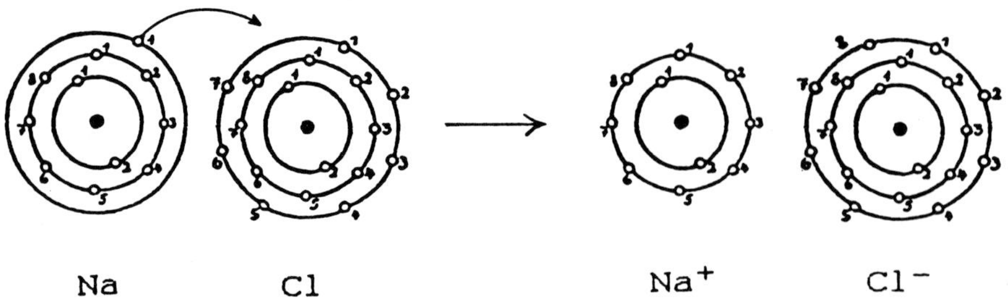

Fig. 3. Exchange of electrons between Cl and Na atoms to form Cl^- and Na^+ ions.

quence, one atom becomes an anion and the other a cation, and the attraction between the two oppositely charged ions forms an ionic bond. Like the gravitational attraction which acts in all directions, this electric force of attraction is non-directional (Figure 3).

Two atoms with relatively high but similar electronegativities may not be able to engage in electron capture between their outer shells, but they may both attain a configuration similar to that of the inert gases if they share electrons between them. Such a situation is called 'covalent bond'. This kind of linkage can correspond to single, double or triple bonds. A double bond occurs if two orbitals from one atom overlap two from the other, so that two pairs of electrons are shared between them. The covalent bond acts in a definite direction and therefore is a directional bond (Figure 4).

In minerals, certain more complex covalent bonds occur quite often, involving different levels of energy, which are better suited to the sharing of electrons. As an example, the diamond structure will be considered. The carbon atom in the neutral state has an electronic configuration $\overset{\uparrow\downarrow}{1s} \mid \overset{\uparrow\downarrow}{2s} \mid \overset{\uparrow}{2p}\overset{\uparrow}{2p}2p$, where the arrows represent electrons with opposite spin. When one carbon atom is linked to another carbon to form the diamond structure, each of these four outside electrons forms a new hybrid orbital of equal energy level, one s and three p orbitals designated by $2(sp^3)$, the properties of which are a mixture of the properties of the original s and p orbitals. The electronic configuration of carbon becomes

$\overset{\uparrow\downarrow}{1s} \mid \overset{\uparrow}{\overline{2(sp^3)}}\ \overset{\uparrow}{\overline{2(sp^3)}}\ \overset{\uparrow}{\overline{2(sp^3)}}\ \overset{\uparrow}{\overline{2(sp^3)}}$. These 'hybrid' orbitals are distributed along four lobes at 109° 28' to each other, the axes of these lobes extending towards the corners of a tetrahedron centred in the carbon atom (Figure 5). These lobes represent the regions where it is more probable to encounter the electrons of the hybrid orbitals $2(sp^3)$,

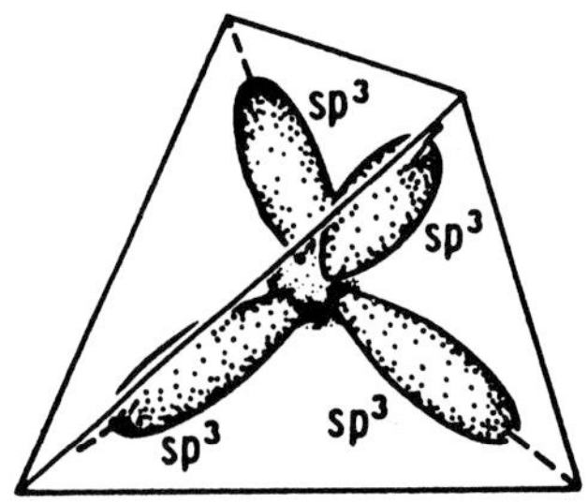

Fig. 5. Schematic illustration of the tetrahedral configuration of the hybrid sp^3 orbitals (after Bloss, 1971).

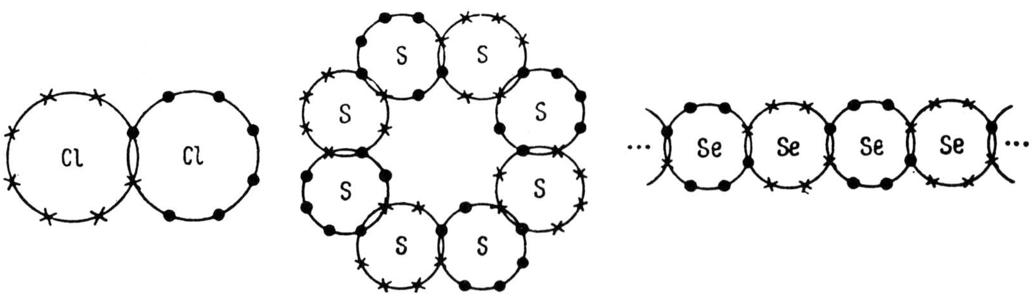

Fig. 4. Very schematic representation of the molecules of chlorine Cl_2 and sulphur S_8, and of a chain of selenium atoms (after Bokii, 1954).

the maximum probability lying along the lobe axes. The linkage between the carbon atoms is such that their orbitals overlap. The tetrahedral coordination of the carbon atoms in the diamond structure is, therefore, related to the distribution of these hybrid orbitals. According to quantum mechanics, the greater the overlap between the orbitals, the stronger the resulting covalent bond. This explains the hardness of diamond.

There are other kinds of hybrid covalent bonds (Table 6). For instance, the hybrid bonds dsp^2, which are distributed along the diagonals of a square, thus representing the square coordination of copper in CuO, and of platinum in PtS.

In a closest packing of atoms of a metallic chemical element the atoms have coordination twelve. Each metallic atom shares its valence electrons with each of its 12 neighbours, and this is the reason why the 'metallic bond' is weaker than the normal covalent bond or the ionic bond. In the metallic packing, the energy levels of the s and p outer orbitals become so even that the electrons easily transfer from an orbital s to a p orbital, and the packing looks like an assemblage of positive charges in an electron ocean (Figure 6).

The metallic bond takes place, typically, between chemical elements of similar low electronegativity. Like the ionic bond, the metallic bond is also non-directional.

The weakest of chemical bonds is the van der Waals bond, which corresponds to small concentrations of positive and negative charges at op-

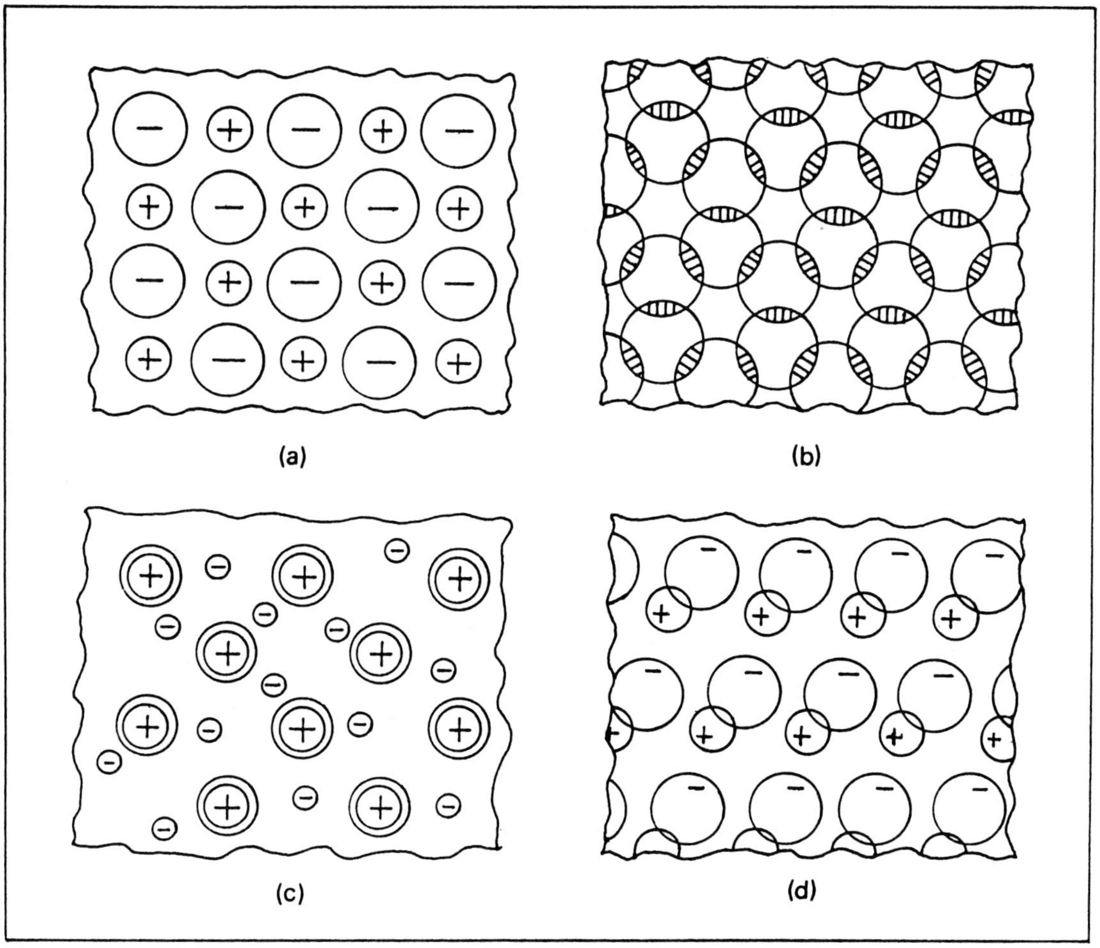

Fig. 6. Very schematic representation of the four main types of bonding forces in crystal structures: (a) ionic, (b) covalent, (c) metallic and (d) van der Waals.

posite poles in covalent bonds of atoms. In molecules where atoms are strongly linked together by covalent bonds, dipoles are generated which bond them together by van der Waals forces.

Some of the chemical bonds between atoms are intermediate among the four basic types of bonds. Of these 'intermediate bonds', the most important one is the so-called 'hydrogen bond', which is intermediate between an ionic and a van der Waals bond. When hydrogen is linked to a strong electronegative atom (fluorine or oxygen) by a covalent bond, its unique electron is located near the region of superposition of the orbitals. Consequently, the part of the hydrogen atom farthest from the overlap consists of an exposed proton, because no inner electron shells exist to shield its exposed nucleus. As a result, the proton attracts negative ions, and this attraction is known as the 'hydrogen bond'.

2.4. The size of atoms

From the determination of a crystal structure the positions of the atoms and the distances between them are obtained. But one does not get information about the size, or zone of influence, of the atoms. To derive the size of the atoms one needs to determine, using other physico-chemical methods, the size of an atom frequently found in crystal structures, then to work out the size of the other atoms from known interatomic distances.

Landé (1920) admitted that the shape of the atoms should be spherical, and that, in most cases, the anions, with added electrons to their shells, should be bigger than the cations, which come about from a loss of electrons. He also assumed that in the LiI structure, which is of the NaCl type, the iodine anions would be nearly in contact, because iodine ions are among the largest anions and litium ions are among the smallest cations (Figure 7). On the basis of these hypotheses, he calculated the size of the iodine anion, namely rI = 2.12 Å, – a figure that yielded both the minimum distance between them and their own diameter. It was then easy to determine the radii of Li^+, Na^+, K^+ and Rb^+, and consequently also of Fl^-, Cl^- and Br^- from data on the structures of the corresponding halogenides.

A more direct way to tackle this problem,

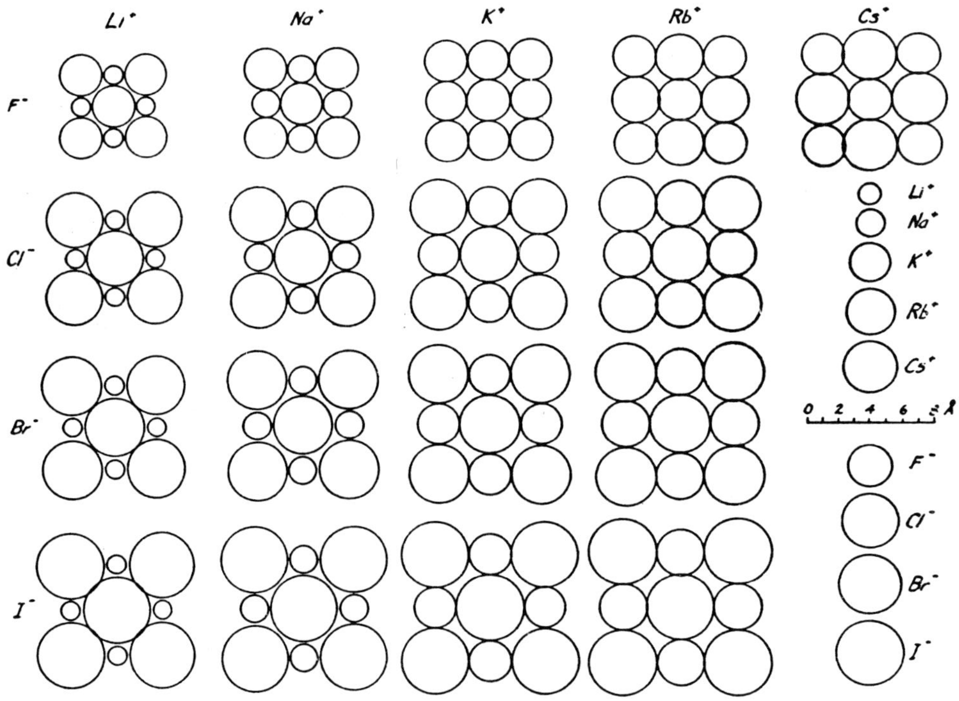

Fig. 7. The arrangement of ions on the cubic face of some alkali halogenide crystals with the sodium chloride structure (after Pauling, 1960).

Table 6. Ionic radii of several chemical elements (after Pauling, 1960)

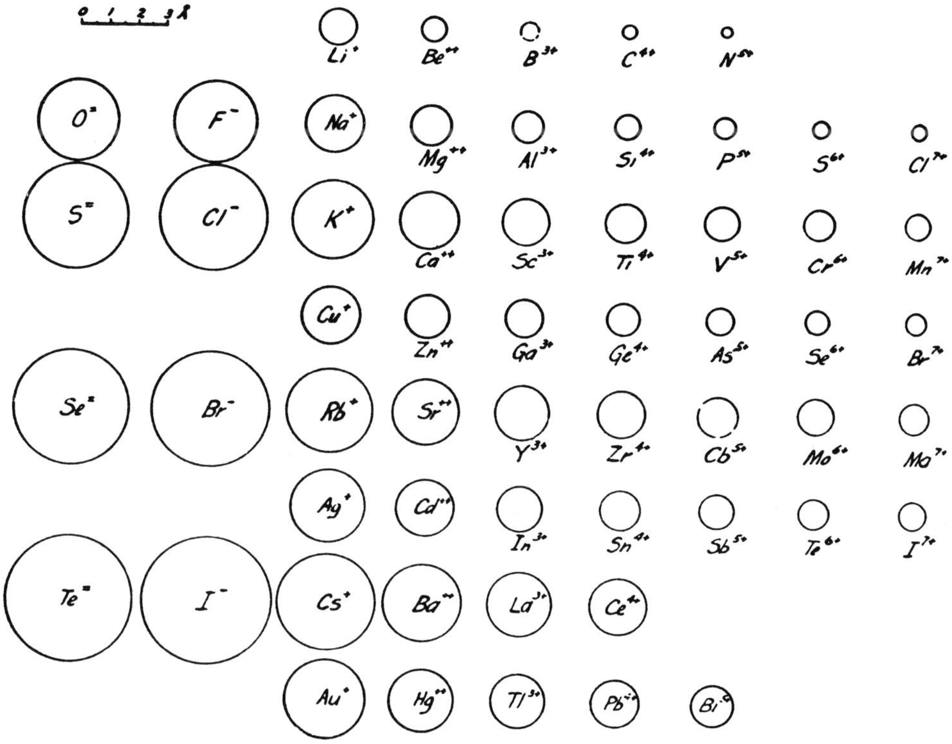

was followed by Wasastjerna (1923) who used molar refraction, which depends on the ionic volume. The measurements and calculations of Wasastjerna gave the value 1.32 Å for the oxygen anion and 1.33 Å for the fluorine anion. The radius of the fuorine ion calculated by Landé was very similar.

The first general table of ionic radii was produced by Goldschmidt (1926), using values for O^{2-} and F^- determined by Wasastjerna and many interatomic distances of well-known structures, together with data gathered for that purpose in his laboratory.

Pauling (1927) and Ahrens (1952) have also dealt with this problem and improved the data on radius size (Table 6).

Admitting sphericity and additivity in the study of ionic radii, we are assuming that the distances between ions are independent of the other forces that act upon them from other parts of the crystal structure. This corresponds to a quite restrictive view of the situation. No doubt more flexibility should be attributed to the ions. An important contribution to this problem has been that of Shannon & Prewitt (1969), who have considered several possible situations, especially the coordination of both cations and anions, and have based their work on approximately one thousand experimental determinations of interatomic distances, considered to be the most precise ones. They have called their figures 'effective' ionic radii.

The other categories of radii (covalent, metallic and van der Waals) have been calculated by several authors, using similar methodologies (Tables 7 and 8). No table of van der Waals radii is presented because it would be of little use in mineralogy.

The knowledge of ionic radii is of great importance in structure determination, particularly in the establishment of the so-called 'trial structure' and, also, in crystal chemistry, to understand the replacement of one ion by another in the same kind of structure (diadochy).

The spherical shape of atoms has been called into question, particularly in alloys and intermetallic compounds. However, instead of the spherical shape, we can always think in terms of a zone (or sphere) of influence (or interaction), related to a certain atom, and which, obviously, oc-

Table 7. Covalent radii of the chemical elements, except for Fr, Ra, Ac and lanthanons and actinons (after Sanderson, 1960)

	IA	IIA	IIIB	IVB	VB	VIB	VIIB	VIIIB			IB	IIB	IIIA	IVA	VA	VIA	VIIA	O
1	H 0.37																H 0.37	He 0.93
2	Li 1.34	Be 0.90											B 0.82	C 0.77	N 0.75	O 0.73	F 0.72	Ne 1.31
3	Na 1.54	Mg 1.30											Al 1.18	Si 1.11	P 1.06	S 1.02	Cl 0.99	Ar 1.74
4	K 1.96	Ca 1.74	Sc 1.44	Ti 1.36	V	Cr	Mn	Fe	Co	Ni	Cu 1.38	Zn 1.31	Ga 1.26	Ge 1.22	As 1.19	Se 1.16	Br 1.14	Kr 1.89
5	Rb 2.11	Sr 1.92	Y 1.62	Zr 1.48	Nb	Mo	Tc	Ru	Rh	Pd	Ag 1.53	Cd 1.48	In 1.44	Sn 1.41	Sb 1.38	Te 1.35	I 1.33	Xe 2.09
6	Cs 2.25	Ba 1.98	La 1.69	Hf	Ta	W	Re	Os	Ir	Pt	Au 1.50	Hg 1.49	Tl 1.48	Pb 1.47	Bi 1.46	Po	At	Rn 2.14

Table 8. Metallic radii of the chemical elements, except for Fr, Ra, Ac and lanthanons and actinons (after Sanderson, 1960)

	IA	IIA	IIIB	IVB	VB	VIB	VIIB	VIIIB			IB	IIB	IIIA	IVA	VA	VIA	VIIA	O
1																		
2	Li 1.549	Be 1.123											B 0.98	C 0.914	N 0.92			
3	Na 1.896	Mg 1.598											Al 1.429	Si 1.316	P 1.28	S 1.27		
4	K 2.349	Ca 1.970	Sc 1.620	Ti 1.467	V 1.338	Cr 1.267	Mn 1.261	Fe 1.260	Co 1.252	Ni 1.244	Cu 1.276	Zn 1.379	Ga 1.408	Ge 1.366	As 1.39	Se 1.40		
5	Rb 2.48	Sr 2.148	Y 1.797	Zr 1.597	Nb 1.456	Mo 1.386	Tc 1.358	Ru 1.336	Rh 1.342	Pd 1.373	Ag 1.442	Cd 1.543	In 1.660	Sn 1.620	Sb 1.59	Te 1.60		
6	Cs 2.67	Ba 2.215	La 1.871	Hf 1.585	Ta 1.457	W 1.394	Re 1.373	Os 1.350	Ir 1.355	Pt 1.385	Au 1.439	Hg 1.570	Tl 1.712	Pb 1.746	Bi 1.70	Po 1.76		

cupies a certain space in the structure.

From Tables 6 to 8 one can realize how great the change in radius is of a chemical element according to its state of chemical bond.

2.5. The coordination of atoms

In a crystal structure the ions tend to surround other ions of opposite electric charge. In general, any atom in a structure is surrounded by or coordinated to other atoms, and the coordination number of an atom, [N], is given by the number of its 'coordinating atoms'. The definition of coordination is not simple and depends on the bonding model, the nature of the problem, and the methods of calculation.

The coordination polyhedron of an atom is the polyhedron that has its vertices coincident with the centres of the coordinating atoms. In structures which contain lone electron pairs and in which volumes ascribable to these lone electron pairs are comparable with the volume of individual atoms, the coordination polyhedra can also be considered to include lone electron pairs.

Several methods have been proposed to define coordination numbers in complex structures. Most provide a 'weighted' coordination number, resulting from an expression for weighting the coordinating atoms, according to their distances from the central atom, e.g., by means of Voronoi polyhedra (the same as Wirkungsbereiche or Dirichlet domains), with or without consideration of atomic radii.

The most common coordination polyhedra occurring in inorganic structures are represented on Figure 8.

For a certain cation-anion relation, it is possible to predict the corresponding coordination from the radius ratio, on the basis of a principle proposed by Goldschmidt (1926), namely: the number of anions surrounding a cation tends to be as big as possible, limited only by the condition that all the anions touch the cation. This means that the structural architecture is fundamentally determined by the interactions cation-anion and not cation-cation or anion-anion.

The minimum values for the most common coordination can be easily calculated and are presented on Figure 9.

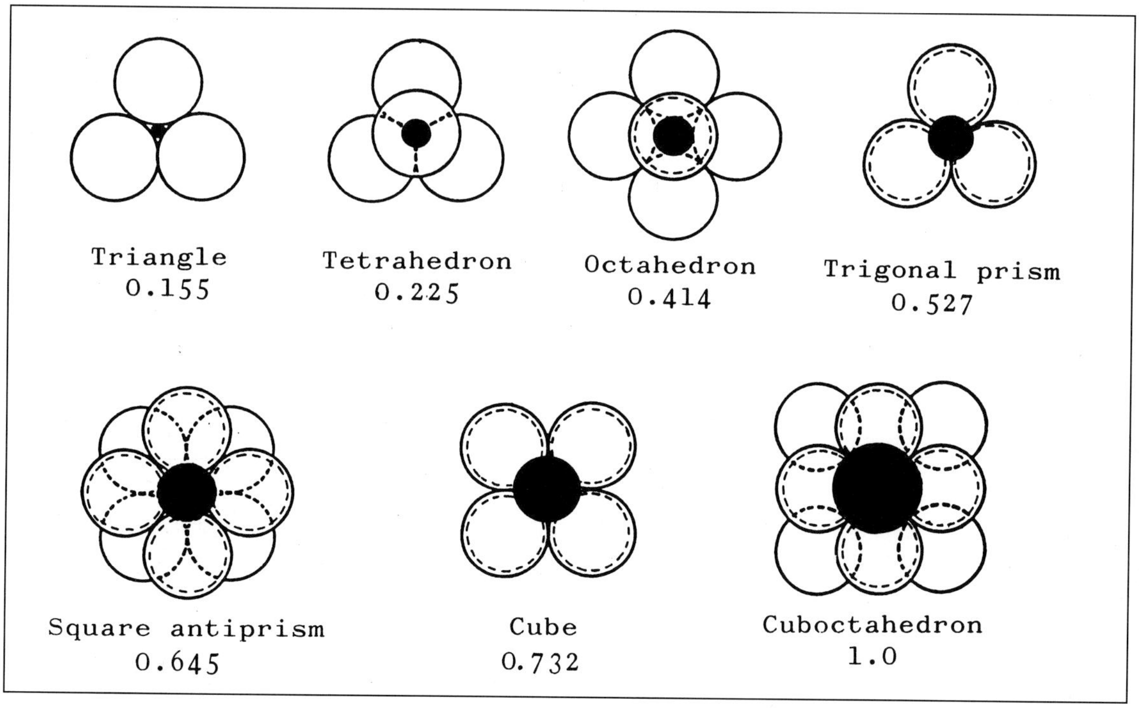

Fig. 9. Some groupings of anions around a cation. The minimum radius ratio (cation/anion), that is, for touching anions, is indicated in each case. The dashed circles mean that they are below the full circles.

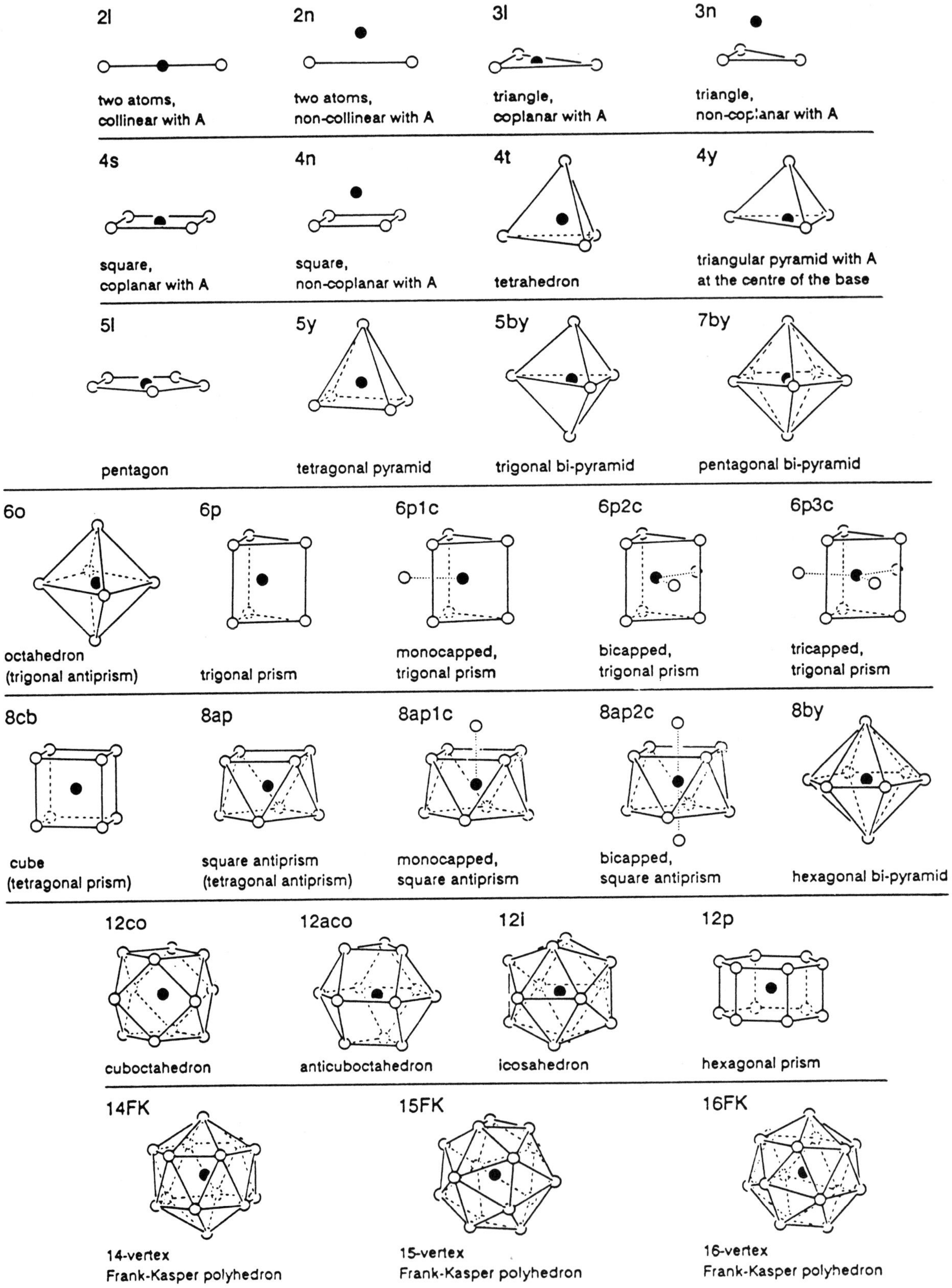

Fig. 8. More commonly occurring coordination polyhedra in inorganic structures (after Parthé, 1990).

Table 9. Limiting radii for the various coordinated configurations.

Radius Ratio	Coordination Number	Configuration
0.15-0.22	3	Triangular
0.22-0.41	4	Tetrahedral
0.41-0.57	6	Octahedral
0.57-0.73	6	Trigonal prismatic
0.73-1	8	Cubic
1	12	Cuboctahedral

A certain coordination is stable if the radius ratio between atoms A and B is such that its value is between the tangential limit of the anions and the upper limiting situation, which corresponds to the next higher coordination (Table 9 and Figure 10). For instance, in the case of octahedral coordination, these values are 0.414 for the tangential limit and 0.527 for the next higher limit, the trigonal prism. However these values are often not rigorously held by mineral structures due to several other factors.

The coordination of a certain atom is also intimately related to the electron configuration of the atom, e.g., to its orbitals. There are hybrid orbitals with a tetrahedral configuration, as is the case of carbon in the diamond structure, others with octahedral, trigonal, bipyramidal configuration, etc. (Table 10). In minerals many chemical elements which form cations have octahedral or tetrahedral orbital distributions, which explains the formation of so many close-packed mineral structures where the principal interstices are tetrahedrally and octahedrally coordinated.

Table 10. Some common hybrid orbitals and their geometrical configurations (after Evans, 1964)

Hybrid	Number of bonds	Distribution of bonds	Ref.*	Examples
sp	2	Linear	(*a*)	Cu^{I}, Ag^{I}, Au^{I}
sp^2	3	Planar to corners of equilateral triangle	(*b*)	—
dsp^2	4	Planar to corners of square	(*c*)	Cu^{II}, Ag^{II}, Au^{III}; Ni^{II}, Pd^{II}, Pt^{II}
sp^3	4	To corners of regular tetrahedron	(*d*)	Cu^{I}, Ag^{I}
d^2sp^3	6	To corners of regular octahedron	(*e*)	Fe^{II}, Fe^{IV}, Co^{II}, Co^{III}, Ni^{II}, Ni^{III}, Pd^{IV}, Pt^{IV}
sp^3d^2	6	To corners of regular octahedron	(*e*)	Ti^{IV}, Zr^{IV}
d^4sp	6	To corners of trigonal prism	(*f*)	Mo^{IV}

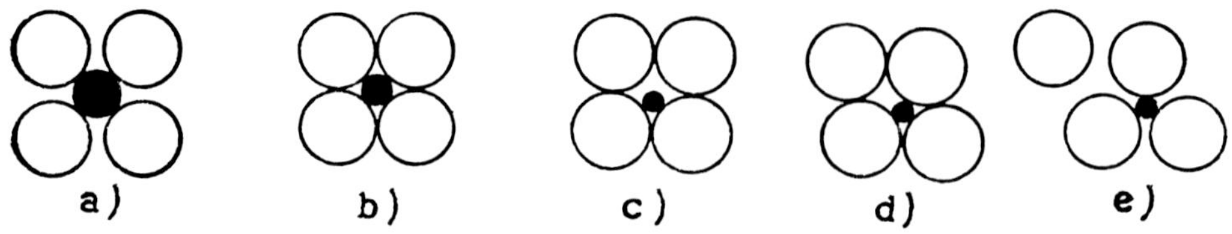

Fig. 10. Scheme illustrating the degree of stability of a coordination polyhedron. (a) Stable: each ion touches only ions of the opposite sign; (b) less stable: anions touch each other; (c), (d), (e) unstable: free movement of cations leads to a reduction in the coordination number (after Bokii, 1954).

CHAPTER 3

The architecture of crystal structures

3.1. The structural units and their 'polymerization' process

The structural units are normally formed by smaller subunits, called building units, which can be dimers (the linear linkage of two atoms), polygons, or polyhedra of atoms. In minerals, structural units may be found to vary a great deal but, fortunately, they are not so different. One may imagine a kind of 'polymerization' (or condensation) process which enables the derivation of complex structural units from simpler ones. The silicate structures offer a good example. Several silicate sheets are related to certain silicate chains, from which they can be derived by a polymerization process. (Liebau, 1956; Belov, 1956) (Figure 11).

This polymerization process was later generalized by Lima-de-Faria & Figueiredo (1976), expressing the polymerization of atoms into groups, of groups into chains, of chains into sheets, and of sheets into frameworks (Figure 12).

The polymerization process of structural units based on linked octahedra, particularly among phosphate structures, has been developed by Moore (1973). He established their hierarchy on this basis and also related the polymerization process of the phosphate structural units with their paragenesis.

3.2. The packing of structural units

The structural units tend to pack as closely as possible and their packing is a kind of 'skeleton' of the mineral structure. The other atoms are called interstitial and occupy the holes in the packing.

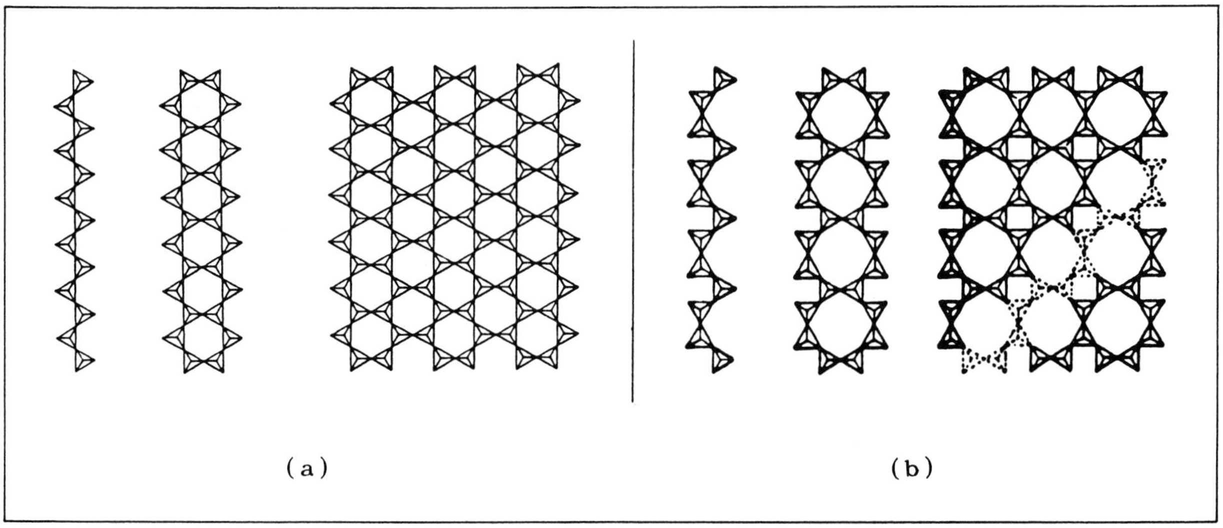

Fig. 11. Examples of silicate structural units related by the 'polymerization' process. (a) a pyroxene chain, giving rise to an amphibole chain, then to a mica sheet; (b) a wollastonite chain, giving rise to a pektolite chain and this to an idealized apophyllite sheet (after Liebau, 1956).

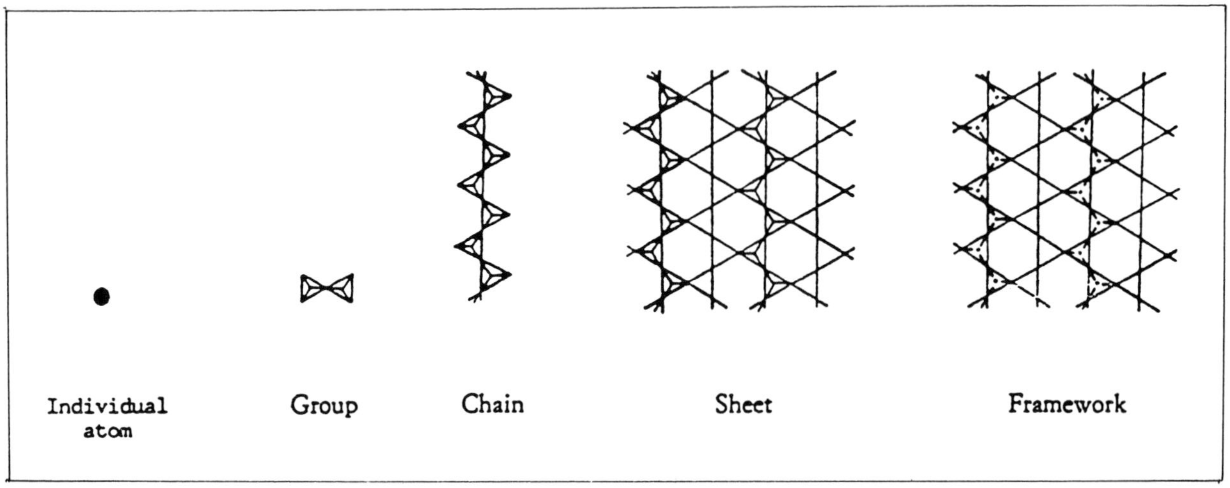

Fig. 12. Generalized polymerization process. An example of the different stages is: Si or O atom → $[Si_2O_7]$group → $[Si_2O_6]$ pyroxene chain → $[Si_2O_3(OH)_2]$sheet → $[Li_2SO_4]$framework. In the case of the framework only the connected unit is depicted; the way the tetrahedra are drawn indicates that they are linked by their top vertices (after Lima-de-Faria, 1986).

When one considers the packing of structural units, it is clear that their shapes and sizes are important characteristics. The packing efficiency of the structural units will be limited by the configuration of the structural units or, in the case of individual atoms, by their electronic configuration which may impose a particular coordination.

3.2.1. Kinds of close packing of individual atoms

In homogeneous close packings the layers are all alike and the way of stacking is the same for all of them.

The well-known cubic closest packing, (Kepler, 1611) the hexagonal closest packing and, the simple cubic packing, the simple hexagonal packing, and the body-centred cubic packing, were all been recognized by Barlow (1883). There are other more open arrangements, not so well-known but important, which exist in many inorganic crystal structures.

If one considers a closest-packed layer of atoms (designated by T, from triangular net), represented by spheres (in projection, by circles of radius R) (Figure 13), with their centres in positions 'A', another closest-packed layer may be stacked over its voids in two possible positions, either 'B' or 'C' (Figure 13). If the second layer is stacked with the centres of its atoms in positions B, see Figure 13, the projections of the voids of this second layer will coincide with positions A and C, defined before. One can then realize that whatever the sequence of the stacking layers, all their relative positions will be projected on positions A, B and C. Therefore, any sequence of layers may be designated by the combination of the letters A,B,C (Pauling, 1928). The hexagonal closest packing is defined by the sequence ABAB...(example: Mg structure), the cubic closest packing by ABCABC...(example: Cu), the double hexagonal closest packing by ABCBABCB...(example: La), the sequence for the Sm structure by ABABCBCACABABC...,etc. (Figure 13).

This way of expressing the various sequences of closest packing may be simplified by using the notation proposed by Ewald & Hermman (1931) for the two possible kinds of layers, 'p' and 'a', which form these sequences. In fact, there are only two kinds of layers in relation to the two adjacent layers: either a symmetry plane exists and they are designated by p, or it does not exist, and they are designated by a. The letters p and a have been replaced by 'h' and 'c' by Belov (1939) and these are presently most used. In the hexagonal closest packing, all the layers are of the kind h, and in the cubic closest packing they are all of the kind c. This was the reason to choose these two letters h and c to designate the two different kinds of layer pure sequences. Considering the examples given above, one can write Th for Mg, Tc for Cu, Tch for La and

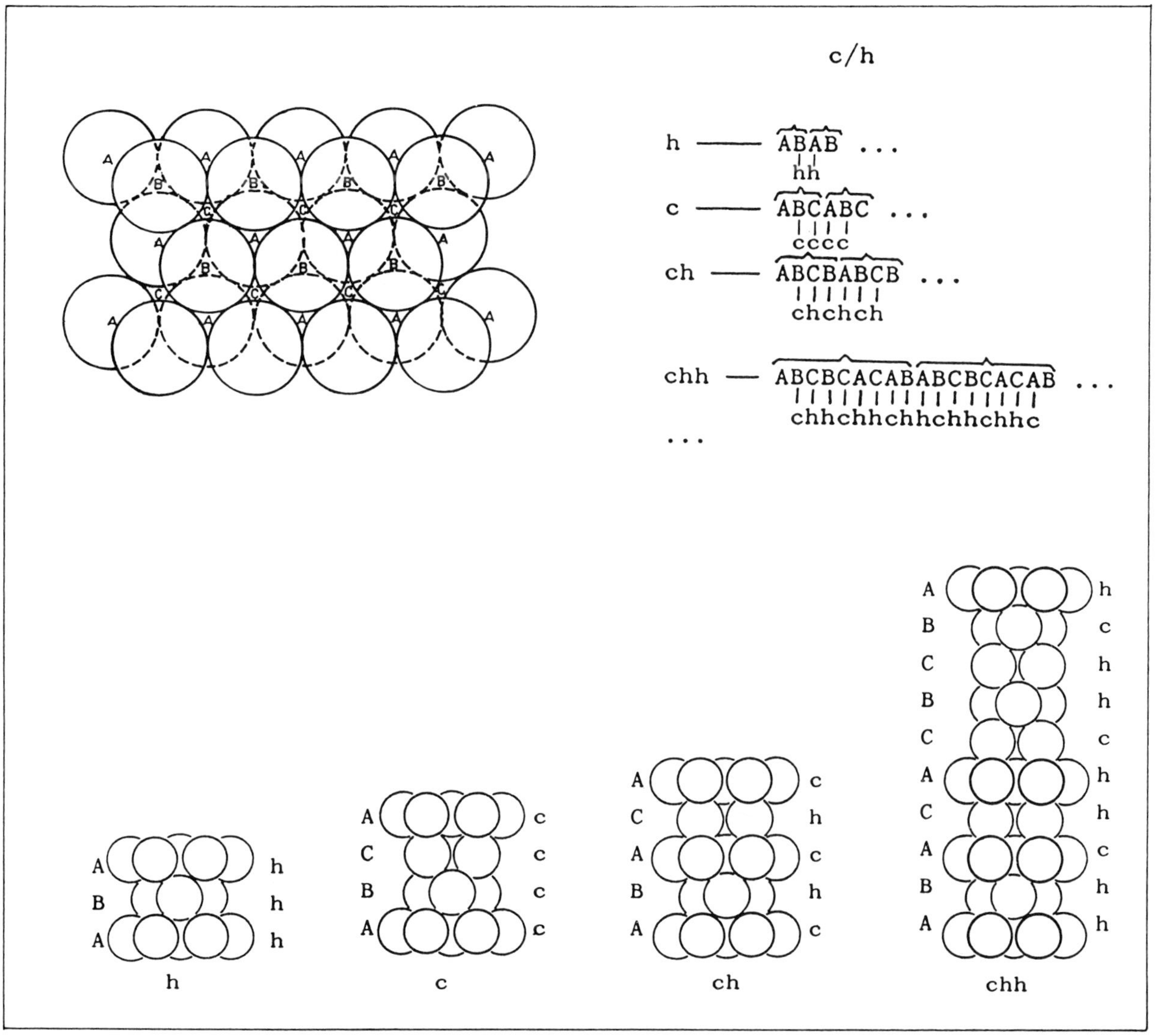

Fig. 13. The closest stacking of two closest-packed layers A and B. Some simple cases of stacking sequences of closest-packed layers, c/h, and the corresponding ball packing drawings.

Tchh for Sm or, in abbreviated form, h, c, ch, and chh, respectively (Figure 13).

The interstices generated in these closest packings are mainly the tetrahedral and octahedral voids. If we consider two layers stacked one over the other, we may notice that all the octahedral voids (r = 0.414R) are at the same level, while the tetrahedral voids (r = 0.225R) are located in two different levels, namely, low and high. A low level tetrahedral void lies in between three spheres with one on the top; a high level tetrahedral void lies on the top of a sphere with three spheres over it (Figure 14).

One can also imagine triangular voids corresponding to the position between three atoms and at the level of the centre of these packing atoms, forming a triangle, as the polygon of coordination. However, these voids can only be occupied by atoms which are very small (r = 0.155R) in relation to the packing atoms of radius R; for this reason they are very rarely occupied in mineral structures, and are not indicated on Figure 14.

In closest packings there are as many octahedral voids as packing atoms, and twice as many tetrahedral voids (Figure 14).

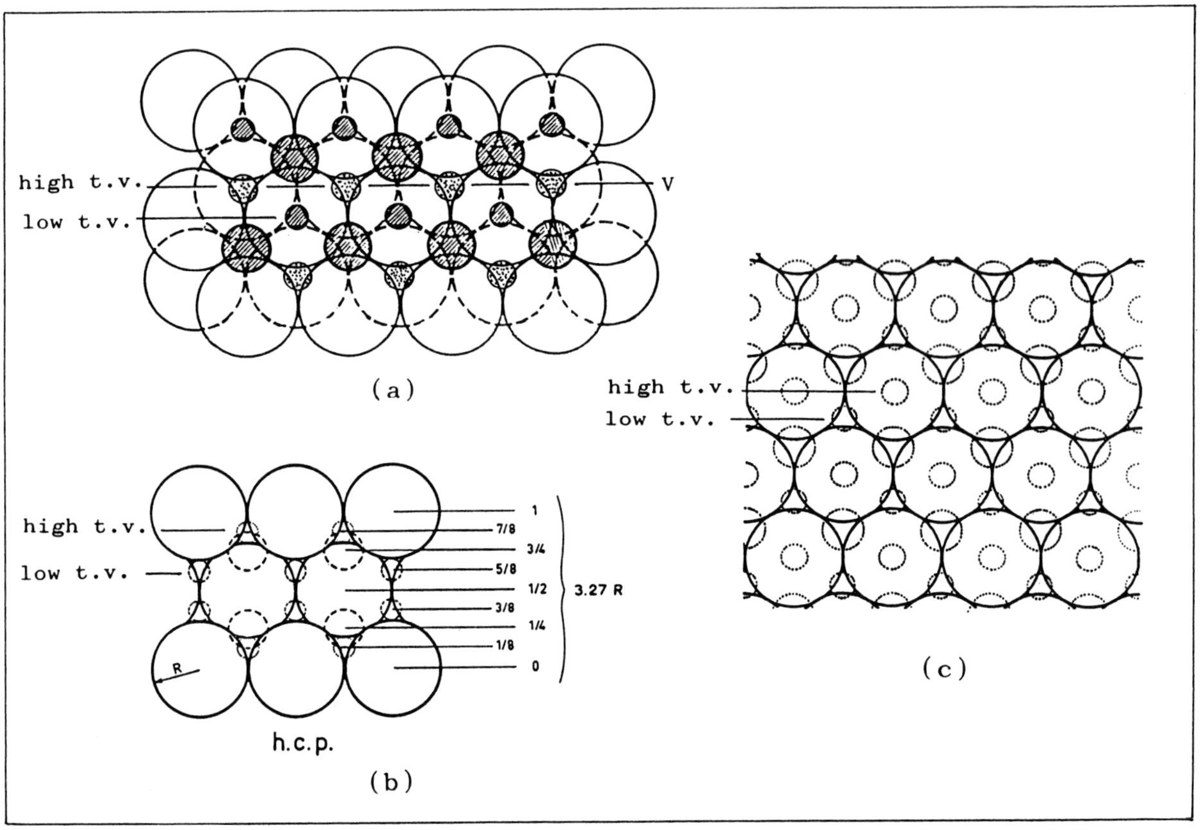

Fig. 14. (a) Generation of octahedral and of tetrahedral (low and high) voids by the closest stacking of two closest-packed layers; (b) cross section along a vertical plane through three layers of a hexagonal closest packing; and (c) representation of the voids (dotted circles) in a closest-packed layer generated by a closest stacking.

The simple hexagonal packing is generated by the stacking of closest-packed layers, not over the holes, but over the atoms themselves (superimposition), each packing atom touching only one packing atom of the layer below. This kind of stacking has, obviously, no alternatives, that is, only one kind of this packing exists, and it generates two categories of interstices: trigonal prismatic (p) (r = 0.527R), and square (s) (r = 0.414R). The structure of AlB_2 is based on a simple hexagonal packing of aluminium atoms, with boron atoms in trigonal prismatic voids (Figure 15).

One may now consider still other ways of stacking the closest-packed layers T, for instance, 'over valleys', that is, in such a way that in each layer the packing atoms touch only two packing atoms of the layer below. On the stacking over valleys, the projections of the centres of the atoms of the second layer may occupy positions α, β or γ (Figure 16). The simple sequences are b = $A\alpha A\alpha$..., v = $A\alpha\beta A\alpha\beta$...and d = $A\alpha\beta\gamma A\alpha\beta\gamma$... (Lima-de-Faria & Figueiredo, 1976). This kind of stacking over valleys generates three types of voids: distorted trigonal anti-prismatic (ap) (r = 0.225R) and two categories of distorted tetrahedral voids (r_1 = 0.291R and r_2 = 0.323R).

An example of a structure with such kind of packing is β-beryllia, one synthetic polymorph of BeO, where the oxygen atoms form a packing designated by Tb and beryllium atoms are located in distorted tetrahedral voids. This packing is also designated by 'tetragonal packing' because of its tetragonal symmetry (Matkovich, Giese & Economy, 1965).

If other kinds of layers are now considered besides the closest layers (T), other close packings may be generated and, some of them, are of great interest. Let us start from the square (or

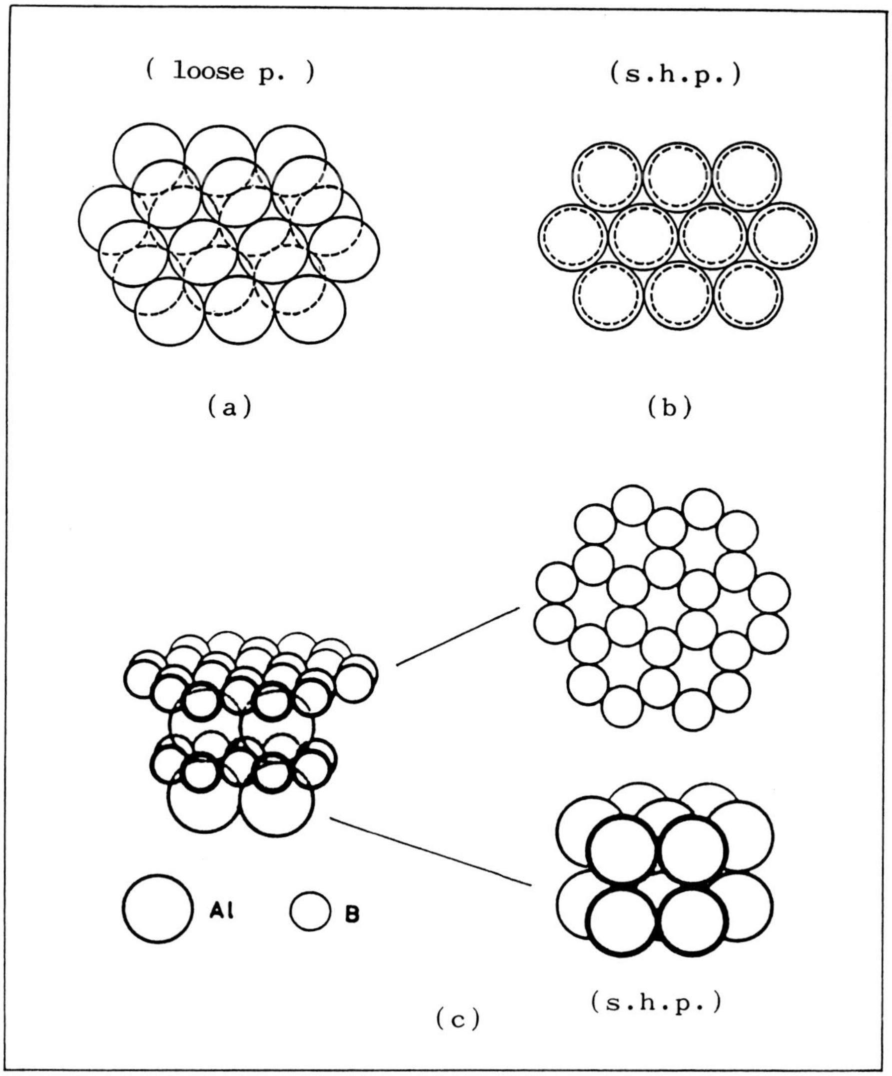

Fig. 15. (a) Closest layers stacked over 'valleys' (loose packings); (b) simple hexagonal packing (s.h.p.); (c) AlB_2 structure; the aluminium atoms form a simple hexagonal packing and the boron atoms occupy all the triangular prismatic voids with a honeycomb pattern (adapted from Laves, 1956).

quadrangular) layers (Q). The stacking of these layers over the holes (position B) with a sequence ABAB... (Figure 16) gives rise to the cubic closest packing, already described (Figure 17). In fact, instead of stacking the closest-packed layers, which are perpendicular to the triad axes, we may obtain the same cubic closest packing by stacking square layers Q, which are parallel to the cubic faces, in a sequence f = ABAB... (Figs. 16 and 17). The cubic closest packing can then be expressed either by T_c or Q_f.

It is handy to realize that the cubic closest packing can be generated in this way, because many cubic or tetragonal structures have unit-cell directions which can be easily marked on the Q layers, but only with difficulty on T layers.

The stacking of Q layers not over holes but over 'valleys', so that each packing atom touches only two atoms of the layer below, generates two main kinds of close packings, one with a sequence $A\alpha A\alpha$... designated by Qb, and another with a sequence $A\alpha B\beta A\alpha B\beta$... called Qd. This kind of stacking generates trigonal prismatic and square interstices (Figure 18). The Qb packing is equal to

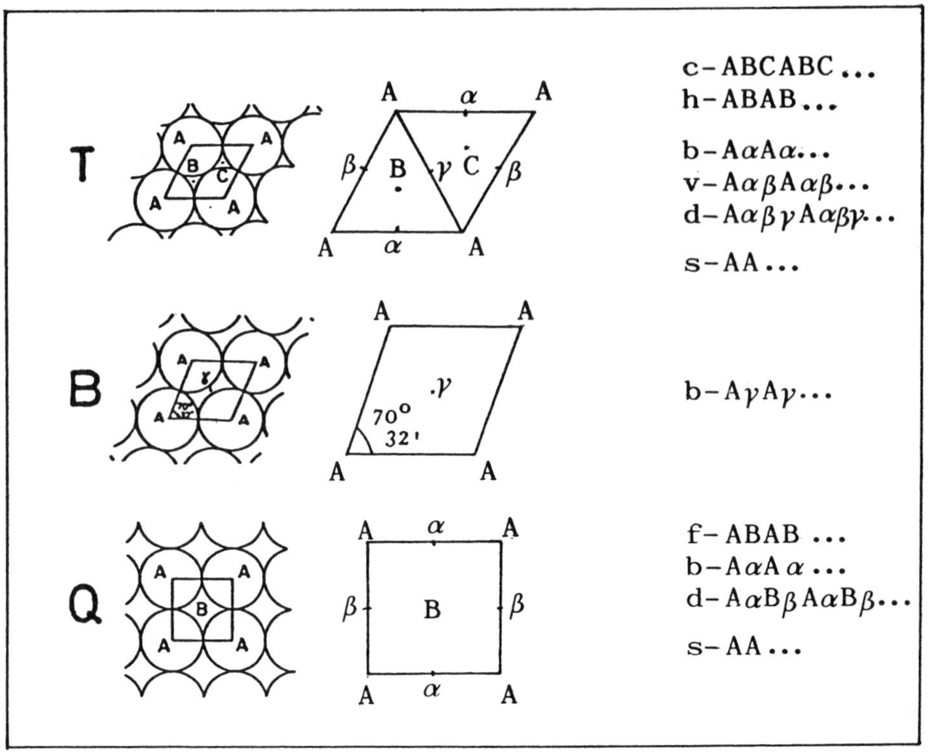

Fig. 16. The main stacking sequences generated by the packing of triangular layers (T), of rhombic layers (B) (which form the body-centred cubic packing), and of square layers (Q) (adapted from Lima-de-Faria & Figueiredo, 1976).

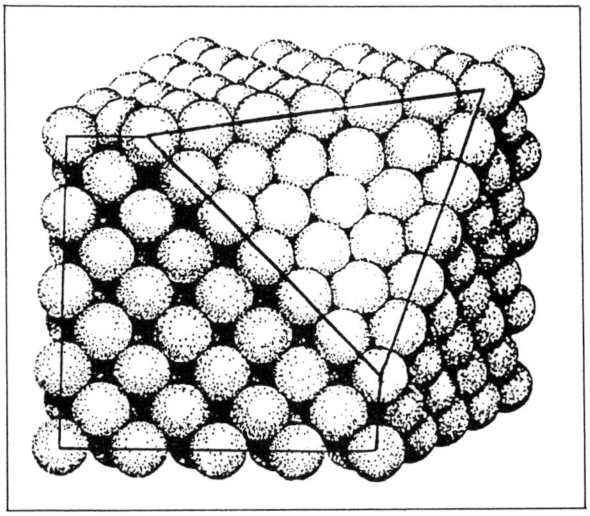

Fig. 17. Cubic closest packing of spheres showing the two kinds of packing layers T and Q, which can generate it (adapted from Zemann, 1969).

the simple hexagonal packing Ts, but Qd is a new close packing. Examples of structures based on a Qd close packing are PNb, where the Nb atoms form the packing and the P atoms occupy square voids, and Si_2Th, where the Th atoms form a Qd packing and the Si atoms occupy triangular prismatic voids.

The stacking of Q layers, where there are superimposition of the atoms, sequence s = AA..., is called simple cubic packing and gives rise to cubic voids and square voids. Examples of structures based on simple cubic packings (Qs), are: fluorite, CaF_2, where F forms the simple cubic packing and Ca occupies cubic voids (see Figure 53 of Chapter 7), and cooperite, PtS, with S forming the simple cubic packing with Pt in square voids. The different voids generated by the various stackings of layers T or Q are represented on Figure 18.

The body-centred cubic packing is another category of close packing and it can be described by the stacking over 'valleys' of rhombic layers designated B layers (Figure 19). These packing layers are parallel to (110) planes of the body-centred cubic packing and are intermediate between the T and Q packing layers. They may be imagined as distor-

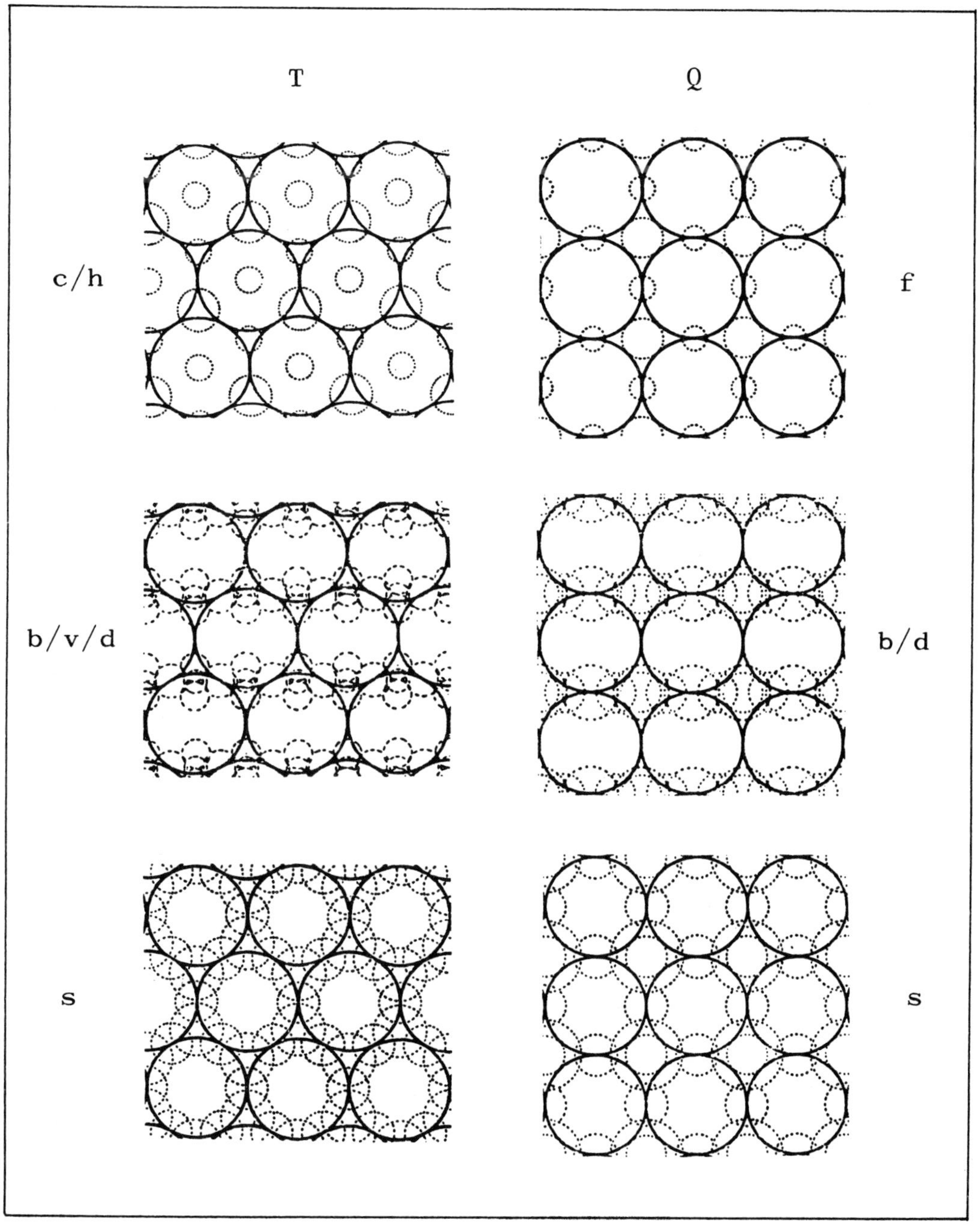

Fig. 18. Representation of the main interstices (dashed circles) generated by the three types of packing layers T and Q, corresponding to the various ways of stacking sequences: over holes, c/h, f; over 'valleys', b, v, d; and by superimposition s (adapted from Lima-de-Faria, 1965b; and Figueiredo & Lima-de-Faria, 1978).

tions of either the triangular layers T (60° → 70° 32'), or the square layers Q (90° → 70° 32') (Figure 16). This kind of packing generates two main categories of voids: distorted octahedral (r = 0.155R) and distorted tetrahedral (r = 0.291R) (Figure 19). Since the voids of both categories are very small, few examples of such packings are found in mineral structures.

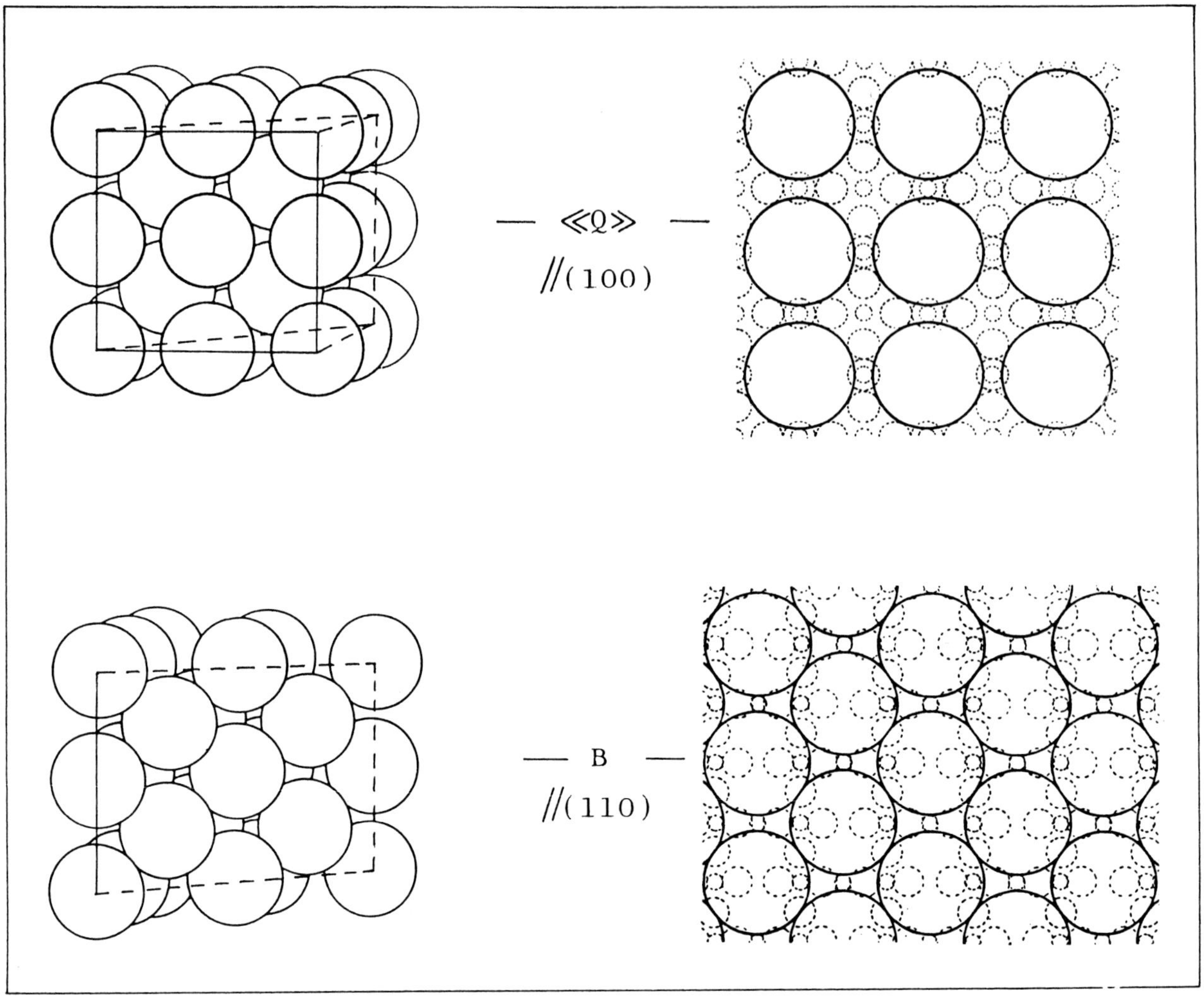

Fig. 19. Packing drawing of the body-centred cubic packing showing two layers: one parallel to the (110) plane (the rhombic layer **B**), and another parallel to the (100) plane (the open square layer «Q») (adapted from Figueiredo & Lima-de-Faria, 1978).

This same body-centred cubic packing, like the cubic closest packing, can be represented by the stacking of other kind of layers, the square *open* layers, «Q» where the spheres are not tangent to each other. Double angular brackets are used to distinguish it from the normal Q layer of the cubic closest packing (Figure 18). Its sequence is also f = ABAB... (Figure 19). The double angular brackets are also used to indicate any derivation from an imaginary packing, that is, only geometrically valid. An example is tellurium, ꝺ [$Te^{«c»}$], where the tellurium atoms form infinite chains but are distributed geometrically in an approximate cubic closest packing.

The less dense close packings which are obtained by stacking closest packed layers over 'valleys' have been called 'loose packings' by Kripyakevich (1973). Similar modes of stacking have been found in structures based on square layers (Q). Therefore, these close packings are included under the same designation of *loose packings*.

The closest-layers T are formed exclusively by triangles of spheres, and the square layers Q by squares of spheres. Continuing this generalization of packings, consideration may now be given to other kind of layers formed by triangles and spheres (Figure 20). Some of these layers are called R^{mn} because they are formed by m rows of

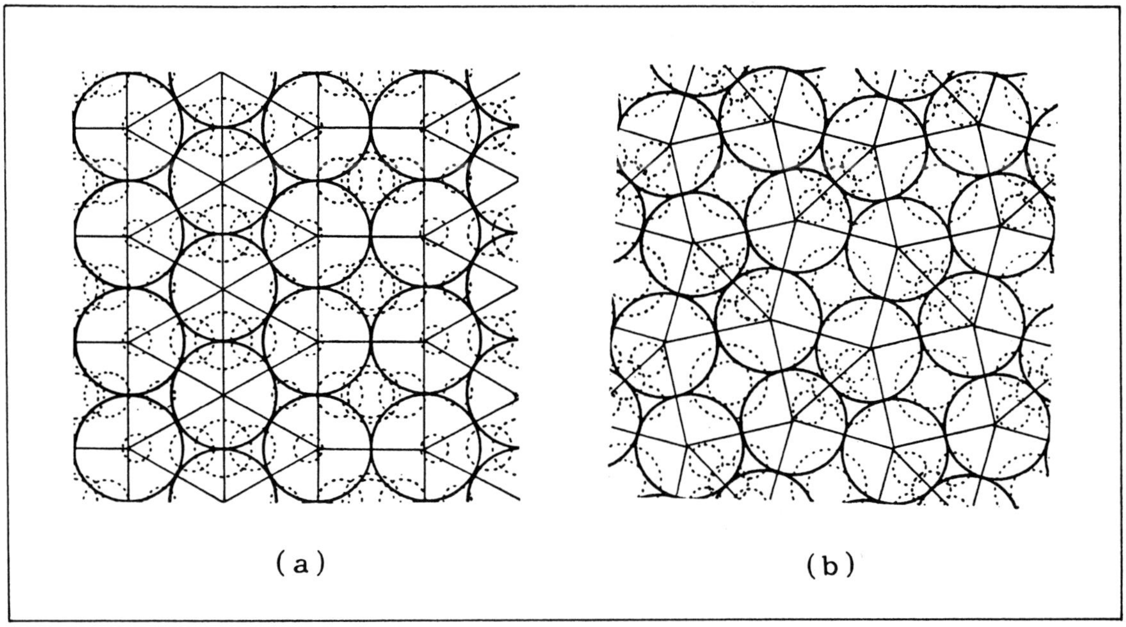

Fig. 20. (a) R^{21} layers with voids (dashed circles) generated by a stacking 'f' (after Figueiredo, 1976) and (b) N^{21} layer with voids generated also by a stacking 'f' (after Figueiredo & Lima-de-Faria, 1977).

triangles and n rows of squares. Lepidocrocite, γ-FeOOH, is a structure where the oxygen ions and the hydroxyls form R^{31} layers, that is, with three rows of triangles followed by one row of squares.

Another kind of layer which also occurs in certain inorganic structures is the one designated by N^{21}, which is formed by interconnected triangles and squares, in the proportion of two triangles for one square (Figure 20). Examples are: $CuAl_2$, where Al atoms form N^{21} layers, with Cu occupying anticubic voids; and $NbTe_4$, with N^{21} layers of Te atoms and Nb occupying anticubic voids.

All the close packings that have been mentioned can be described in terms of the stacking of layers of atoms. However, there are structures with a high value of packing efficiency but that cannot be decomposed into layers. The packing has a three directional character, as in garnet, $Ca_3^{do}Al_2^{o}Si_3^{t}[O_{12}]^{*}$. This kind of homogeneous close packing is indicated by an asterisk.

So far, discussion has been restricted to packings where the layers are all alike and the kind of stacking is the same. However, continuing this generalization, one may also consider heterogeneous packings formed by different ways of stackings or by different layers of atoms.

There are packings formed by the same kind of layers but stacked in different ways. An example is molybdenite, MoS_2, which is formed by slabs of two T layers stacked in 's' (simple hexagonal), these slabs being stacked together in the closest way, with a *h* sequence. On the whole, the complete stacking is (2Ts)h of the S atoms, with Mo atoms in trigonal prismatic voids (Figure 21).

Matlockite, PbFCl, is built of two kinds of square packing layers but differing by the relative size of their atoms. The square layers formed by the larger packing atoms, designated by Q^1, correspond to Cl atoms and the layers formed by the smaller packing atoms, Q^2, correspond to fluorine. The packing Q^2Q^1 is built of alternate Q^2 and Q^1 layers, stacked together as close as possible, designated by *f*, with Pb atoms in the voids with coordination nine (Figure 22).

There are other kinds of heterogeneous packings which correspond to interpenetrated slabs of homogeneous closest packings (c/h) or of other close packings (e.g., simple hexagonal and closest packings, Ts/h). They pertain to the so-called recombination structures. Examples are: galenobismutite, $Pb^{[7]}Bi_2^{[6/7]}[S_4]^{c/h}$, and apatite, $Ca_5^{p}P_3^{t}[O_{12}(OH, F)]^{Ts/h}$.

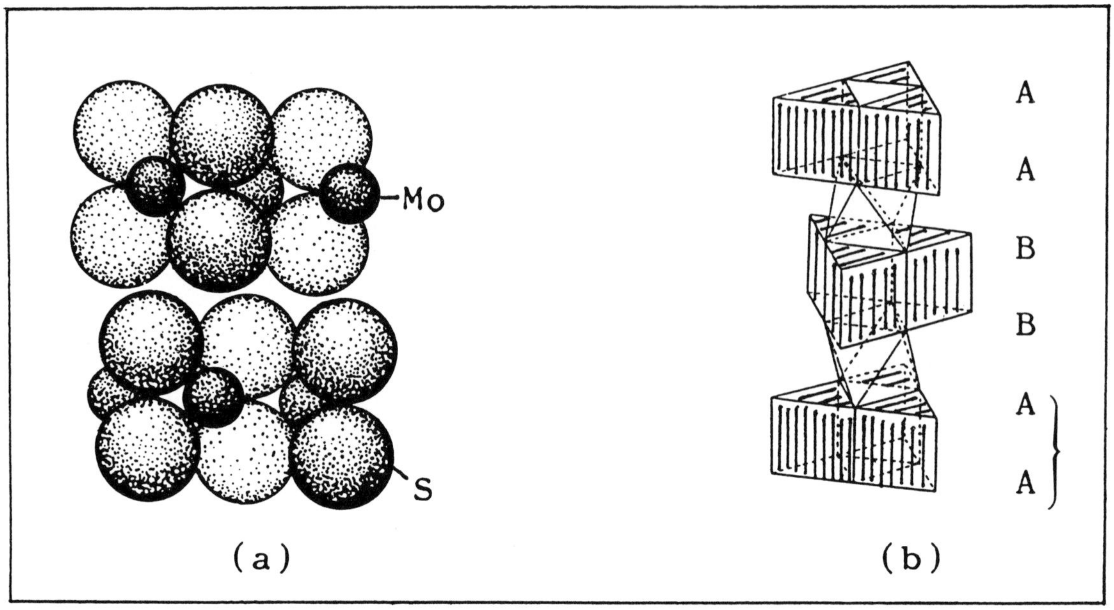

Fig. 21. Representation of the molybdenite structure, MoS_2, with a heterogeneous close packing (2Ts)h of sulphur atoms with Mo in trigonal prismatic voids: (a) close-packed description, and (b) polyhedral drawing (after Wyckoff, 1963, and adapted from Kostov, 1968, respectively).

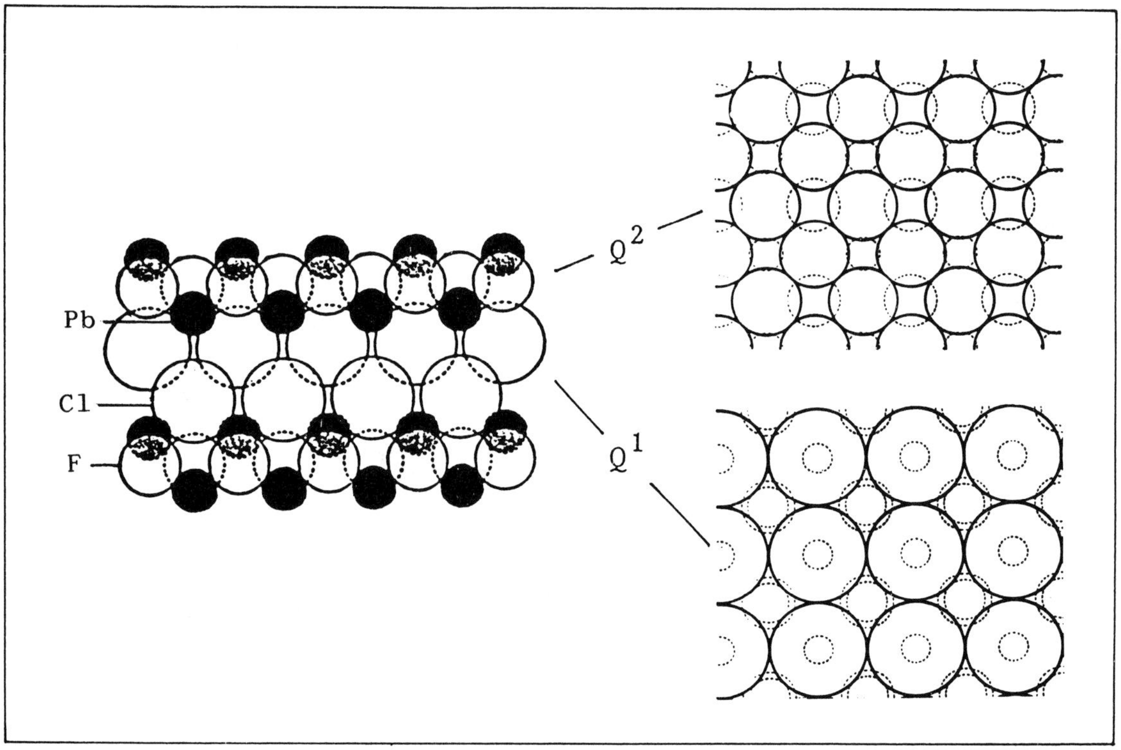

Fig. 22. (a) Packing drawing of the matlockite structure, PbFCl, projected along one of the *a* axis (adapted from Wells, 1962), (b) The corresponding Q^2 and Q^1 packing layers with the voids (dashed circles) generated by the 'f' stacking (adapted from Figueiredo & Lima-de-Faria, 1991b).

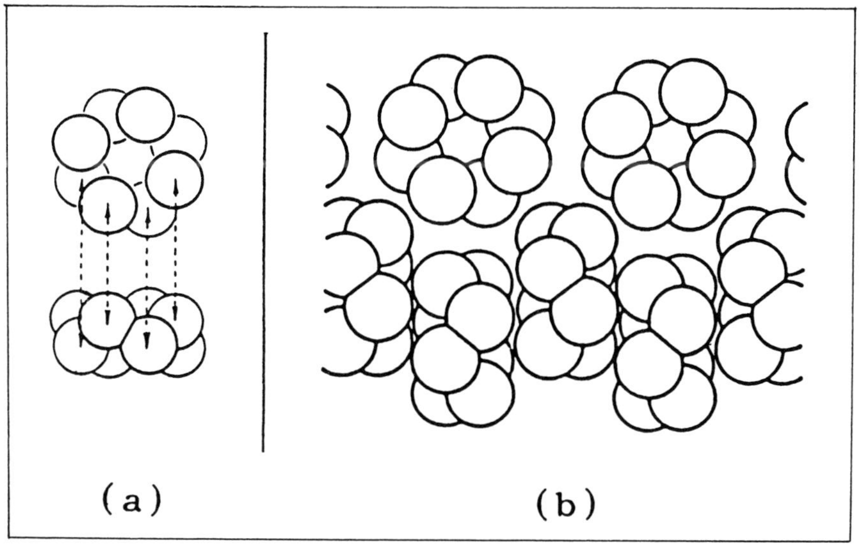

Fig. 23. (a) Representation of the S_8 molecules of the α-form of sulphur (after Kostov, 1968). (b) Packing drawing showing how the S_8 molecules are packed together (after Bunn, 1964).

For a more detailed description of the various close packings, their corresponding layers and voids, see Lima-de-Faria & Figueiredo (1990a and b) and Figueiredo & Lima-de-Faria (1991a and b).

3.2.2. The packing of groups, chains and sheets

The other kinds of structural units also tend to pack together as closely as possible. When the structural units are groups, they tend to behave as organic molecules. The molecules tend to stack in such a way that the bumps in one are inserted in the hollows of another molecule. A very complete study of the packing of molecules in organic structures has been made by Kitaigorodskii (1955, 1961), based either on finite molecules, infinite molecular chains, or infinite molecular sheets. This study may be applied to the corresponding categories of inorganic structural units.

Let us consider the packing of groups. An example is the α-form of sulphur, which is built of S_8 groups (Figure 23).

As an example of infinite chains, we chose the pyroxenes, which pack in a close way (Figure 24). The infinite chains do not pack side by side (Figure 24b), but alternately up and down, giving rise to a better fit (Figure 24c).

For sheet structures many examples could be given, such as muscovite, kaolinite and talc, the 'skeletons' of which are based on silicate sheets, $[Si_2{}^tO_5]$, packed together in a dense way.

3.2.3. The connectivity in frameworks

In framework structures, the structural unit is the whole framework. However, the framework may be imagined subdivided in parts, the so-called 'connected units' (Lima-de-Faria & Figueiredo, 1976), so as to facilitate its description and representation. In this way a framework may be decomposed into sheets, or chains, or even groups, which by connection give rise to the whole framework. Whenever possible, the framework is decomposed into layers, because this is the best way which leads to a simpler representation. Examples of frameworks built of connected sheets are: cristobalite and tridymite, which may be considered formed by silicate puckered sheets connected in 'c' or 'h' close packing, respectively. The connected units also tend to link together in a close way, as if they were structural units.

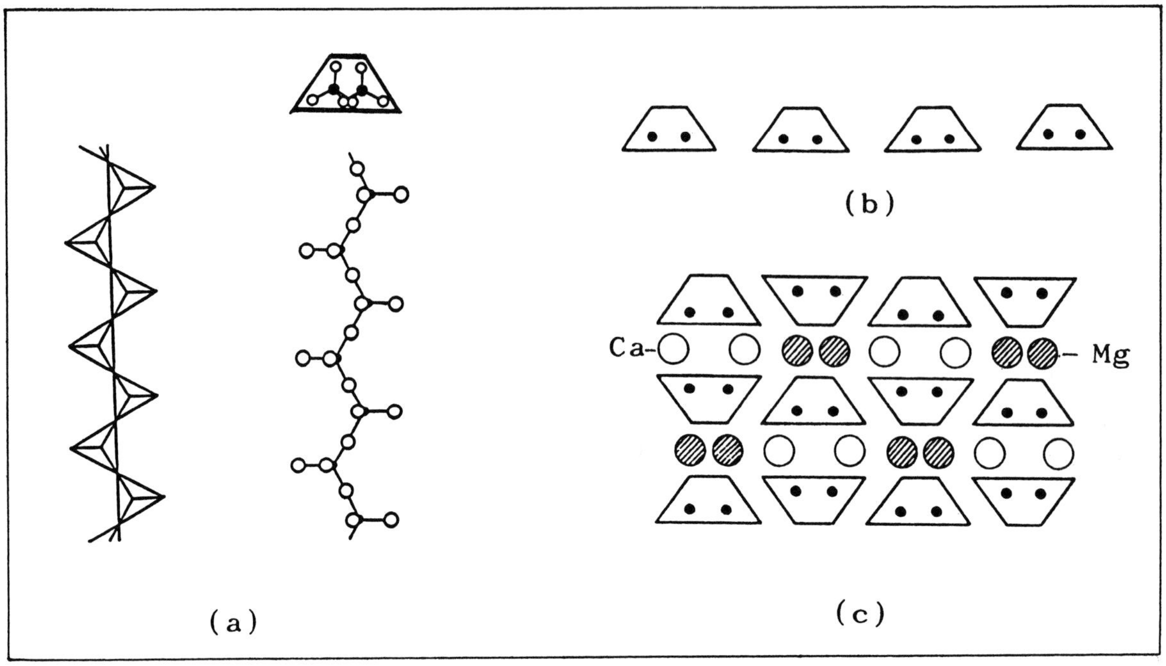

Fig. 24. (a) Two schematic representations of the silicate SiO_3 pyroxene chain and its cross section (after Kostov, 1968). (b) Imagined packing of the pyroxene chains in a non-close way: chains just packed side by side. (c) Representation of the real structure of diopside, $CaMgSi_2O_6$, showing the way these chains pack together in the closest way (after Lima-de-Faria & Figuiredo, 1990b).

CHAPTER 4

The stability of crystal structures

4.1. The general conditions of stability

Mineral structures are possibly the most stable among inorganic crystal structures. Bragg (1964) said:

> A mineral must be very stable in order that it may exist for so long, and so must have a structure of minimum energy.

In fact, the general condition of stability for a physical system corresponds to a minimum free energy G (of Gibbs),

$$G = U + pV - TS$$

where U is the internal energy of the system, p and V the pressure and volume, and T and S the temperature and entropy, respectively. When this principle is applied to crystal structures, it is concluded that, for stability at a given pressure and temperature, the internal energy of a structure should be a minimum, and its entropy, a maximum. A minimum internal energy corresponds to a minimum volume; this means that the atoms in a crystal structure will tend to be as close as possible. The maximum entropy should not be interpreted as a maximum disorder, but as a maximum homogeneity of the atomic distribution. A periodic orderly distribution may possibly be the most favourable situation. According to Buerger (1971)

> a random assemblage is one of greater potential energy than an ordered one.

The measure of the internal energy, U, of a crystal structure has been the object of several studies. The theoretical studies of Madelung (1918) and of Born & Landé (1918) were made on alkali halides, and afterwards developed by Born (1919), Haber (1919), Kapustinskii (1933) and Fersman (1935). This enabled theoretical values to be established for simple ionic structures, which are very nearly the experimental values. However, satisfactory results have only been obtained for relatively simple ionic structures.

4.2. The extension of Laves principles to minerals

According to Laves (1956, 1963), what governs the stability of the alloy structures are three main principles: the 'space filling principle' (tendency of the atoms to optimize space filling), the 'high symmetry principle' (tendency to form arrangements of high symmetry), and the 'connection principle' (tendency of the atoms to form connections of high dimension). However, he admitted that other physico-chemical factors, such as the bond factor and the temperature factor, may also be important.

The space filling principle is in complete agreement with the minimum free energy condition, which implies a minimum volume of the atomic distribution. The high symmetry principle possibly expresses the equilibrium of forces reaching each atom; thus, for atoms of the same kind, it would also mean equal interatomic distances.

Certain structures, like pyroxenes, amphiboles, and micas, may be considered intermediate cases between their symmetrical and their packing analogues (Lima-de-Faria, 1988a). Belov (1951) was the first to consider these two extreme ideal cases in micas (Figure 25a). For micas with Si_2O_5 layers, the possible SiO_4 tetrahedral distribution in a closest-packed layer can only have triangular shaped rings (Figure 25b), and the layer with hexagonal rings corresponds to the symmetrical analogue and does not fit in any closest-packed layer (Figure 25c). Moreover, Kitaigorodskii (1955) has also emphasized the importance of these two tendencies for layer stacking in molecular structures:

The nature of the layer stacking, like the

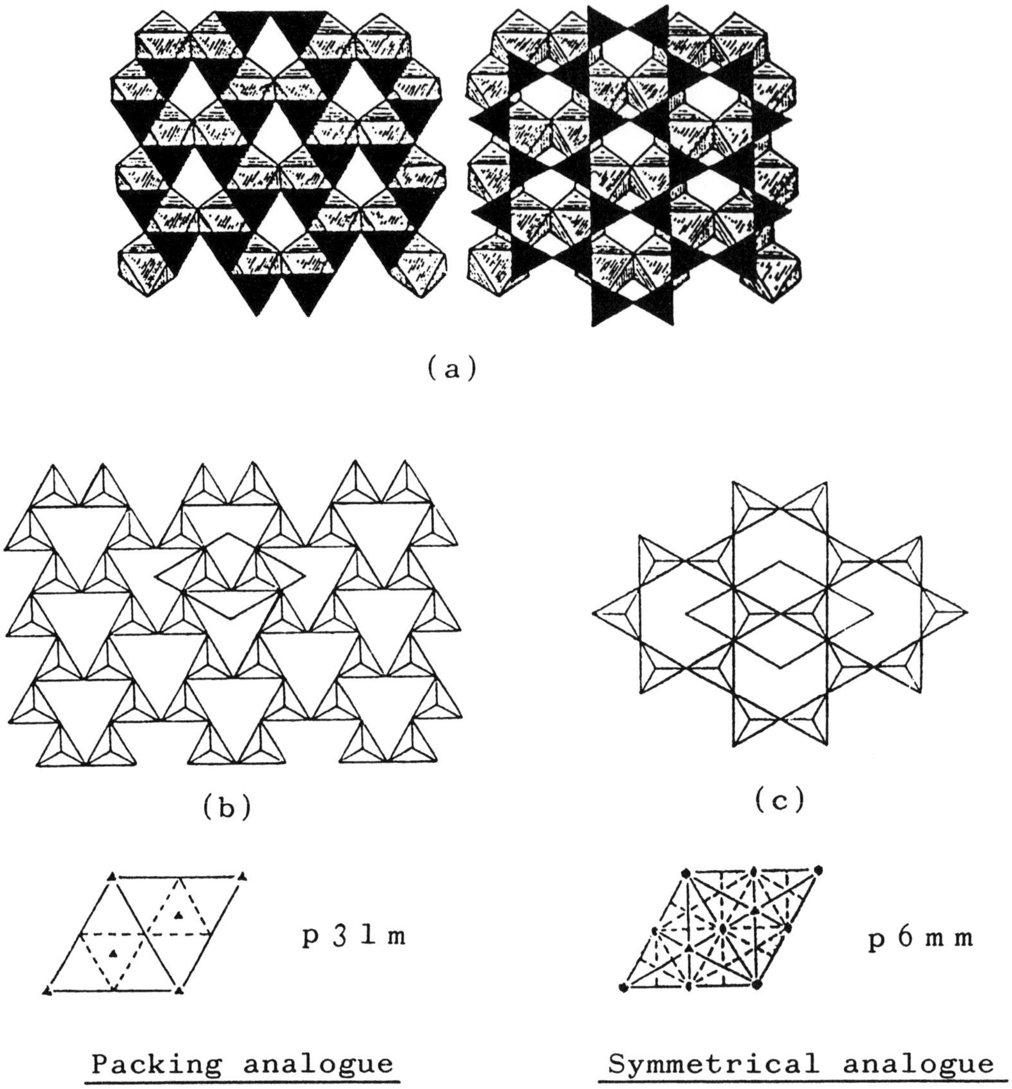

Fig. 25. (a) Extreme ideal models of mica (after Belov, 1951); (b), (c) Tetrahedral sheets based on triangular shape rings, and on hexagonal shape rings, to which correspond the plane groups p31m and p6mm, respectively (after Lima-de-Faria, 1988a).

molecular packing in a layer, is determined by the closest packing requirements and by the tendency for the symmetry of the molecular arrangement to be as high as possible.

Consequently, the Laves principles of stability can possibly be extended to mineral structures by considering that *there is a tendency for the formation of highly symmetrical structural units, and also a tendency for the close packing of such structural units.* These two tendencies may conflict with each other, and give rise to intermediate situations between the two extreme ideal analogues (Lima-de-Faria, 1988a).

4.3. Stability rules

4.3.1. Rules governing the polyhedral constitution of inorganic crystal structures (Pauling rules)

One of the ways to look at structures is to consider the coordination polyhedra formed by the anions around the cations, and how they are linked to each other. Pauling (1929) has proposed five rules for ionic structures, which govern the linkage of these polyhedra (whether linked by corners, edges or faces), in order to determine the possible stable structures.

First rule:

A coordination polyhedra of anions is formed about each cation, the cation-anion distance equalling the sum of their characteristic packing radii and their radius ratio determining both the nature of the coordination polyhedron and, therefore, the coordination number of the cation.

In other words, the maximum number of large ions X that can be simultaneously tangent to a smaller ion A is geometrically dependent on the value of R_A/R_X, their radius ratio.

Second rule:

The strength of an electrostatic bond may be defined as a cation's valence charge divided by its coordination number. Pauling's second rule states that

an ionic structure will be stable to the extent that the sum of the strengths of the electrostatic bonds that reach an anion (from adjacent cations) equals the charge of that anion but with opposite sign.

In NaCl, for example, each Cl^- anion is surrounded by 6 Na^+ neighbours, each of which directs a bond of strength +1/6 towards it. Consequently, the negative unit of charge of a Cl^- anion is completely neutralized by the six +1/6 bonds from the adjacent Na^+ ions; no portion of the positive charge from more distant cations is required to neutralize the Cl^- anion charge. Pauling's second rule calls for the neutralization of charge around an anion in an ionic structure to be sharply localized, that is, at short range distance. Hence, the rule is often known as the electrostatic valence principle.

Third rule:

The sharing of edges and particularly of faces, by two anion polyhedra, decreases the stability of an ionic crystal structure.

This is particularly true if the cation within each anion polyhedron is highly charged, has a small coordination number, and has a radius ratio in respect to the anions that is near the lowest limit tolerated by the anion polyhedron. This third rule might be acceptably restated as: highly charged cations prefer to maintain as large a separation as possible within a structure and to have anions intervening between them so as to screen one from the other. Either feature decreases the potential energy of a crystal by minimizing the repulsive forces existing between neighbour cations.

To illustrate Pauling's third rule we may consider the tetrahedron that four O^{2-} ions typically form around a Si^{4+} ion in silicate crystals. In such crystal structures, these tetrahedra often share corners, but not edges or faces. Two reasons dictate this, namely, (i) the cation-to-cation distance is the greatest for corner sharing and the least for face sharing, and (ii) the O^{2-} ions positioned between one cation from the adjacent ones screen most effectively for corner sharing and least effectively for face sharing.

Fourth rule:

In a crystal structure containing different cations, those of high valency and small coordination number tend not to share polyhedron elements with each other.

This 'fourth rule' is merely an extension of the third one and the considerations that justified the former also justify the latter.

Fifth rule:

The number of essentially different kinds of constituents in a crystal tends to be small.

If the simplifying assumption is adopted that most crystal structures consist of a closest packing of the large ions, or a variation thereof, with the smaller ions in the interstices, then the fifth rule may be restated as: the number of types of interstitial sites present within a periodically regular packing of anions tends to be small. This is also called the rule of parsimony.

4.3.2. The bond-valence rules (Brown rules)

To predict the structure and properties of complex structures the bond-valence method can be of considerable help. Basic to the method is the prediction of both atomic valences and bond lengths, as achieved by solving a model that is based on a network of chemical bonds.

The fundamental concept of bond-valence stems from the assumption that the valence of an atom is distributed among the bonds that such an atom forms. Because bond-valences are found to correlate well with a number of parameters, namely with bond lengths, experimental values of the latter can be used to estimate values of the former. These may then be of assistance in the evaluation of crystal structures.

In ionic networks, every bond has a Lewis acid (cation) at one end and a Lewis base (anion) at the other end. Brown (1981) established two empirical rules for compounds with acid-base networks:

Stoichiometric rule:

The total valence of the Lewis acids is equal to the total valence of the Lewis bases.

Valence rule:

The sum of bond valences at each atom is equal to the atomic valence.

This rule is closely related to Pauling's electrostatic valence rule.

When identification is needed of elements that cannot be easily distinguished by X-ray diffraction, resorting to experimental values of atomic valences may prove quite useful. Si and Al are two such elements, because they frequently occur in a similar environment in a given crystal and they have nearly the same X-ray scattering factors. However, an atom can be readily identified as Si if its atomic valence is found to be 4.0 and as Al if its atomic valence is found to be 3.0. Should the atomic valence be found to be 3.60, the conclusion would be drawn, to a first approximation, that the site is occupied 60% by Si and 40% by Al.

4.3.3. *Rules governing the layer organization of inorganic crystal structures*

The distribution in space of the structural units, which plays a fundamental role in the constitution of crystals, is a three dimensional problem (3D), though in most cases it can be formally decomposed into a two-plus-one dimensional problem (2D + 1D), that is, into the organization of the structural units in layers, and the way the layers stack together (Lima-de-Faria & Figueiredo, 1976). The systematic derivation of simple inorganic close-packed structures based on the stacking of equal or alternate layers (Lima-de-Faria, 1965b; Lima-de-Faria & Figueiredo, 1969) and the use of condensed models have shown that the constitution and the stacking of the layers obey certain rules (Lima-de-Faria, 1978).

In the representation of layered structures by condensed models (Lima-de-Faria, 1965a) the structures are sliced into atomic layers formed by the packing atoms and the interstitial atoms which are immediately above them, so that the interstitial atoms which are below it will pertain to the layer located underneath. The slice formed by the packing layer and the corresponding interstitial atoms, as defined, is called 'constructive layer', to distinguish it from the packing layer alone, which involves just the packed atoms (Lima-de-Faria, 1978). The stacking of the constructive layers generates, completely, the crystal structure.

Many inorganic structures are built of the stacking not only of equal packing layers but also of equal constructive layers; this means that the distribution pattern of the interstitial atoms (Lima-de-Faria, 1965a) is also the same in the constructive layers. This fact should be a very difficult guess with the sole help of the usual drawings and models of crystal structures; it only became apparent through the systematic use of condensed models.

The concept of constructive layers does not apply exclusively to the close packing of atoms, that is, to close-packed structures, but may be extended to structures based on the packing of other structural units such as groups, chains or sheets, and even to framework structures. The problem is to find the proper plane direction in the structure which leads to a simple description, and that in general coincides with the plane direction of the highest density of atoms. Most of the inorganic structures admit such a description by layers (Lima-de-Faria & Figueiredo, 1976).

The rules which apply to constructive layers (in what follows called simply 'layers') of inorganic crystal structures (Lima-de-Faria, 1978) are:

1. Inorganic structures are in general built up of equal layers, in certain cases of two alternate layers, and in a very few cases of more than two different layers.

This rule expresses the simplicity of the architecture of crystal structures. It is related to the relatively small values of the lattice parameters and to the symmetry within the unit cell (Figure 58).

2. The way the layers stack together is, in general, the same for all the layers.

This expresses a short-range interaction in the structure, which seems to be related to local electrical neutrality.

3. In each layer the atoms of the same kind tend to be crystallographically equivalent, that is, to oc-

cupy a minimum number of sets of equivalent positions.

This rule expresses the fact that equal atoms tend to have the same environment, and corresponds to Pauling's fifth rule.

4. The distribution pattern of the atoms of the same kind (either interstitial or packed) tends to be as symmetrical and homogeneous as possible.

This rule expresses electrical equilibrium and is related to the symmetry principle (Laves, 1956) and the vector equilibrium principle (Loeb, 1970). Only in rare cases (certain particular proportions of interstitial atoms) is this rule not completely satisfied (Figueiredo, 1973).

5. The distribution pattern of the atoms of the same kind (either interstitial or packed) tends to be as distant as possible from the corresponding distribution pattern of the adjacent layers.

This rule expresses a minimum of electrostatic energy and corrresponds to the 'distant distribution rule' (Lima-de-Faria, 1965b). Related statements were presented by Iida (1957) for magnetic oxides, which he called 'assumption II', and by Radoslovich (1963) for certain silicate structures relative to the interstitial atoms and under the designation of 'cation-avoidance rule'. Pauling's third rule is again another form of expressing this statement, but in terms of the sharing of corners, edges and faces of coordination polyhedra.

These rules are especially useful in the domain of close-packed structures. It should also be noted that the rules presented here express tendencies, and do not take into account certain competing factors, such as the bond factor, which may lead to other equilibrium compromises.

CHAPTER 5

Structure and properties

5.1. Physical properties and crystal structure

There is an intimate relation between the internal structure of minerals and their properties. Povarennykh (1972) said:

> Polymorphic species (diamond and graphite, calcite and aragonite, kyanite and sillimanite) provide the best evidence that the structure type and bond distribution are the decisive factors as regards properties[...]; only a consideration of structural features (bond distribution and strength) can give a full conception of all the major properties.

Certain physical properties are more sensitive to the geometrical arrangement of the atoms in space, that is, to symmetry; for example, elasticity, thermoconductivity, electroconductivity and thermoexpansion (Figure 26). Others, such as cleavage, habit, hardness and twinning, are more related to the kind of structural units and their packing.

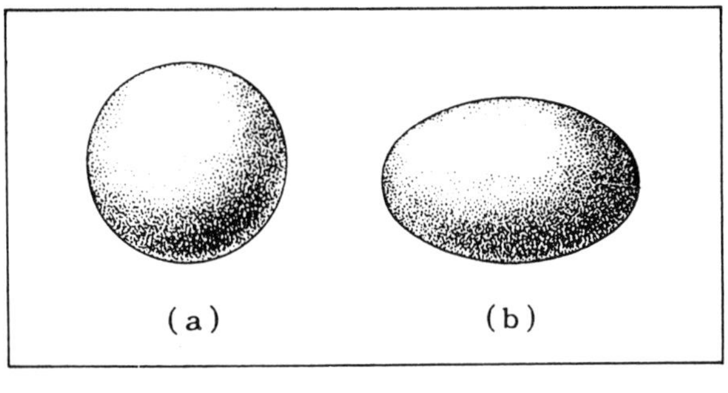

Fig. 26. Schematic representation of the anisotropic thermoexpansion of a crystal of low symmetry: (a) Cut sphere of the crystal; (b) The sphere takes the shape of an ellipsoid by the action of heat (after Chaskolskaia, 1959).

Hardness is related to the surface resistance of the sample to abrasion. The facility in breaking the bonds depends on the behaviour of the bonds under the effect of an external perturbation, and it is related to the symmetrical character of the electronic orbitals involved in the bonds. For instance, in diamond the crystal is built up of atoms linked by strong covalent bonds (sp^3 orbitals) which have a strong directional character. The displacement of an atom, or its removal, therefore requires the breaking of strong bonds, and this is why diamond is very hard.

The arrangement of the atoms in a crystal stucture is normally different in different directions and, as a consequence, the hardness may also depend on the particular face and direction under consideration. These differences are often rather small; however in the case of kyanite, Al_2SiO_5, the hardness may vary between 4 and 7 for the (100) face, depending on the direction.

The density of crystals depends on the atomic weight of the chemical elements and also on the degree of close packing of their atoms; normally, it will decrease from close-packed to framework structures.

The shape of crystals, their habit, depends on the relative rates of growth in the different directions. There is consequently a definite relation between structure and habit, though there are often obscure additional factors, such as the conditions of formation, which may play an important role. The kind of structural units seems to be the main factor. In sheet structures the habit is normally platy or leafy, in chain structures it is prismatic or acicular, and in close-packed, group and framework structures it is generally isometric, that is, the crystal develops more or less equally in all directions. However, in group structures, groups of certain particular shapes, like long sticks or flat disks, may give rise to platelike or needlelike crystals, respectively. In fact needleshaped molecules pack parallel like a bundle of sticks, and additional ones are added to their sides quite quickly, generating

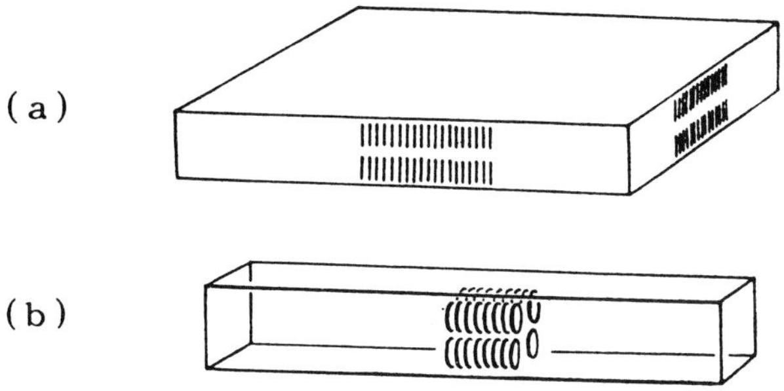

Fig. 27. (a) Needleshaped molecules, packed parallel, give platelike crystals, while (b) flat molecules, packed parallel, give needlelike crystals (after Bunn, 1964).

platelike crystals. Small flat groups tend to pack together more quickly in a direction perpendicular to the plane of the groups, forming needlelike crystals (Figure 27).

It is clear that crystals will break more easily along the plane surfaces containing weak bonds, and therefore the bond strength distribution will define the cleavage directions. Cleavage will easily reveal chain or sheet structures because it will be parallel to one or to two directions, respectively. Structures belonging to close-packed, group or framework structures will normally show cleavage along three directions.

Structures having a common geometric arrangement may however show different directions of cleavage due to different bonds between the atoms. Examples are diamond and sphalerite, ZnS. In diamond the cleavage is along the (111) plane, involving the smallest number of bonds per unit area, but in sphalerite it is along the (110) plane because the fragments formed have to be electrically neutral (Figure 28).

Optical properties are essentially symmetry dependent. The optical indicatrix is related to the symmetry, and the behaviour of crystals under polarized ligth is easily predictable on this basis. Optical properties are also greatly influenced by chemical composition. But the refractive index, birefringence and optical sign depend greatly on the structural character. The refractive index depends on the packing density and is normally low for framework structures. Birefringence and optical sign reveal minerals with chain or sheet structures if high or very high, respectively.

Twinning is related to interfaces of common atomic arrangement. The twinning of aragonite (Figure 29) is explained by its structure. The section of the crystal between the broken lines, the so-called twinning plane (Figure 29), may be regarded as belonging either to the original crystal on the left, or to the twin crystal on the right, since it conforms to both.

Every property may reflect, to a certain extent, the constitution of a structure, but it is by considering all the properties of a structure that its character may be revealed.

Epitaxy is an oriented overgrowth of two different crystalline substances, and is very dependent

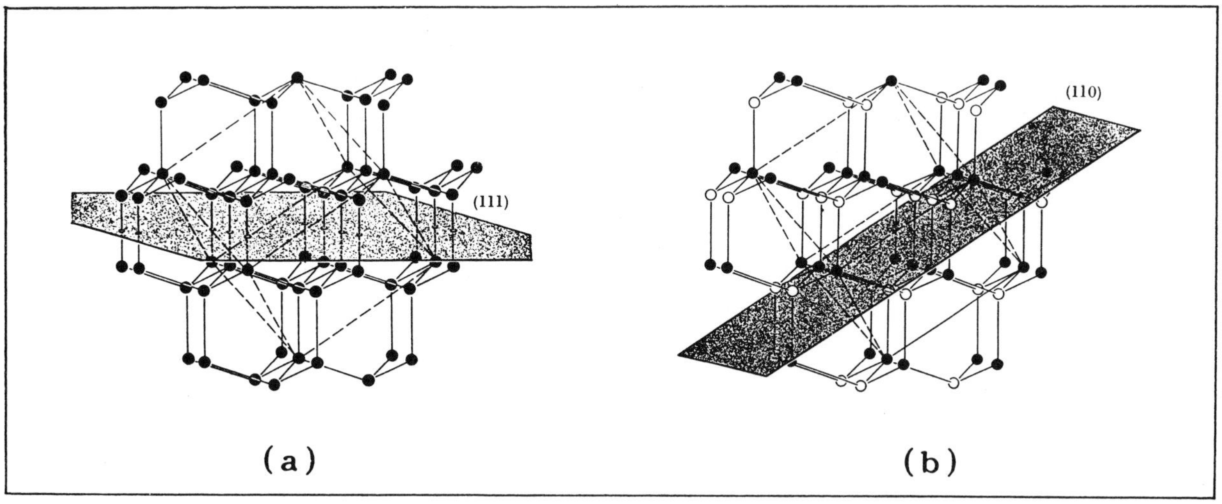

Fig. 28. (a) Structure of diamond showing cleavage plane (111), and (b) structure of sphalerite showing cleavage plane (110) (after Zemann, 1969).

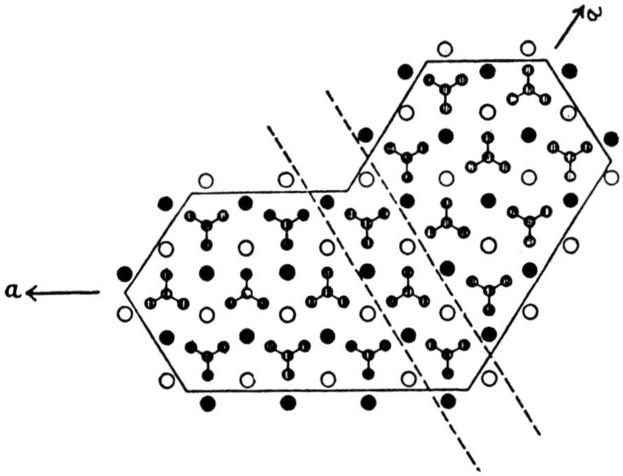

Fig. 29. The twinning of aragonite. The structure between the broken lines is consistent with the symmetry scheme in either individual (after Bragg & Claringbull, 1965).

on the crystal structures involved. Although the crystals related by epitaxy have different structures, there will be structural planes where there may exist a good fit between the two individuals. As an example we may mention the epitaxial growth of albite over microcline (Figure 30).

Epitaxy is of great importance in thin-film technology, and in particular in the manufacture of semiconductors.

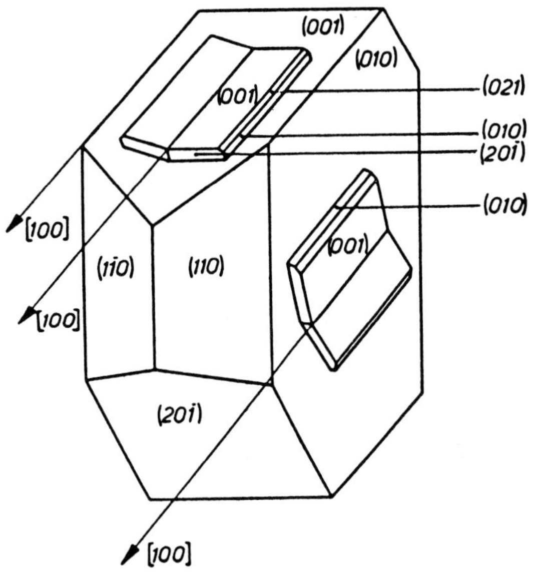

Fig. 30. Example of parallel growth in epitaxy: albite over microcline (after Kern & Gindt, 1958).

5.2. The importance of crystal structure in phase transformations

5.2.1. Topotaxy

The study of phase transformations of minerals, that is, of their behaviour under heat, pressure and certain atmospheric conditions, may reveal important characteristics of their structures. When we break a mineral the cleavage reveals certain directions of weak bonds and, in a similar way, when a crystal is heated the first parts to collapse and transform are also regions of weak bonds.

The transformations of mineral structures by heat may give rise to new phases with random orientation in relation to the original mineral, to a certain preferred orientation, or to a well-oriented phase. In this last case the transformation is called 'topotatic' and the oriented phenomenon 'topotaxy'.

When the transformation of a mineral by heat is topotatic and irreversible, instead of using a temperature and a time which completely transform the mineral, one should use successive heat treatments at a relatively moderate temperature in order to obtain a kind of 'movie' of the transformation. This technique may reveal intermediate stages of the transformation which will help in interpreting its mechanism. Also, instead of powder X-ray photographs to follow the transformation, which do not give information on the relative orientation of the original and transformed structures, it is better to use single-crystal X-ray photographs.

As an example of a topotatic transformation, the goethite → hematite dehydration may be described. The orientation relationship of the original goethite to the formed hematite is [001] of hematite // [100] goethite , and [110] hematite // [010] of goethite (Francombe & Rooksby, 1959). The crystal-chemical equation for this irreversible transformation is

$$\underset{\text{goethite}}{2Fe^{o}[O(OH)]^{h}} \longrightarrow \underset{\text{hematite}}{Fe^{o}_{2}[O_{3}]^{h} + H_{2}O}$$

where 'h' means hexagonal closest packing, and 'o' octahedral coordination.

A convenient temperature for studying this dehydration is 350 °C because successive and relatively short heat treatments at this temperature

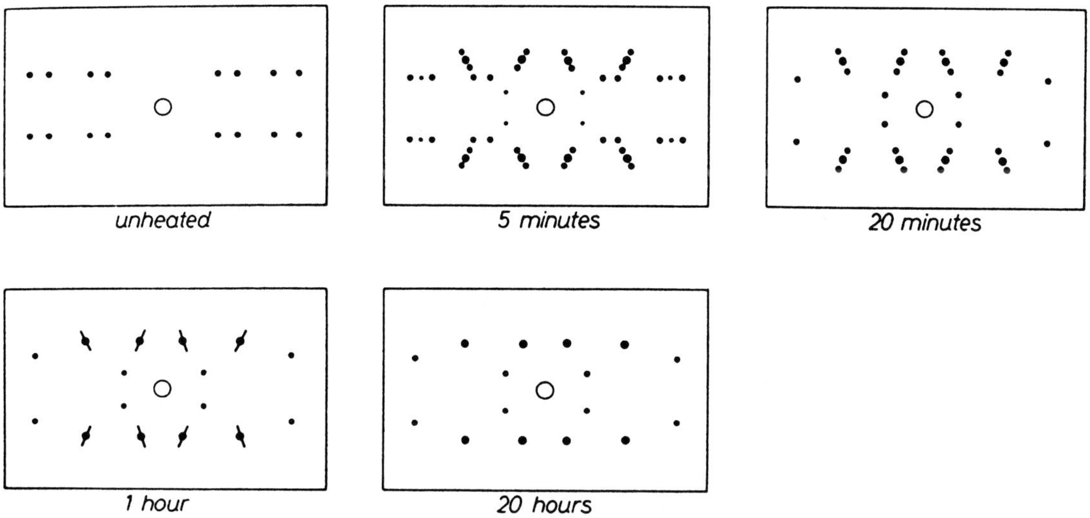

Fig. 31. Sketches of X-ray oscillation photographs of a sample of original goethite, unheated, then heated at 350 °C for several periods of time (oscillation about the 'a' axis) (after Lima-de-Faria, 1963).

may produce a good 'movie' of the transformation. After each treatment, a single-crystal X-ray photograph may be taken to indicate the stage of the transformation (Figure 31).

On the second sketch one can notice, together with the spots of the original goethite, already those of hematite with satellites; on the third sketch, the goethite spots have disappeared; on the fourth, the satellites transform to diffuse regions; only on the last sketch is goethite completely transformed to hematite.

The satellites (Figure 32) observed in the goethite-hematite dehydration correspond to intermediate stages of the transformation, and were very important for the interpretation of its mechanism (Lima-de-Faria, 1963).

The structure of goethite is based on a hexagonal closest packing of O and OH ions, with Fe ions in octahedral voids. Hematite is also based on a hexagonal closest packing of oxygen ions with iron ions in octahedral voids. Due to the similarity of the packing in both goethite and hematite, this hexagonal closest packing tends to be preserved during the transformation, and the iron ions change their positions from a row pattern to a honeycomb distribution within the close packing. The mechanism is inhomogeneous (Brindley, 1961, Ball & Taylor, 1961), that is, certain regions of the crystal are preserved while others, where the water splits off, are completely destroyed, giving rise to micro cracks in the crystal. The preservation of certain parts of the close packing, where the hematite structure is developed, explains the topotatic character of this transformation (Lima-de-Faria, 1963).

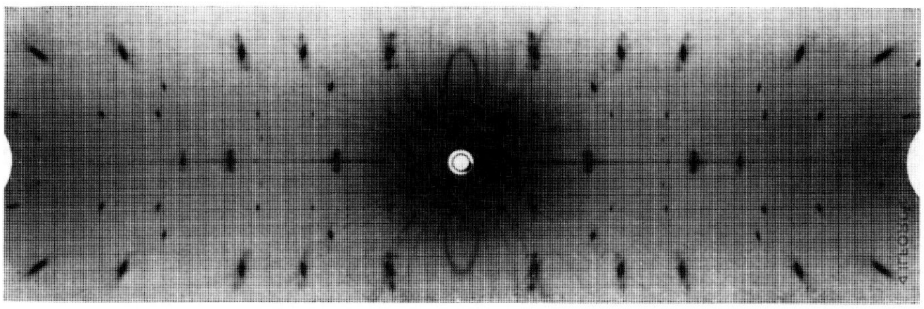

Fig. 32. X-ray rotation photograph of transformed goethite corresponding to a treatment of 300 °C for 19 hours, showing sattelites (FeK radiation, rotation along the 'a' axis of goethite) (after Lima-de-Faria, 1963).

Topotaxy involving structures based on closest packing can be studied in a more general way (Lima-de-Faria, 1967). The possible transformations of structures based on the two fundamental closest-packed arrangements, viz. cubic closest packing (c.c.p.) and hexagonal closest packing (h.c.p.) are: (a) h.c.p. → h.c.p., (b) c.c.p. → c.c.p., (c) h.c.p. → c.c.p. and (d) c.c.p. → h.c.p. A tendency for the preservation of the type of packing has been pointed out by several authors (e.g., Ervin, 1952; Dent Glasser, Glasser & Taylor, 1962; Brindley, 1963). If the original structure and the transformed structure have the same type of packing, this is preserved throughout the transformation, the small ions only moving inside the packing. Moreover, when there is a change in the type of packing, the tendency for the preservation of the closest-packed layers determines the possible orientations.

The simplest way of preserving the closest-packed layers is for the transformation to proceed by slipping of the layers over one another. Because there are only two possible ways of placing one layer over another, the movement of a layer is restricted to the second possible position. This movement can consist of a single slip along one out of three different directions (spheres will move along valleys, that is, from one hole to the nearest hole) all giving rise to the same final state, and therefore being equivalent. To describe the movements of layers in these transformations, it is convenient to consider the (110) section of the c.c.p. structure and the section of the h.c.p. structure parallel to $(2\bar{1}\bar{1}0)$ (Christian, 1951) (Figure 33). The advantage of these sections is that the spheres corresponding to the sequences ABAB... and ABCABC... all lie on the plane of the figure. With the help of packing models of plastic balls it can easily be seen that to

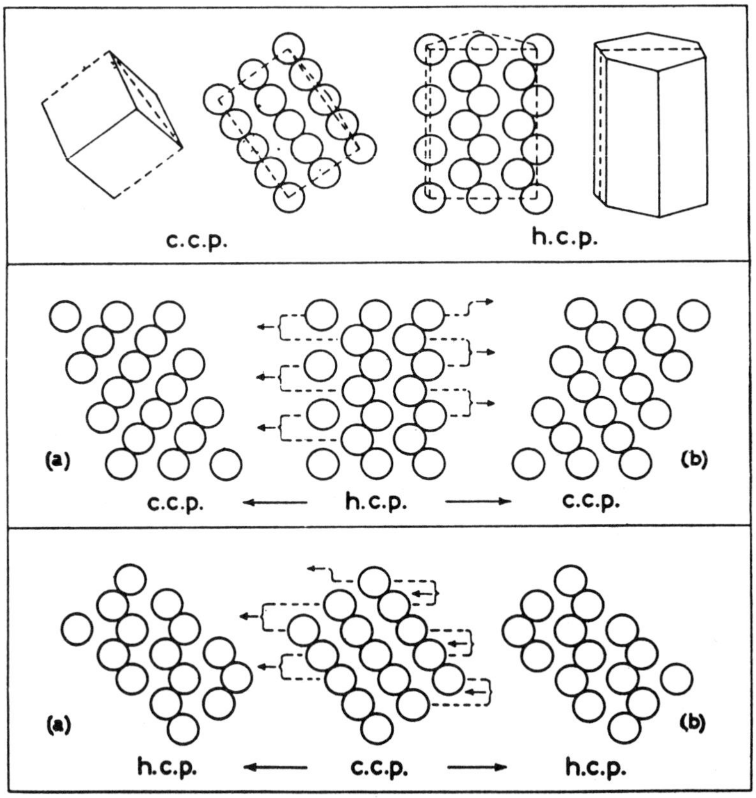

Fig. 33. Mechanism of h.c.p. → c.c.p. transformation and *vice versa*. The top part of the figure shows schematically the sections of close packing models of plastic balls, appropriate for the description of the mechanism of these transformations: (110) section of c.c.p. and $(2\bar{1}\bar{1}0)$ section of h.c.p. Two different movements are represented, corresponding one to the left-hand arrows and the other to the right-hand arrows, leading respectively to structures (a) and (b). The layers slip by pairs, as locked together (after Lima-de Faria, 1967, and adapted from Christian, 1951).

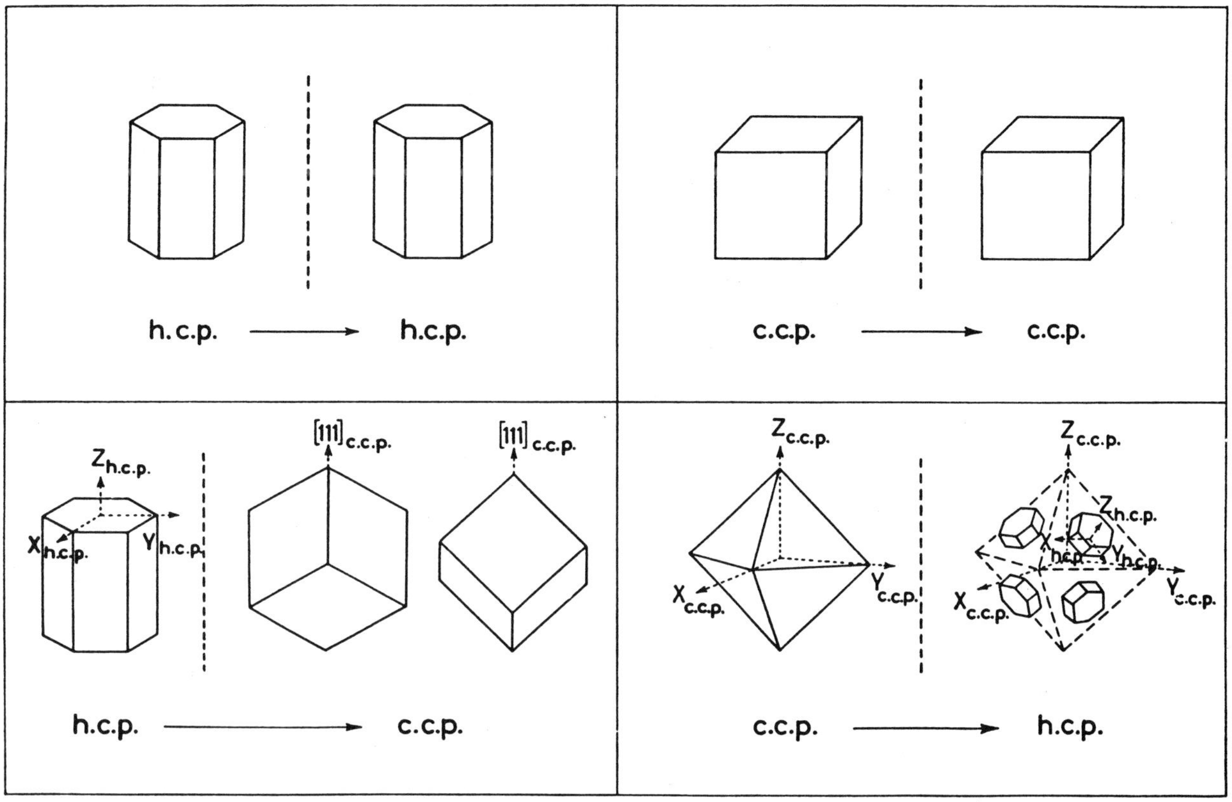

Fig. 34. Normal orientations derived in transformations involving c.c.p. and h.c.p. (after Lima-de-Faria, 1967).

change from h.c.p. to c.c.p. the layers have to slip by pairs, as if locked together, in the $[01\bar{1}0]$ direction, and that all the pairs slip in the same direction (Christian, 1951) (Figure 33, middle section). The necessity for the layers to move in pairs is the consequence of the single new position that can be obtained by slipping, as pointed out above. For instance, if the third layer slips relative to the layer below, which has already slipped, the resulting position of this third layer will be the same as initially, the sequence ABA being maintained. Two possible c.c.p. orientations can be derived from h.c.p. as movements of pairs AB or BA take place (Figure 33), thus twinning is obtained in this transformation. Since a unique set of closest-packed planes exists in the h.c.p. structure, only two possible c.c.p. orientations are obtained in the transformation h.c.p. → c.c.p.

To transform the c.c.p. into h.c.p. the layers have again to be moved by pairs, slipping in the $[11\bar{2}]$ direction and the two possible orientations derived for h.c.p. are parallel, not giving rise to twinning (Figure 33, lower part). However, because there are four equivalent sets of close-packed planes in the c.c.p. structure, four different orientations are obtained in the transformation (Lima-de-Faria, 1967). These results are summarized in Figure 34.

In Figure 34, lower section, the various packing orientations developed in transformations involving change from h.c.p. into c.c.p., and *vice versa*, are shown with their packing axial relationships. From the figure many particular transformations can be predicted if the packing characteristics of both the original and transformed structures are known.

5.2.2. Transformation twins: Packing and interstitial twinning

Apart from the twinning that comes from closest packing, twinning can also arise from different possible distributions of the interstitial ions within the closest packing. This is the case of the hematite

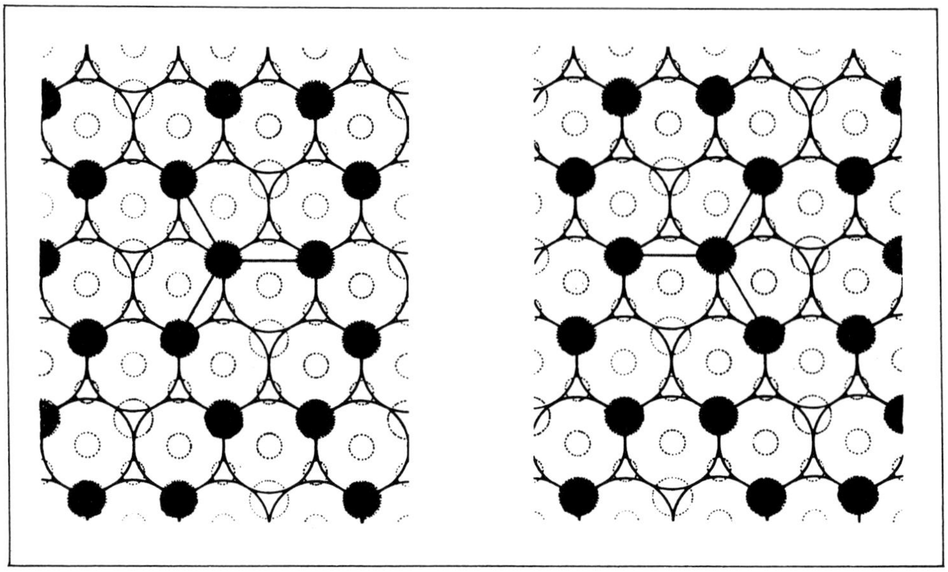

Fig. 35. The two possible different orientations of nucleation of the honeycomb iron pattern inside the hexagonal closest packing.

derived from heated goethite. In fact the X-ray oscillation photographs show a plane of symmetry which does not exist in single crystals of hematite with this same orientation (Figure 35). The reason is that two differently oriented hematites develop during the transformation, because the honeycomb pattern of the iron ions in hematite may develop in two possible positions inside the hexagonal closest packed layers.

Equal amounts of hematite crystallites are formed with the two orientations because these are equally probable, which is confirmed by the equal intensities of the spots (Lima-de-Faria, 1963).

Consequently, when considering *transformation twins*, one should distinguish between *packing twinning*, which derives from displacements of packing layers, and *interstitial or internal twinning*, derived from more than one possible distributions of the interstitial atoms inside the packing.

Another kind of twinning also arising from phase transformations was described by Lonsdale (1966).

5.2.3. The measure of symmetry in crystal structures and its application to phase transformations

When dealing with phase transformations, it is often needed to compare the symmetries ot two structures, that is, to judge which has higher symmetry. In fact, according to Megaw (1973),

> pseudo-symmetric structures are always likely to undergo transition to high-symmetry form

and

> a high-symmetry structure tends to have higher entropy (and therefore lower free energy),

and again

> the higher-temperature phase is not necessarily characterized by the higher symmetry.

The measure of symmetry is a problem that has been ignored or disguised until now, and many crystallographers may have different opinions about it or even not to know how to tackle it. To a certain extent, it has been treated under the heading of group-subgroup relations, though in an implicit manner (see, for instance, Vol. A of the International Tables of Crystallography, 1983, 1987); however, it would certainly be useful if this problem were dealt with explicitly.

Admittedly, this is not a simple problem, and a recent proposal for measuring the symmetry of a crystal structure, as put forward by the author (1988b, 1991), may involve aspects hitherto not considered. However, should such a proposal provoke a wide discussion of this problem, it would certainly prove very useful.

What is crystallographic symmetry? Symmetry is essentially repetition. When a motif is repeated several times within a pattern, we say that a certain kind

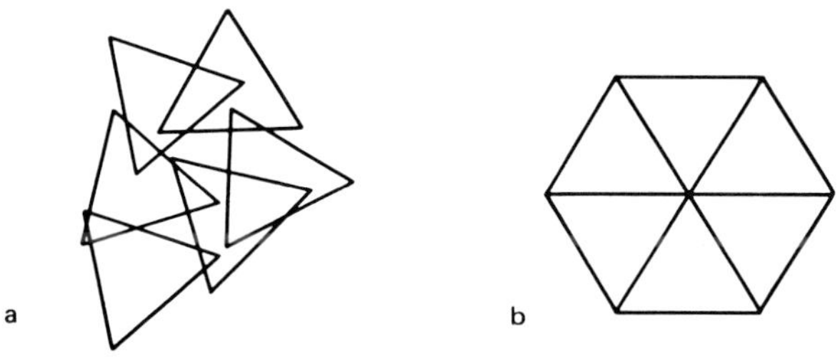

Fig. 36. Asymmetrical (a) and symmetrical (b) figures (after Shubnikov & Koptsik, 1972)

of symmetry exists. However, symmetry is not only repetition; it has to be produced in a regular way. In fact, according to Shubnikov & Koptsik (1972),

> a symmetrical figure must have, in addition to geometric equality of its parts (a), identical arrangement of the parts (b)

(Figure 36).

According to Helen Megaw (1973),

> An object possesses certain symmetry if after the application of a particular operation it looks exactly as it did before and continues to do so however often the operation is repeated. For instance, a hexagonal prism looks exactly the same after rotation through 60°.

Therefore, as Fedorov (1901) said,

> symmetry is the property of geometrical figures to repeat their parts, or more precisely, it is the property of figures in different positions to bring them in coincidence with the figures in the initial positions.

Although symmetry is essentially repetition in a certain regular way, repetition by itself cannot be a measure of symmetry. In fact, if we compare a cube and an hexaoctahedron, the number of repeated faces in the cube is 6, and that in the hexaoctahedron is 48, but we cannot say that the hexaoctahedron has higher symmetry than the cube, because they have exactly the same symmetry.

What defines the crystallographic symmetry of these two forms is the group of elements of symmetry which is the same for both, and what should measure the symmetry is not the actual repetition of the faces but the maximum number of repetitions these elements of symmetry may generate. This number measures the 'power of repetition' or 'symmetry capacity' of these elements of symmetry, and corresponds to the multiplicity of the general form (Lima-de-Faria, 1988b). The maximum number of repetitions can obviously occur when the face is in a general position, giving rise to the general form (in this case the hexaoctahedron).

We can then propose that the crystallographic symmetry of a pattern should be measured by the capacity of symmetry, or power of repetition of its elements of symmetry.

A pattern may be finite (a figure or a form), or infinite (in two or three dimensions). In the case of a figure or a form (e.g., a crystal in the morphological sense), the symmetry capacity of its elements of symmetry corresponds to the multiplicity of the general form, or the 'order' of the point group (Shubnikov & Koptsik, 1974; Vol. A of Int. Tables, 1983). If we consider the 10 two-dimensional crystallographic point groups and the 32 three-dimensional crystallographic point groups, and arrange them according to their 'order', we obtain the results presented in Tables 11 and 12, respectively. These tables correspond to the group-subgroup relations shown on Figure 10.3.1, and Figure 10.3.2, of Vol. A of Int. Tables, 1983 edition, (pages 774 and 775, respectively).

For the two- and three-dimensional infinite patterns we have to order them by the multiplicity of the general position of the corresponding crystallographic space groups (Lima-de-Faria, 1988b, 1991). We obtain Table 13 for the 17 crystallographic plane groups and Table 14 for the 230 crystallographic space groups. The latter table enables the ordering of crystal structures with respect to symmetry. Although there are several tables in Vol. A of Int. Tables involving the multiplicity of the space groups, no table could be found there to correspond to either Table 13 or Table 14.

Applying our definition to the comparison of two infinite patterns corresponding to the plane groups *pm* and *cm* (plane groups chosen for reasons of simplification), it is found that the multiplicity of *pm* is 2 and that of *cm* is 4. Therefore the symmetry of *cm* is higher than that of *pm* (Figure 37). Moreover, not only the multiplicity is higher in *cm* but more elements of symmetry are present, namely the glide planes.

Group-subgroup relations when applied to these examples may give contradictory results, depending on the sense and mechanism of the transformation (Lima-de-Faria, 1991).

Figure 38 represents the original plane group *pm*

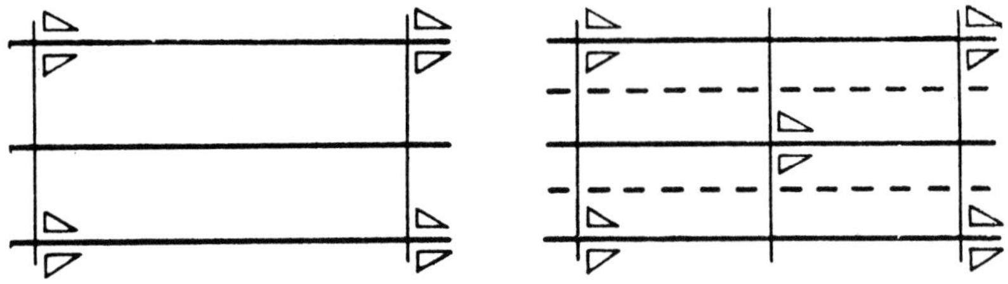

pm cm

Fig. 37. Comparison of the symmetries *pm* and *cm* (after Lima-de-Faria, 1991).

and the composed plane group *cm*. By considering that, in certain positions, the motif has slightly changed (full triangles), the composed plane group will show repeated distances twice the original ones. According to group-subgroup relations the formed plane group *cm* has lost the infinitely repeated translations a_1 and b_1, and has only translations at $2a_1$ and $2b_1$. On the other hand, certain mirror planes have been converted into glide planes. Consequently, *pm* has higher symmetry than *cm*.

However, one can imagine a transition which should lead precisely to the opposite conclusion. If one starts with the space group *cm* (Figure 39), one can transform it by a similar process into space group *pm*, as indicated in Figure 39. The infinitely repeated translations are maintained but the glide planes are lost. Consequently *pm* is a subgroup of *cm*, and *cm* has higher symmetry than *pm*. The fact that this result is in complete contradiction with the one presented in Figure 38 clearly expresses the ambiguity of the measure of symmetry based on group-subgroup relations.

Such contradiction might possibly be resolved if in Figure 38 one would compare the actual unit cells (double lined) of the original and the composed structures. Then it should be obvious that there is an increase of symmetry (the gain of the glide planes), rather than a loss, which leads to the conclusion that *cm* has higher symmetry than *pm*. What seems important is to consider the plane groups independently of their unit-cell size, that is, of their density of symmetry. Should this be admitted, the answer would be the same in both cases,

Table 11. Hierarchy of the 10 two-dimensional point-groups (after Lima-de-Faria, 1991)

Order	Oblique	Rectangular	Square	Hexagonal	
1	1				
2	2	m			
3				3	
4		2 mm	4		
6				6	3 m
8			4 mm		
12				6 mm	

Table 12. Hierarchy of the 32 three-dimensional point-groups (after Lima-de-Faria, 1991)

Order	Triclinic	Monoclinic	Orthorhombic	Trigonal	Tetragonal	Hexagonal	Cubic
1	1						
2	$\bar{1}$	2 | m					
3				3			
4		2/m	mm2 | 222		4 | $\bar{4}$		
6				3m $\bar{3}$ | 32		6 | $\bar{6}$	
8			mmm		4/m | 4mm $\bar{4}$2m | 422		
12				$\bar{3}$m		6/m | 6mm $\bar{6}$/2m | 622	23
16					4/mmm		
24						6/mmm	m$\bar{3}$ $\bar{4}$3m | 432
48							m$\bar{3}$m

Table 13. Hierarchy of the 17 plane groups (after Lima-de-Faria, 1991)

Multiplicity (*)	Oblique	Rectangular	Square	Hexagonal
1	p1			
2	p2	pm | pg		
3				p3
4		cm | p2mm p2mg | p2gg	p4	
6				p3m1 | p31m p6
8		c2mm	p4mm | p4gm	
12				p6mm

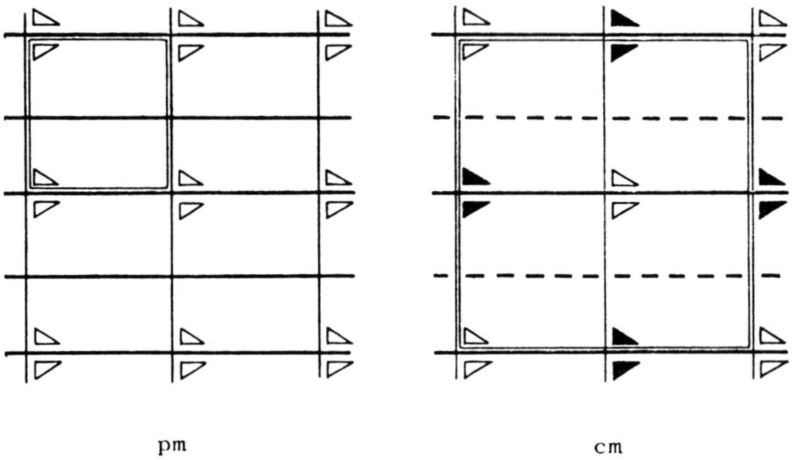

Fig. 38. The transformation of a pattern with *pm* symmetry into a pattern with symmetry *cm* (after Lima-de-Faria, 1991).

namely *cm* > *pm* (with respect to symmetry), and the ambiguity would vanish.

According to these ideas, *density of symmetry* of a crystal structure should be defined as the number of general equivalent positions per unit volume, that is, the ratio of the multiplicity of the general position of a space group to the volume of its unit-cell. The density of symmetry may also be important in phase transitions, but should be considered separately from the symmetry itself.

It is interesting to notice that certain crystal structures with orthorhombic symmetry may have higher symmetry than others with tetragonal, trigonal or even hexagonal symmetry. An example is Cmmm (multiplicity 16) with higher symmetry than P6mm and P$\bar{3}$ml (both with multiplicity 12) and P$\bar{4}$m2 (multiplicity 8).

It may now be realized that in the goethite-hematite transformation there is an increase of symmetry from Pbnm of goethite (multiplicity 8) to R$\bar{3}$c of hematite (multiplicity 18).

The proposed measure of symmetry is independent of the size of the unit cell, therefore also of the density of symmetry. It is clear that an amplified pattern has the same symmetry as the original one (Figure 40).

On the contrary, the measure of symmetry based on group-subgroup relations intermixes symmetry with density of symmetry, which possibly gives rise to the mentioned ambiguity in group-subgroup relations. Some crystallographers would admit solving this ambiguity by using the concept of index of

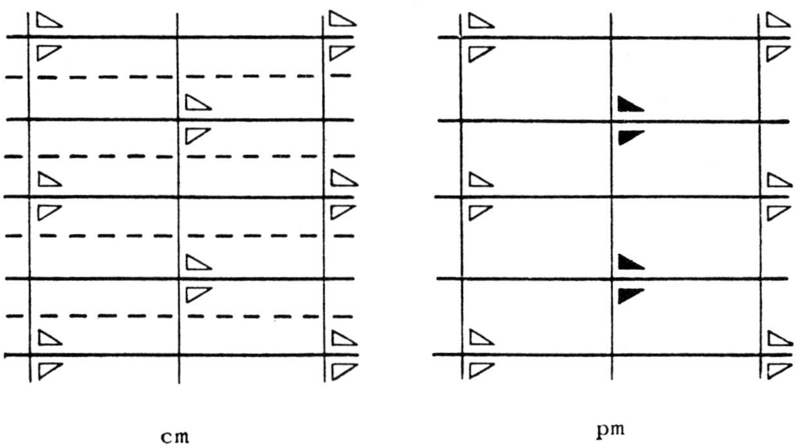

Fig. 39. The transformation of a pattern with *cm* symmetry into another pattern with symmetry *pm* (after Lima-de-Faria, 1991).

Table 14. Hierarchy of the 230 space groups (after Lima-de-Faria, 1991)

Multi-plicity*	Triclinic	Monoclinic		Orthorhombic		Trigonal		Tetragonal		Hexagonal		Cubic	
1	$P1$												
2	$P\bar{1}$	$P2$	$P2_1$										
		Pm	Pc										
3						$P3$	$P3_1$						
						$P3_2$							
4		$C2$	Cm	$P222$	$P222_1$			$P4$	$P4_1$				
		Cc	$P2/m$	$P2_12_12$	$P2_12_12_1$			$P4_2$	$P4_3$				
		$P2_1/m$	$P2/c$	$Pmm2$	$Pmc2_1$			$P\bar{4}$					
		$P2_1/c$		$Pcc2$	$Pma2$								
				$Pca2_1$	$Pnc2$								
				$Pmn2_1$	$Pba2$								
				$Pna2_1$	$Pnn2$								
6						$P\bar{3}$	$P312$			$P6$	$P6_1$		
						$P321$	$P3_112$			$P6_2$	$P6_3$		
						$P3_121$	$P3_212$			$P6_4$	$P6_5$		
						$P3_221$	$P3m1$			$P\bar{6}$			
						$P31m$	$P3c1$						
						$P31c$							
8		$C2/m$	$C2/c$	$C222_1$	$C222$			$I4$	$I4_1$				
				$I222$	$I2_12_12_1$			$I\bar{4}$	$P4/m$				
				$Cmm2$	$Cmc2_1$			$P4_2/m$	$P4/n$				
				$Ccc2$	$Amm2$			$P4_2/n$	$P422$				
				$Abm2$	$Ama2$			$P42_12$	$P4_122$				
				$Aba2$	$Imm2$			$P4_12_12$	$P4_222$				
				$Iba2$	$Ima2$			$P4_22_12$	$P4_322$				
				$Pmmm$	$Pnnn$			$P4_32_12$	$P4mm$				
				$Pccm$	$Pban$			$P4bm$	$P4_2cm$				
				$Pmma$	$Pnna$			$P4_2nm$	$P4cc$				
				$Pmna$	$Pcca$			$P4nc$	$P4_2mc$				
				$Pbam$	$Pccn$			$P4_2bc$	$P\bar{4}2m$				
				$Pbcm$	$Pnnm$			$P\bar{4}2c$	$P\bar{4}2_1m$				
				$Pmmn$	$Pbcn$			$P\bar{4}2_1c$	$P\bar{4}m2$				
				$Pbca$	$Pnma$			$P\bar{4}c2$	$P\bar{4}b2$				
								$P\bar{4}n2$					
9						$R3$							
12						$P\bar{3}1m$	$P\bar{3}1c$			$P6/m$	$P6_3/m$	$P23$	$P2_13$
						$P\bar{3}m1$	$P\bar{3}c1$			$P622$	$P6_122$		
										$P6_222$	$P6_322$		
										$P6_422$	$P6_522$		
										$P6mm$	$P6cc$		
										$P6_3cm$	$P6_3mc$		
										$P\bar{6}m2$	$P\bar{6}c2$		
										$P\bar{6}2m$	$P\bar{6}2c$		

Table 14. Continued

16				F222	Fmm2			I4/m	$I4_1/a$				
				Fdd2	Cmcm			I422	$I4_122$				
				Cmca	Cmmm			I4mm	I4cm				
				Cccm	Cmma			$I4_1md$	$I4_1cd$				
				Ccca	Immm			$I\bar{4}m2$	$I\bar{4}c2$				
				Ibam	Ibca			$I\bar{4}2m$	$I\bar{4}2d$				
				Imma				P4/mmm	P4/mcc				
								P4/nbm	P4/nnc				
								P4/mbm	P4/mnc				
								P4/nmm	P4/ncc				
								$P4_2/mmc$	$P4_2/mcm$				
								$P4_2/nbc$	$P4_2/nnm$				
								$P4_2/mbc$	$P4_2/mnm$				
								$P4_2/nmc$	$P4_2/ncm$				
18						$R\bar{3}$	R32						
						R3m	R3c						
						$R\bar{3}c$							
24										P6/mmm	P6/mcc	I23	$I2_13$
										$P6_3/mcm$	$P6_3/mmc$	$Pm\bar{3}$	$Pn\bar{3}$
												$Pa\bar{3}$	P432
												$P4_132$	$P4_232$
												$P4_332$	$P\bar{4}3m$
												$P\bar{4}3n$	
32				Fmmm	Fddd			I4/mmm	I4/mcm				
								$I4_1/amd$	$I4_1/acd$				
36						$R\bar{3}m$							
48												F23	$Im\bar{3}$
												$Ia\bar{3}$	I432
												$I4_132$	$I\bar{4}3m$
												$I\bar{4}3d$	$Pm\bar{3}m$
												$Pn\bar{3}n$	$Pm\bar{3}n$
												$Pn\bar{3}m$	
96												$Fm\bar{3}$	$Fd\bar{3}$
												F432	$F4_132$
												$F\bar{4}3m$	$F\bar{4}3c$
												$Im\bar{3}m$	$Ia\bar{3}d$
192												$Fm\bar{3}m$	$Fm\bar{3}c$
												$Fd\bar{3}m$	$Fd\bar{3}c$

* of the general position.

Fig. 40. (a) An example of a pattern with *p2mm* plane symmetry; (b) amplification of (a). Both have obviously the same symmetry.

the space group, but this also involves the size of the unit-cell and therefore the density of symmetry.

Another sensitive point seems to be the relations between centred and primitive cells (Koptsik & Belov, 1977). For certain mathematical crystallographers, centred and primitive cells may have different crystallographic multiplicities, but the same multiplicity from a mathematical point of view.

Anyhow, with regard to point groups, there is complete agreement between the group-subgroup theory and the present proposal. The only concern is in space groups divergence. This proposal corresponds to the natural extension of the definition of the measure of symmetry from point groups to space groups. In fact, to the multiplicity of the general form, or order of the point groups, corresponds the multiplicity of the general position in space groups.

As another application of these concepts, one may now return to the ideal models of mica (Chapter 4). The symmetrical model with plane symmetry p6mm has higher symmetry (multiplicity 12) than the packing model with plane symmetry p31m (multiplicity 6).

CHAPTER 6

Representation of crystal structures

6.1. Kinds of representation

There are several ways of representing crystal structures. The most common representations can be exemplified for the structure of olivine. In Figure 41 four main different ways of representing this structure are shown: (a) the projection of the atoms on one side of the unit cell, (b) the distribution of the atoms in perspective within the unit cell as a kind of ball and spoke model, (c) the coordination polyhedra description, and (d) the atomic packing.

The representation (a) indicates the exact position of the atoms in the structure, though it gives no global visual idea of their distribution in space; (b) shows the coordination of the atoms, in terms of lines of valence, according to the chemical approach; (c) depicts the coordination polyhedra and the way they are linked together; and (d) shows the relative size of the atoms and their packing.

The crystal structure models that we see in many mineralogical museums are still of the (b) type (ball and spoke), because they correspond more directly to the chemical classification of minerals. However, nowadays, there is a tendency in literature to

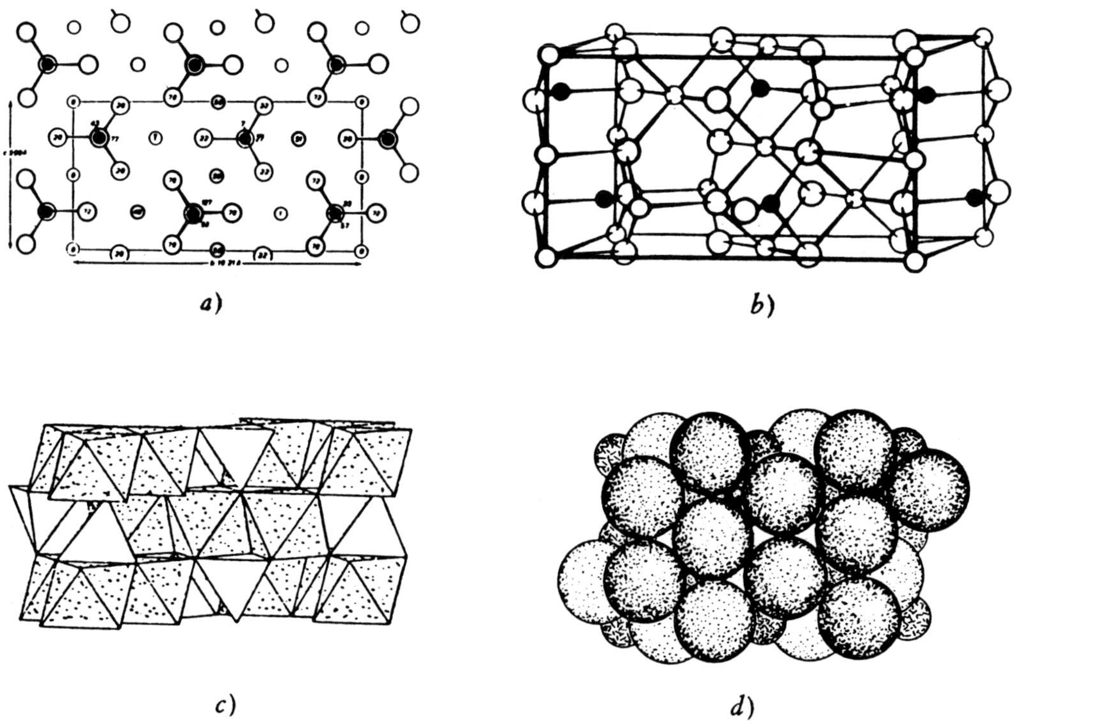

Fig. 41. Four different ways of representing the structure of olivine: (a) the projection of the atoms on one side of the unit cell (after Bragg & Claringbull, 1965); (b) the distribution of the atoms in perspective within the unit-cell (after Povarennykh, 1972); (c) the coordination polyhedra description (after Zoltai, 1975) and (d) the atomic packing description (after Wyckoff, 1965).

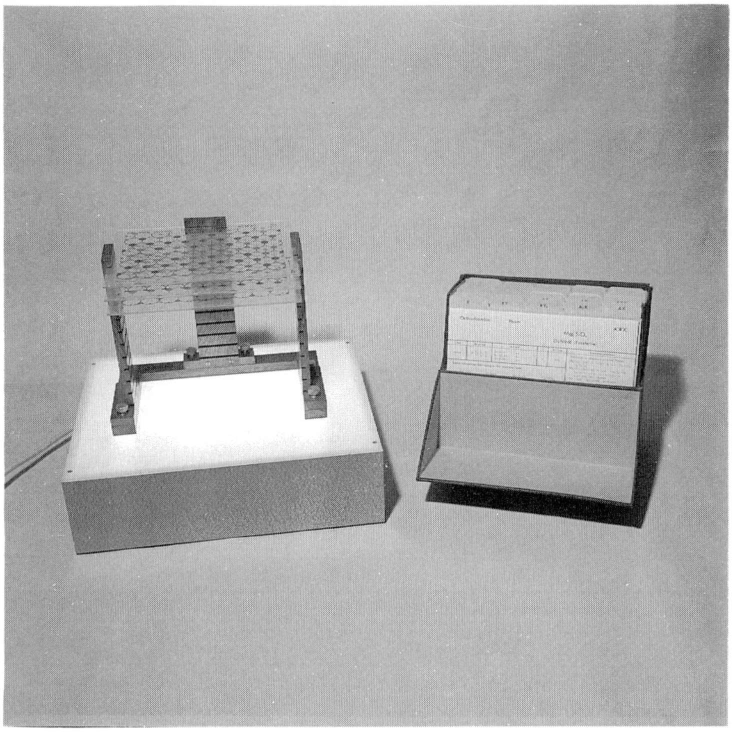

Fig. 42. A condensed model with its supporting rack placed on a viewing screen. Also shown is a small index box containing condensed model sheets corresponding to several inorganic structure types (after Lima-de-Faria, 1986).

use the (c) representation, with coordination polyhedra, possibly because it gives a better idea of the whole structure and also because it enables to think in terms of the Pauling rules of stability.

To these kinds of representation another one should be added which is based on the layer description of crystal structures, the so-called 'condensed models' (Lima-de-Faria, 1965a). This will be treated in detail in the next section.

There are no representations of crystal structures which may be considered useless. Instead of being opposed they are complementary. Each has advantages and disadvantages.

6.2. Layer description and the condensed models of crystal structures

The layer description stems from slicing the crystal structure into atomic layers, and many authors have considered this kind of description of crystal structures (e.g., Smirnova, 1956; Iida, 1957; Frank & Kasper, 1958; Kripyakevich, 1963).

In the condensed model technique (Lima-de-Faria, 1965a) the atomic layers are formed by the packing atoms and the interstitial atoms which are immediately above them; they are represented by transparent plastic sheets (cellulose acetates), placed in an appropriate supporting rack (Figure 42). A viewing screen may be used underneath to improve the visibility of the model. The condensed model gives an idea of how a structure is built by layers of atoms, and shows how these layers stack together in order to produce the structure.

Although the condensed models were invented to study the mechanism of phase transformations, namely the dehydration of goethite into hematite (Lima-de-Faria, 1963), it was soon realized that they were a suitable tool for the description and relationship of inorganic structures, and for the study of crystal chemistry in general.

Coming back to the crystal structure of olivine (forsterite), Mg_2SiO_4, it might be noted that the oxygen ions form a hexagonal closest packing while magnesium ions occupy octahedral voids and silicon ions tetrahedral voids. To build its con-

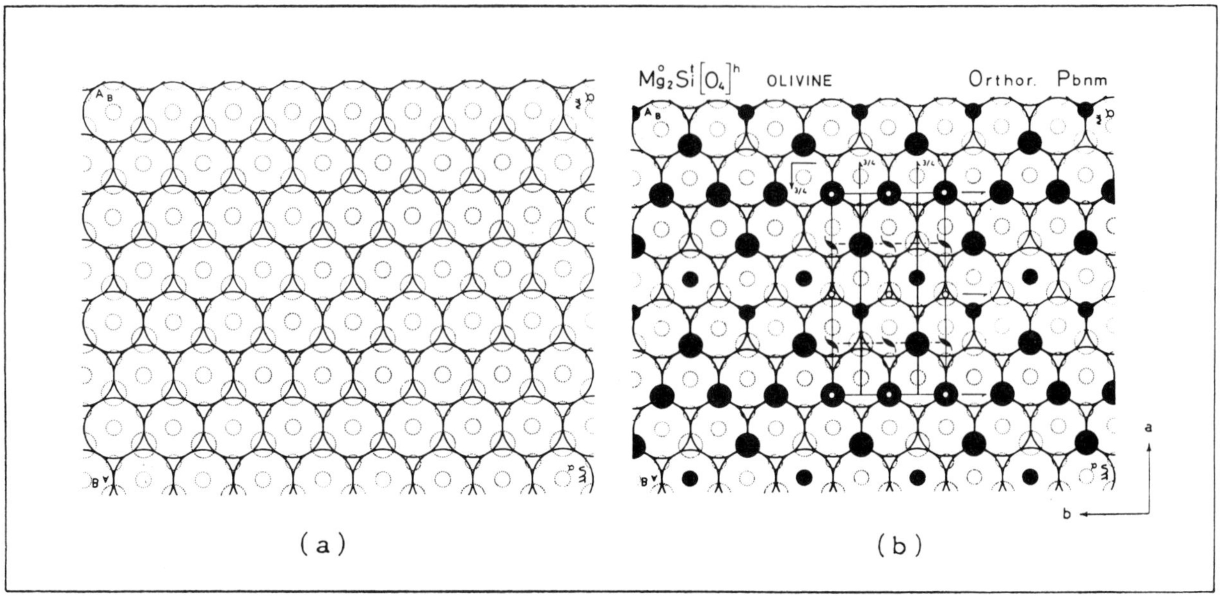

Fig. 43. (a) Standard sheet for building condensed models of structures based on closest packing, with all the octahedral and tetrahedral voids marked (small dotted circles); (b) first sheet of the condensed model of olivine (forsterite); the large open circles represent the oxygen ions while Mg occupying octahedral voids and Si tetrahedral voids correspond to the full circles.

densed model one should start with a hexagonal closest-packed sheet, where the oxygen atoms are represented by large open circles, and all the octahedral and tetrahedral voids generated by the layer to be placed above are indicated by small dotted circles (Figure 43a). Once the octahedral and tetrahedral voids that are occupied are painted black, the first sheet of the condensed model of olivine is obtained (Figure 43b).

Mg ions (the larger full circles) and Si ions (the smaller full circles) occupy 1/2 of the octahedral voids and 1/8 of the tetrahedral voids, respectively, within the closest-packed layer. In the olivine condensed model the sheets are all alike, only displaced in relation to one another. The condensed model of a crystal structure is formed by the minimum number of sheets necessary to get a repetition of the first sheet, and should also contain the unit-cell. The condensed model of olivine is formed by just three sheets (Figure 44). Because the sheets normally contain the projections of more than one unit cell, these models show certain atomic patterns that the other techniques do not. In the case of olivine, the zig-zag patterns of Mg ions and of Si ions may be noticed. The zig-zag of Mg ions is located over and midway between the two zig-zags of Mg ions pertaining to the layer below, that is, the cations in one layer are as distant as possible from the cations on the adjacent layers, as stability demands. The same applies to the zig-zag pattern of Si ions.

The transparent sheets can be kept in an envelope and a large collection of models may be organized and kept in a small index box (Figure 42). This is the reason why these models have been called 'condensed models'. Another advantage of this kind of model is the possibility of visualizing and easily marking the elements of symmetry of the space group (Lima-de-Faria, 1966). In the case of olivine, one can notice the mirror planes, the glide planes, the screw axes, and the centres of symmetry acting on all the O, Mg and Si atoms (Figure 43b).

The problem of slicing a structure into layers is not so easy as may be imagined and, whenever more than one simple layer description can be envisaged, the most appropriate one has to be found. As already seen, the structure of halite, NaCl, can be represented in two different ways, namely, by densest layers T parallel to the (111) planes (Figure 18), or by less dense layers, Q, parallel to the cubic faces (Figure 45). Of these two possible descriptions, the one parallel to the cubic faces is the more appropriate because one can mark on the sheets, quite easily, the elements of symmetry of the cubic space group of halite (Figure 45).

Many other examples could be mentioned like AuCu (Figure 46).

An important characteristic of condensed models is their idealized geometry. The fact was borne in mind that there are hundreds of crystal structures based on slightly distorted closest packings, either cubic or hexagonal, of the larger atoms (normally the anions), with small cations in the octahedral and tetrahedral voids. Thus, instead of producing a specific and exact layer to each of the structures, the decision was made to standardize the problem by drawing idealized 'standard sheets'. These could adequately reproduce any closest packing: cubic (c), hexagonal (h), or mixed sequences (c/h), with all the octahedral and tetrahedral voids marked on them (Lima-de-Faria, 1965a).

The shape of the standard sheet for condensed models was chosen rectangular because this was considered the most suitable (Lima-de-Faria, 1965a; Figueiredo & Lima-de-Faria, 1977). The radius of the circles representing the packing atoms was chosen as 1 cm, and the spacing between the perspex plates in the supporting rack was taken as 1.63 cm, corresponding to the ideal distance between the layers of closest packing for T layers. Accordingly, additional minor distortions will accompany other packings, with different layers or different ways of stacking.

The second aspect to be considered for the appropriate standardization was the minimum number of standard sheets necessary to figure out the different packings (Lima-de-Faria, 1965a; Figueiredo & Lima-de-Faria, 1978).

Because the plastic sheets are transparent and both faces are indistinguishable, one may take advantage of sheet rotations in order to produce, with the same sheet, several different stacking positions. To take the maximum advantage of the two diad axes of the rectangular sheet, the layer pattern has to be located asymmetrically with respect to the centre of the sheet, and all the possible layer stacking positions have to be taken into consideration. The maximum number of standard sheets necess-

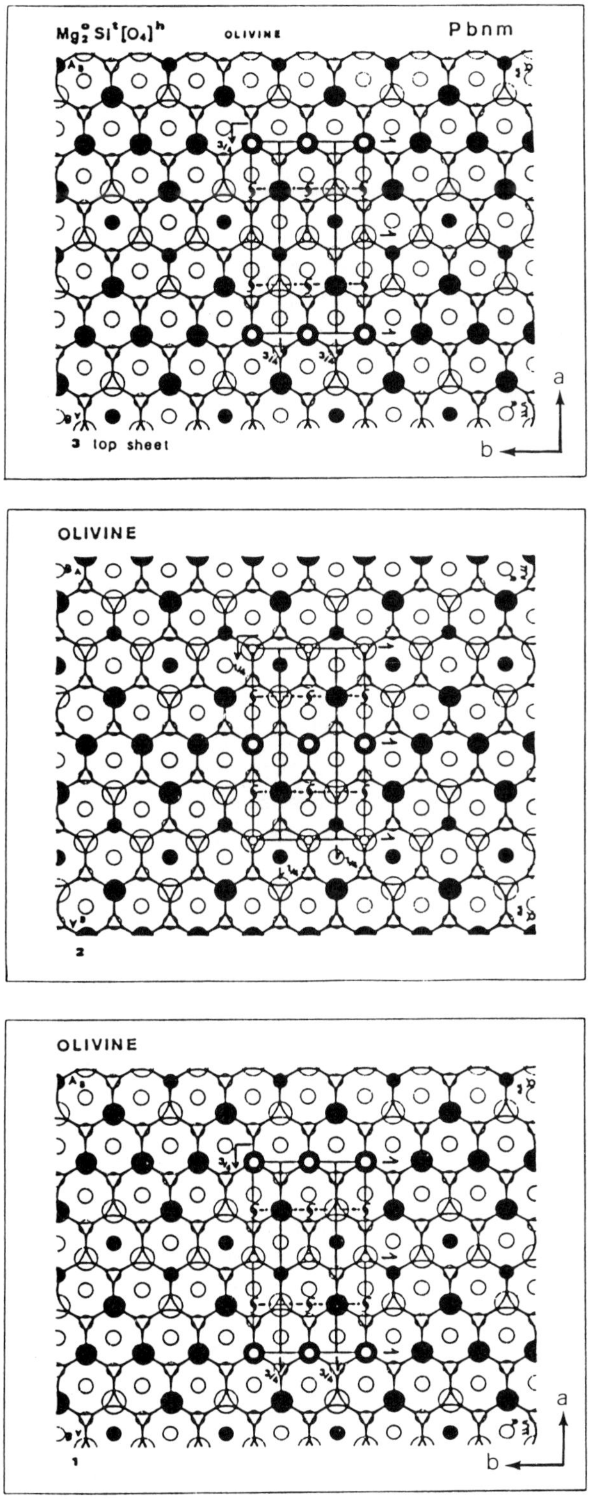

Fig. 44. Condensed model of olivine (forsterite), $Mg_2{}^{o}Si^{t}[O_4]^{h}$, formed by three transparent plastic sheets. The oxygen ions are represented by the large circles and form a hexagonal closest packing (h). The magnesium atoms are placed in octahedral voids (o) and form a zig-zag pattern, the silicon atoms being placed in tetrahedral voids (t) (smaller full circles). The three sheets display the same pattern, and are only alternately shifted in relation to one another.

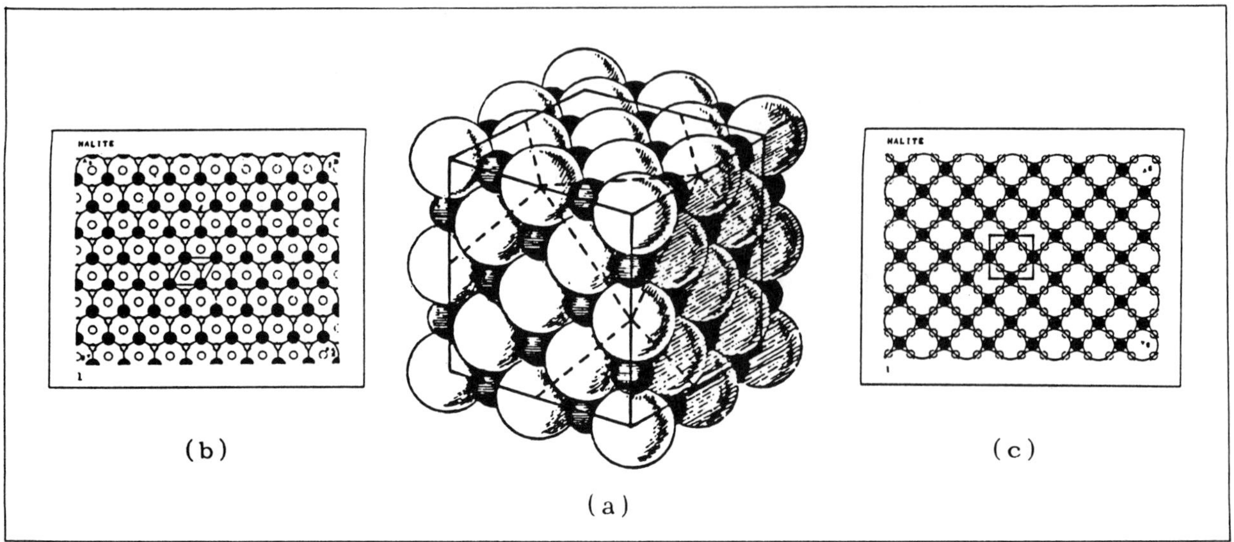

Fig. 45. The structure of halite, NaCl: (a) packing drawing (adapted from Barlow, 1883); (b) and (c) the two alternative layers for condensed model representation. The unit cell marked on (b) refers to the T layer and not to NaCl.

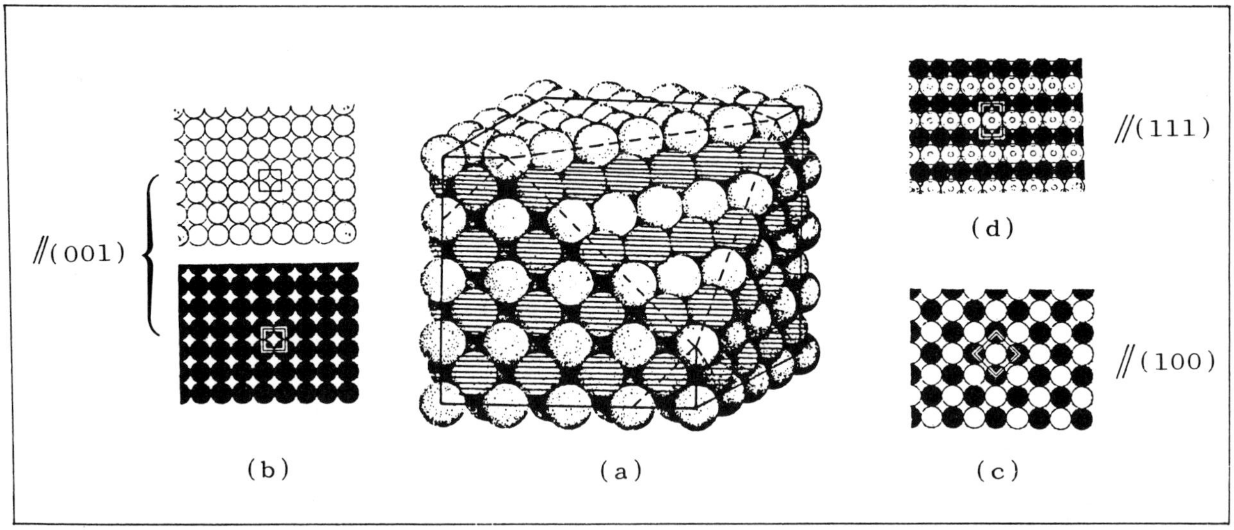

Fig. 46. The structure of AuCu: (a) packing model, showing various possible descriptions; (b) layers parallel to the cubic face (001) which are perpendicular to the vertical tetrad axis, and which corresponds to alternate layers of Au atoms (open circles) and of Cu atoms (hatched circles); (c) parallel to (100), and (d) parallel to (111).

ary to represent the closest, the loose, and the simple packings can then be easily derived.

For instance, in the case of square layers (Q), if the centre of the sheet is located over the middle of the small square formed by positions A, α, B, β, only one sheet is needed to generate by rotation all possible positions A, B, α, and β (Figure 47).

If one starts with the packed spheres in position A, position α is generated by rotating the standard sheet about axis I, and position β is obtained by rotating it about axis II; position B is produced by the combination of both rotations, or by rotation of the standard sheet about an axis perpendicular to its plane and passing through the same centre.

Letters are inscribed on each corner of the sheets (Figure 47) to report the above listed stacking pos-

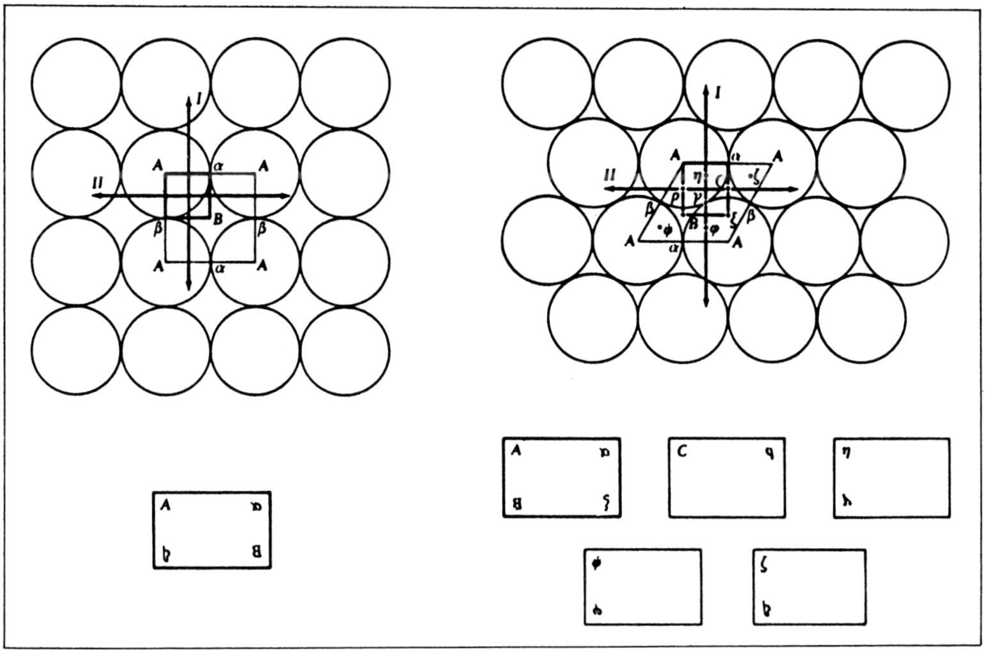

Fig. 47. Positions of square (Q) and triangular (T) layers with the projections of all possible stacking positions of regular stacking modes: over holes, over valleys and over atoms. The location of the sheet centre in relation to the layer pattern for position A is defined by the intersection of the two coplanar rotation axes I and II of the rectangular sheet. The sheets necessary to generate all possible stacking positions are also schematically represented. Some of the letters are printed upside down or reversed in order to achieve their normal reading when placed, by rotation, at the left hand corner of the sheet (adapted from Figueiredo & Lima-de-Faria, 1978).

itions. A letter identifies a stacking position only when placed at the upper left-hand corner, which is chosen as a reference. To bring a letter to the reference position, a suitable rotation should be imposed on the sheet. This is the reason why, in condensed model sheets, some of the letters are printed upside down or reversed, in order to achieve their normal reading positions when placed, by rotation, at the left hand corner of the sheet. A standard sheet placed in a given position can then be specified by the letters labelling the four corners taken in a clockwise sequence, starting with that placed at the reference corner, and enclosed within square brackets. For instance, the first sheet schematically represented in Figure 47 corresponds to position A, and is specified by [AαBβ].

If the pattern of all the interstices defined by the stacking of the layer placed above is added to the pattern of the packing atoms, then instead of one sheet, four standard sheets differing only by the marked interstices, and corresponding to the possible ways of stacking will be generated (Figure 48).

To simplify the building of a condensed model, the letters indicating the stacking positions are supplemented by smaller letters, as subindices, indicating the position of the layer to be placed immediately above. Again, each letter pair is printed in such a way as to appear properly written after the rotation of the standard sheet that brings it to the upper left-hand corner (Figure 48).

To represent cubic closest packing by Q layers (Qf, see Chapter 17, namely Figure 16 and Figure 18) only one sheet is needed, [$A_B\alpha_\beta B_A\beta_\alpha$]. Simple cubic packing also calls for one sheet only, [$A_A\alpha_\alpha B_B\beta_\beta$]. However, for the representation of loose packing sequences, corresponding to the stacking over 'valleys', two sheets are necessary, namely [$A_\alpha\alpha_A B_\beta\beta_B$] and [$A_\beta\alpha_B B_\alpha\beta_A$]. These last two sheets differ in the relative orientation of the rows of voids, which may be parallel either to the shorter or to the longer side of the rectangular sheet.

To derive the complete standard sheets with letter pairs at their corners, it is important to notice that when complementing the pattern of the packed atoms with interstices, satisfying a given stacking situation for the reference position (letter at the up-

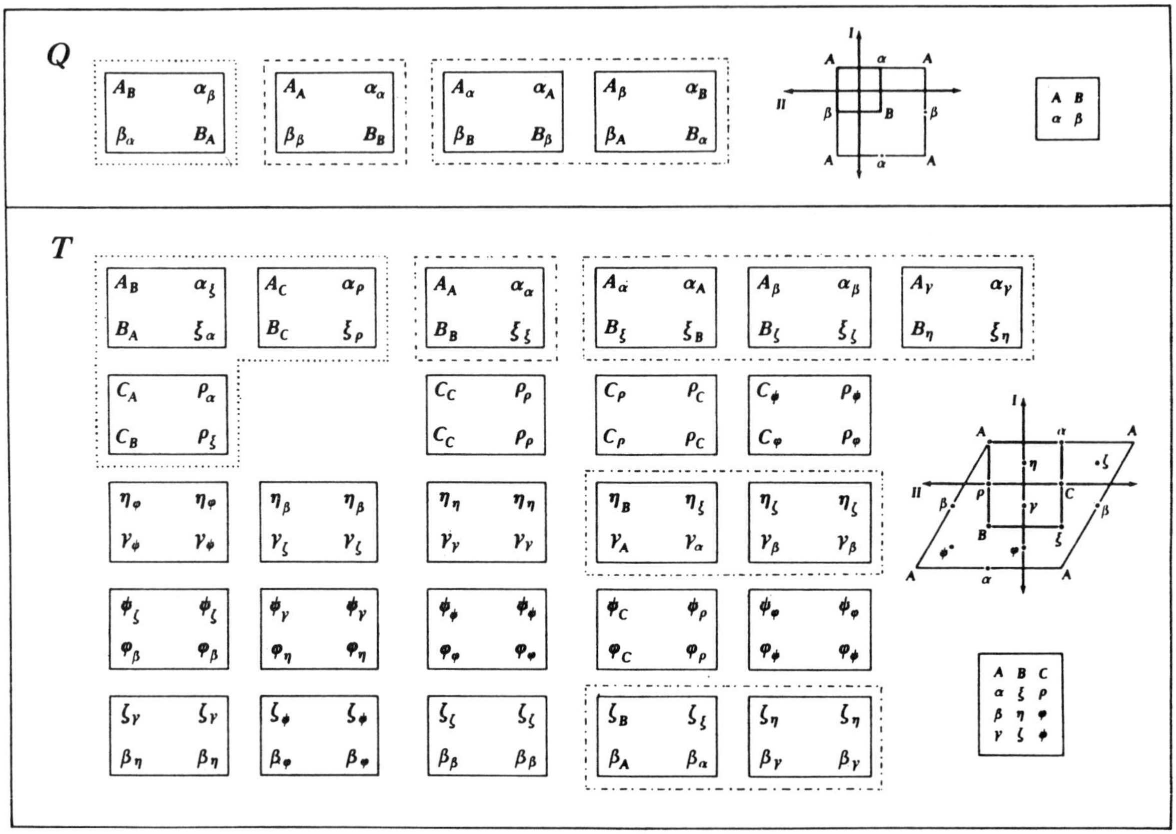

Fig. 48. Schematic representation of the minimum set of complete standard sheets of Q and T layers, which enables the construction of all possible sequences of regular stacking modes. The sheets necessary to produce only pure stacking sequences are enclosed by dotted lines for close stacking, by dashed lines for simple stacking, and by dotted and dashed lines for loose stacking. Unlike real model sheets, and just for the readers' convenience, the schematic sheets in this figure show, at their four corners, letters that are in their normal reading positions (after Figueiredo & Lima-de-Faria, 1978).

per left-hand corner), a similar stacking situation is attained by the remaining positions (letters at the other corners). Consequently, in a complete standard sheet, the letters grouped in pairs at each corner are related by the same kind of stacking. It is therefore easy to derive schematically the complete standard sheets because one only needs to imagine the superimposition of the incomplete sheets according to each situation, and combine the letters of the superimposed corners by pairs.

Figure 48 represents a full scheme of the standard sheets required to build any close packing based on square layers and involving one of the three modes of stacking mentioned above. It can be seen that all the sheet corners are different, and that they cover all the possible stacking situations. A similar process of derivation applies to the packings based on the stacking of T layers (Figure 48) (Figueiredo & Lima-de-Faria, 1978).

For reasons of simplicity, in all the layers and stackings considered up to now, the standard dimension R = 1 cm has been kept for the circles representing the packing atoms and the same supporting rack has been used, in spite of the minor distortions already mentioned, arising from small differences in the interlayer distances. However, as such difference is rather big for the (100) layers, «Q», of the body-centred cubic packing, the radius of the circles representing the packing atoms on these layers was increased (R = 1.41 cm) to avoid great distortions.

The condensed model technique may also be applied to structures based on more complex close packings such as the packings of R^{21} or N^{21} layers,

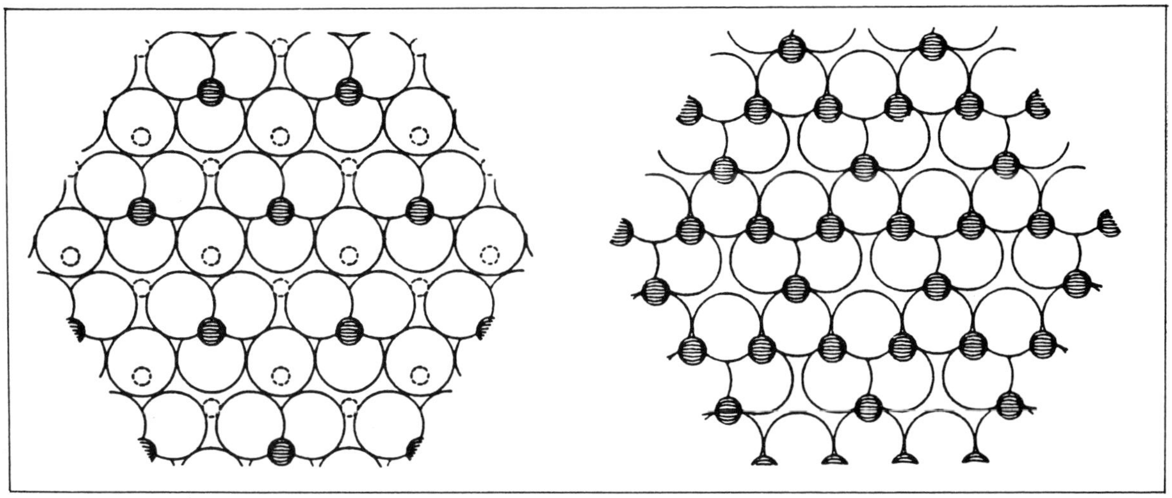

Fig. 49. Two alternate layers of the structure of spinel, $Al_2{}^oMg^t[O_4]^c$, drawn by the computing plotter. The small dotted circles represent Mg atoms in tetrahedral voids, and the larger hatched circles correspond to Al atoms in octahedral voids (after Langlet, 1975, 1976).

and Q^2Q^1 packings, or even other structures which are not based on close packings but that can be sliced in layers. Examples are pyroxenes and amphiboles (Figueiredo & Lima-de-Faria, 1983), micas (Figueiredo, 1979a, 1986) and feldspars (Figueiredo, 1981).

6.3. Computing programs for layer description

The problem of cutting a certain structure into layers, and then searching for the densest layers, which normally yield the simplest description in equal layers, can be very much simplified by the use of computers. Computing programs for slicing a structure along any (hkl) direction and for the determination of the densest atomic layers have already been worked out by Langlet (1975) as the PRSH and PRCM programs, respectively (Figure 49).

In the cases of closest packings and of simple loose packings it is relatively easy to calculate the radii of the spheres which will fit into the interstices between two adjacent layers, stacked according to a certain stacking mode; however, in more complex cases this calculation becomes very difficult. Another computing program (void program) for the calculation of the sizes of such voids has also been worked out (Langlet, Figueiredo & Lima-de-Faria, 1977). Figure 50a represents a standard sheet of the condensed model of a N^{21} layer, that is, a layer built of circles forming triangles and squares in a 2 to 1 proportion, corresponding to the stacking which displaces the unit-cell of the layer pattern from the origin to its middle point, the so-called 'f' stacking. The large open circles represent the packing atoms, and the dotted circles, the various voids. In Figure 50b the same kind of layer is shown, but drawn by a computing plotter, as a result of the application of the void program. The great similarity between Figs. 50a and 50b clearly shows the large potentialities of computing methods to solve such kind of problems and to help drawing condensed model sheets.

6.4. Representation of real structures by ideal structures: Packing and symmetrical analogues of crystal structures

Since Kepler (1611) several authors have understood the great importance of the tendency for close packing in the mineral structures. In this respect, Belov declared in 1947:

> In spite of the variety of the mineral crystalline world, the whole 'mineralogical game' just reduces to various modes of filling gaps in uniform close packing with various corresponding patterns.

Robert Hooke (1665) also said:

> [...] had I time and opportunity, I could make

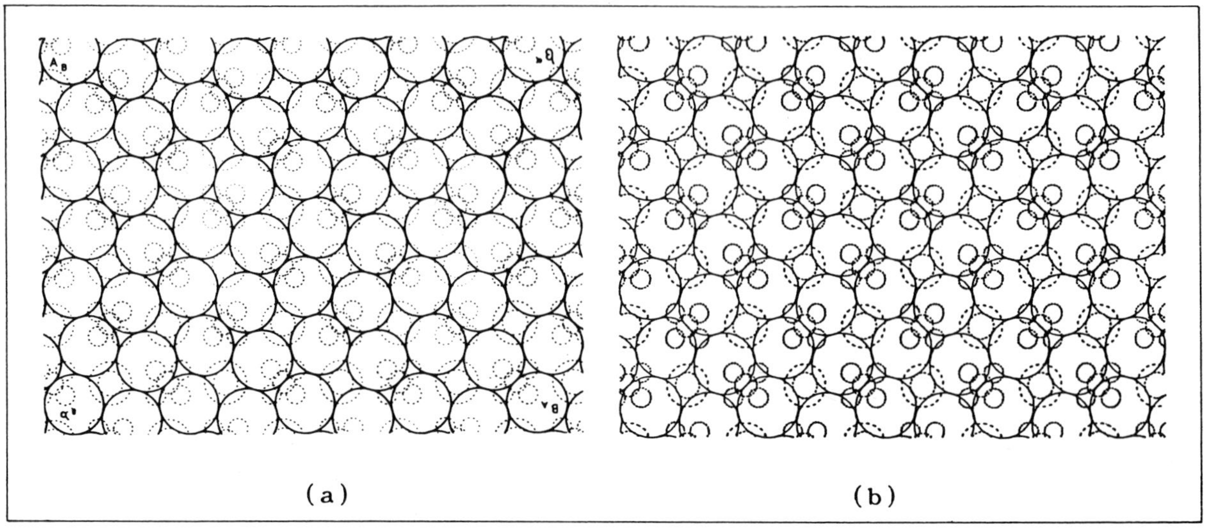

Fig. 50. (a) Condensed model standard sheet of N^{21} layers with stacking 'f' (after Figueiredo & Lima-de-Faria, 1977); (b) same layer N^{21}, with same stacking 'f', drawn by computing plotter as a result of the application of the void program (after Langlet, Figueiredo & Lima-de-Faria, 1977).

probable, that all these regular Figures that are so conspicuously various and curious, and do so adorn and beautifie such multitudes of bodies, as I have above hinted, arise only from three or four several positions or postures of Globular particles,[...]

(quoted by Burke, 1966) (Figure 51). More recently Moore (1992) wrote:

> I believe the principles of close packing are among the most fundamental and useful of chemical crystallographic concepts;

and added:

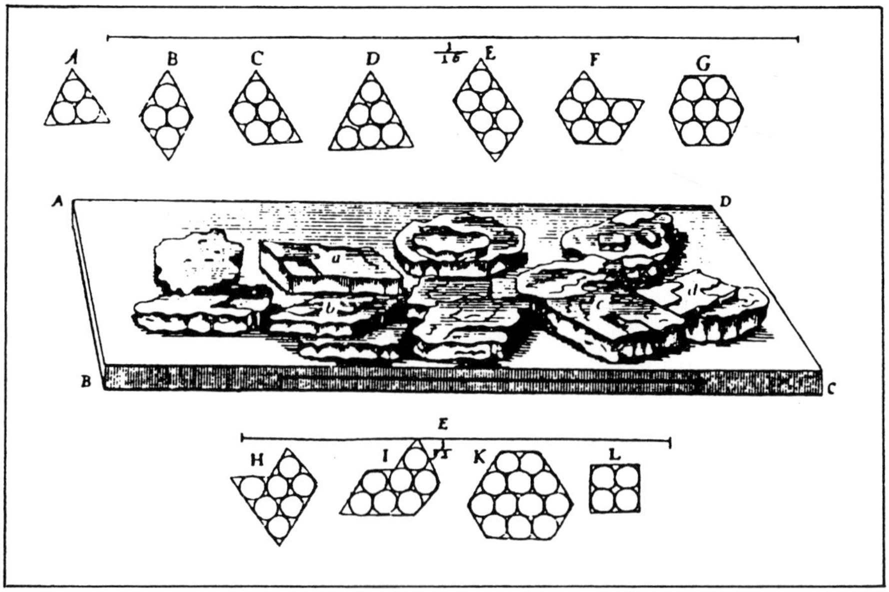

Fig. 51. Hooke's drawing emphasizing the packing of spherical particles to account for the natural shape observed in crystals (after Hooke, 1665, quoted by Burke, 1966).

Crystallography is an intensely hierarchial science. One could offer in fact infinitely many representations of crystal structures. However highest in the hierarchy for mineral structures is representation of a structure based on principles of close- packing.

When we represent a structure based on the close packing of the larger atoms, with smaller atoms in the interstices, in the majority of cases we are adopting an ideal representation. The model of the structure based on an exact close packing is an *ideal model* of the real structure, which normally corresponds to a slightly distorted ideal packing. The condensed models are appropriate for such representation because they are ideal models.

When the structural units are not individual atoms, but groups, chains, sheets or frameworks, the structures can also be directly related to ideal models. The stability conditions reveal two tendencies, namely: (i) for highly symmetrical structural units, and (ii) for close packing of the structural units. Consequently, the *real structures are normally intermediate between two different types of ideal models* (Lima-de-Faria, 1988a). Well-known examples are the pyroxenes, amphiboles and micas (see Figure 25).

The ideal symmetrical structure, which is based on structural units of higher symmetry, may have lower symmetry, due to the distribution pattern of the interstitial atoms. But to layers formed by the structural units alone, that is, without considering the interstitial atoms, correspond a certain symmetry called the 'intrinsic symmetry' of the structure (Figueiredo, 1979a). For reasons of simplicity, the symmetry of the layers containing the structural units, or of the atomic layers forming the structures, is here evaluated by the plane of symmetry of the corresponding model sheets. For instance, the intrinsic symmetry of ideal structures based on sheets of hexagonal rings of tetrahedra is higher than the intrinsic symmetry of the ideal structures based on sheets of triangular rings which correspond to the closest packing (see Figure 25). In fact, the plane group symmetry of the ideal symmetrical sheets is p6m, and that of the packing layers is p31m (Figueiredo, 1979a), the multiplicity of p6m being twelve and that of p31m, six (see Chapter 5). To emphasize the two stability tendencies, for high symmetry and for close packing, these ideal models are called *symmetrical analogues* and *packing analogues*, respectively. Other names have been proposed by other authors to express the second ideal category, the packing analogues, such as ideal close-packed or symmetrical packed model (Zoltai, 1975; Zoltai & Stout, 1984), and close-packed or atomic analogue (Figueiredo, 1977, 1979a).

The structure of a pyroxene such as diopside, $CaMgSiO_6$, can be regarded as intermediate between a structure built of ideal symmetrical tetrahedral chains, with a straight spine (Figure 52a,b), and the corresponding closest-packed structure of tetrahedral silicate chains with a zigzag shaped spine, based on the cubic closest packing of the oxygen atoms (Thompson, 1970; Papike, Prewitt, Sueno & Cameron, 1973) (Figure 52c,d). It is interesting to notice that the packing analogue of diopside is not a mere speculation, but it corresponds to a real structure type, namely, cobalt germanate, $CoGeO_3$ (Papike, Prewitt, Sueno & Cameron, 1973).

Although the symmetrical analogue and the packing analogue have the same overall symmetry (C2/c), they have different intrinsic plane symmetries with respect to the sheets formed by their tetrahedral silicate chains, that of the symmetrical analogue being higher than that of the packing analogue. In fact, the plane symmetry of the sheets formed by the tetrahedral chains of the symmetrical analogue is p2mg (multiplicity 4) and that of the packing analogue is pg (multiplicity 2) (Figure 52).

Many mineral structures possibly have their corresponding symmetrical analogues and packing analogues, and the search for them will certainly contribute to a better understanding of the stability of mineral structures. It is the kind of representation which is chosen that in many cases makes difficult the search for the ideal analogues of a certain mineral structure. A great part of the diversity and complexity of mineral structures is, possibly, only apparent.

DIOPSIDE

Symmetrical analogue	Packing analogue
$Ca^{8}Mg^{o}[Si^{t}_{2}O_{6}]^{II\overset{\uparrow\uparrow}{\bar{c}x}}_{\infty 1}$	$Ca^{o}Mg^{o}Si^{t}_{2}[O_{6}]^{c}$

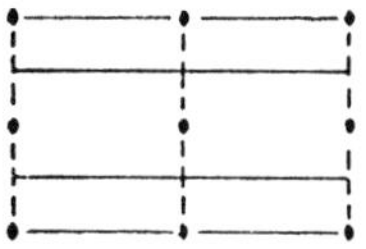

p 2 *m g*

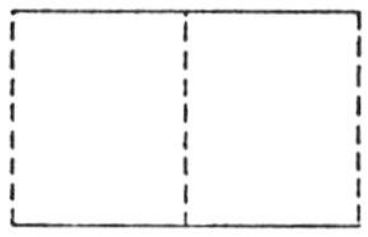

p g

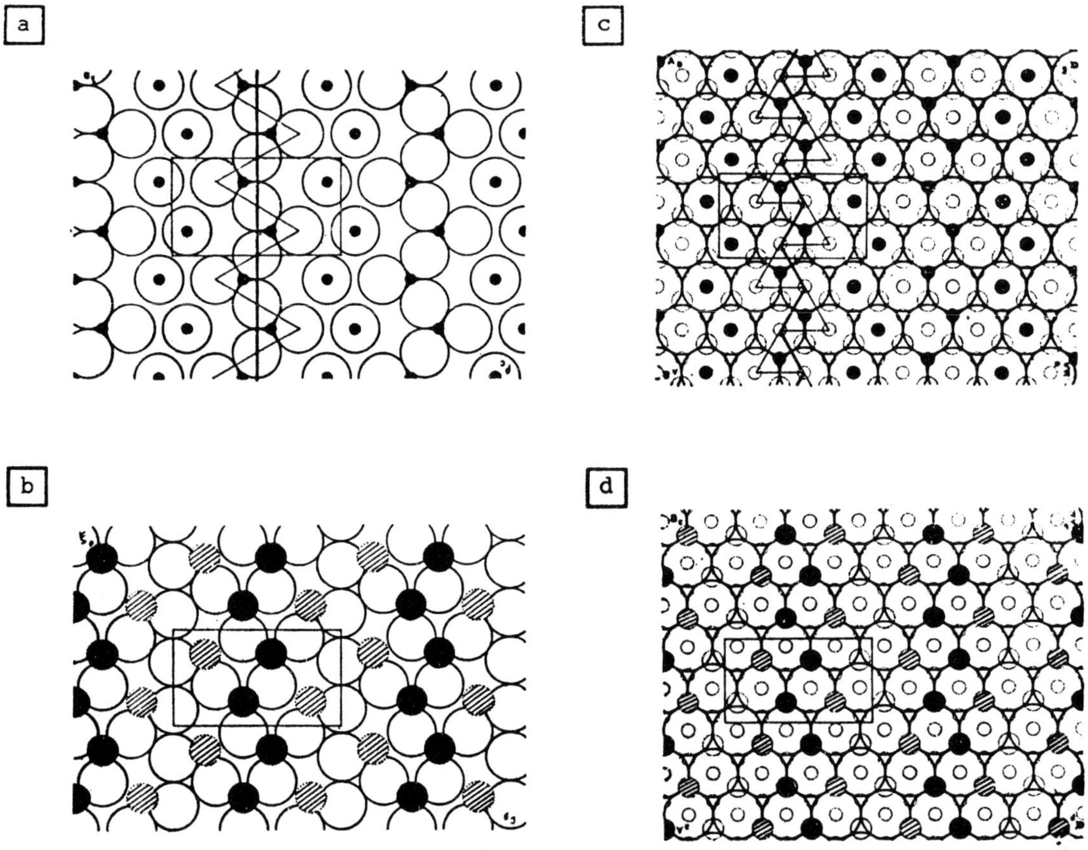

Fig. 52. Alternate sheets of the condensed models of diopside: (a) and (b) of the symmetrical analogue, (c) and (d) of the packing analogue (identical to the cobalt germanate structure). Small full circles represent Si atoms, large full and hatched circles represent Mg and Ca, respectively. The dotted circles, small and large, correspond to non-occupied tetrahedral and octahedral voids, respectively. The superscript on the top right-hand side of the square brackets in the structural formulas characterizes the kind of structural unit and its packing (after Lima-de-Faria, 1988a).

CHAPTER 7

Nomenclature of crystal structures

Structural nomenclature is an important and very useful scientific tool if established appropriately. It will enable the structural characteristics of crystal structures to be seen at a glance and it will also facilitate the establishment of a number of relationships, in particular between structures with similar atomic arrangements.

7.1. Degrees of similarity among mineral structures. Concept of structure type

There are several ways to establish relationships among crystal structures to which different degrees of similarity correspond. Two crystal structures are similar when a correspondence exists between the structural arrangements of their atoms. For example, thorianite, ThO_2, has a structure similar to that of fluorite, CaF_2, these two structures having the same symmetry (space group), and the corresponding atoms occupying the same equivalent positions; they differ only by the parameters of their unit cells. In such cases, the compounds are called isotypic, that means, they belong to the same structure type. In this example the equivalent positions are invariant, and the atomic arrangement is exactly the same.

Even if equivalent positions are invariant, a change in radius ratio of the corresponding atoms may bring about a very different atomic arrangement, at least in what concerns anion and cation coordination. This is the case with CaF_2 and Li_2O structures (Lima-de-Faria & Figueiredo, 1969). In the Li_2O structure, the lithium ions are quite small (radius 0.60Å) as compared to the oxygen ions (radius 1.40Å), therefore this structure can be considered a cubic closest packing of oxygens with lithium ions occupying all the tetrahedral voids (Figure 53). On the other hand, the fluorine ions (radius 1.36Å) are much larger than the calcium ions (radius 0.99Å), consequently the fluorite structure should be regarded not as a cubic closest packing of calcium ions with fluorine ions in tetrahedral voids, but rather as a simple cubic packing of fluorine ions with calcium ions in cubic voids. No doubt these two structures correspond to different structure types. Some investigators, noting only that cations are merely switched, call them 'antistructures'.

When the equivalent positions are not invariant and the radius ratio between corresponding atoms also changes, the problem may become more complicated. Variation in the atomic parameters may give rise to quite different coordinations of the corresponding atoms, and the atomic arrangements are no longer similar. Examples of this situation were given by Wells (1962), for the relationship among $LiNiO_2$, $NaHF_2$ and $CsICl_2$, and in a more general way by Figueiredo (1976a) for AB compounds having the same symmetry, P4/nmn, and the same occupied equivalent positions, for instance, with A atoms in (2a) and B atoms in (2c). Several values of these equivalent positions and different radius ratios between atoms A and B may give rise to at least five different kinds of atomic arrangements (Figure 54).

These and other crystal-chemical considerations, such as bond strength distribution, bond character, electronegativities assigned to the atoms, and electronic states, may still complicate the definition of isotypy. A few attempts to clarify this problem have been presented (Kripyakevich, 1963; Buerger, 1967; Lima-de-Faria & Figueiredo, 1976); more recently in a report of a Subcommittee of the Crystallographic Commission on Crystallographic Nomenclature of the International Union of Crystallography (Lima-de-Faria, Hellner, Liebau, Makovicky & Parthé, 1990). In

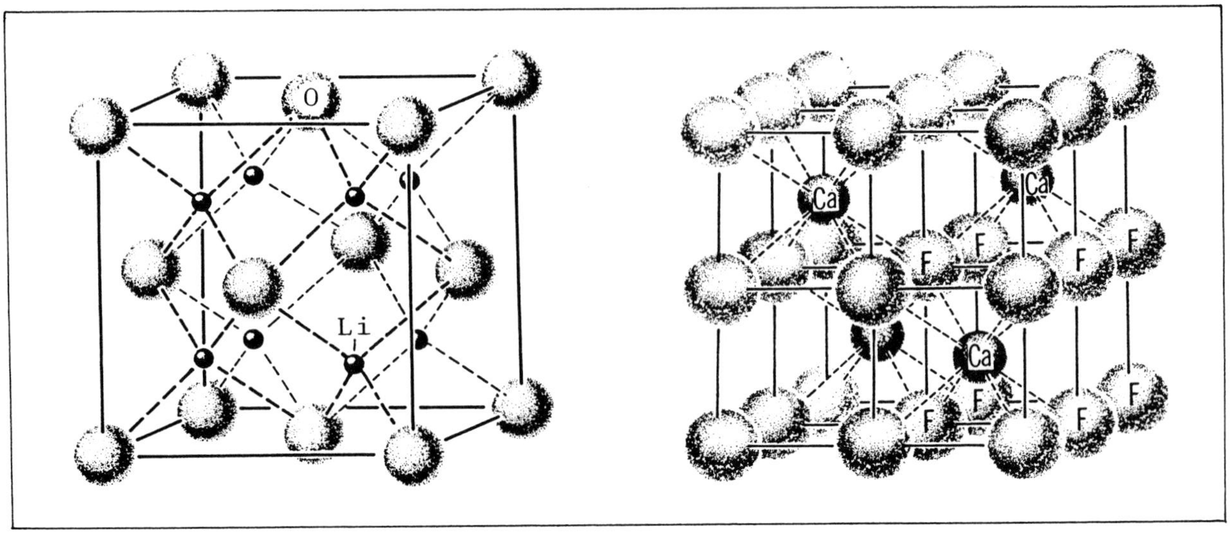

Fig. 53. Packing drawings of the Li_2O and CaF_2 structures, with the packing atoms quite apart, in order to better see the distribution of the smaller interstitial atoms (adapted from Bloss, 1971).

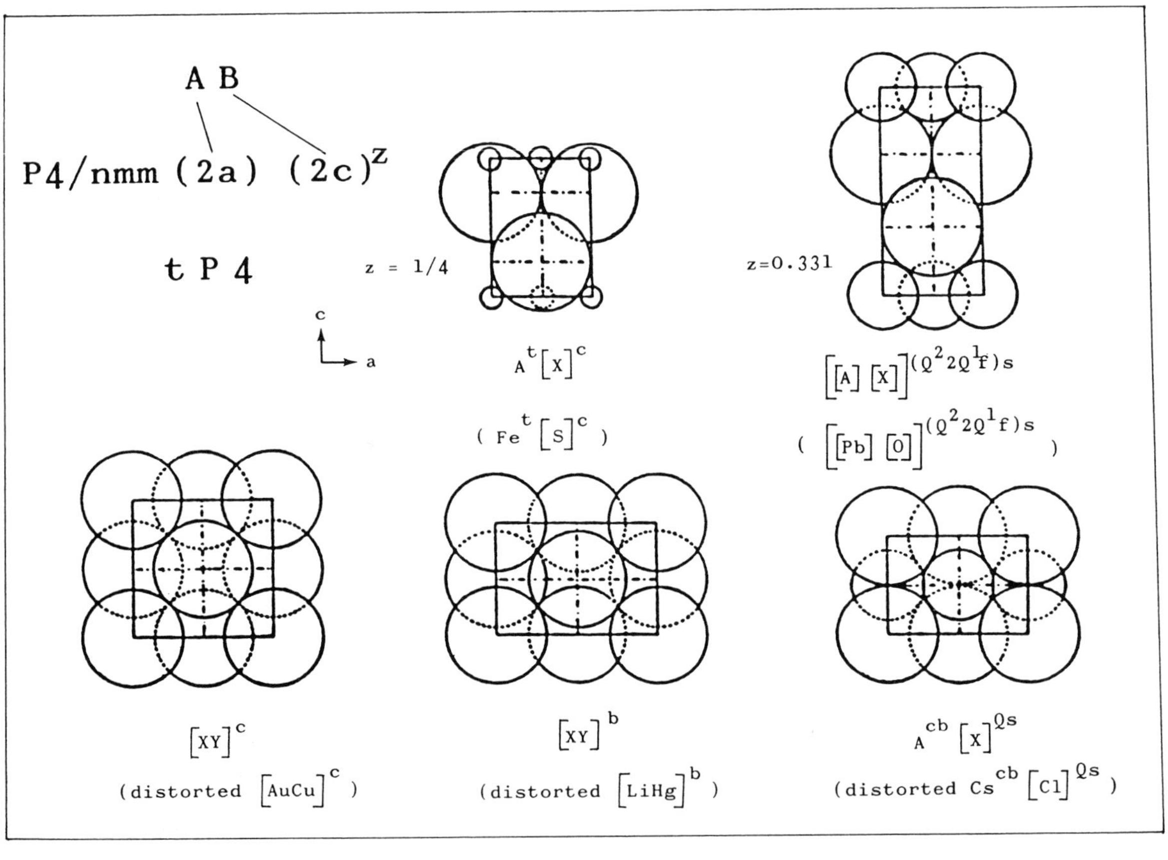

Fig. 54. Five different structural arrangements for an AB compound corresponding to the symmetry P4/nmm, and atoms A and B in the same equivalent positions, namely, A in (2a) and B in (2c) but different values of z, different values of radius ratio rA/rB and different c/a ratio. X represents atoms B larger than A, and similar in size to Y (adapted from Figueiredo, 1976a).

that report, three main degrees of similarity between crystal structures have been defined:

(1) Two structures are *isopointal* if they have the same space group, or belong to a pair of enantiomorphous space groups, and the corresponding atoms occupy the same equivalent positions, either fully or partially at random (Ex.: FeS_2 and CO_2, and the examples considered in Figure 54).

(2) Two structures are *isoconfigurational* if they are isopointal, and their geometrical properties, such as axial ratios and interaxial angles, are similar (Ex.: CaF_2 and Li_2O, in the strict geometrical sense).

(3) Two structures are *crystal-chemical isotypic*, or belong to the same structure type if they are isoconfigurational and the corresponding atoms have similar physical/chemical characteristics (Ex.: CaF_2 and ThO_2; NaCl and MgO; Mg_2SiO_4, olivine, and Al_2BeO_4, chrysoberyl). The word 'similar' arises from the inherent difficulty in defining a priori limits for similarity of geometrical configuration or physical/chemical characteristics.

Consequently, crystal chemical isotypism may be defined in different ways according to the physical/chemical characteristics considered, as further discussed in the report referred to above. However, in this work the following definition will be adopted: *two structures are isotypic if they are isoconfigurational and have the same structural units packed in the same way.*

Other close relations between structures may be considered, which are called homeotypic. Two structures are *homeotypic* if one or more of the conditions required for isotypism are relaxed, such as identity of the space groups, similarity of axial ratios and interaxial angles, values of the adjustable positional parameters, coordinations of atoms, same site occupancy allowing given sites to be occupied by different atomic species.

Some of the important homeotypic structures are the so-called *distortion* and *substitution derivatives* (Buerger, 1947). In the case of distortion derivatives, the atomic arrangement suffers a slight distortion. Examples are NaCl, halite, space group Fd3m, and FeSi, fersilicite, space group $P2_13$; and $TiCaO_3$, ideal perovskite, space group Fd3m, and $CuKF_3$, with subgroup symmetry. The other cases are the substitution derivatives where a certain atom in the basic structure is replaced by two or more atoms, but always in the same equivalent positions. Examples are $Cu^tSb^t[S_4]^c$, famatinite, and $Cu^t_2Fe^tSn^t[S_4]^c$, stannite, both substitution derivatives of $Zn^t[S]^c$, sphalerite, where Zn is replaced by Cu and Sb, and by Cu, Fe and Sn, respectively.

One can also imagine the so-called 'coalescent derivatives' which are derived from close-packed structures by an imaginary 'coalescence process' of some of their packing or interstitial atoms, e.g., calcite, where three oxygen and one carbon atoms form a CO_3 group within a hexagonal closest packing of oxygens, and O_2Rb_2, with dumbbells of Rb_2 derived by the coalescence of two Rb atoms in prismatic voids of a simple hexagonal packing of oxygen atoms (Figure 55). (Lima-de-Faria, 1978b).

Apart from the homeotypic structures, many other interesting closely related structures may be considered such as: polytypic structures ($C^t[Si]^{c/h}$ polytypes), 'interstitial' (or 'stuffed') derivatives ($Ti^o[CaO_3]^c$ of the basic structure $[AuCu_3]^c$), and 'recombination' structures (cannizzarite; olivine-norbergite homologous series). More information on such structures may also be found in the mentioned report (Lima-de-Faria, Hellner, Liebau, Makovicky & Parthé, 1990).

7.2. Structural formulas

7.2.1. General considerations

An appropriate scientific notation for a structure should obey certain fundamental characteristics: it should be as simple and self-explanatory as possible, and it should be kept as close as possible to the way chemists write their formulas in order to facilitate the transfer from chemical to crystal-chemical nomenclature. On the other hand, so as to obtain appropriate structural formulas, not overloaded but containing sufficient information, one must select the most relevant structural characteristics, as inferred from the definition of structure type and from the structural classification used.

Although symmetry may account for some physical properties, it is not necessarily an expression of the kind of arrangement of atoms, since slight changes in the positions of the atoms may give rise to quite different symmetries. Conversely, the same symmetry and equivalent positions may

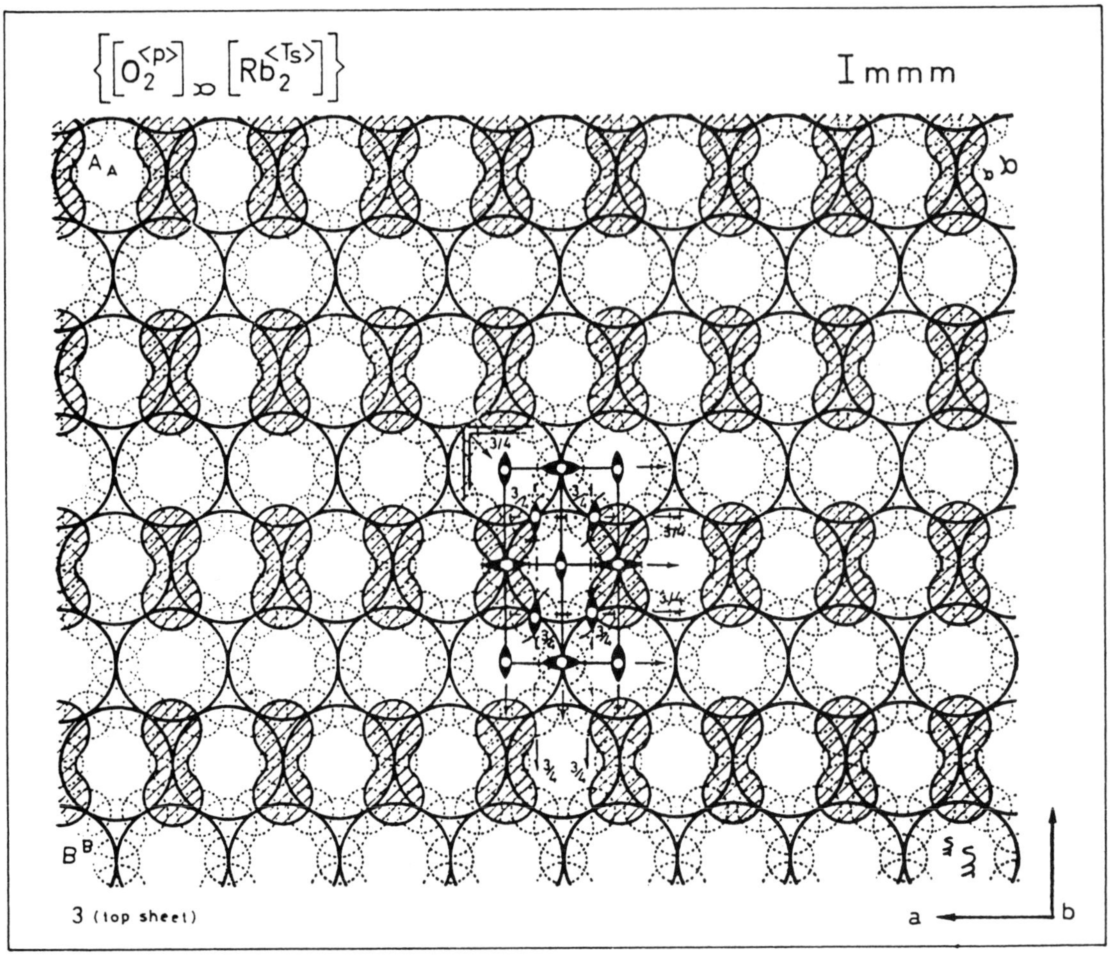

Fig. 55. Top sheet of the condensed model of rubidium peroxide. The small dotted circles represent prismatic and square voids of the simple hexagonal packing formed by the rubidium atoms, and where some of the prismatic voids are occupied by oxygen atoms, coalesced in pairs (after Lima-de-Faria, 1978b).

comply with quite different arrangements of atoms, as was seen before. Therefore, if one wants to express mainly the kind of arrangement of the atoms in the structure, an indication of symmetry can be omitted in structural formulas. However, chemical composition, the coordination of atoms, the category and constitution of the structural units and the way they pack together all are important structural features which should figure in structural formulas.

Several notations have been proposed for structural formulas, and among them the pioneer works of Niggli (1945) and of Machatschki (1947) may be distinguished.

7.2.2. Notation for the coordination of atoms

The definition of coordination and the most commonly observed types of coordination polyhedra in mineral structures have already been discussed (Chapter 2).

As an improvement on earlier notations (Machatschki, 1947; Donnay, Hellner & Niggli, 1964; Lima-de-Faria & Figueiredo, 1976), a specific set of symbols for coordination polyhedra has been proposed by Lima-de-Faria, Hellner, Liebau, Makovicky & Parthé (1990). Such coordination symbols are added as trailing superscripts to the symbols used for the chemical elements in

Table 15. Symbols for common coordination polyhedra (adapted from Lima-de-Faria, Hellner, Liebau, Makovicky & Parthé, 1990). The symbol 'tr' is suggested as an alternative for [3l], and 'sq' for square coordination in order to avoid confusion with symbol 's' which means stacking by superposition

Coordination polyhedron around atom *A*	Complete symbol	Simplified symbols		
Single neighbour	[1*l*]		[1]	
Two atoms collinear with atom *A*	[2*l*]		[2]	
Two atoms non-collinear with atom *A*	[2*n*]			
Triangle coplanar with atom *A*	[3*l*]		[3]	tr
Triangle non-coplanar with atom *A*	[3*n*]			
Triangular pyramid with atom *A* in the centre of the base	[4*y*]		[4]	
Tetrahedron	[4*t*]	[*t*]		*t*
Square coplanar with atom *A*	[4*l*]*	[*s*]		*s* **
Square non-coplanar with atom *A*	[4*n*]			
Pentagon coplanar with atom *A*	[5*l*]		[5]	
Tetragonal pyramid with atom *A* in the centre of the base	[5*y*]			
Trigonal bipyramid	[5*by*]			
Octahedron	[6*o*]	[*o*]	[6]	*o*
Trigonal prism	[6*p*]	[*p*]		*p*
Trigonal antiprism	[6*ap*]	[*ap*]		*ap*
Pentagonal bipyramid	[7*by*]		[7]	
Monocapped trigonal prism	[6*p*1*c*]			
Bicapped trigonal prism	[6*p*2*c*]		[8]	
Tetragonal prism	[8*p*]			
Tetragonal antiprism	[8*ap*]			
Cube	[8*cb*]	[*cb*]		*cb*
Anticube	[8*acb*]	[*acb*]		*acb*
Dodecahedron with triangular faces	[8*do*]	[*do*]		*do*
Hexagonal bipyramid	[8*by*]			
Tricapped trigonal prism	[6*p*3*c*]		[9]	
Cuboctahedron	[12*co*]	[*co*]	[12]	*co*
Anticuboctahedron (twinned cuboctahedron)	[12*aco*]	[*aco*]		*aco*
Icosahedron	[12*i*]	[*i*]		*i*
Truncated tetrahedron	[12*tt*]			
Hexagonal prism	[12*p*]			
Frank–Kasper polyhedra with:				
14 vertices	[14*FK*]		[14]	
15 vertices	[15*FK*]		[15]	
16 vertices	[16*FK*]		[16]	

* Also [4*s*]. ** or sq

crystal-chemical formulas and, preferably, they are placed between square brackets.

Two levels of symbols are proposed, namely, complete and simplified:

(1) Each complete symbol gives the total number of atoms coordinated to a central atom A, and the type of coordination polyhedron, indicated by lower-case letters. The symbols for the most common coordination polyhedra are listed in Table 15; for example, for fluorite, CaF_2, one has: $Ca^{[8cb]}F_2^{[4t]}$.

Some of the notations adopted when creating coordination symbols and which can be used in further developments are as follows:

[Nl] denotes a N-sided coplanar (collinear for N<3) coordination polygon around atom A;

[Nn] denotes a N-sided non-coplanar coordination polygon around A;

[Np] denotes a N/2-sided coordination prism around A;

[Ny] denotes a (N-1)-sided coordination pyramid around A;

[Nby] denotes a (N-2)-sided coordination bipyramid around A;

(2) The simplified symbol requires only the coor-

dination number [N], without specifying the polyhedron type. On the other hand, for most common coordination polyhedra, a simplified letter notation can be used as a trailing superscript with or without square brackets ('t' for tetrahedron, 'o' for octahedron, 'cb' for cube, etc., as in Table 15); for example,

$$Ca^{[8]}F_2^{[4]} \text{ or } Ca^{[cb]}F_2^{[t]} \text{ or } Ca^{cb}F_2^{t}$$

The notation must be able to describe coordination by different sets of atoms, or coordination at different distances, or self-coordination and coordination polyhedra composed of several distinct atomic species, in addition to giving the shape of coordination polyhedra and/or the number of coordinating atoms. The notation should also be flexible and able to express either the complete coordination or only the limited amount of information desired.

For normal oxycompounds, a simple coordination notation such as that for perovskite, $CaTiO_3$:

$$Ca^{[12co]}Ti^{[6o]}O_3,\ Ca^{[12]}Ti^{[6]}O_3,\ Ca^{[co]}Ti^{[o]}O_3 \text{ or } Ca^{co}Ti^{o}O_3$$

will always be interpreted as coordination of Ca and Ti by oxygen.

However, in the general case such simplification results in ambiguity of interpretation. The coordination of atom A in the compound $A_a\ B_b\ C_c$ for such a case is written:

$$A^{[m,n;p]}$$

where 'm' and 'n' denote the numbers of atoms B and of atoms C (always in the sequence they are presented in the formula), respectively, which are coordinated to atom A. These coordination numbers are separated by commas; the self-coordination number 'p', of A by atoms A, follows a semicolon. The coordination of atom B is written

$$B^{[m',n';p']}$$

where *m'*, *n'* and *p'* denote the numbers of atoms A, C and B around atom B, respectively. Likewise for atom C.

For example, a more complete information on perovskite reads:

$$Ca^{[8cb,12co;6o]}\ Ti^{[8cb,6o;]}\ O_3^{[4l,2l;8p]}$$

The crystal-chemical formula can be simplified as follows:

$$Ca^{[8,12;6]}\ Ti^{[8,6]}\ O_3^{[4,2;8]}$$

and, if only information on coordination by oxygen atoms is required, it can be further simplified to

$$Ca^{[,12]}\ Ti^{[,6]}\ O_3^{[;8]} \text{ or } Ca^{[12]}\ Ti^{[6]}\ O_3^{[;8]} \text{ or still } Ca^{co}Ti^{o}O_3, \text{ as above.}$$

7.2.3. Notation for the structural units and their packing

The constitution of a structural unit expresses its extensional and geometrical 'structure', i.e., the way the structural unit is built from its subunits, which may be dimers, polygons, polyhedra or any other clusters. Some of the constitutional aspects are concerned only with the structural unit as a whole, whereas other aspects are concerned only with the way each subunit is linked to other subunits. The former include dimensionality, multiplicity, branchedness and periodicity, and the latter linkedness and connectedness.

The *dimensionality* is the number of dimensions in which a structural unit has infinite extension. It is zero for individual atoms and finite groups, and one, two or three for infinite chains, sheets and frameworks, respectively. The corresponding symbols to be used in a crystal-chemical formula are $\overset{0}{\infty}$, $\overset{1}{\infty}$, $\overset{2}{\infty}$, and $\overset{3}{\infty}$, the well-known Machatschki symbols.

The following specific symbols may be used for 0-dimensional structural units:

individual: {a}

group: {g} { ring: {r} or ○; chain fragment: {f} or ∧; cage: {k} or ⊗ }

group: {g}	ring:	{r} or ○
	chain fragment:	{f} or ∧
	cage:	{k} or ⊗

Examples are: $Cs_{2\wedge}[S_6]$, $Na_{4\otimes}[Si_4]$, $Cu_6\{r\}[Si_6O_{18}] \bullet 6H_2O$.

For details of this notation see Parthé (1980).

If dimensionality is the only information expressed, the $\overset{n}{\infty}$ and the pictorial symbols may be used without curly brackets. Otherwise, curly brackets are compulsory in order to avoid ambiguity.

Due to the fact that many mineral structures are based on close packings of individual atoms, the symbol {a} may be omitted for reasons of simplicity; only for the other categories of structural units the dimensionality symbol is compulsory. If the packing is mixed, that is built of atoms of different kinds but similar in size, one may include the various packing atoms, together with their proportion, inside square brackets. For instance, in perovskite, one may write $Ti^{o}[CaO_3]^{c}$. Moreover, if the packing is heterogeneous, that is based on atoms with different sizes, then each kind of atom should be written within square brackets. For $MgCu_2$ (Friauf-Laves phase), one should write

$$[[Mg][Cu_2]]$$

In the case of group structures, e.g., ring, chain fragments and cage structures, the number of atoms of each chemical element within square brackets must be equal to the number of atoms of each chemical element in the finite group. For instance, calcite, $Ca\{g\}[CO_3]$.

The *multiplicity* of a structural unit is the number of single subunits, e.g., polyhedra, single rings, single chains or single layers, which are linked to form a complex structural unit of the same dimensionality.

With regard to *branchedness*, finite structural units and single chains are called unbranched if they do not contain any subunits that are linked to more than two other subunits. They are called branched if they do.

The *periodicity* of a structural unit of infinite extension is the number of subunits, excluding branches, within one repeat unit of the infinite chain from which the strucutral unit can be generated by successive linking.

The main constitutional aspects concerned only with the way each subunit is linked to the other subunits are linkedness and connectedness.

The *linkedness* is the number L of peripheral atoms shared by two subunits.

The *connectedness* of a subunit is the total number 's' of adjacent subunits with which it shares atoms, irrespective of its linkedness with a particular adjacent subunit.

The specific values L_1, L_2, etc., of linkedness L and/or connectedness 's' of a subunit are written within 'Japanese brackets' $\ulcorner\lrcorner$ as trailing superscripts to its central atom, by analogy with the coordination symbols. The first entries in the Japanese brackets are the different values of L_n, separated from the value of *s* by a semicolon. The general formula for a structural unit with only one kind of subunit then reads

$$[A_m^{\ulcorner L_1, L_2 ..; s \lrcorner} B_n]$$

For example, SiO_2 exists in a number of polymorphs having different values of linkedness and connectedness of the SiO_4 tetrahedra. In fibrous silica the SiO_4 tetrahedra form infinite chains with tetrahedra linked by their edges; the corresponding formula is

$$\overset{1}{\infty} [Si^{[4t]\ulcorner 2;2 \lrcorner} O_2]$$

For details of these concepts, see Liebau (1982, 1985).

A notation has also been proposed to emphasize the 'polymerization' (or 'condensation') process of the structural units (Lima-de-Faria & Figueiredo, 1976), in which the number of silica tetrahedra in a group silicate is designated by a Roman numeral, with a superscript to differenciate it from other groups with the same number of silica tetrahedra (Figure 56).

The symbol of a chain derived from a certain group by polymerization is formed by the symbol of the group plus another superscript, to distinguish different chains derived from the same group. In this way the symbol of the structural unit indicates immediately what is the structural subunit from which it derives (Figure 56).

A framework is in itself an infinite structural unit, but it can be imagined subdivided in parts which can be infinite sheets, infinite chains, or finite groups. These subunits are called 'connected units', and the framework may be considered a condensation of such 'connected units'.

The packing of the structural units shall now be considered. Normally, the packing of structural units (either individual atoms or more complex units) is indicated by layer description, that is, by stating the type of layer formed by the structural units, and the way the layers stack together.

However, as a matter of convenience, a few simplifications are adopted in the notation. Because cubic and hexagonal close packings are so common in close-packed and group structures, use is made of symbols c and h instead of Tc and Th, respectively. Also, for the body-centred cubic packing, b is substituted for Bb, while the other pack-

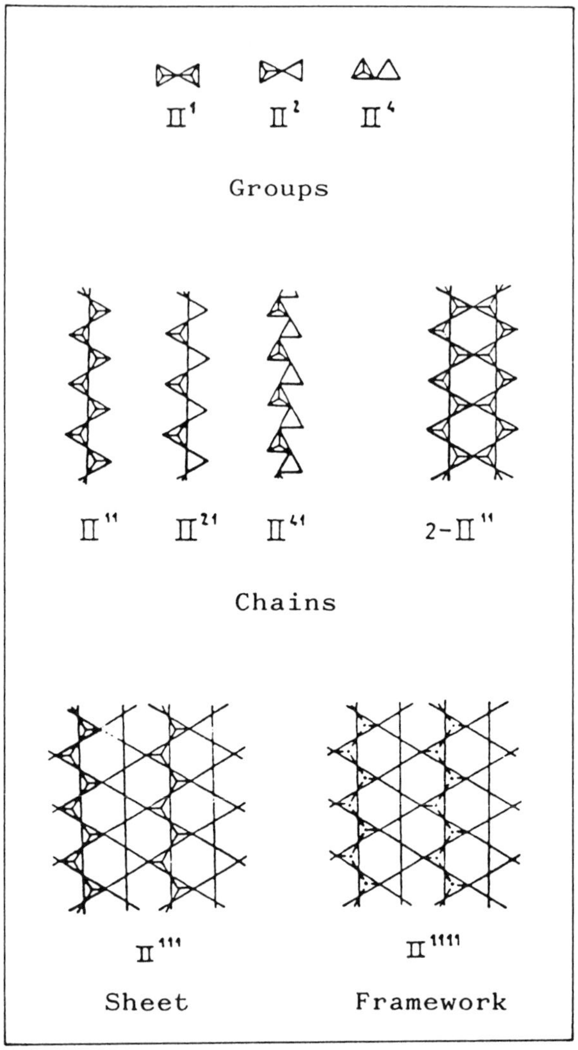

Fig. 56. Nomenclature of structural units expressing the 'polymerization' process. The number of subunits of a group is expressed by a Roman numeral (adapted from Lima-de-Faria, 1986).

ings with b stacking, like Tb and Qb, are written in their complete form.

When no other packing information is provided but the close packing (closest, loose and simple packings), symbols for the latter may be given as trailing superscripts to the square brackets which contain the structural unit. In this case, angle brackets are not compulsory, for instance, $[ABC]^c$. Any other packing information, particulary the packing (or stacking) symbolism used by individual authors should be given in angle brackets on the line, e.g., [ABC] < >.

If packing information is to be given for a set of atoms which does not constitute a structural unit, the symbol should be placed within vertical bars followed by the packing information:

$$|ABC|^c \text{ or } |ABC| < >$$

The symbolism for the packing in close-packed structures calls for a few definitions and the use of a small set of numbers and letters. When dealing with complex layers of the kind found in the packing of group, chain and sheet structures, rather than introducing new symbols, one may extend previous ones while generalizing their meaning. Complex layers which normally correspond to plane directions of higher density of atoms are symbolized, their unit cells are marked, and their ways of stacking indicated by stacking vectors (Figure 57).

To describe the packing in *group structures*, the layer with highest density of groups is selected, the geometrical centres of the groups are considered as if they were individual atoms, and the symbolism for atomic packings is applied. However, for groups whose shape is far from spherical, an uncertainty remains regarding orientation relative to the plane of the layers, and the same packing description may correspond to various assemblages of the groups.

In *chain structures*, it is also necessary to choose the plane direction along which the chains are packed with highest density and mark on them their unit cell.

In *framework structures*, where the structural units extend in three dimensions, the concept of packing has no meaning. The packing of structural units is replaced by the connectivity of connected units, but the symbolism is the same as is used for the other categories of structures, accounting for the possibility to consider a framework derived from the connectivity of infinite sheets, infinite chains or finite groups.

7.2.4. *General scheme for the structural formulas*

To build the structural formula of a mineral structure one has to know first its chemical composition. The atoms belonging to the structural unit should be written within square brackets [ ], information on the packing should follow between angle brackets < >, then information on the constitution of the structural unit as a whole should be added within curly brackets < > or else be expressed by

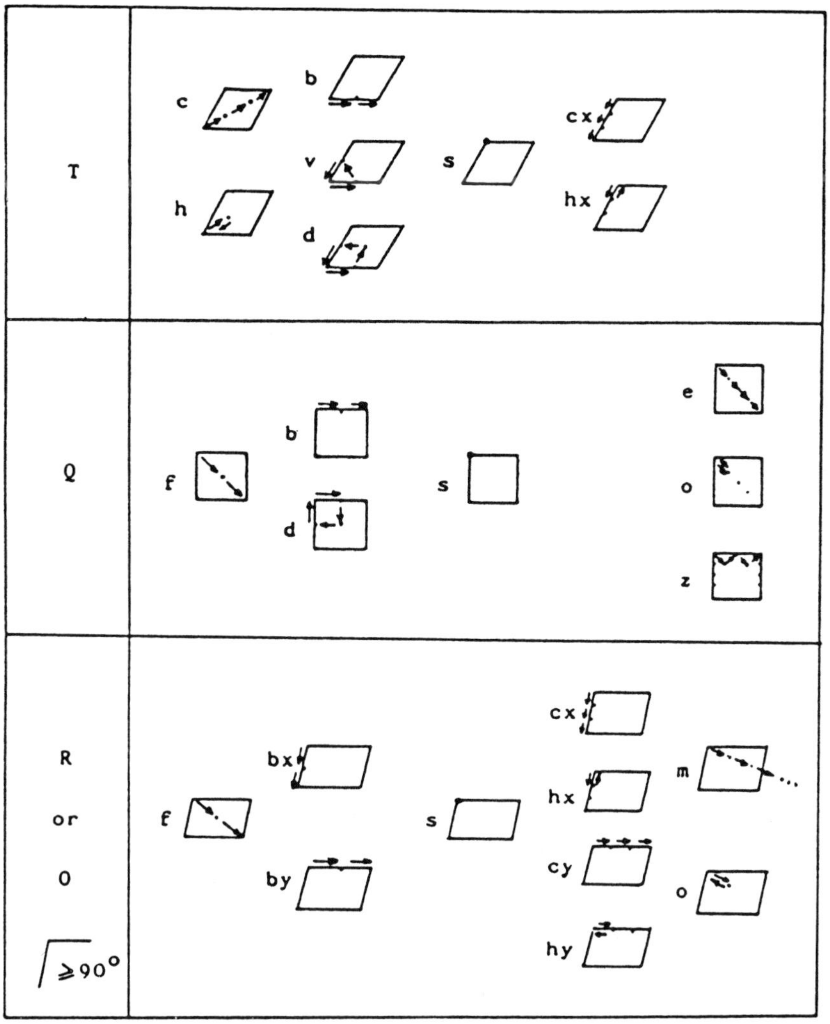

Fig. 57. Generalized stacking symbols. R and O mean rectangular and oblique unit cell, respectively (adapted from Lima-de-Faria & Figueiredo, 1976).

trailing superscripts to the chemical elements or subunits inside the structural unit. Ex.: cristobalite, SiO_2,

$${}^{3}_{\infty}[Si^{[4t]\ulcorner 1;4 \lrcorner}O_2]$$

Information concerning interstitial atoms and/or interstitial groups of atoms is placed before the structural unit. Coordination is expressed, in general, within small square brackets as trailing superscripts to the chemical symbols, the written sequence being preferably from high to low coordination.

The structural formula for a compound A_aB_b-$C_cD_dE_eF_fG_g$ should thus read:

$$A_a^{[\alpha]}\,B_b^{[\beta]}\,C_c^{[\gamma]}\,D_d^{[\delta]}\quad\{\quad\}\quad[E_e^{[\epsilon]}\,F_f^{[\zeta]}\,G_g^{[\eta]}]\;<\;>$$

interstitial atoms — structural unit — constitution of structural unit — packing of structural unit

Table 16. Examples of different ways and criteria of writing the crystal-chemical formulas of some mineral and other inorganic structures (adapted from Lima-de-Faria, Hellner, Liebau, Makovicky & Parthé, 1990)

Cu	${}^{0}_{\infty}[Cu]\langle c\rangle$	$[Cu]^{c}$	${}^{0}_{\infty}[Cu^{[12co]}]$
NaCl	$Na^{[6]}{}^{0}_{\infty}[Cl^{[6]}]$	$Na^{o}[Cl]^{c}$	${}^{3}_{\infty}[Na^{[6o]}Cl^{[6o]}]$
SiO_2 (cristobalite)	${}^{3}_{\infty}[Si^{[4t]}O_2]$	${}^{3}_{\infty}[Si^{[4t]\lceil 1;4\rfloor}O_2]$	${}^{3}_{\infty}[Si^{t}O_2]$
FeS_2 (pyrite)	$Fe^{[6o]}\{{}^{0}_{\infty}\}[S_2^{[3;(1+2)]}]$	$Fe^{[6o]}\,{}^{\wedge}[S_2^{[(3;1)t]}]$	$Fe^{o}\{g\}[S_2]^{c}$
$(Mg,Fe)_2SiO_4$ (olivine)	$(Mg,Fe)_2^{[6o]}\{{}^{0}_{\infty}\}[Si^{[4t]}O_4]$		$(Mg,Fe)_2^{o}\,Si^{t}\,[O]_4^{h}$
$MgAl_2O_4$ (spinel)	${}^{3}_{\infty}[Mg^{[4t]}Al_2^{[6o]}O_4^{[1,3;12co]}]$		$Al_2^{o}Mg^{t}[O]_4^{c}$
$KAl_3Si_3O_{10}(OH)_2$ (muscovite)	$K^{[6+6]}\{{}^{2}_{\infty}\}[Al_2^{[6o]}\{{}^{2}_{\infty}\}[(Al_{0.5}Si_{1.5})^{[4t]\lceil 1;3\rfloor}O_5]_2(OH)_2]$		
		$K^{[6]}Al_2^{o}\,{}^{2}_{\infty}[Al^{t}Si_3^{t}\,O_{10}](OH)_2$	
LaP_2 (HT form)	$La_4\{{}^{0}_{\infty}\}[P_2^{[;1]}P^{[;2]}]\{{}^{0}_{\infty}\}[P_2^{[;1]}P_3^{[;2]}]$		$La_4\,{}^{\wedge}[P_3]{}^{\wedge}[P_5]$
$Ca_3Si_2O_7$ (rankinite)	$Ca_3\{{}^{0}_{\infty}\}[Si_2^{[4t]\lceil 1;1\rfloor}O_7]$		

If several distinct structural units are present, each should be considered separately with its information in curly brackets followed by that in square brackets. Ex.:

$$A_a^{[\alpha]}\,B_b^{[\beta]}\,C_c^{[\gamma]}\,[\{\ \}\,[D_d^{[\delta]}\,E_e^{[\epsilon]}]\,\{\ \}\,[F_f^{[\zeta]}G_g^{[\eta]}]]\ \langle\ \rangle$$

The packing information within angle brackets describes the way the two different structural units pack together.

In accordance with IUPAC (1990) rules, the valency state of each atom is expressed immediately after its chemical symbol by a Roman numeral in parentheses (e.g., $Fe^{(III)}$), a superscripted Roman numeral (e.g., Fe^{III}), or by a superscripted Arabic numeral followed by the + or − sign (e.g., Fe^{3+}).

The hierarchy of bonds leads to a hierarchy of structural units when several degrees of bond strength may be discerned in a structure. This often translates into weaker bond-strength units incorporating previous more strongly bonded units, and can be expressed by multiple brackets, with the central brackets referring to the structural unit having the strongest bonds. Ex.:

$$A_a^{[\alpha]}\,B_b^{[\beta]}\,C_c^{[\gamma]}\,[D_d^{[\delta]}\,\{\ \}\,[E_e^{[\epsilon]}\,F_f^{[\zeta]}G_g^{[\eta]}]\langle\ \rangle]\ \langle\ \rangle$$

The same structural formula may correspond to more than one polymorph of a certain compound, e.g., α and β quartz. This may happen only when the symmetry and/or the distribution pattern of the interstitial atoms vary in the structures. In such cases, the designation of the particular form has to be included in the structural formula in order to avoid ambiguity.

The structural formula can be used with any amount and any selection of structural information depending on the purpose of the study. Examples are given in Table 16.

7.2.5. The importance of structural formulas for the relationship of crystal structures

The widely adopted use of the chemical formula alone to ascribe the crystal structures makes their relationship very difficult. The use of superscripts and other symbols added to the chemical formula, expressing their structural characteristics, makes this relationship much easier. From table 17, where the structures of some binary compounds are indicated by their chemical formula alone, and by the corresponding structural formula, one can recognise the importance of the structural nomenclature.

Authors would certainly gain by writing the

Table 17. Chemical formulas and corresponding structural formulas of some binary compounds

A	AB	AB_2	AB_3
Cu	UPb	$MoPt_2$	$CuPt_3$
Mg	PtCu	$CdTi_2$	$AuCu_3$
Pa	AuCu	$ZrGa_2$	$TiAl_3$
Pu	AuCd	$HfGa_2$	$ZrAl_3$
W	LiRh	$ZrSi_2$	$SbAg_3$
Po	NaCl	$TaPt_2$	$SnNi_3$
	NiAs	TiO_2	$TiCu_3$
	ZnS (c)	CTi_2	$AlCl_3$
	ZnS (h)	$CdCl_2$	MoO_3
	FeS	PbO_2	PdF_3
	CuTi	$CaCl_2$	NNi_3
	CW	CdI_2	$TiCl_3$
	PV	NFe_2	BiI_3
	NW	Li_2O	OTi_3

A	AB	AB_2	AB_3
$[Cu]^{c}$	$[UPb]^{c}$	$[MoPt_2]^{c}$	$[CuPt_3]^{c}$
$[Mg]^{h}$	$[PtCu]^{c}$	$[CdTi_2]^{c}$	$[AuCu_3]^{c}$
$[Pa]^{Tb}$	$[AuCu]^{c}$	$[ZrGa_2]^{c}$	$[TiAl_3]^{c}$
$[Pu]^{Td}$	$[AuCd]^{c}$	$[HfGa_2]^{c}$	$[ZrAl_3]^{c}$
$[W]^{\ll Q \gg f}$	$[LiRh]^{h}$	$[ZrSi_2]^{c}$	$[SbAg_3]^{h}$
$[Po]^{Qs}$	$Na^{o}[Cl]^{c}$	$[TaPt_2]^{c}$	$[SnNi_3]^{h}$
	$Ni^{o}[As]^{h}$	$Ti^{o}[O_2]^{c}$	$[TiCu_3]^{h}$
	$Zn^{t}[S]^{c}$	$C^{o}[Ti_2]^{c}$	$Al^{o}[Cl_3]^{c}$
	$Zn^{t}[S]^{h}$	$Cd^{o}[Cl_2]^{c}$	$Mo^{o}[O_3]^{c}$
	$Fe^{t}[S]^{c}$	$Pb^{o}[O_2]^{h}$	$Pd^{o}[F_3]^{h}$
	$[CuTi]^{Tb}$	$Ca^{o}[Cl_2]^{h}$	$N^{o}[Ni_3]^{h}$
	$C^{p}[W]^{Ts}$	$Cd^{o}[I_2]^{h}$	$Ti^{o}[Cl_3]^{h}$
	$P^{p}[V]^{Ts}$	$N^{o}[Fe_2]^{h}$	$Bi^{o}[I_3]^{h}$
	$N^{p}[W]^{Ts}$	$Li_2^{t}[O]^{c}$	$O^{o}[Ti_3]^{h}$

structural formulas of mineral structures, because it would impel them to summarize their structural results in an explicit way (structural category, coordination of the atoms, etc.).

Structural formulas are also important to express the relationship among parental structures. On Table 18 various structures closely related to the ideal perovskite structure are represented.

7.3. Structure type symbolism

There is a difference between the structural formula and the symbol of a mineral structure type. The structural formula of a mineral structure is the chemical formula of the compound with the addition of superscripts, parentheses, etc., expressing its various structural characteristics. The symbol of a structure type represents a certain kind of structural arrangement and may be based on various notations such as letters and numbers, chemical formulas, lattice complexes, or structural formulas (Table 19).

A general notation for structure types was first proposed by Ewald and Hermann in 1931. 'A' for the chemical elements, 'B' for simple binary compounds AB, 'C' for AB_2, and so on. Within each category they were distinguished by an ordering number. Ex.: copper symbol A1; diamond A4; NaCl, halite, B1; PtS, cooperite, B17; MoS_2, molybdenite, C7. This notation has not been very widespread, possibly because of its lack of structural information. Consequently, structure types are frequently referred to by the best known substance or earliest analysed compound having that structure type. Hence about two hundred compounds having the NaCl arrangement are regarded as pertaining to the NaCl structure type. The American Society for Testing and Materials (1957) proposed a nomenclature for ordering alloy phases, which expressed the chemical formula of a representative substance followed by the number of atoms per unit cell, then the Bravais lattice type codified in a literal symbol. Improvements on this nomenclature were proposed by Schubert (1964) for elements and binary compounds, and by Pear-

Table 18. An example of the efficiency of the structural nomenclature. The perovskite family, showing the corresponding basic and derivative structures: by distortion (dist.), substitution (subs.), insertion (inser.), subtraction (subtr.) and displacement of packing layers (disp.)

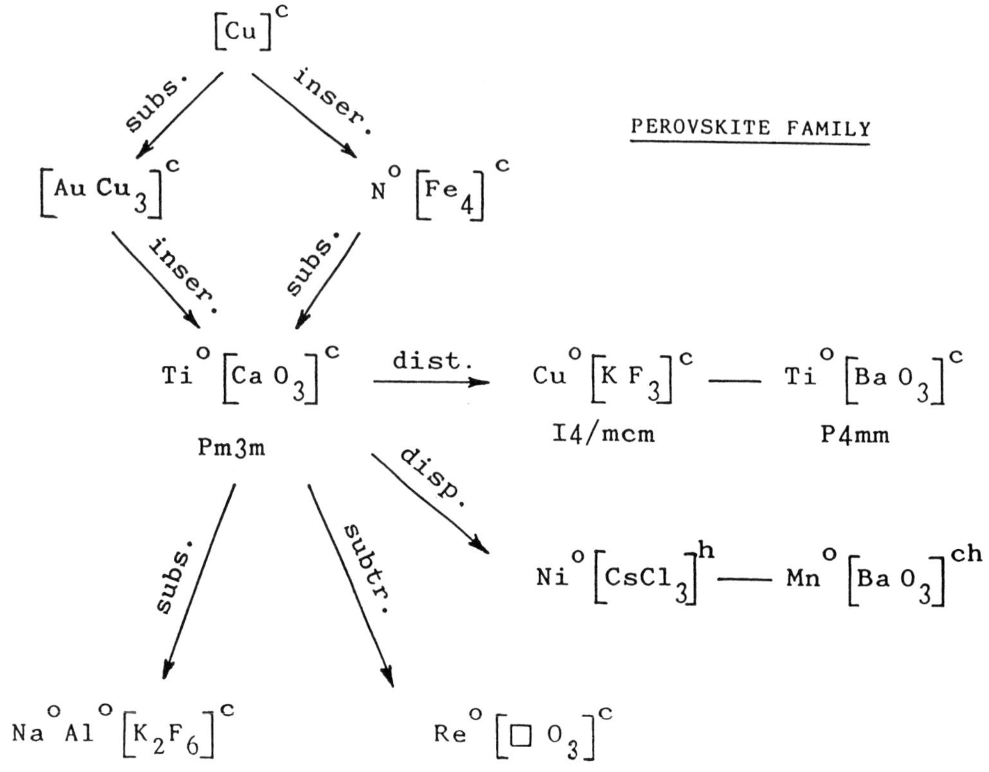

son (1967) for metals and alloys. However, such a nomenclature is ambiguous because the same Pearson symbol may pertain to several structure types. Ex.: oP8 (orthorhombic primitive, eight atoms per unit cell) corresponds to SiTi, BFe, CuS and MnP, which are different structure types. Although this symbolism is simple and useful, especially for ordering structure types in a list for structure determination purposes, it is of no use when expressing the relationship between crystal structures.

A more structurally oriented nomenclature has been attempted by Hellner (1965) and by Donnay, Hellner & Niggli (1966), who proposed descriptive symbols based on lattice complexes. For instance, to the copper structure corresponds the F lattice complex and to ZnS, sphalerite, the symbol F + F''. However, this symbolism becomes very complicated for less symmetrical structures and it seems of little practical use.

Offering still more importance to atomic arrangement than to symmetry characteristics, Lima-de-Faria (1965b) proposed a symbolism for close-packed structure types based on general structural formulas. For instance, the NaCl structure type, symbol B1 in the Ewald-Hermann notation, was represented by the symbol A^oX^c, where X^c meant a large packing atom X in cubic close packing(c), and A^o meant an interstitial atom A in an octahedral void (o). However, this notation called for the addition of superscripts to distinguish among structure types with the same general structural formula but with different distribution patterns of interstitial atoms. For instance, TiO_2, rutile, and α-PbO_2 would be represented by the same symbol A^oX^h, while corresponding to different structural arrangements, namely, one with rows of Ti atoms and the other with Pb forming a zig-zag pattern.

To solve this ambiguity it was proposed that the structural formula of a certain substance be used instead of the general structural formula. Such symbolism, in addition to its own descriptive value, would also have the advantage of being in agree-

Table 19. Various symbolisms proposed for the structure types

	Ewald and Hermann SB(1931)	Hellner (1965)	Lima-de-Faria (1965)	Lima-de-Faria and Figueiredo (1976)
Cu	A1	F or (c) H	X^c	$[Cu]^c$
Mg	A3	E or (h) H	X^h	$[Mg]^h$
Na Cl	B1	F + F'	$A^o X^c$	$Na^o[Cl]^c$
Zn S (sphalerite)	B3	F + F''	$A^t X^c$	$Zn^t[S]^c$
Al_2MgO_4 (spinel)	H11	F'''_{222} + D, T'	$A_2^o B^t X_4^c$	$Al_2^o Mg^t [O_4]^c$
Mg_2SiO_4 (olivine)	H12	(h)n C_{22}+00 1/2 $I2_{xy}$, A_{211}1/4 1/4 F	$A_2^o B^t X_4^h$	$Mg_2^o Si^t [O_4]^h$

ment with the custom of making use of well-known compounds when representing structure types. Thus, the structural formula of the most representative substance, or Megaw's aristotype (Megaw, 1973) has been proposed (Lima-de-Faria & Figueiredo, 1976) for the corresponding structure type. The B1 structure type would then be represented by the symbol $Na^o[Cl]^c$. On Table 19 are presented the various symbolisms that have been proposed for the structure types.

CHAPTER 8

Systematics of minerals on structural grounds

8.1. Choice and presentation of data

Approximately three thousand and five hundred mineral species are known and the structures of many of them have been determined. In a first approach to a structural classification of minerals, it would be impracticable to appropriately treat all of them; consequently, the necessity was felt to reduce their number and to define a convenient domain of mineral species to be considered in this book. We chose the most common minerals, and a few others were added which fulfilled some links in the structural classification, or had a certain particular structural interest.

For the choice of the most common minerals the list was adopted that is given in Kostov's 'Mineralogy' and which includes approximately 300 mineral species. Because emphasis on the various kinds of structural arrangements is important, mineral species have been grouped according to mineral structure types. In this way, the 300 minerals are reduced to approximately 190 mineral structure types. With the addition of 40 mineral structure types, a total of 230 different structure types are described.

The mineral chosen to define the mineral structure type was the most representative mineral species, the so-called aristotype in the sense of Megaw (1973). There are certain mineral species to which several names correspond, but the name given in Hey's 'Chemical Index of Minerals' has been adopted (Ex.: sphene, rather than titanite). There are also some mineral species which pertain to an homologous series, and then the extreme cases are mentioned just below the series name (Ex.: the olivine series: forsterite-fayalite).

The structural data presented include the structural formula, the space group, the cell parameters, the number of structural formulas per unit cell (Z), and the atomic positions of the corresponding chemical elements.

The rules proposed by the Nomenclature Subcommittee Report (Lima-de-Faria, Hellner, Liebau, Makovicky & Parthé, 1990) for the structural formulas were adopted, with a few alterations. Main differences are the location of the structural units at the end, and of all the interstitial atoms in the beginning.

The space group and the atomic positions were taken from Wyckoff's 'Crystal Structures' unless a more recent structure determination was known to give substantially different or unambiguous results. The symbolism presented for the space group in some cases is not the standard but that used by the authors, in order not to change the description of the atomic positions. The experimental errors in the atomic positions are omitted for reasons of simplicity.

The population corresponding to the same mineral structure type, that is, the mineral species having the same structure (isotypic), are indicated by their names and corresponding structural formulas. Underlined mineral names correspond to the most common minerals of Kostov list. Due to the lack in many books of a clear definition of isotypy, many minerals that were designated isotypic have not been included in the population. The conviction is felt that in the future many other mineral species will be added, substantially increasing their population. The population is followed, when it is the case, by the distortion and/or substitution derivatives. Only when a derivative is sufficiently important is it described on a separate page.

Several complementary representations of the same mineral structure type are given, in order to enable a better view and understanding of the structure. These representations may include: packing drawing and corresponding projection of the unit cell content, polyhedral representation, ball and spoke description, layer dercription (condensed model) or any other suitable structural pictorial information, like special coordination polyhedra of some of the chemical elements. Among the various kinds of representations particular attention is given to the close-packing description, based on condensed models, because the general tendency in mineral structures is toward the densest packing.

The original figures are, in many cases, modified, adapted or completed for continuity and to avoid certain misunderstandings. They are often completed with the chemical symbols of atoms, when these are not explicit in the original, and the axes of the unit cell may have been changed in order to produce a uniform description of the structures, and facilitate the comparison among the various representations.

The figures are referred to their original authors, whenever possible. If a figure is taken from a textbook where the original is not mentioned in full, then it is also indicated that it is quoted by the author of the book.

The properties of the mineral representing the structure type had to be selected due to shortage of space and lack of uniform information. The main properties chosen are: habit, cleavage (cleav.), fracture (fract.), twinning (twin.), hardness (hard.), density (dens.), colour, transparence (transp.), refractive index/reflectance (refr./reflect.), birefringence (birefr.), lustre, streak, melting point (melt. p.), and the packing efficiency under the designation of 'close packing index' (CPI).

The packing efficiency of a structure deserves special treatment. It refers to the density of packing in the structure, and is determined by the ratio of the volume occupied by the packing atoms within the unit cell and the volume of the unit cell. It is called here 'close packing index' (CPI). The maximum packing efficiency would be 1 if parallelepipedic shaped atoms could be imagined to occupy the whole space of the unit cell, and 0.74 if spherical atoms were considered corresponding to the closest packing. Normally, the packing atoms are anions, but in some cases they may also be cations, e.g. in perovskite, $Ti^{o}[CaO_3]^{c}$, where Ca^{++} and $O^{=}$ are the packing ions. It was realized that several authors give different definitions of the packing efficiency, some considering only the anions and others the anions and the influence of the expansion effect of cations located in tetrahedral or octahedral voids. This situation and the use of different values of atomic radii have given rise to different values of the packing efficiency for the same mineral structure. Exs.: Rutile Ti^{o} $[O_2]^{h}$: 72 for Zoltai & Stout (1984), 0.74 for Giacovazzo, Monaco, Viterbo, Scordari, Gilli, Zanotti & Catti (1992). and 6.6 for Berry & Mason (1959); perovskite Ti^{o} $[CaO_3]^{c}$: 68 for Zoltai & Stout (1984), and 0.62 for Giacovazzo et al. (1992). Though we think that this is a very important subject, we decided to use in the present approach the values of the 'Symmetrical Packing Index' (SPI) established by Zoltai & Stout, and leave the clarification of this problem for a future opportunity.

The properties refer to the mineral that represents the structure type, and in the case of homologous series, the values of the extreme members of the series are indicated whenever possible. The sources of all this information are indicated in the references in a simplified manner, the complete references being given in the bibliography at the end of the book.

The description of the structure type is often brief in order to enable all the structural data to be included in one page. In a very few cases two pages have been used to describe the same mineral structure type.

8.2. Ordering of the main categories of mineral structure types

The mineral structure types are ordered according to the five main categories of crystal structures, namely: close-packed, group, chain, sheet and framework structures. Within each main structural category the mineral structure types are organized from the simplest to the most complex, with respect to the structural formula or structural arrangement. In certain cases, the insertion (or stuffed) derivatives are placed near the corresponding basic structure type. Although this contradicts the principle of hierarchy of formula simplicity, it facilitates the relationship among closely related mineral structure types.

The close-packed structures are divided in two main subcategories: homogeneous and heterogeneous, and these may or may not admit a layer description. The various subdivisions are distinguished by their packing characteristics.

A) *Layered homogeneous close-packed structures*, based on the packing of equal layers stacked in the same way; they correspond to structures based on the various well-known packings: closest packings, cubic body-centred packing, simple cubic packing, etc.

A.1. Closest packings

A.1.1. Without occupied interstitial atoms:

Ex.: copper $[Cu]^{c}$

A.1.2. With occupied octahedral voids:

Ex.: halite $Na^{o}[Cl]^{o}$

A.1.3. With occupied tetrahedral voids:

Ex.: wurtzite $Zn^{t}[S]^{h}$

A.1.4. With occupied octahedral and tetrahedral voids:

Ex.: olivine $Mg^{o}{}_{2}Si^{t}[O_4]^{h}$

A.2. Cubic body-centred packings:

Ex.: iron $[Fe]^{Bb}$

A.3. Simple cubic and simple hexagonal packings

Ex.: fluorite $Ca^{cb}[F_2]^{Qs}$

A.4. Packings based on R-layers with a pattern formed by rows of triangles and rows of squares, or based on N-layers formed by other patterns of interconnected triangles and squares.

Exs.: andalusite $Al^{o}Al^{[5]}Si^{t}[O_5]^{R^{21}_{by}}$

by vysotskite $Pd^{sq}[S]^{N^{21}_{s}}$

B) *Non-layered homogeneous close-packed structures*, based on a three-dimensional network (*) of atoms packed in a close way.

Ex.: garnet $Ca^{do}Al_2{}^{o}Si_3{}^{t}[O_{12}]^{*}$

C) *Layered heterogeneous close-packed structures*, based on different layers and/or different kinds of stacking.

C.1. Packings based on equal layers but with different kinds of stacking

Ex.: molybdenite $Mo^{p}[S_2]^{(2Ts)h}$

C.2. Packings based on different layers

Ex.: matlockite $Pb^{[9]}[[F][Cl]]^{(Q^2,2Q^1f)f}$

D) *Non-layered heterogeneous close-packed structures*, based on interpenetrated parts (blocks, rods or slabs) of homogeneous close packings

Exs.: galenobismutite $Pb^{[7]}Bi_2{}^{[6/7]}[S_4]^{c/h}$

apatite $Ca_5{}^{p}\ P_3{}^{t}[O_{12}(OH,F)]^{Ts/h}$

The group, chain, and sheet structures may also be divided in homogeneous and heterogeneous, as regards their formation by one only or several kinds of structural units. The homogeneous structures can be considered simple or composite as their structural units consist of one only or a number of building units (dimers, polygons or polyhedra). The structures may still be subdivided, according to the shape of their building units into linear, polygonal and polyhedral.

When a structure is built up of several structural units of different dimensionalities it is classified according to the structural unit of highest dimensionality.

The classification of the mineral structures considered in this work, and according to the principles referred to above, is summarized on Table 20.

A number (24) of mineral structures pertaining to the most common (Kostov's list) could not be clearly classified; for a few of them, a guess of their probable structural category is indicated. The most numerous structures in this unclassified category are sulphates, such as thenardite $Na_2{}^{o}S^{t}O_4$, and kieserite $Mg^{o}(H_2O)S^{t}O_4$. With more structural information and the help of computer programs they will come to be appropriately classified.

8.3. Final remarks

The 230 mineral structure types that have been described are distributed among the following structural categories:

84 Close-packed structure types including 255 mineral structures

29 Group structure types with 73 mineral structures

21 Chain structure types with 30 mineral structures

26 Sheet structure types with 38 mineral structures

Table 20. Scheme of the structural classification of minerals

- Close-packed
  - homogeneous
    - layered —— (equal layers and same stacking)
    - non-layered — (three directional packings)
  - heterogeneous
    - layered —— (different layers and/or different stackings)
    - non-layered — (interpenetrated parts of homogeneous close packings)
- Group / Chain / Sheet
  - homogeneous
    - simple ——
      - linear
      - polygonal
      - polyhedral
    - composite
  - heterogeneous
- Framework
  - simple ——
    - linear
    - polygonal
    - polyhedral
  - composite

46 Framework structure types with 65 mineral structures
24 Non-classified structures with 41 mineral structures

This summary clearly shows the predominance of close-packed structures, and confirms the stability principle that expresses the tendency for the densest packing.

Another fact that strikes one's attention is the great number of close-packed structures which admit a simple layer description, that is, which are based on equal layers or on two different alternate layers. This may be related to the small values of the unit-cell parameters, (less than 15Å, for most mineral structures) with the consequence that any plane direction of high density of atoms repeats at very short distances (Figure 58).

It is also important to notice the simple and symmetrical character of the atomic layers (condensed model sheets) in which most of the close-packed structures can be decomposed. Such layers normally correspond to the most symmetrical distribution of the interstitial atoms within the packing layers. This fact shows that the stability of atomic layers is also an important factor in crystal structures. Stability in two dimensions is possibly related to the process of crystal growth by addition of atomic layers over one another.

One of the goals of crystal chemistry is the systematic derivation of crystal structures. The results mentioned above point to a systematic derivation based on the determination of the most symmetrical distributions of the interstitial atoms within the packing layers, carried out by application of adequate stacking rules (Lima-de-Faria, 1965b; Lima-de-Faria & Figueiredo, 1969).

This is a first attempt to present a detailed systematics of minerals based on structural features, and much work will be needed in order to reveal all its potentialities. The author is aware of severe difficulties that still exist in the application of the structural classification, though they are but a natural characteristic of any new development in science. Spronsen (1969) said:

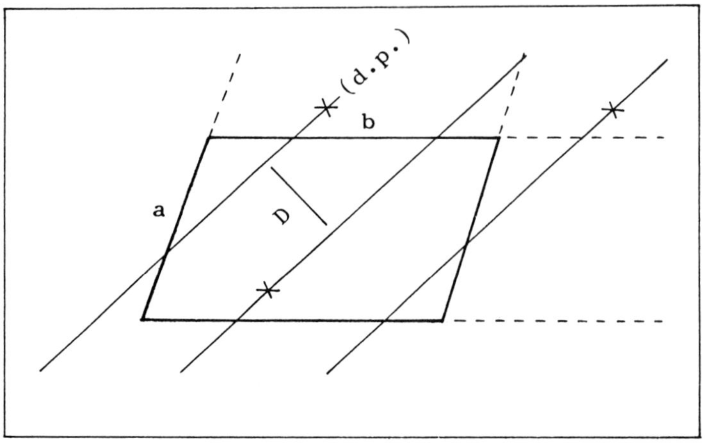

Fig. 58. The densest atomic plane direction is marked (d.p.) on a plane unit cell with parameters 'a' and 'b'. If 'a' and 'b' are small, the distance D between two densest planes is also small, and will only accommodate a few atoms in between. Normally, there will be space for no more than one or two atomic layers.

There are three distinguishable main stages in the history of the periodic system, namely that of initiation, that of phenomenological development and that of theoretical development.

Such stages seem not to be confined to the periodic system of chemical elements; most likely they are main stages in every historic development in scientific work. In what regards the structural classification of minerals, and of inorganic structures in general, no doubt we are at the beginning, that is, just at their initiatial stage.

8.4. Descriptive charts of mineral structure types

8.4.1. Close-packed structures

COPPER $[Cu]^{c}$

a = 3.61496 Å (at 18°C) Cu (4a)

Fm3m Z = 4

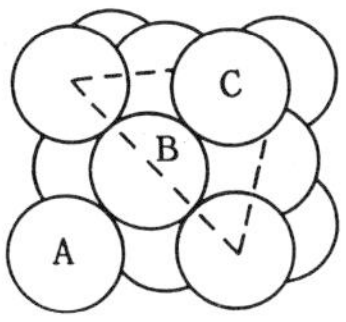

Fig. 1

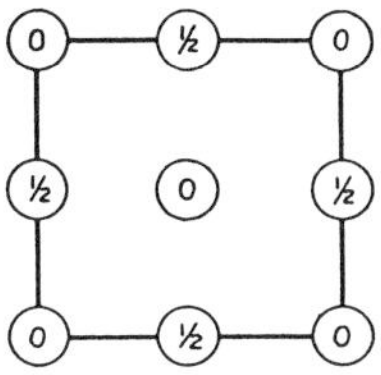

Fig. 2

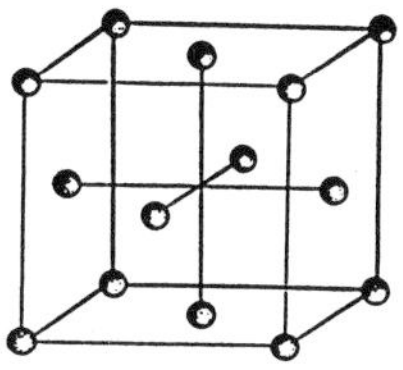

Fig. 3

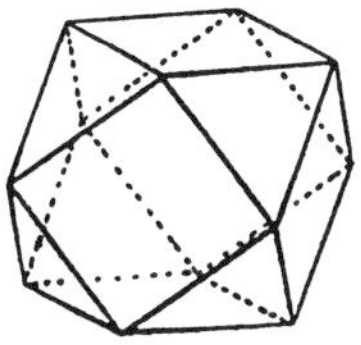

Fig. 4

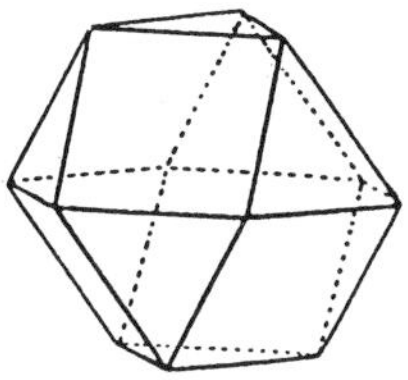

Fig. 5

Properties

Habit	Cleav.	Fract.	Twin.	Hardn.	Dens.	Colour	Transp.
cubic, dodecahedral, arborescent	none	hackly	(111)	2.5-3	8.95	light rose to copper--red	opaque

Refr. index/Reflect.		Birefr.	Luster	Streak	Melt.p.	CPI
n=0.641	81%		metallic	metallic, shining	1083° C	(SPI) 74

Population

Silver	$[Ag]^{c}$
Gold	$[Au]^{c}$
Electrum	$[(Ag, Au)]^{c}$
Kongsbergite	$[(Ag, Hg)]^{c}$
Platinum	$[Pt]^{c}$
Palladium	$[Pd]^{c}$
Iridium	$[Ir]^{c}$
Nickel	$[Ni]^{c}$
Lead	$[Pb]^{c}$

Distortion derivatives

Mercury	$[Hg]^{c}$	R3m (at <-39°C)
Indium	$[In]^{c}$	I4/mmm

References

Kostov (1968) 94,95.

Wyckoff (1963) Vol. 1, 7-11.

Povarennykh (1972) 116, 192--195.

Palache et al. (1944) Vol. 1, 99.

Zoltai + Stout (1984), 375.

Description

The copper atoms form a cubic closest packing (c). The coordination of the copper atoms in relation to each other is therefore 12, and the corresponding polyhedron is a cuboctahedron (Fig. 4 and 5).

Figures

Fig. 1. Packing model of Cu structure. The direction of the closest packed layers is shown by a triangle (dashed lines).

Fig. 2. Unit cell projection parallel to (001) (after Wyckoff, 1963, vol. 1).

Fig. 3. Ball and spoke model of the copper structure (after Povarennykh, 1972).

Figs 4. and 5. Coordination polyhedron of Cu (cuboctahedron in two different orientations).

Fig. 6. Condensed model of the copper structure formed by three transparent sheets of cubic closest packing layers parallel to (001). The large open circles are the copper atoms, small dotted circles correspond to all possible voids, octahedral and tetrahedral, in c.c.p., which in this case are all empty.

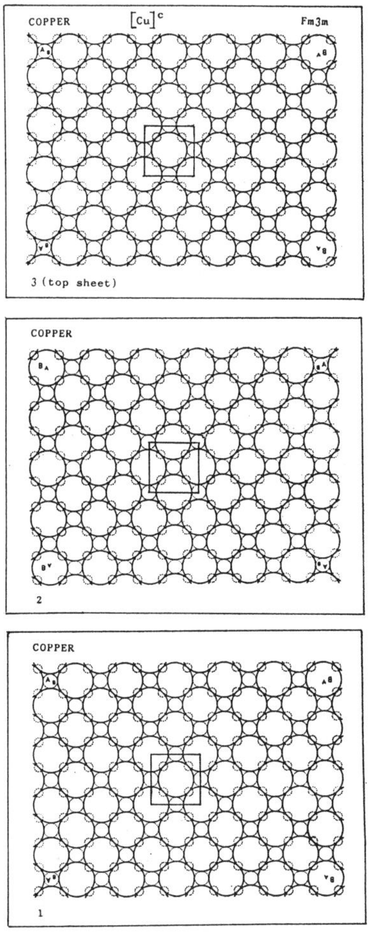

Fig. 6

IRIDOSMINE $[Os, Ir]^h$

$P6_3/mmc$

a = 2.63 Å (Os, Ir) (2c)

c = 4.24 Å c/a = 1.62

Z = 2

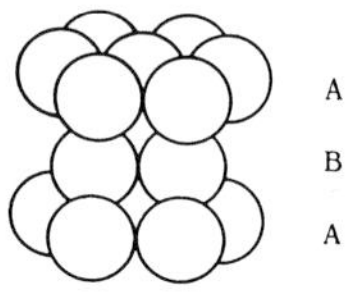

Fig. 1

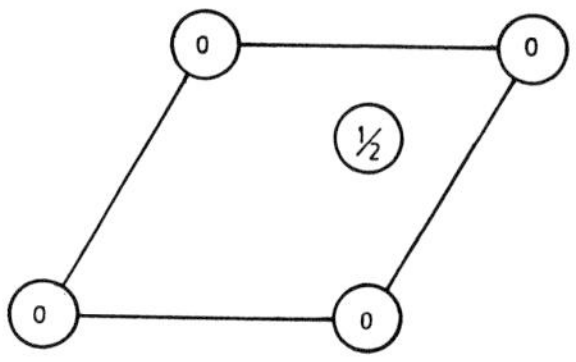

Fig. 2

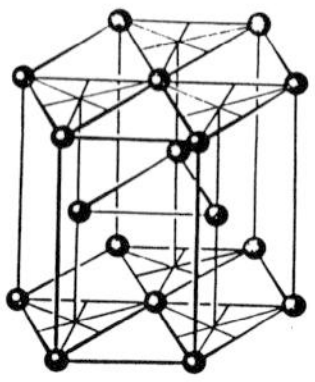

Fig. 3

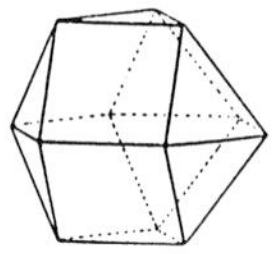

Fig. 4

Properties

Habit	Cleav.	Fract.	Twin.	Hardn.	Dens.	Colour	Transp.
tabular triangular plates	perfect (0001)			6-7	19-21	tin--white	opaque
Refr. index/Reflect.		**Birefr.**		**Luster**	**Streak**	**Melt.p.**	**CPI**
66%				metallic	grey		(SPI) 74 (for h.c.p.)

Distortion derivatives

Zinc $[Zn]^h$ $P6_3/mmc$ but c/a=1.86
a = 2.664
c = 4.945 (c/a ideal = 1.63)

Figures

Fig. 1. Packing model of the iridosmine structure.

Fig. 2. Unit cell projection parallel to (0001) (after Wyckoff, 1963, Vol. 1).

Fig. 3. Ball and spoke model of the iridosmine structure (after Povarennykh, 1972).

Fig. 4. Polyhedron representing the twelve coordination of the Os or Ir packing atoms.

Fig. 5. Condensed model of the iridosmine structure, based on closest packed layers stacked on a sequence A,B,A,B,... (h sequence). The large open circles represent the Os and Ir atoms; small dotted circles correspond to all possible octahedral and tetrahedral voids in h.c.p., which in this case are empty.

Description

The Os and the Ir atoms form together an hexagonal closest packing (h), and are distributed at random within the packing. Any Os or Ir atom is surrounded by 12 packing atoms, and the corresponding coordination polyhedron is represented on Fig. 4. Compare it with that of copper.

Iridosmine pertains to the $[Mg]^h$ inorganic structure type.

References

Kostov (1968) 90.
Wyckoff (1963) Vol. 1, 8-13.
Povarennykh (1972) 117, 192, 194.
Palache et al. (1944) Vol. 1, 111-113.
Roberts et al. (1974) 304.
Zoltai + Stout (1984) 147.
Strunz (1982) 95, 98.

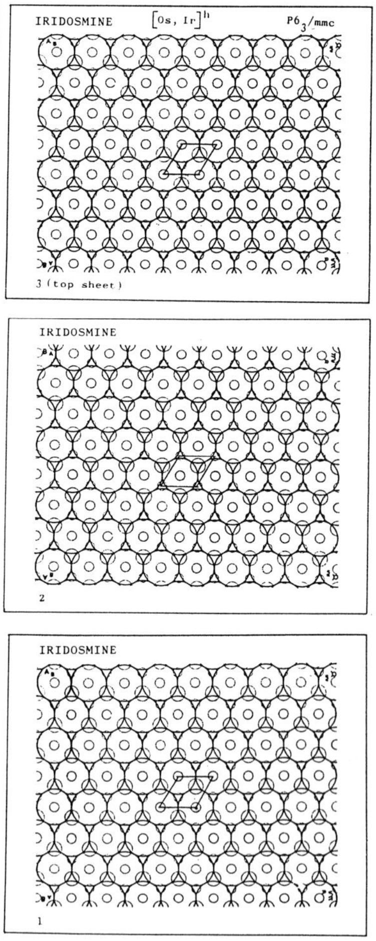

Fig. 5

HALITE $Na^{o}[Cl]^{c}$

Fm3m

a = 5.62779 Å (at 18°C)

Z = 4

Cl (4a)

Na (4b)

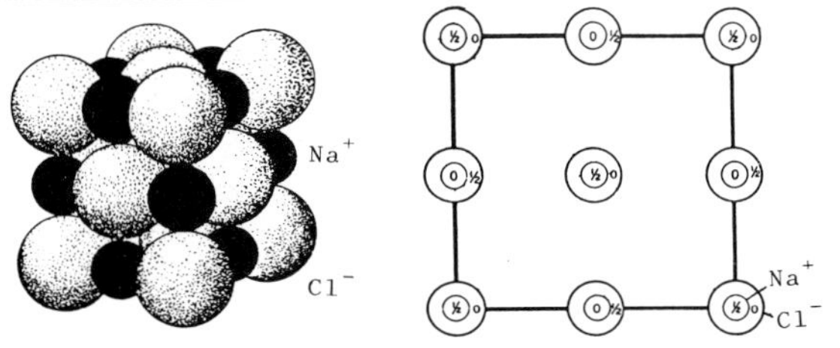

(a) Fig. 1 (b)

Fig. 2

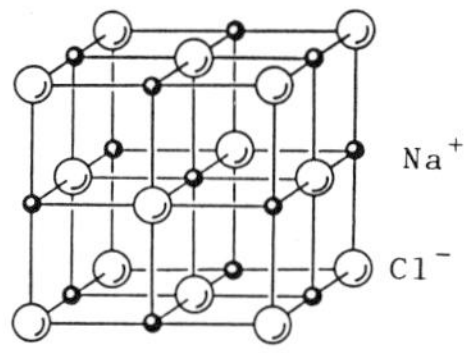

Fig. 3

Properties

Habit	Cleav.	Fract.	Twin.	Hardn.	Dens.	Colour	Transp.
cubic, massive	perfect (001)	conchoidal		2.0	2.168	colourless, white, variable	transparent

Refr. index/Reflect.	Birefr.	Luster	Streak	Melt.p.	CPI
n = 1.5446		vitreous	colourless, white	800°C	(SPI) 74

Population

Villiaumite	$Na^{o}[F]^{c}$
Sylvine	$K^{o}[Cl]^{c}$
Cerargyrite	$Ag^{o}[(Br,Cl)]^{c}$
Periclase	$Mg^{o}[O]^{c}$
Bunsenite	$Ni^{o}[O]^{c}$
Wüstite	$Fe^{o}[O]^{c}$
Manganosite	$Mn^{o}[O]^{c}$
Lime	$Ca^{o}[O]^{c}$
Monteponite	$Cd^{o}[O]^{c}$
Oldhamite	$Ca^{o}[S]^{c}$
Alabandite	$Mn^{o}[S]^{c}$
Galena	$Pb^{o}[S]^{c}$
Clausthalite	$Pb^{o}[Se]^{c}$
Altaite	$Pb^{o}[Te]^{c}$
Osbornite	$Ti^{o}[N]^{c}$

Figures

Fig. 1.(a) Packing model of the halite structure and (b) unit cell projection parallel to (001).

Fig. 2. Polyhedral representation of the halite structure (after Kostov, 1968).

Fig. 3. Ball and spoke model of the halite structure (after Kostov, 1968).

Fig. 4. Condensed model of the halite structure. It is based on three standard sheets of cubic closest packing parallel to (001) planes. Cl^- (large open circles) form the packing, and Na^+ (small black circles) occupy all the octahedral interstices . The small dotted circles represent all the possible tetrahedral voids, which, in this case, are not occupied.

References

Kostov (1968) 155, 198,199.
Wyckoff (1963) Vol. 1, 85-94.
Palache et al. (1951) Vol. 2, 4, 5.
Ingerson (1955) 351.
Zoltai + Stout (1984) 401.
Strunz (1982) 123.
Povarennykh (1972) 49.

Description

The structure is formed by the closest packing of Cl^- anions in cubic closest packing with Na^+ cations occupying all the octahedral interstices.

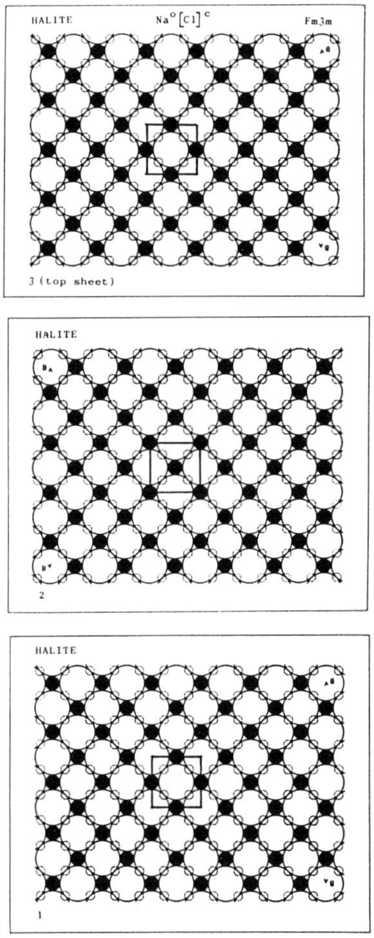

Fig. 4

NICCOLITE $Ni^{o}[As]^{h}$

$P6_3/mmc$

a = 3.602 Å Ni (2a)
c = 5.009 Å As (2c)
Z = 2

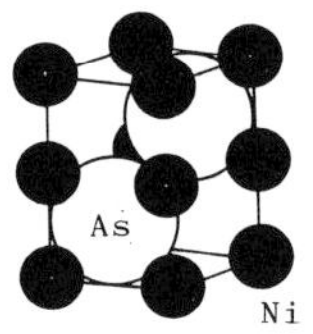

Fig. 1

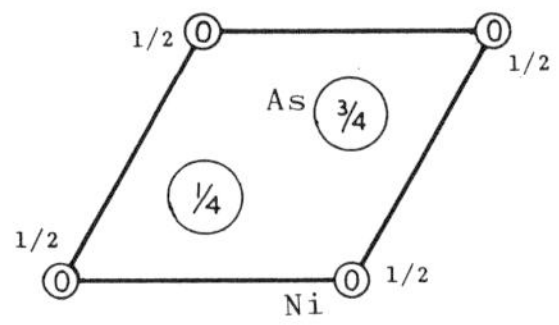

Fig. 2

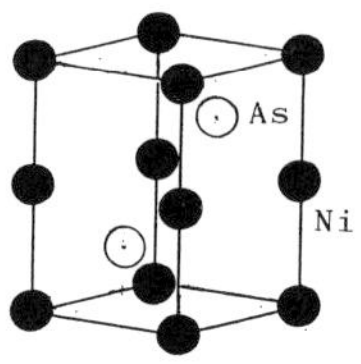

Fig. 3

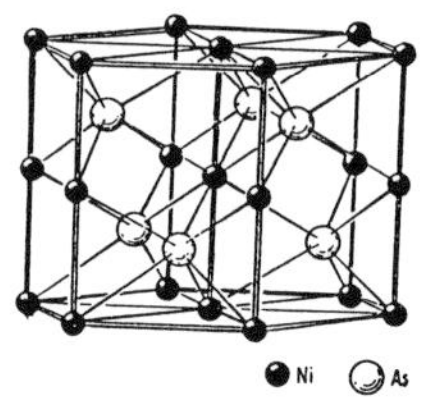

Fig. 4

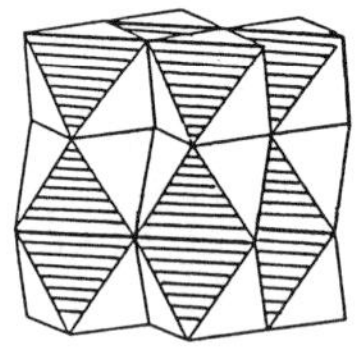

Fig. 5

Properties

Habit	Cleav.	Fract.	Twin.	Hardn.	Dens.	Colour	Transp.
massive, reniform arborescent	none	uneven		5-5.5	7.784	pale copper-red	opaque

Refr. index/Reflect.		Birefr.	Luster	Streak	Melt.p.	CPI
n_{ω} = 2.11	56%	(-)	metallic	pale brownish-black	970°C	(SPI) 74 (h.c.p.)
n_{ϵ} = 1.80						

Population

Langisite	$Co^{o}[As]^{h}$
Breithauptite	$Ni^{o}[Sb]^{h}$
Jaipurite	$Co^{o}[S]^{h}$
Achavalite	$Fe^{o}[Se]^{h}$
Freboldite	$Co^{o}[Se]^{h}$
Imgreite	$Ni^{o}[Te]^{h}$
Kotulskite	$Pd^{o}[Te]^{h}$
Niggliite	$Pt^{o}[Sn]^{h}$
Pyrrhotine	$(Fe^{II}_{1-x},Ni^{III}_{2/3x})^{o}[S]^{h}$
Sederholmite	$(Ni^{II}_{1-x},Ni^{III}_{2/3x})^{o}[Se]^{h}$

Distortion derivatives

Modderite $Co^{o}[As]^{h}$ Pbna

a=5.97 Å b=5.16 Å c=3.52 Å

Figures

Fig. 1. Packing model of the niccolite structure.

Fig. 2. Unit cell projection parallel to (0001).

Fig. 3. Niccolite structure.

Fig. 4. Ball and spoke model of the niccolite structure (after Bokii, 1954).

Fig. 5. Polyhedral representation of the niccolite structure.

Fig. 6. Condensed model of the niccolite structure. It is based on three standard sheets of hexagonal closest packing: As large open circles, Ni small black circles. The small dotted circles represent all the tetrahedral voids, which, in this case, are not occupied.

Description

The structure is formed by the hexagonal closest packing of As atoms with Ni atoms occupying all the octahedral voids. The packing is quite distorted (c/a=1.38, c/a ideal = 1.633).

References

Kostov (1968) 120,121.
Wyckoff (1963) Vol. 1, 122-124.
Palache et al.(1944) Vol. 1, 236,237.
Zoltai+Stout (1984) 387.
Ingerson (1955) Table II.
Povarennykh (1972) 195, 203, 210, 220, 225.
Bokii (1954) 103.

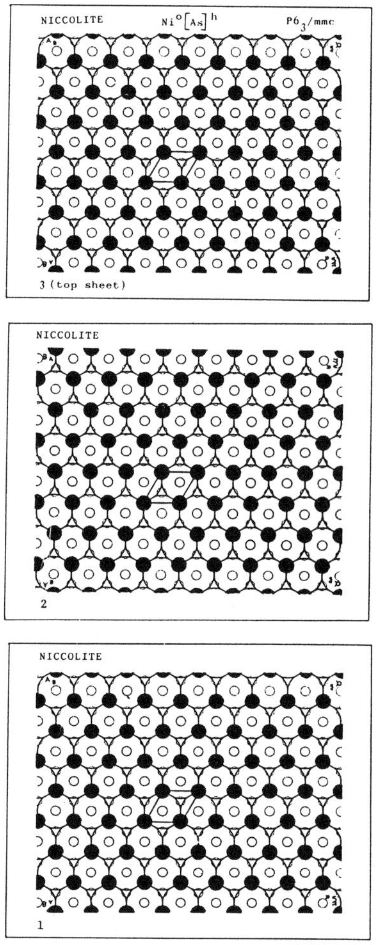

Fig. 6

HYDROPHILITE

$Ca^{o}\left[Cl_2\right]^{h}$

P n n m

a = 6.24 Å
b = 6.43 Å
c = 4.20 Å
Z = 2

Ca (2a)
Cl (4g) u = 0.275, v = 0.325

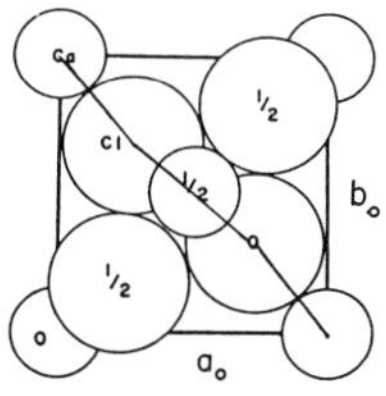

Fig. 1

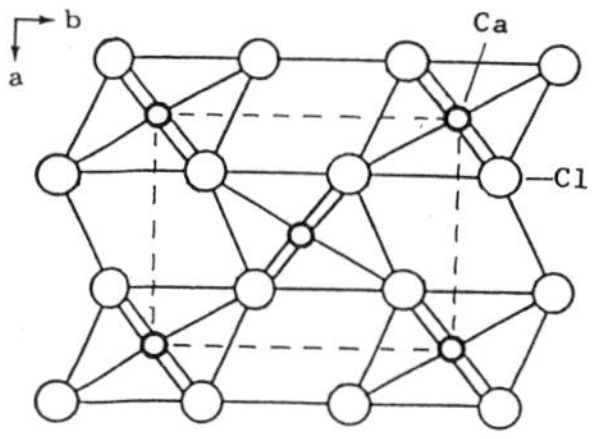

Fig. 2

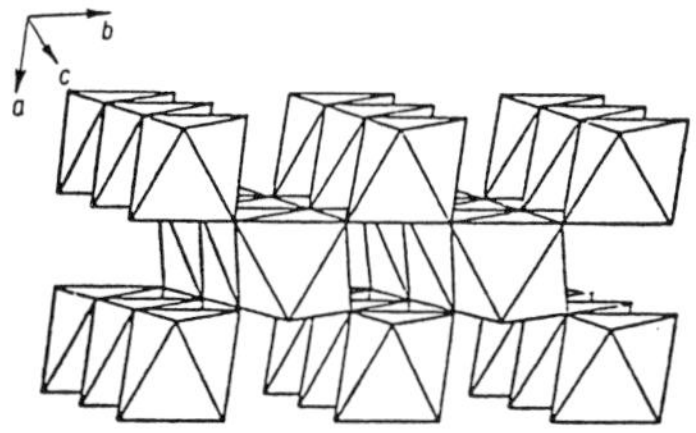

Fig. 3

Properties

Habit	Cleav.	Fract.	Twin.	Hardn.	Dens.	Colour	Transp.
massive	perfect prismatic		polysyn-thetic on a prism		2.22	white	transparent to trans-lucent

Refr. index/Reflect.	Birefr.	Luster	Streak	Melt.p.	CPI
n_α = 1.600 n_β = 1.605 n_γ = 1.603	(+)			754° C	

Figures

Fig. 1. Hydrophilite structure projected along the c axis (after Kyckoff, 1963, Vol. 1).

Fig. 2. Hydrophilite structure projected along the c axis (adapted from Wells, 1962).

Fig. 3. Polyhedral description of the hydrophilite structure (after R. Burns+ V. Burns, 1979).

Fig. 4. Condensed model of the hydrophilite structure. Layers parallel to (100), showing the octahedral occupancy of the Ca atoms in parallel rows.

Description

The chlorine atoms form a hexagonal closest packing, and the Ca atoms are located in octahedral holes, with a row pattern: one filled row and another empty, alternately. This structure is related with rutile, differing mainly by the symmetry.

The distortion of the real structure in relation to the ideal hexagonal closest packing is small: b (ideal) = = 2× 1.732 R (R = radius of the packing atom, which in this case is Cl^-), and a (ideal) = 2× 1.633 R, consequently b/a (ideal) = 1.06, and b/a (real) = 1.03

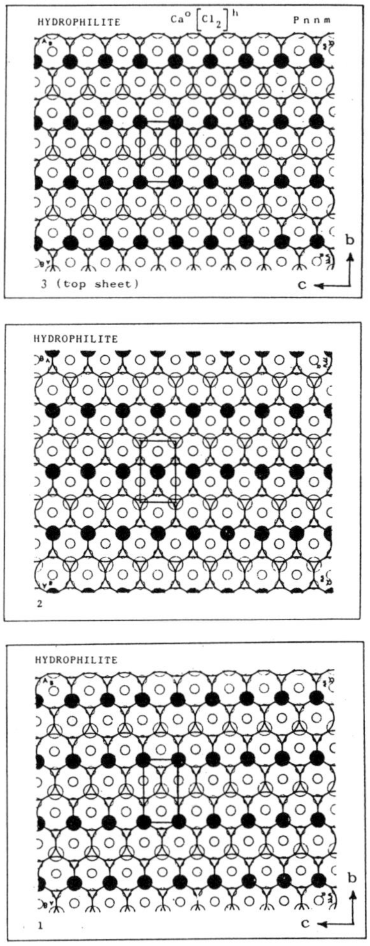

Fig. 4

References

Wyckoff (1963) Vol. 1, 252.

Povarennykh (1972), 643.

Wells (1962) 125.

R. Burns + V. Burns (1979) 13.

Palache et al. (1944) Vol. 2, 41,42.

Ingerson (1955) 351.

Roberts et al. (1974) 291.

RUTILE

$Ti^{o}[O_2]^{h}$

P 4/m n m

a = 4.59 Å
c = 2.96 Å
Z = 2

Ti (2a)
O (4f) u = 0.3053

(a)

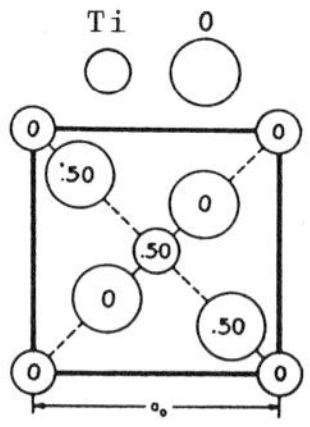
(b)

Fig. 1

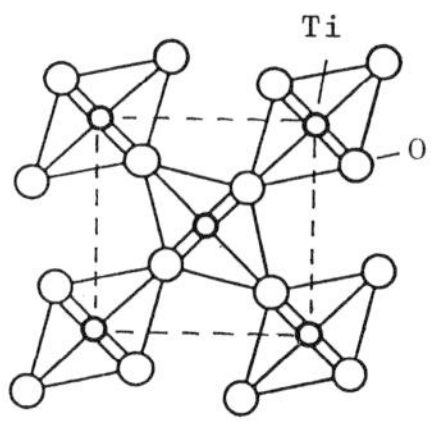

Fig. 2

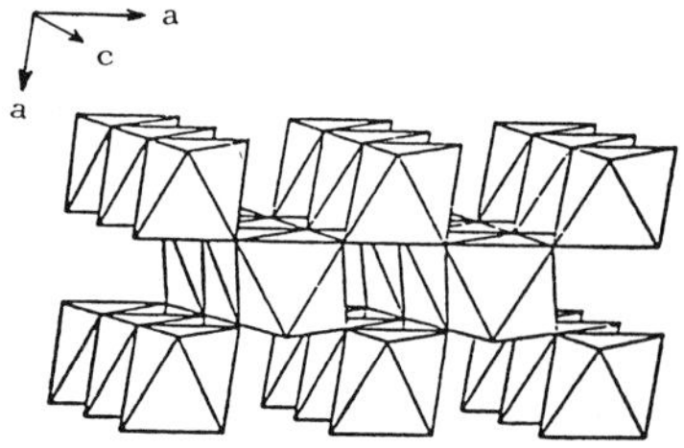

Fig. 3

Properties

Habit	Cleav.	Fract.	Twin.	Hardn.	Dens.	Colour	Transp.
prismatic, acicular [001]	good (100) (110)	subcon-choidal	(011)	6-6.5	4.23	redish brown	transparent to trans-lucent

Refr. index/Reflect.	Birefr.	Luster	Streak	Melt.p.	CPI
n_ω = 2.90 n_ϵ = 2.61	(+)	submetallic, adamantine	light brown	1825° C	(SPI) 72

Population

Pyrolusite	$Mn^{o}[O_2]^{h}$
Cassiterite	$Sn^{o}[O_2]^{h}$
Plattnerite	$Pb^{o}[O_2]^{h}$
Stishovite	$Si^{o}[O_2]^{h}$
Sellaite	$Mg^{o}[F_2]^{h}$

Description

The oxygens are approximately arranged in hexagonal closest packing, and the titanium atoms occupy a row pattern like in hydrophilite. The rutile structure can be derived from that of hydrophilite by a certain distortion (compare figures 1 and 2 of hydrophilite with Fig.1b and 2 of rutile), however rutile has higher symmetry. The condensed model is similar to that of hydrophilite, however the rutile structure is more apart from the ideal h.c.p. (b/a ideal = 1.06 corresponds to 1.00 in rutile).

Figures

Fig. 1. (a) Packing representation of rutile structure, (b) projection along the c axis (after Wyckoff, 1963. Vol. 1).

Fig. 2. Rutile structure projected along the c axis (after Wells, 1962).

Fig. 3. Polyhedral representation of the rutile structure (after R. Burns + V. Burns, 1979).

Fig. 4. Condensed model of the rutile structure. The small black circles represent Ti atoms in octahedral voids.

References

Kostov (1968) 242,243.
Wyckoff (1963) Vol. 1, 250-252.
Palache et al. (1944) Vol. 1, 555-558.
Zoltai+Stout (1984) 408.
Wells (1962) 125.
R. Burns+V. Burns (1979) 13.
Ingerson (1955) 351.
Povarennykh (1972) 297, 666.

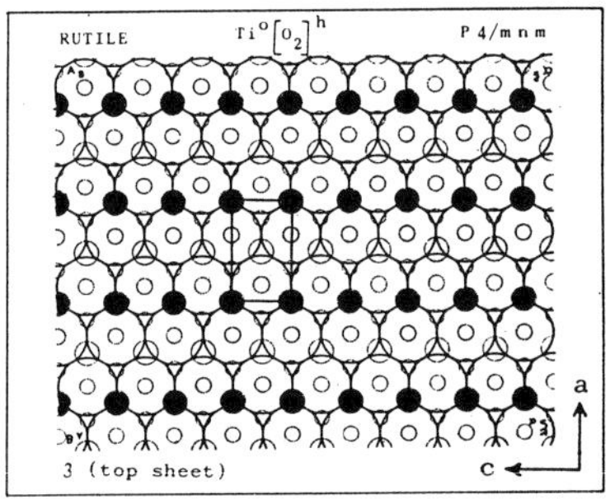

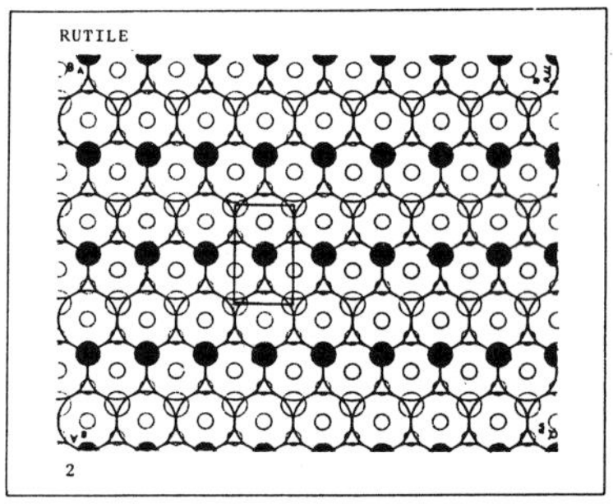

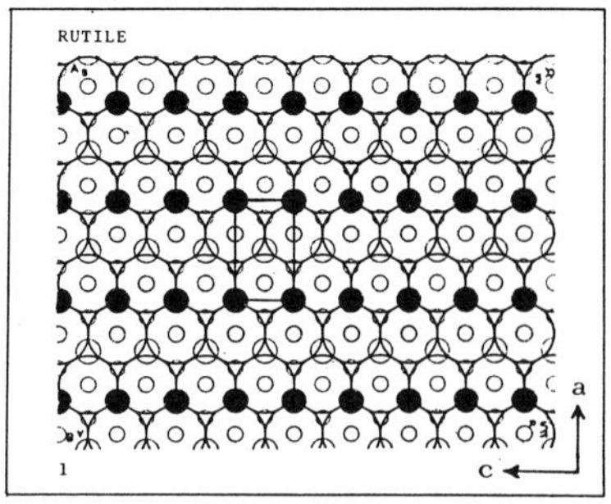

Fig. 4

MANGANITE

$Mn^{o}[O(OH)]^{h}$

$B\,2_1d$

a = 8.88 Å	Mn (8f) x = 0.0047, y = -0.0119, z = 0.2401
b = 5.25 Å	O (8f) x = 0.1218, y = 0.1250, z = 0.0011
c = 5.71 Å	(OH)(8f) x = 0.1239, y = 0.1250, z = 0.5016
β = 90°	
Z = 8	

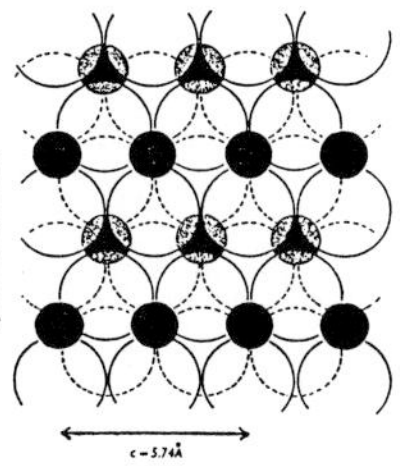

Fig. 1

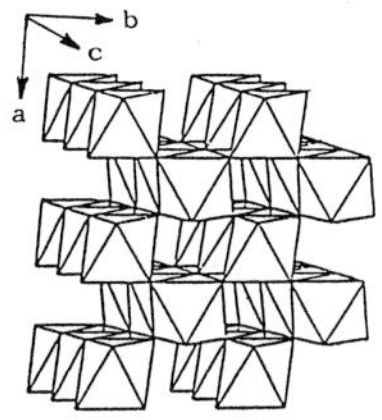

Fig. 2

Properties

Habit	Cleav.	Fract.	Twin.	Hardn.	Dens.	Colour	Transp.
prismatic [001]	perfect (010) good (110)	sectile	(011)	4	4.33	dark black steel--grey	opaque

Refr. index/Reflect.		Birefr.	Luster	Streak	Melt.p.	CPI
$n\alpha$ = 2.24, $n\beta$ = 2.24, $n\gamma$ = 2.53	15%	(+)	submetallic	redish brown		(SPI) 59

Figures

Fig. 1. Schematic representation of two hexagonal closest packing layers of the manganite structure (after Bragg+ Clarinbull, 1979).

Fig. 2. Polyhedral representation of the manganite structure (adapted from R. Burns + V. Burns, 1979).

Fig. 3. Condensed model of the manganite structure. The large open circles represent OH and O atoms, and the small black circles Mn atoms in octahedral voids. Notice the row pattern of the Mn atoms.

Description

Manganite is a distorted and packing substitution derivative of rutile. It is also very related to arsenopyrite (a group structure). Manganite is the γ-form of Mn O (OH), the α-form beeing groutite, with a structure equal to diaspore.

References

Kostov (1968) 232,233.

Dachs (1963) 303-326.

Bragg + Claringbull (1965) 124.

Palache et al. (1944) Vol. 1, 646-648.

R. Burns + V. Burns (1979) 13.

Zoltai + Stout (1984) 421.

Strunz (1982) 217.

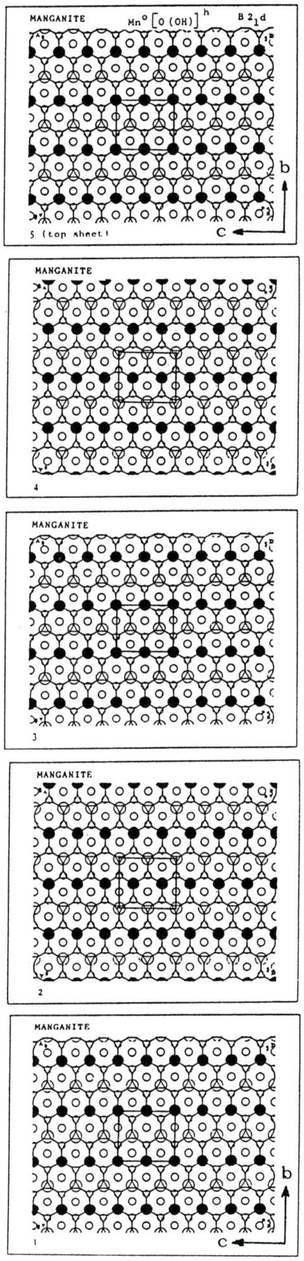

Fig. 3

ANATASE

(Octahedrite)

$Ti^{o}[O_2]^{c}$

$I\,4_1/amd$

a = 3.785 Å
c = 9.514 Å
Z = 4

Ti (4a)
O (8e) u = 0.2066

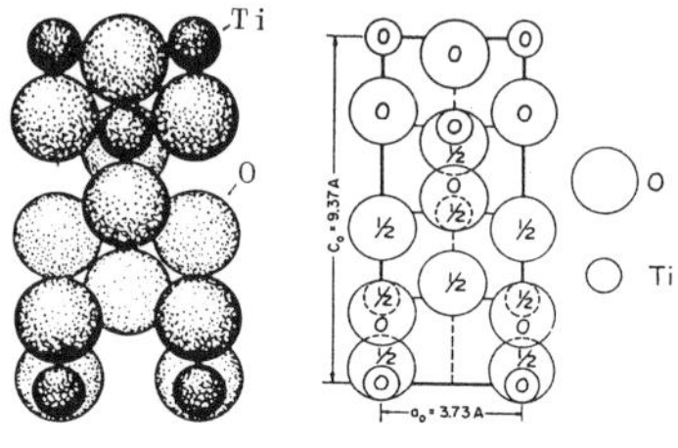

(a) (b)

Fig. 1

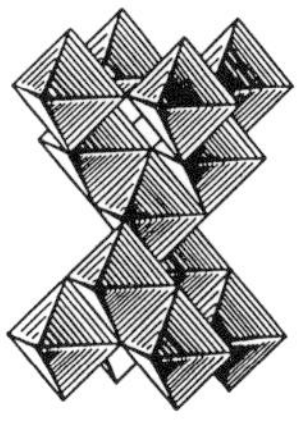

Fig. 2

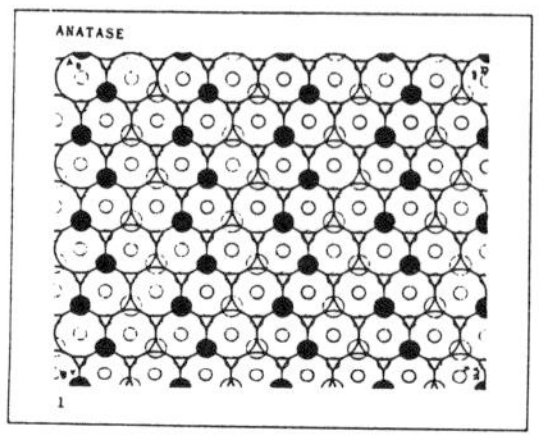

Fig. 3

Properties

Habit	Cleav.	Fract.	Twin.	Hardn.	Dens.	Colour	Transp.
pyramidal, tabular (001)	perfect (001) (111)	subconchoidal	rare (112)	5.5-6	3.90	brown, variable	transparent to translucent

Refr. index/Reflect.	Birefr.	Luster	Streak	Melt.p.	CPI
n_ω = 2.561 n_ϵ = 2.488	(-)	adamantine, metallic	colourless yellow		(SPI) 65

Figures

Fig. 1. (a) Packing representation of the anatase structure, and (b) projection on (010) of the unit-cell content (afer Wyckoff, 1963, Vol. 1).

Fig. 2. Polyhedral representation of the anatase structure (after Povarennykh, 1972).

Fig. 3. First sheet of the condensed model corresponding to a description based on closest packed layers; all the closest packed layers are alike, showing the zig-zag pattern of the occupied octahedral interstices by Ti atoms. The unit cell is inclined in relation to this layer.

Fig. 4. Condensed model of anatase structure according to the other alternative way of describing the cubic packing: square layers (Q) parallel to the (001) planes showing a square pattern of the occupied octahedral interstices. The condensed model is in agreement with Fig. 2.

Description

In the anatase structure the oxygens form a cubic closest packing, and the Ti atoms are located in octahedral voids. On a layer description parallel to (001) the Ti atoms show a square pattern, but in a description parallel to the closest packed layers the Ti atoms show a zig-zag pattern.

References

Kostov (1968) 242-245.

Wyckoff (1963) Vol. 1, 253,256.

Povarennykh (1972) 293.

Palache et al. (1944) Vol. 1, 583,584.

Zoltai+Stout (1984) 408.

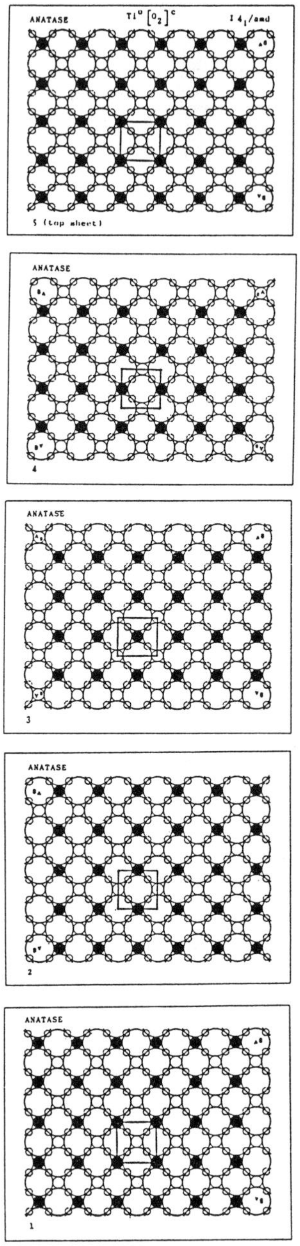

Fig. 4

TAPIOLITE $Ta^{o}_{2}\ Fe^{o}\ [O_6]^{h}$

$P4_2/mnm$

a = 4.67 Å
c = 9.14 Å
Z = 2

(for the isotypic str. $Sb^{o}_{2}\,Zn^{o}\,[O_6]^{h}$)

Ta (4e) u = 1/3
Fe (2a)
O_I (4f) u = 0.305
O_{II} (8j) u = 0.303, v = 0.328

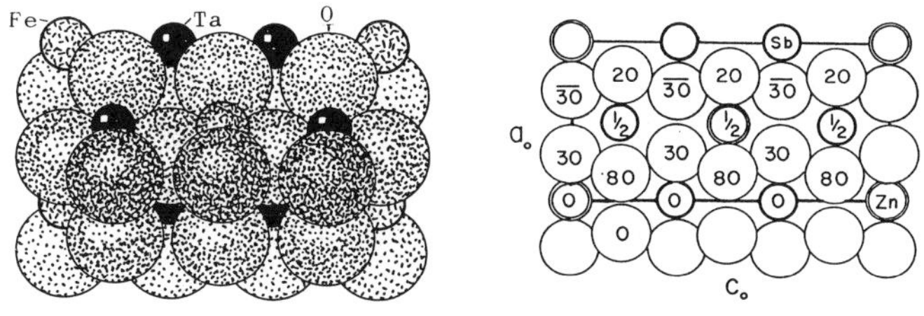

(a) (b)

Fig. 1

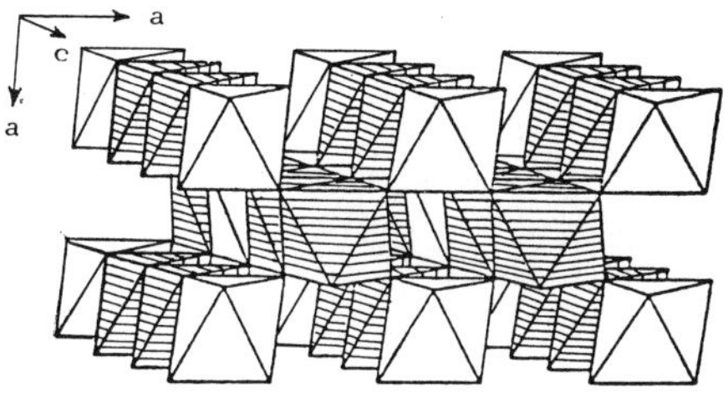

Fig. 2

Properties

Habit	Cleav.	Fract.	Twin.	Hardn.	Dens.	Colour	Transp.
short prismatic, equant	none	uneven to subconchoidal	(013)	6-6.5	7.90	black	transparent in splinters

Refr. index/Reflect.	Birefr.	Luster	Streak	Melt.p.	CPI
n_ω= 2.27 n_ϵ= 2.42	(+)	subadamantine to submetallic	cinnamon brown		

Population

Tripuhyite	$Sb^{o}_{2}\ Fe^{o}\ [O_6]^{h}$
Byströmite	$Sb^{o}_{2}\ Mg^{o}\ [O_6]^{h}$
Ordoñezite	$Sb^{o}_{2}\ Zn^{o}\ [O_6]^{h}$
Mossite	$(Ta,Nb)^{o}_{2}\ Fe^{o}\ [O_6]^{h}$

Description

The oxygens form a hexagonal closest packing and the tantalum and iron atoms occupy one half of the octahedral voids, with a row pattern. The structure is a substitution derivative of the rutile structure. It pertains to the $Sb^{o}_{2}\ Zn^{o}\ [O_6]^{h}$ inorganic structure type.

Figures

Fig. 1. (a) Packing drawing of the tapiolite structure seen along an a axis, and (b) $Sb^{o}_{2}\,Zn^{o}\,[O_6]^{h}$ structure isotypic with tapiolite, projected along an a axis (after Wyckoff 1965, Vol. 3).

Fig. 2. Polyhedral representation of the tapiolite structure (adapted from R. Burns + V. Burns, 1979).

Fig. 3. Condensed model of the tapiolite structure. Layers parallel to (100), showing a row pattern, like in rutile.

References

Kostov (1968) 232, 237, 247.
Wyckoff (1965) Vol. 3, 360-362.
Palache et al. (1944) Vol. 1, 775,776.
R. Burns + V. Burns (1979) 13.
Povarennykh (1972) 301,302.

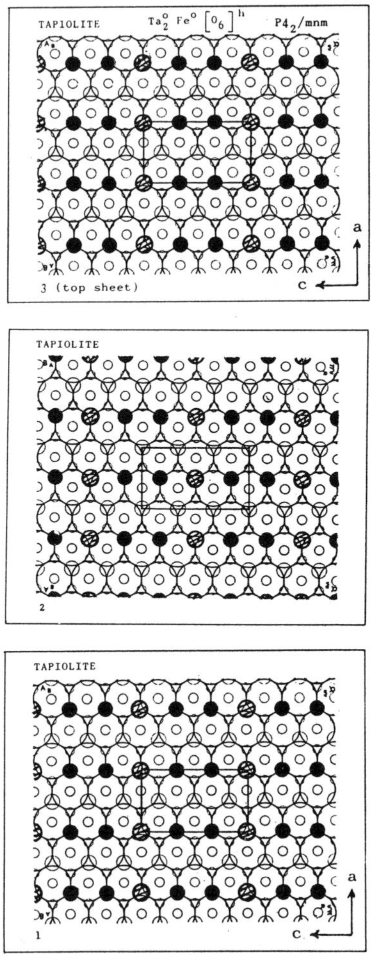

Fig. 3

WOLFRAMITE $(Fe,Mn)^o W^o [O_4]^h$

[Ferberite (Fe) - – Hübnerite (Mn)] P 2/ c

a = 4.70 Å
b = 5.69 Å
c = 4.93 Å
$\beta = 90^o$
Z = 2

Fe (2f) u = 0.653
W (2e) u = 0.180
O_I (4g) x = 0.22, y = 0.11, z = -0.04
O_{II} (4g) x = 0.26, y = 0.38, z = 0.39
(for the isotypic str. $Ni^oW^o [O_4]^h$)

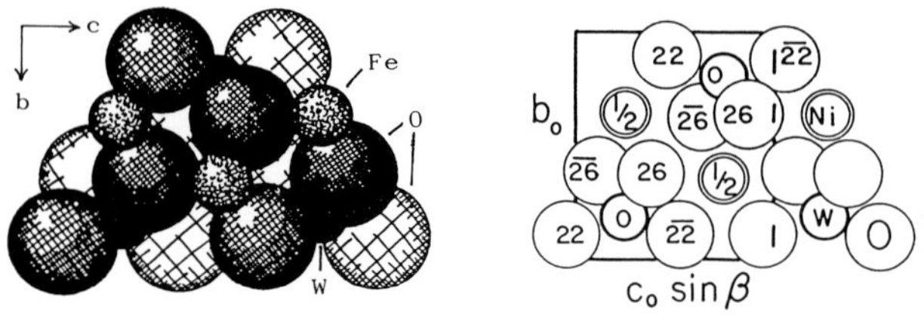

(a) (b)

Fig. 1

Fig. 2

Properties

Habit	Cleav.	Fract.	Twin.	Hardn.	Dens.	Colour	Transp.
short prismatic, tabular	perfect (010)	uneven	(100)	4 - 4.5	7.60-7.29	yellowish brown to black	opaque

Refr. index/Reflect.	Birefr.	Luster	Streak	Melt.p.	CPI
n_α = 2.31-2.17	(+)	subme-	nearly		(SPI)
n_β = 2.40-2.22		tallic	black		44-41
n_γ = 2.46-2.30	$2V = 79^o - 73^o$	resinous	to brown		

Population

Sanmartinite $Zn^o W^o [O_4]^h$

Basic structure

Ixiolite $(Fe,Sn)^o Ta^o [O_4]^h$

Pbcn

Figures

Fig. 1.(a) Packing drawing of the wolframite structure seen along the a axis. The oxygen atoms are large and cross - - line shaded. The small black circles represent wolframium atoms and the small hook shaded circles represent iron atoms. (b) $Ni^o W^o [O_4]^h$ structure, which is isotypic with wolframite, projected along the a axis (after Wyckoff, 1965, Vol. 3),

Fig. 2. Polyhedral representation of the wolframite structure in a setting analogous to that of the brookite structure. The line shaded octahedral correspond to the iron coordinated atoms (after Povarennykh, 1972).

Fig. 3. Condensed model of the wolframite structure. Small black circles represent iron atoms in octahedral voids, and the dashed small circles represent the wolframium atoms.

Description

The oxygens form a hexagonal closest packing, and the iron and wolframium atoms are located in one half of the octahedral voids. The iron and the wolframium atoms form a zig-zag pattern like in brookite. The layers parallel to (100) containing the iron atoms alternate with the layers containing the wolframium atoms. It is a distortion derivative of ixiolite, and a substitution derivative of α- $Pb^o [O_2]^h$.

References

Kostov (1968) 483.
Wyckoff (1965) Vol. 3, 41-43.
Povarennykh (1972) 135, 302,303.
Palache et al. (1944) Vol. 1, 1065,1066.
Zoltai+Stout (1984) 446.

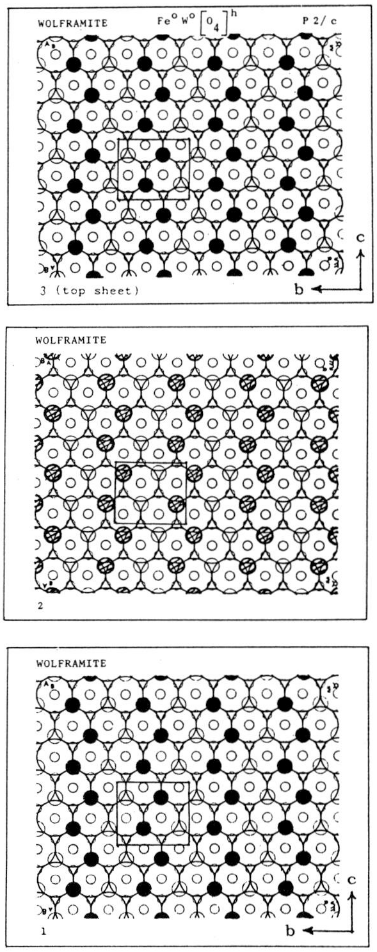

Fig. 3

COLUMBITE

$Nb^{o}_{2}\ Fe^{o}[O_6]^{h}$

P b c n

a = 14.238 Å	Fe(4c)	u	= 0.350
b = 5.730 Å	Nb(8d)	x	= 0.163
c = 5.082 Å		y	= 0.175
Z = 4		z	= 0.750
	O_I(8d)	x	= 0.090
		y	= 0.095
		z	= 0.083
	...		

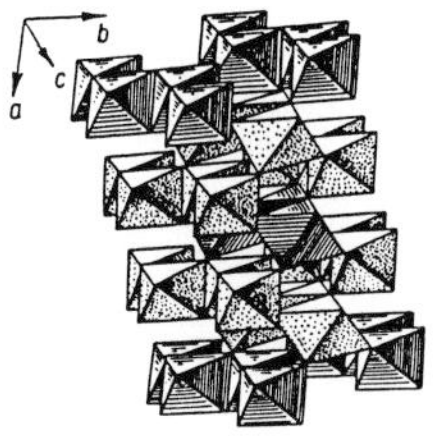

Fig. 1

Properties

Habit	Cleav.	Fract.	Twin.	Hardn.	Dens.	Colour	Transp.
short prismatic	distinct (001)	subconchoidal	(012)	6-6.5	5.20--7.95	iron-black brownish--black	translucent to opaque

Refr. index/Reflect.		Birefr.	Luster	Streak	Melt.p.	CPI
n_α = 2.44	17%	(+)	submetallic	brown		(SPI) 68
n_β = 2.32						
n_γ = 2.38		2V = 75°				

Population

Tantalite	$Ta^{o}_{2}Fe^{o}[O_6]^{h}$
Magnocolumbite	$Nb^{o}_{2}Mg^{o}[O_6]^{h}$

Figures

Fig. 1. Polyhedral representation of the columbite structure in a setting analogous to that of brookite (after Povarennykh, 1972).

Fig. 2. Condensed model of the columbite structure. Small black circles represent iron atoms in octahedral voids, and the dashed small circles represent niobium atoms.

Description

The oxygens form a hexagonal closest packing, and the iron and niobium atoms are located in half of the octahedral voids, forming a zig-zag pattern, like in anatase and brookite. The layers containing the iron atoms alternate with pairs of layers containing the niobium atoms. It is a substitution derivative of wolframite, or still of α- $Pb^{o}[O_2]^{h}$.

References

Kostov (1968) 247.
Wyckoff (1965) Vol. 3, 362,363
Povarennykh (1972) 303.
Palache et al. (1944) Vol. 1, 780-782.
Zoltai+Stout (1984) 409, 412.

Crystallographic data (continued)

O_{II} (8d)	x = 0.410		O_{III}(8d)	x = 0.750
	y = 0.100			y = 0.080
	z = 0.083			z = 0.070

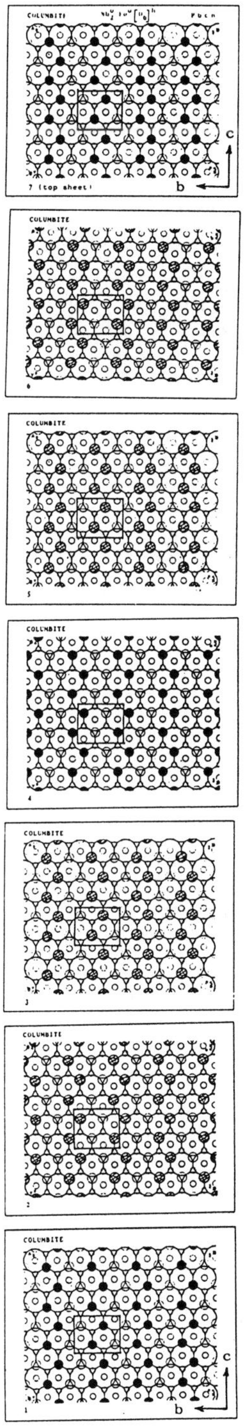

Fig. 2

RAMSDELLITE

$Mn^{o}[O_2]^{h}$

Pbnm

a = 4.533 Å
b = 9.27 Å
c = 2.866 Å
Z = 4

Mn (4c)	u	=-0.022
	v	= 0.136
O_I (4c)	u	= 0.17
	v	=-0.23
O_{II} (4c)	u	=-0.21
	v	=-0.033

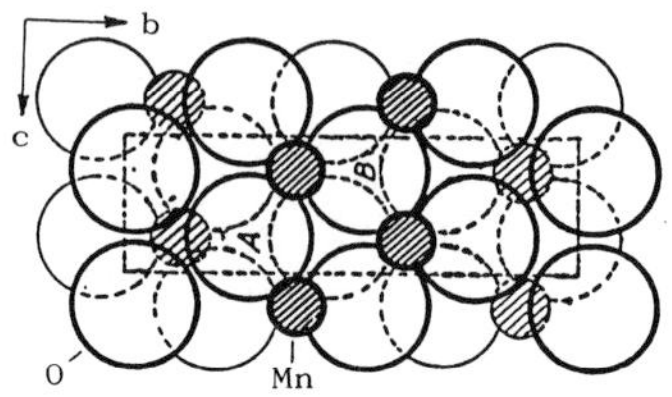

Fig. 1

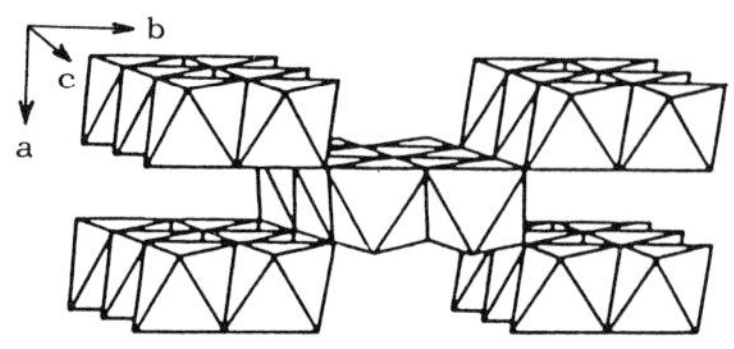

Fig. 2

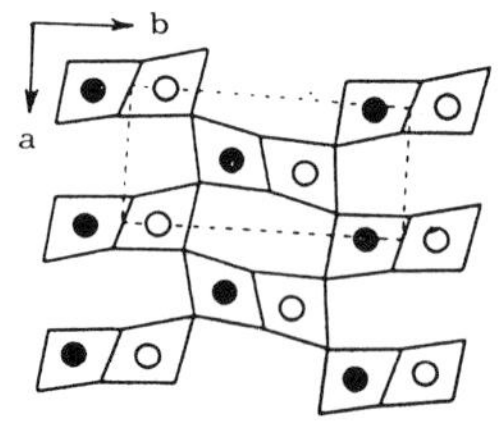

Fig. 3

Properties

Habit	Cleav.	Fract.	Twin.	Hardn.	Dens.	Colour	Transp.
fibers (001) massive	on three pinacoids			2 - 4	4.37-4.83	steel--grey, iron-black	opaque

Refr. index/Reflect.	Birefr.	Luster	Streak	Melt.p.	CPI
		brillant metallic	dull black		

Population

Paramontroseite $V^{o}[O_2]^{h}$

Description

Is based on a hexagonal closest packing of oxygens, with one half of the octahedral voids occupied by manganese atoms, forming a double row pattern. Noticed that the double rows on the adjacent layers are placed over the empty rows, that is in between the double rows of the adjacent layers (distant distribution rule, Lima-de-Faria, 1965a).

References

Kostov (1968) 236,237.
Wyckoff (1963) Vol. 1, 290-293.
Wells (1962) 558.
R. Burns + V. Burns (1979) 15 .
Roberts et al. (1974) 509.
Povarennykh (1972) 298.
Lima-de-Faria (1965a) 361.

Figures

Fig. 1. Projection along the a axis of the ramsdellite structure (after Wells, 1962).

Fig. 2. Polyhedral representation of the ramsdellite structure viewed along the c axis (after R. Burns + V. Burns, 1979).

Fig. 3. Structural scheme of ramsdellite structure (after Kostov, 1968).

Fig. 4. Condensed model of the ramsdellite structure. Small black circles represent the manganese atoms in octahedral voids.

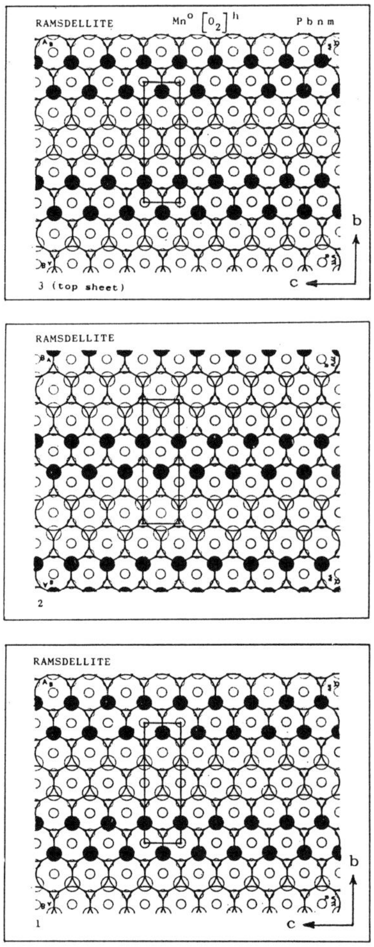

Fig. 4

BROOKITE $Ti^{o}[O_2]^{ch}$

P b c a

a = 9.184 Å
b = 5.447 Å
c = 5.145 Å
Z = 8

Ti	(8c)	x = 0.1290, y = 0.0972, z = -0.1371
O_I	(8c)	x = 0.0101, y = 0.1486, z = 0.1824
O_{II}	(8c)	x = 0.2304, y = 0.1130, z = -0.4629

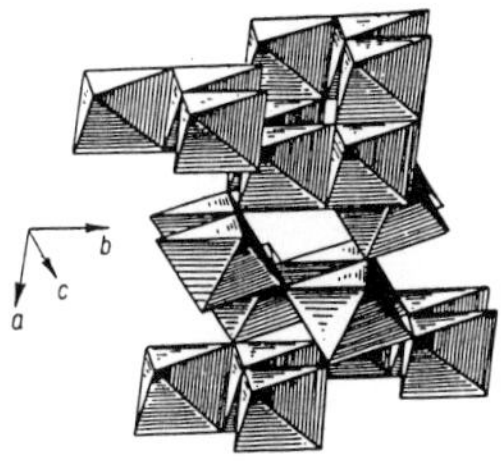

Fig. 1

Properties

Habit	Cleav.	Fract.	Twin.	Hardn.	Dens.	Colour	Transp.
tabular (100) elongated	indistinct (210)	subconchoidal	uncertain	5.5-6	4.14	hair-brown, yellowish brown	transparent in small fragments

Refr. index/Reflect.	Birefr.	Luster	Streak	Melt.p.	CPI
$n_\alpha = 2.5831$ $n_\beta = 2.5843$ $n_\gamma = 2.7004$	(-) 2V = 0-30°	metallic, adamantine	uncoloured to greywish		(SPI) 69

Figures

Fig. 1. Polyhedral representation of the brookite structure (after Povarennykh, 1972).

Fig. 2. Condensed model of the brookite structure. Notice the zig-zag pattern of the titanium atoms, like in anatase.

Description

The oxygens form a closest packing with a ch sequence, where the Ti atoms are located in one half of the octahedral voids. On the closest packed layers, which are parallel to (100), the Ti atoms form a zig-zag pattern.

References

Kostov (1968) 245.

Povarennykh (1972) 135.

Wyckoff (1963) Vol. 1, 254-256.

Zoltai+Stout (1984) 408.

Palache et al. (1944) Vol. 1, 588-590.

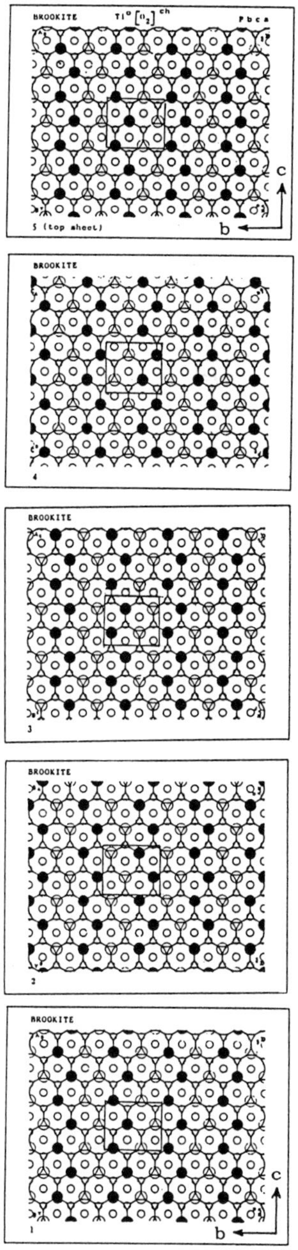

Fig. 2

DIASPORE $Al^{o}[O(OH)]^{h}$

P b n m

a = 4.396 Å (at 25°C)
b = 9.426 Å
c = 2.844 Å
Z = 4

Al	(4c)	u = -0.0451, v = 0.1446
H	(4c)	u = -0.4095, v = -0.0876
O_I	(4c)	u = 0.2880, v = -0.1989
O_{II}	(4c)	u = -0.1970, v = -0.0532

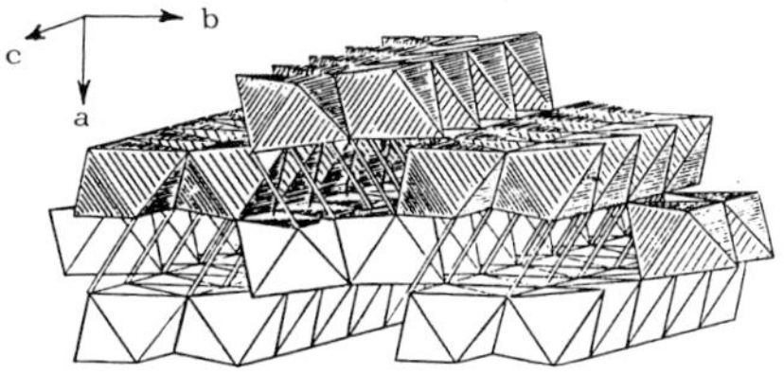

Fig. 1

Properties

Habit	Cleav.	Fract.	Twin.	Hardn.	Dens.	Colour	Transp.
platy, acicular, massive	perfect (010)	conchoidal		6.5-7	3.3-3.5	white, greyish white	transparent

Refr. index/Reflect.	Birefr.	Luster	Streak	Melt.p.	CPI
n_α = 1.68-1.71 n_β = 1.71-1.72 n_γ = 1.73-1.75	(+) 2V = 85°	brilliant, pearly, vitreous	white, yellow		CPI 57

Population

Goethite	$Fe^{o}[O(OH)]^{h}$
Groutite	$Mn^{o}[O(OH)]^{h}$
Bracewellite	$Cr^{o}[O(OH)]^{h}$
Montroseite	$V^{o}[O(OH)]^{h}$

Description

It is formed by the hexagonal closest packing of oxygens and hydroxyls, with aluminium atoms occupying one half of the octahedral holes, in a double row pattern. It can be considered as a substitution derivative of ramsdellite. More detailled structural work has enable the location of the hydrogen atoms, and it was found out that they are in two coordination (linking two oxygens). This hydrogen distribution in the diaspore structure is responsable for its stability. On the contrary the non existence of such positive charges in between the double rows on ramsdellite explains its metastable equilibrium.

Figures

Fig. 1. Polyhedral representation of the diaspore structure. The double lines indicate O-H-O bonds (after Ewing, 1935, quoted by Wells, 1962).

Fig. 2. Condensed model of the diaspore structure. The small black circles represent aluminium atoms in octahedral voids, and the black points hydrogen atoms.

References

Kostov (1968) 219.
Wyckoff (1963) Vol. 1, 291.
Wells (1962) facing page 556.
Palache et al. (1944) Vol. 1, 675-677.
Zoltai+Stout (1984) 422.
Povarennykh (1972) 317.
Ewing (1935) 203.

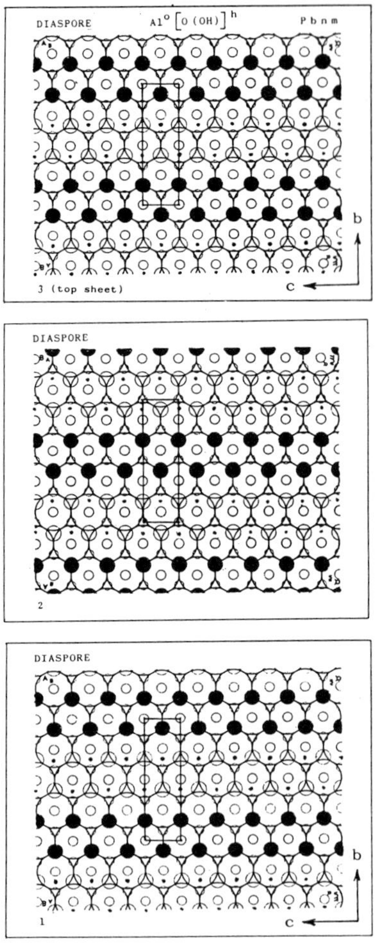

Fig. 2

CHLOROMAGNESITE $Mg^{o}[Cl_{2}]^{c}$

R $\bar{3}$ m

a_R = 6.22 Å (hex. descrip.)
α = 33° 36'
Z_R = 1
a_H = 3.596 Å
c = 17.589 Å
Z_H = 3

Mg (3a)
Cl (6c) u ≃ 0.25

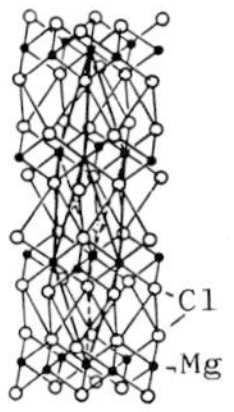

Fig. 1

Fig. 2

Properties

Habit	Cleav.	Fract.	Twin.	Hardn.	Dens.	Colour	Transp.
hexagonal plates	perfect (0001)			1 - 1.5	2.325	colourless, white	

Refr. index/Reflect.	Birefr.	Luster	Streak	Melt.p.	CPI
n_ω = 1.675 n_ϵ = 1.59	(-)				

Population

Scacchite $Mn^{o}[Cl_{2}]^{c}$

Lawrencite $Fe^{o}[Cl_{2}]^{c}$

Description

It is based on the closest packing of Cl^- ions, with Mg^{2+} in octahedral voids. Two different closest packed layers alternate, one with all the octahedral voids occupied, and another completely empty. It is isotypic with $Cd^{o}[Cl_{2}]^{c}$.

References

Kostov (1968) 195,

Wyckoff (1963) Vol. 1, 270-272.

Povarennykh (1972) 56, 644.

Bokii (1954) 130.

Palache et al. (1951) Vol. 2, 41.

Roberts et al. (1974) 127.

Figures

Fig. 1. Ball and spoke representation of the chloromagnesite structure (after Bokii, 1954).

Fig. 2. Polyhedral representation of the chloromagnesite structure (adapted from Povarennykh, 1972).

Fig. 3. Condensed model of the chloromagnesite structure. Packing layers parallel to (0001) planes. Large open circles represent Cl^-, and the small black circles Mg^{2+}, in octahedral voids. The structure is formed by alternate completely filled and empty layers in relation to the octahedral magnesium ions.

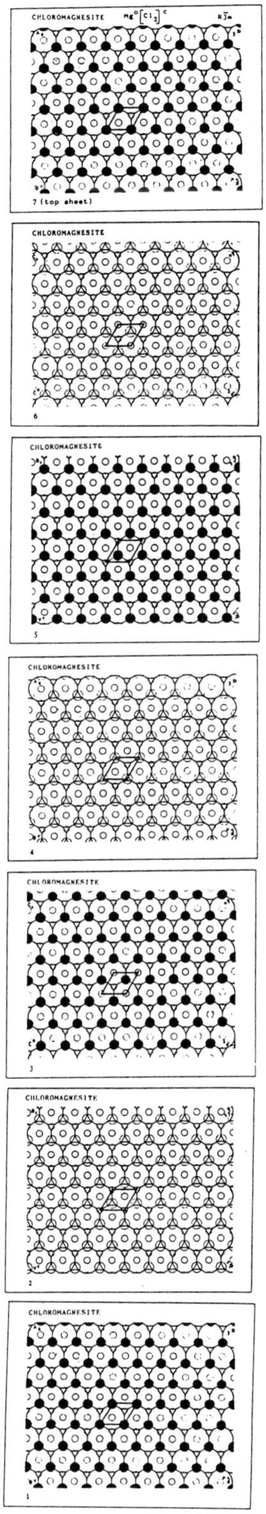

Fig. 3

MELONITE $Ni^{o}[Te_2]^{h}$

$P\bar{3}m1$

a = 3.861 Å
c = 5.297 Å
Z = 1

Ni (1a)
Te (2d) u = 1/4

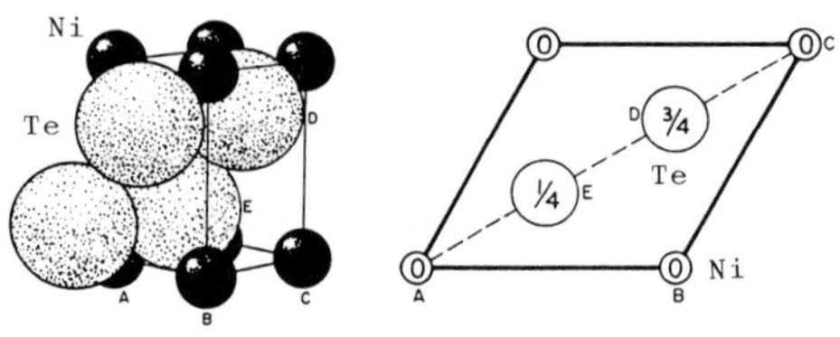

(a) (b)

Fig. 1

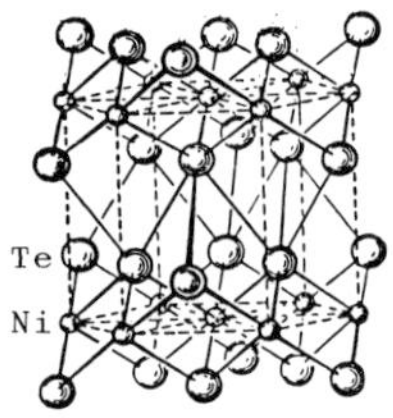

Fig. 2

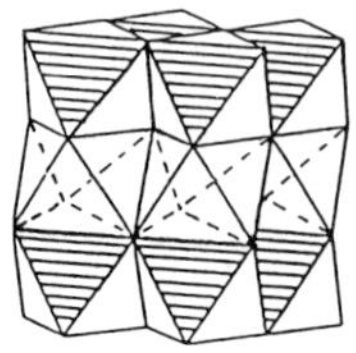

Fig. 3

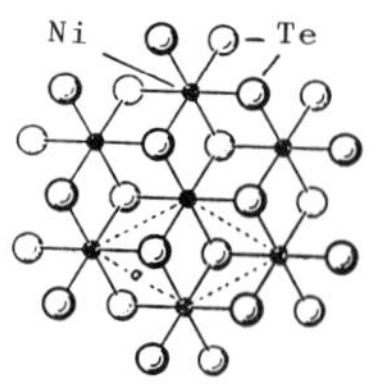

Fig. 4

Properties

Habit	Cleav.	Fract.	Twin.	Hardn.	Dens.	Colour	Transp.
platy foliated	perfect (0001)			1 - 1.5	7.35	redish white	opaque

Refr. index/Reflect.	Birefr.	Luster	Streak	Melt.p.	CPI
		metallic	dark grey		

Population

Moncheite	$Pt^{o}[Te_2]^{h}$
Merenskyite	$Pd^{o}[Te_2]^{h}$
Berndtite	$Sn^{o}[S_2]^{h}$
Kitkaite	$Nd^{o}[Se\,Te]^{h}$
Brucite	$Mg^{o}[(OH)_2]^{h}$
Pyrochroite	$Mn^{o}[(OH)_2]^{h}$
Portlandite	$Ca^{o}[(OH)_2]^{h}$

References

Kostov (1968) 118.
Wyckoff (1963) Vol. 1, 267-269.
Bokii (1954) 130.
Ramdohr + Strunz (1980) 550.
Palache et al. (1944) Vol. 1, 341.
Roberts et al. (1974) 391.
Povarennykh (1972) 214, 259, 324.

Figures

Fig. 1. (a) Packing model of the melonite structure, and (b) unit cell projection parallel to (0001) (after Wyckoff,1963, Vol. 1).

Fig. 2. Ball and spoke representation of the melonite structure (adapted from Bokii, 1954).

Fig. 3. Polyhedral representation of the melonite structure.

Fig. 4. Projection on (0001) of a double packing layer. The oxygens are at two different levels (Ramdohr + Strunz, 1980).

Fig. 5. Condensed model of the melonite structure. The large open circles represent the Te packing atoms, and the small black circles the Ni atoms located in octahedral voids. The small dotted circles represent the other possible voids, octahedral and tetrahedral, which are not occupied. The packing is quite distorted (c/a = 1.37, c/a ideal = 1.633).

Description

It is based on the hexagonal closest packing of the Te atoms with Ni atoms occupying a half of the octahedral interstices. It is formed by two alternate layers, one where the octahedral voids are completely filled, and another layer completely empty.

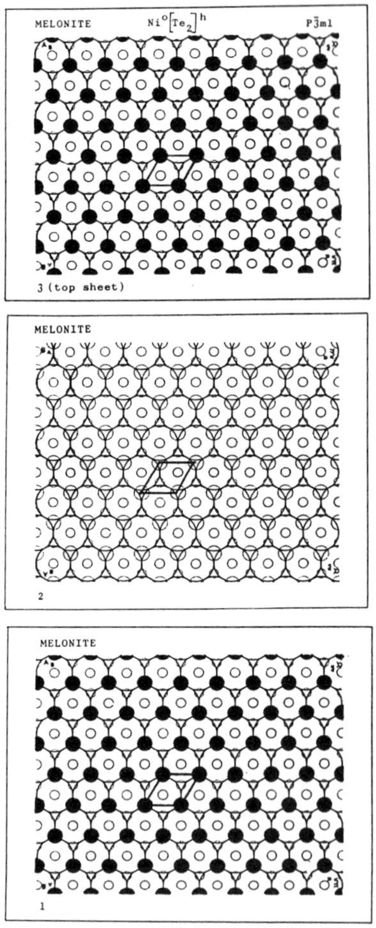

Fig. 5

CORUNDUM

$Al_2^o\left[O_3\right]^h$

$R\bar{3}c$

a_R = 5.128 Å
α = 55° 20'
Z_R = 2
a_H = 4.76280 Å
c = 13.00320 Å
Z_H = 6

(hex. description)

Al (12c) u = 0.3520
O (18e) v = 0.306

Fig. 1

Fig. 2

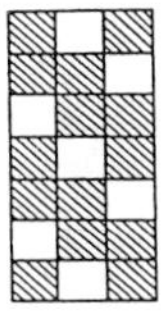

Fig. 3

Properties

Habit	Cleav.	Fract.	Twin.	Hardn.	Dens.	Colour	Transp.
steep-pyramidal botryoidal foliated	none	uneven conchoidal	$(10\bar{1}1)$	9	4.0-4.1	blue, variable, colourless	transparent, translucent

Refr. index/Reflect.	Birefr.	Luster	Streak	Melt.p.	CPI
n_ω = 1.768 n_ϵ = 1.760	(-) 2V= 58°	adamantine, vitreous	uncoloured	2050°C	(SPI) 70

Population

Hematite	$Fe_2^o\left[O_3\right]^h$
Eskolaite	$Cr_2^o\left[O_3\right]^h$
Karelianite	$V_2^o\left[O_3\right]^h$

Ruby and sapphire are red and blue varieties of corundum, respectively.

Description

The oxygens form an hexagonal closest packing with aluminium atoms occupying 2/3 of the octahedral voids, with a honeycomb pattern, which is the most symmetrical distribution for the proportion 2/3.

Figures

Fig. 1. Polyhedral representation of the corundum structure (after Povarennykh, 1972).

Fig. 2. Layer of AlO_6 octahedra parallel to the closest packed layers, with a honeycomb pattern (after Bokii, 1954).

Fig. 3. Elevation of the corundum structure (after Kostov, 1968).

Fig. 4. Condensed model of the corundum structure. The large open circles represent oxygens, and the small black circles the aluminium atoms, which form a honeycomb pattern.

References

Kostov (1968) 216,217, 249.
Wyckoff (1964) Vol. 2, 6,7.
Povarennykh (1972) 53, 270.
Palache et al. (1944) Vol. 1, 521.
Zoltai + Stout (1984) 407.
Ingerson (1955) 351.
Bokii (1954) 143.

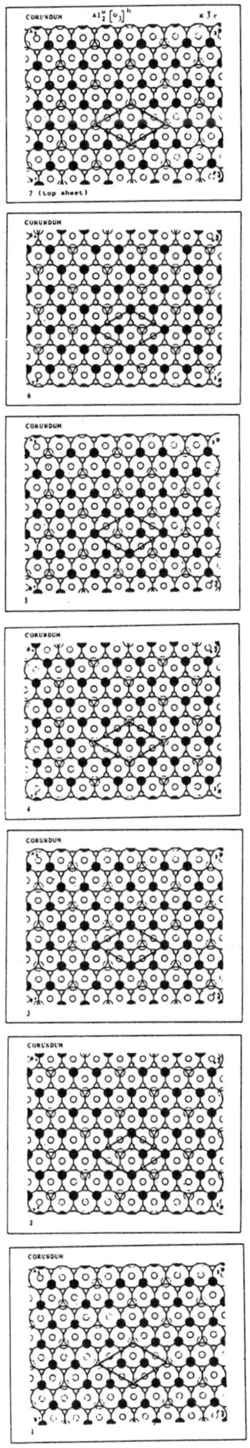

Fig. 4

ILMENITE

$Fe^{o}\ Ti^{o}\ [O_3]^{h}$

$R\bar{3}$

a_R = 5.538 Å	(hex. description)		
α = 54° 41'	Fe(6c)	u = 0.358	
Z_R = 2	Ti(6c)	u = 0.142	
a_H = 5.082 Å	O(18f)	x = 0.305	
c = 14.026 Å		y = 0.015	
Z_H = 6		z = 0.250	

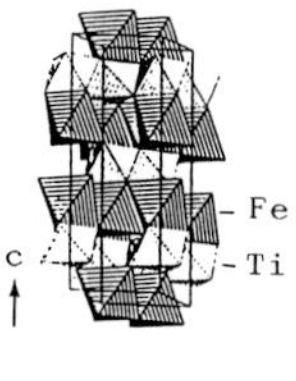

Fig. 1

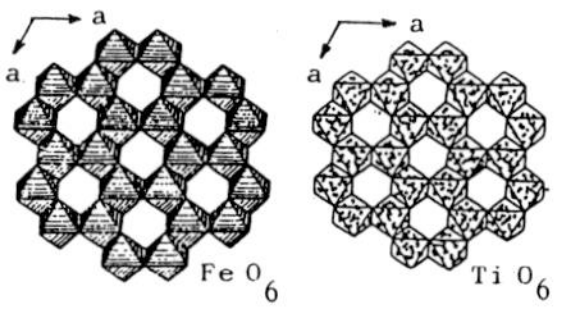

Fig. 2

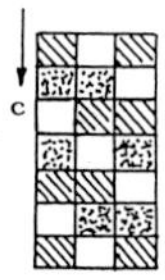

Fig. 3

Properties

Habit	Cleav.	Fract.	Twin.	Hardn.	Dens.	Colour	Transp.
thick prismatic, tabular, rhombohedral	none	conchoidal, subconchoidal	(0001) (10$\bar{1}$1)	5 - 6	4.72	iron--black	opaque

Refr. index/Reflect.		Birefr.	Luster	Streak	Melt.p.	CPI
$n_\omega \gg 2.7$ $n_\epsilon \gg 2.7$	18%	(-)	metallic, submetallic	black, brownish		

Population

Pyrophanite	$Mn^{o}\ Ti^{o}[O_3]^{h}$
Melanostibite	$Mn^{o}\ (Fe, Sb)^{o}[O_3]^{h}$

Figures

Fig. 1. Polyhedral representation of the ilmenite structure (Finger + Hazen, 1991).

Fig. 2. Layers of Fe O_6 and Ti O_6 octahedra parallel to the closest packed layers, with a honeycomb pattern (after Bokii, 1954).

Fig. 3. Elevation of the ilmenite structure.

Fig. 4. Condensed model of the ilmenite structure. The large open circles represent the oxygen atoms, the small black circles correspond to iron atoms, and the small dashed circles to titanium atoms.

Description

It is an interstitial substitution derivative of corundum. The oxygen atoms form an hexagonal closest packing and Fe and Ti occupy 2/3 of the octahedral voids, with a honeycomb pattern, in alternate layers.

References

Kostov (1968) 249.
Wyckoff (1964) Vol. 2, 420,421.
Finger+Hazen (1991) facing p. 566.
Palache et al. (1944) Vol. 1, 535,536.
Zoltai + Stout (1984) 407.
Povarennykh (1972) 279.
Bokii (1954) 143.

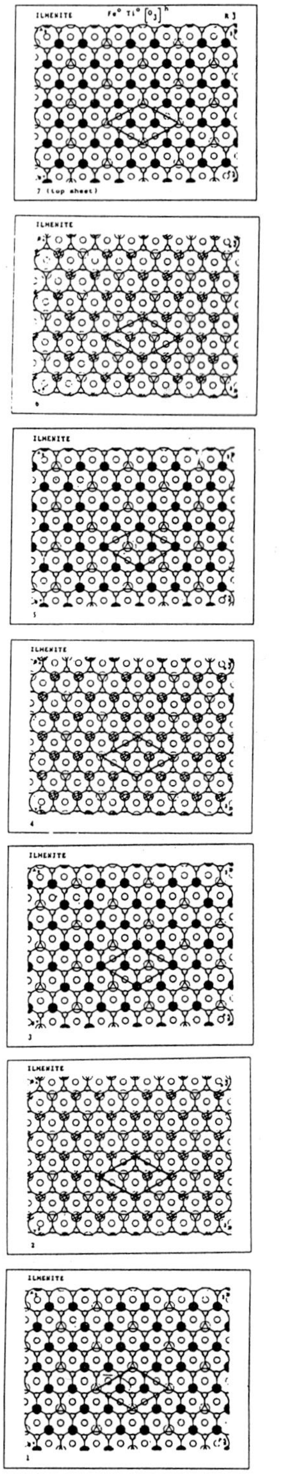

Fig. 4

TELLURBISMUTH

(Tellurobismuthite)

$Bi_2^o [Te_3]^{chh}$

$R\bar{3}m$

$a_R = 10.3$
$\alpha = 24^o\ 10'$
$Z_R = 1$
$a_H = 4.39$ Å
$c = 30.60$ Å
$Z_H = 3$

(hex. description)

Bi (6c) u = 0.399

Te_I (6c) u = 0.792

Te_{II} (3a)

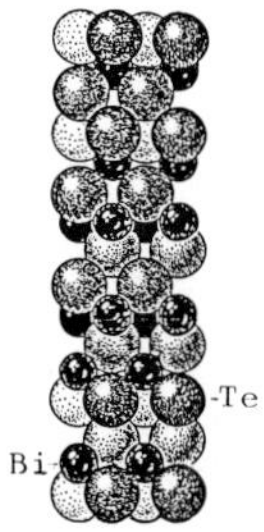

Fig. 1

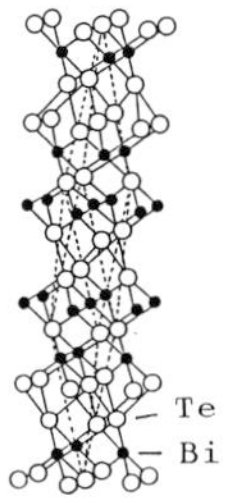

Fig. 2

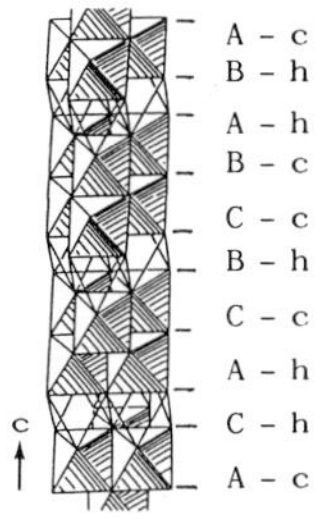

Fig. 3

Properties

Habit	Cleav.	Fract.	Twin.	Hardn.	Dens.
irregular plates, foliated masses	perfect (0001)			1.5 - 2	7.815

Refr. index/Reflect.	Birefr.	Luster	Streak
		metallic	pale lead- -grey

Colour	Transp.	Melt.p.	CPI
pale lead- -grey	opaque		

Figures

Fig. 1. Packing model of the tellurbismuth structure.

Fig. 2. Ball and spoke representation of the tellurbismuth structure (adapted from Povarennykh, 1972).

Fig. 3. Polyhedral representation of the tellurbismuth structure (adapted from Kostov, 1968).

Fig. 4. Condensed model of the tellurbismuth structure. The large open circles are the Te atoms (packing atoms), and the small black circles the Bi atoms in octahedral voids.

Description

The structure is based on a chh closest packing of Te atoms with Bi atoms in octahedral voids. Two completely filled layers with Bi alternate with one empty.

References

Kostov (1968) 163.

Wyckoff (1964) Vol. 2, 30,31.

Povarennykh (1972) 214,215.

Palache et al. (1944), Vol. 1, 160.

Fig. 4

TETRADYMITE

$Bi^{o}_{2}\left[Te_{2}S\right]^{chh}$

R $\bar{3}$ m

a_R = 10.31 Å
α = 24°10'
Z_R = 1
a_H = 4.316 Å
c = 30.01 Å
Z_H = 3

(hex. description)
Bi (6c) u = 0.392
Te (6c) u = 0.788
S (3a)

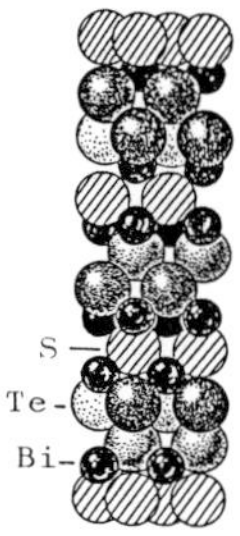

Fig. 1

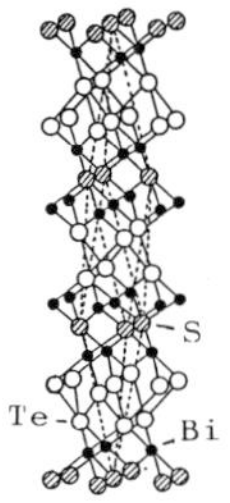

Fig. 2

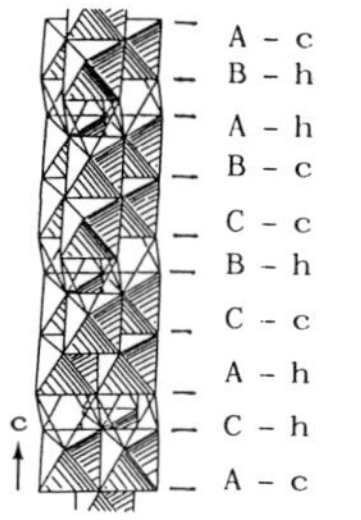

Fig. 3

Properties

Habit	Cleav.	Fract.	Twin.	Hardn.	Dens.
acutely pyramidal, foliated, massive	perfect (0001)		$(01\bar{1}8)$	1.5 - 2	7.3

Refr. index/Reflect.	Birefr.	Luster	Streak
48%		metallic, splendent	pale steel- -grey

Colour	Transp.	Melt.p.	CPI
pale steel- -grey	opaque	600°C	

Population

Kawazulite $Bi^{o}_{2}\left[Te_{2}\,Se\right]^{chh}$

Figures

Fig. 1. Packing model of the tetradymite structure.

Fig. 2. Ball and spoke representation of the tetradymite structure (adapted from Povarennykh, 1972).

Fig. 3. Polyhedral representation of the tetradymite structure (adapted from Kostov, 1968).

Fig. 4. Condensed model of the tetradymite structure. Notice that the only difference between this model and that of tellurbismuth is the packing layers where the hatched packing atoms correspond to the S atoms, sandwiched between the Bi atoms (small black circles).

Description

The structure is a packing substitution derivative of tellurbismuth. It is based on a chh closest packing of Te and S atoms, with Bi atoms in octahedral voids.

References

Kostov (1968) 163.

Wyckoff (1964) Vol. 2, 29--31.

Palache et al. (1944) Vol. 1, 161-164.

Povarennykh (1972) 214,215.

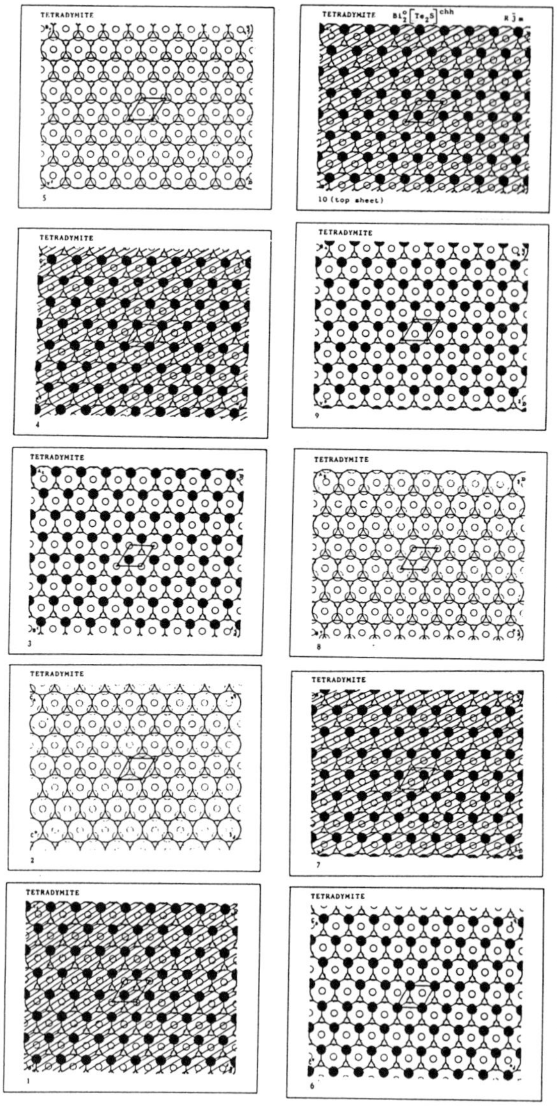

Fig. 4

PEROVSKITE

$Ti^{o}[Ca\,O_3]^{c}$

P b n m

			ideal structure
a = 5.37 Å	Ti (4a)		P m3m
b = 5.44 Å	Ca (4c)	u = 0.00, v = 0.030	
c = 7.64 Å	O_I (4c)	u = 0.037, v = 0.482	a = 3.84 Å — Ca (1a)
Z = 4	O_{II} (8d)	x = −0.268, y = 0.268, z = 0.026	Z = 1 — Ti (1b), O (3c)

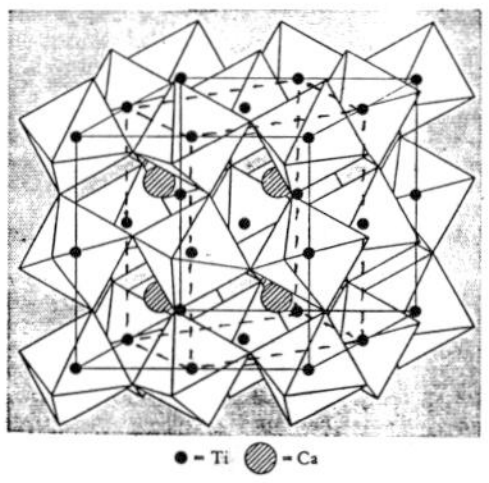

Fig. 1

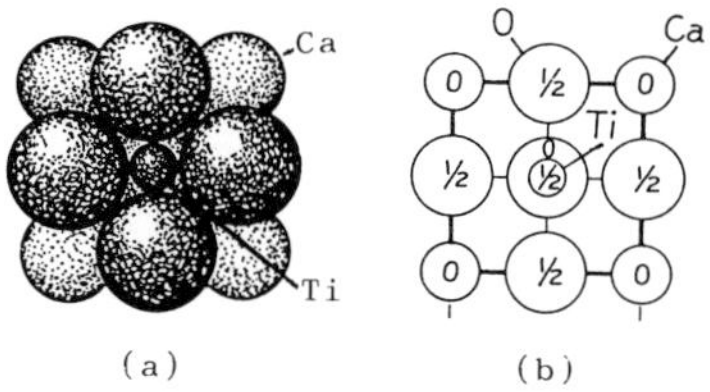

Fig. 2

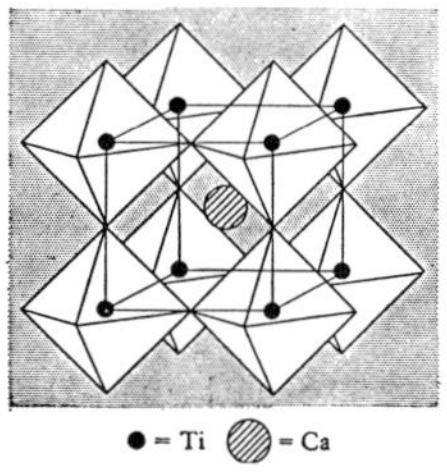

Fig. 3

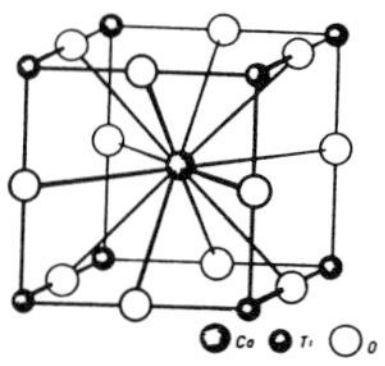

Fig. 4

Properties

Habit	Cleav.	Fract.	Twin.	Hardn.	Dens.	Colour	Transp.
cubic	imperfect (001)	uneven, subcon-choidal	(111)	5.5	4.01	brown black	transparent in thin splinters

Refr. index/Reflect.	Birefr.	Luster	Streak	Melt.p.	CPI
$n_\alpha = 2.34$ $n_\beta = 2.34$ $n_\gamma = 2.34$	(+) 2V= 90°	adamantine to metallic	colourless, greyish	1970°C	(SPI) 69

Population

Real

Latrappite $(Ti,Nb)^{o}[(Na,Ca)O_3]^{c}$

Ideal

Loparite $Ti^{o}[(Na,Ce)O_3]^{c}$

Igdloite $Nb^{o}[Na\,O_3]^{c}$

Figures

Fig. 1. Polyhedral description of the real perovskite structure. The orthorhombic unit cell is marked with dashed lines. The unit cell marked with continuous lines was proposed by Náray-Szabó 1943, $P2_1/m$, $a \approx b \approx c = 7.62$ Å, $\beta = 90°$, Z = 8 (adapted from Deer, Howie and Zussman, 1962, Vol. 5).

Fig. 2. (a) Packing drawing of the ideal perovskite structure, and (b) projection along the a axis of the ideal perovskite structure (after Wyckoff, 1964, Vol. 2).

Fig. 3. Polyhedral description of the ideal perovskite structure (after Náray-Szabó, 1943, quoted in Deer, Howie and Zussman, 1962, Vol. 5).

Fig. 4. Ball and spoke model of the ideal perovskite structure with a different choice of origin: instead of calcium a titanium atom (after Povarennykh, 134). Compare with Fig. 5.

Fig. 5. Condensed model of the perovskite structure. Real an ideal unit cells are marked with continuous and dashed lines, respectively. The dash and dotted unit cell corresponds to the Náray-Szabó monoclinic structure. The large open circles represent oxygen atoms, the large lined circles Ca atoms, and the small black circles Ti atoms.

Description

The real perovskite structure is orthorhombic and based on a cubic closest packing of the oxygen and calcium atoms. Ti atoms occupy 1/4 of the octahedral voids with a square pattern. Perovskite heated above 900°C corresponds to the ideal cubic structure. A monoclinic structure has also been assigned by Náray-Zsabó (Fig. 1).

References

Kostov (1968) 251.

Wyckoff (1964) Vol. 2, 390,391, 401, 410.

Deer, Howie and Zussman (1962) Vol. 5, 49, 55.

Náray-Szabó (1943) 202.

Povarennykh (1972) 134, 295.

Palache et al. (1944) Vol. 1, 730,731.

Zoltai+Stout (1984) 417.

Ingerson (1955) 351.

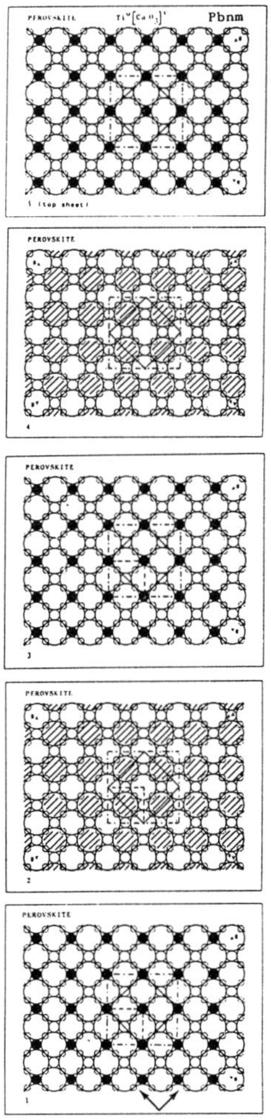

Fig. 5

ELPASOLITE $Al^{o}\ Na^{o}\ [K_2\ F_6]^{c}$

P a 3

a = 8.11 Å
Z = 4

Al(4a)
Na(4b)
K (8c) u= 0.25
F(24d) x= 0.22 y= 0.03 z= 0.01

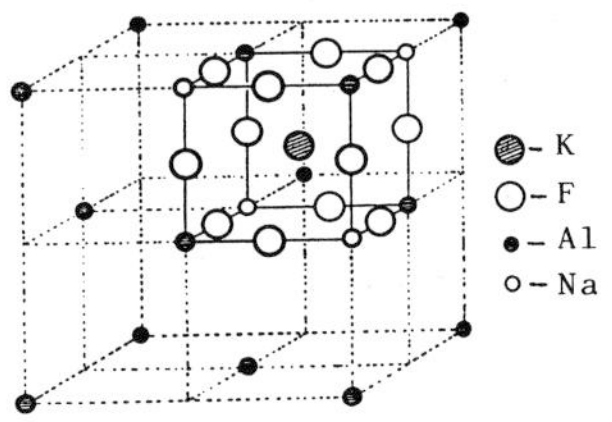

Fig. 1

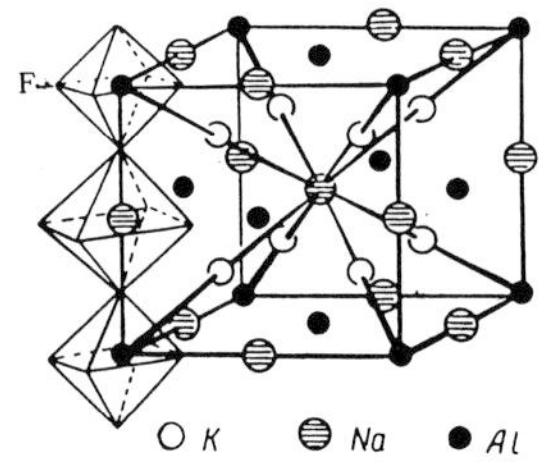

Fig. 2

Properties

Habit	Cleav.	Fract.	Twin.	Hardn.	Dens.	Colour	Transp.
massive	none	uneven		2.5	2.995	colourless	transparent to translucent

Refr. index/Reflect.	Birefr.	Luster	Streak	Melt.p.	CPI
n = 1.376		vitreous to greasy			

Figures

Fig.1. Relation between elpasolite and perovskite structures (see Fig. 4 of perovskite). The elpasolite structure is only fully represented in an octet of the large cube (after Wells, 1962).

Fig. 2. Ball and spoke model of the elpasolite structure (after Povarennykh, 1972). Compare with Fig. 1 of perovskite.

Fig. 3. Condensed model of the elpasolite structure. The large open circles represent fluorine atoms and the lined large circles potassium atoms. The small black circles correspond to aluminium, and the small circles to sodium.

Description

The elpasolite structure is a interstitial substitution derivative of the perovskite structure.

References

Kostov (1968) 192,193.

Wyckoff (1965) Vol. 3, 374,375.

Wells (1962) 369,370.

Povarennykh (1972) 663.

Palache et al (1951) Vol. 2,114.

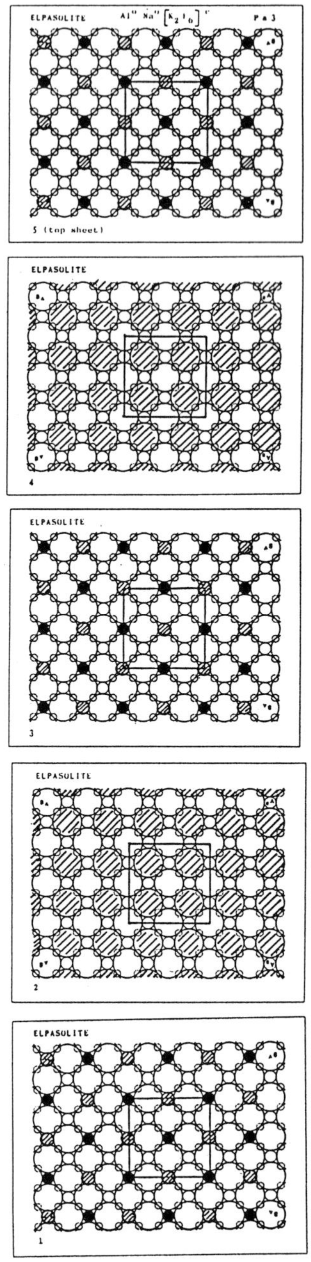

Fig. 3

CRYOLITE

$Na^{o}\,Al^{o}\,[Na_2\,F_6]^{c}$

$P\,2_1/n$

$a = 5.46$ Å
$b = 5.61$ Å
$c = 7.80$ Å
$\beta = 90^{o}\ 11'$
$Z = 2$

Al (2a)

Na_I(2b)

Na_{II}(4c) x = 0.50, y = -0.055, z = 0.24

F_I (4c) x = 0.065, y = 0.06, z = 0.22

F_{II} (4c) x = -0.29, y = 0.16, z = 0.03

F_{III}(4c) x = 0.15, y = 0.28, z = -0.06

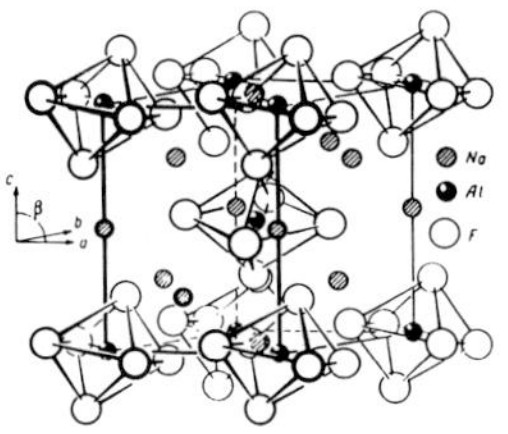

Fig. 1

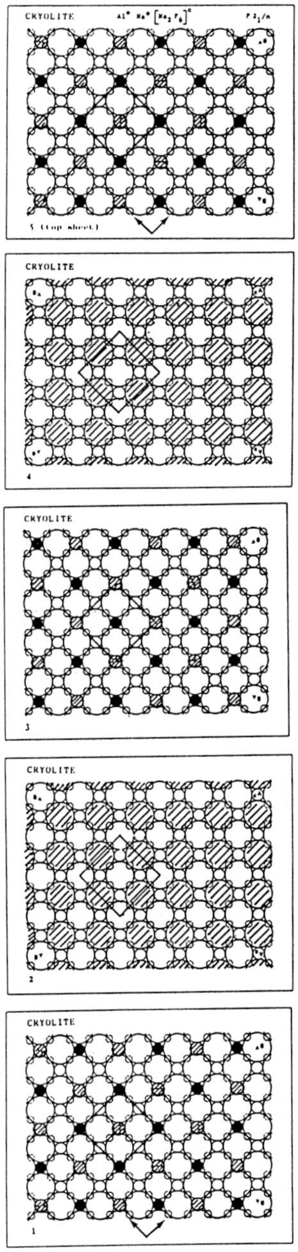

Fig. 2

Properties

Habit	Cleav.	Fract.	Twin.	Hardn.	Dens.	Colour	Transp.
granular, lamellar	none	uneven	(001), (110), (101)	2.5	2.96	colourless, variable	transparent to trans-lucent

Refr. index/Reflect.	Birefr.	Luster	Streak	Melt.p.	CPI
$n_\alpha = 1.3385$	(+)	vitreous	white	1020°C	(SPI)
$n_\beta = 1.3389$	2V = 43°				60
$n_\gamma = 1.3396$					

Figures

Fig. 1. Drawing of the cryolite structure (after Povarennykh, 1972).

Fig. 2. Condensed model of the cryolite structure. Large open circles represent fluorine atoms, and large lined circles certain sodium atoms (those within the square brackets). Small black circles correspond to aluminium atoms, and the small lined circles to the other sodium atoms.

Description

The fluorine and certain sodium atoms form a cubic closest packing and the other sodium atom and aluminium atoms occupy 1/4 of the octahedral voids.

It may be considered as a substitution + distortion derivative of perovskite, or as a distortion derivative of elpasolite.

References

Kostov (1968) 192,193.

Wyckoff (1965) Vol. 3, 382,383.

Povarennykh (1972) 666,667.

Palache et al (1951) Vol. 2, 110-112.

Zoltai + Stout (1984) 403.

HIERATITE

$Si^{o}[K_2F_6]^{c}$

F m 3 m

a = 8.133 Å

Z = 4

Si (4a)

K (8c)

F (24e) u = 0.215

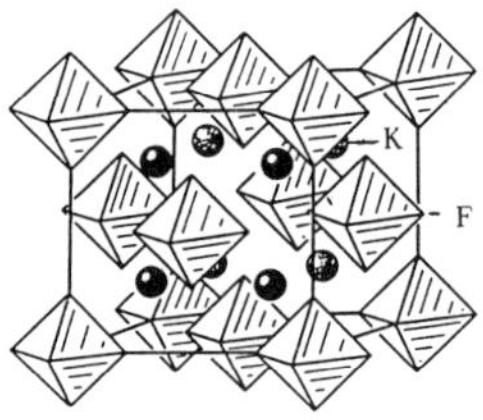

Fig. 1

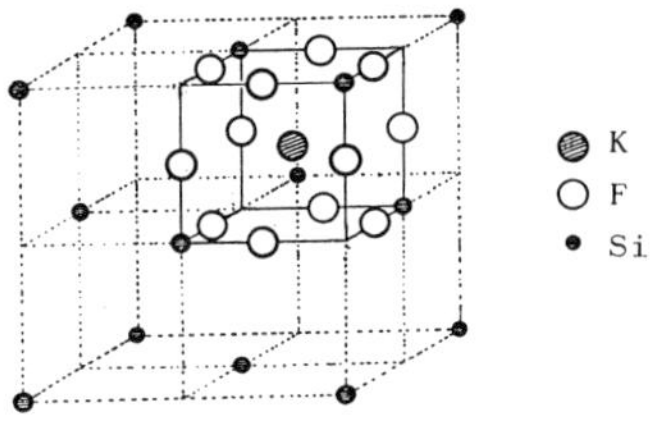

Fig. 2

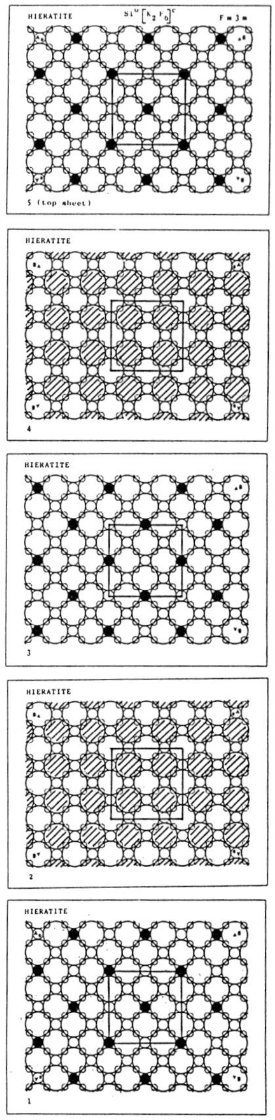

Fig. 3

Properties

Habit	Cleav.	Fract.	Twin.	Hardn.	Dens.	Colour	Transp.
cuboc-tahedral, octahedral	perfect (111)			2.5	2.665	colourless, white, grey	transparent
Refr. index/Reflect.		**Birefr.**		**Luster**	**Streak**	**Melt.p.**	**CPI**
n = 1.340				vitreous			

Figures

Fig. 1. Polyhedral representation of the hieratite structure (after Kostov, 1968).

Fig. 2. Unit cell content of the hieratite structure; only one octet is completely represented (adapted from Wells, 1962).

Fig. 3. Condensed model of the hieratite structure. Large open circles represent fluorine atoms, large lined circles represent potassium, and small black circles Si atoms in octahedral voids, which is a rare coordination for Si.

Description

The fluorine and the potassium atoms form a mixed cubic closest packing, and the silicon atoms occupy one eighth of the octahedral voids. This structure is isotypic with $Pt^{o}[K_2Cl_6]^{c}$. It may also be imagined as a subtraction interstitial derivative of the perovskite structure (compare Fig. 4 with the corresponding Fig. 4 of perovskite).

References

Kostov (1968) 186, 192.

Povarennykh (1972) 665.

Wyckoff (1965) Vol. 3, 339-342.

Wells (1962) 369.

Palache et al. (1951) Vol. 2, 103, 104.

HOLLANDITE $Mn_8^o[Ba_2 O_{16}]^{chh}$

I 4 / m

a = 9.84 Å
c = 2.858 Å
Z = 1
Ba (2b)

Mn (8h) u = 0.348, v = 0.167
O_I (8h) u = 0.153, v = 0.180
O_{II} (8h) u = 0.542, v = 0.167

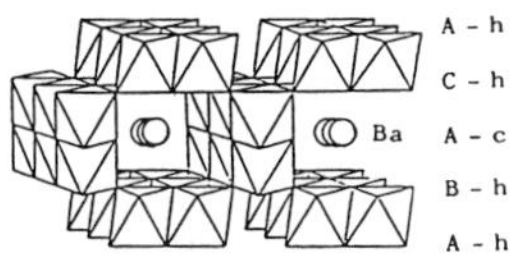

Fig. 1

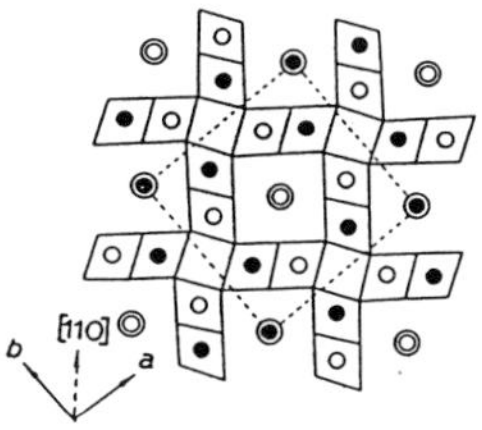

Fig. 2

Properties

Habit	Cleav.	Fract.	Twin.	Hardn.	Dens.	Colour	Transp.
short prismatic, massive	distinct prismatic			6	4.95	silvery grey, black	opaque

Refr. index/Reflect.	Birefr.	Luster	Streak	Melt.p.	CPI
22.5%		metallic shining	black		

Population

Cryptomelane $Mn_8^o[K_2 O_{16}]^{chh}$

Coronadite $Mn_8^o[Pb_2 O_{16}]^{chh}$

Substitution derivative

Priderite $Fe_2^8 Ti_6^8[K_2 O_{16}]^{chh}$

Figures

Fig. 1. Polyhedral description of the hollandite structure (adapted from R. Burns + V. Burns, 1979).

Fig. 2. Projection of the hollandite structure along the c axis (after Kostov, 1968).

Fig. 3. Condensed model of the hollandite structure. The large open circles represent oxygen atoms, and the large lined circles barium atoms. The small black circles correspond to manganese atoms located in octahedral voids. The unit cell is inclined in relation to the packing layers.

Description

Is based on a closest packing of type chh of oxygen and barium atoms, with manganese atoms occupying 4/9 of the octahedral interstices. The closest packed layers are parallel to (110). The structure is related to ramsdellite and hydrophilite.

References

Kostov (1968) 237.
Povarennykh (1972) 141, 305,306.
Wyckoff (1965) Vol. 3, 494-496.
R. Burns + V. Burns (1979) 16.
Palache et al. (1944) Vol. 1, 743.

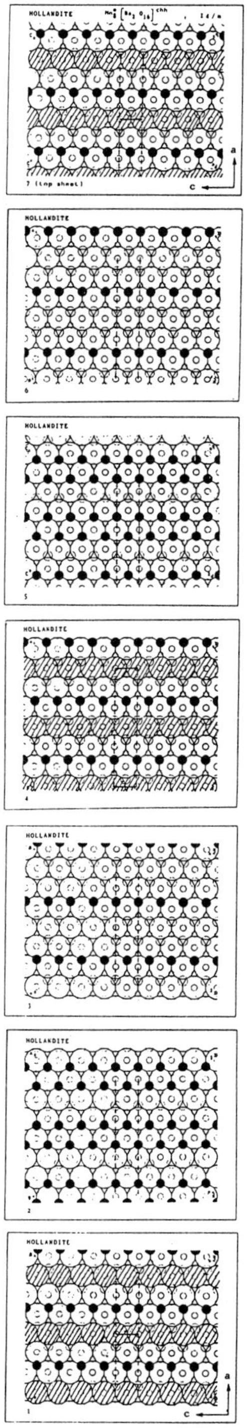

Fig. 3

PSILOMELANE A 2/m

$Mn^{o}_{5}\left[Ba\,(H_2O)\,O_{10}\right]^{cchh}$

a = 9.56 Å	Mn_I (2b)		(Ba,H_2O) (4i) x = 0.124, y = 0, z = 0.246
b = 2.88 Å	Mn_{II} (4i)	x = 0.265, y = 0, z = 0	O_I (4i) x = 0.168, z = 0.572
c = 13.85Å			
β = 92° 30'	Mn_{III} (4i)	x = 0.488, y = 0, z = 0.335	O_{II} (4i) x = 0.421, z = 0.092
Z = 2			...

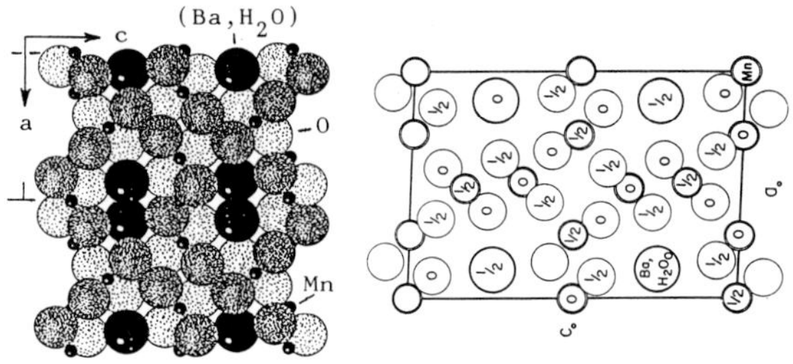

(a) Fig. 1 (b)

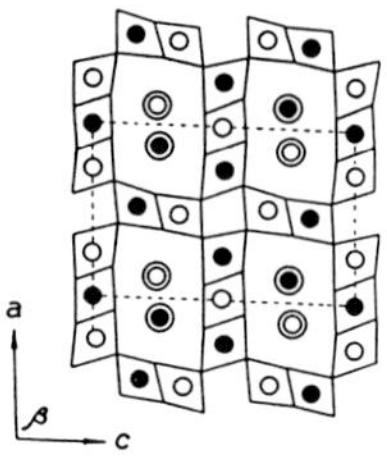

Fig. 2

Properties

Habit	Cleav.	Fract.	Twin.	Hardn.	Dens.	Colour	Transp.
massive, botryoidal, reniform		conchoidal		5 - 6	4.71	black, steel--grey	opaque

Refr. index/Reflect.	Birefr.	Luster	Streak	Melt.p.	CPI
22 - 24 %		submetallic	brownish black		(SPI) 73

Figures

Fig. 1. (a) Packing drawing of the psilomelane structure, and (b) projection along the b axis of the unit cell content (adapted from Wyckoff, 1965, Vol. 3).

Fig. 2. Structural scheme of the structure of psilomelane (after Kostov, 1968).

Fig. 3. Polyhedral description of the psilomelane structure (adapted from R. Burns + V. Burns, 1979).

Fig. 4. Condensed model of the psilomelane structure. The large open circles represent oxygen atoms, and the large lined circles Ba and H_2O. The small black circles correspond to the Mn atoms.

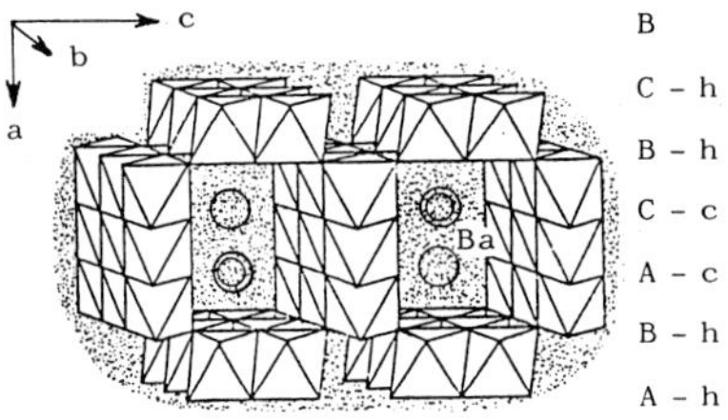

Fig. 3

Description

Is based on a closest packing of type cch of barium , oxygen atoms and water molecules, with Mn atoms occupying 5/12 of the octahedral voids. Psilomelane is also considered as having P222 symmetry, with a = 9.45 Å, b = 13.90 Å, c = 5.72 Å, Z = 2 (quoted by Strunz, 1982).

Crystallographic data (continued)

O_{III}	(4i)	x = 0.394, z = -0.232
O_{IV}	(4i)	x = 0.346, z = 0.422
O_V	(4i)	x = 0.080, z = -0.075

References

Kostov (1968) 236-238.
Povarennykh (1972) 141, 318, 319.
Wyckoff (1965) Vol. 3, 484,485.
Zoltai + Stout (1984) 419.
R. Burns + V. Burns (1979) 18.
Palache et al. (1944) Vol. 1, 669.
Strunz (1982) 201.

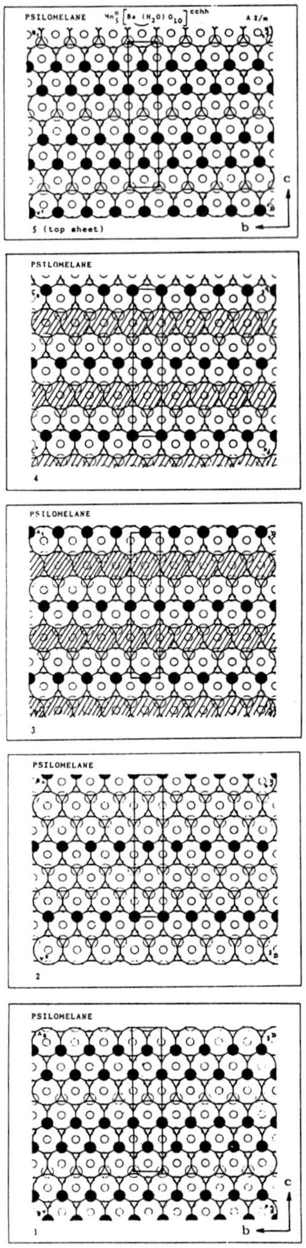

Fig. 4

DIGENITE $Cu_2^t[S]^c$

a = 5.57 Å Cu(8c)

Z = 4 S (4a)

F m 3 m

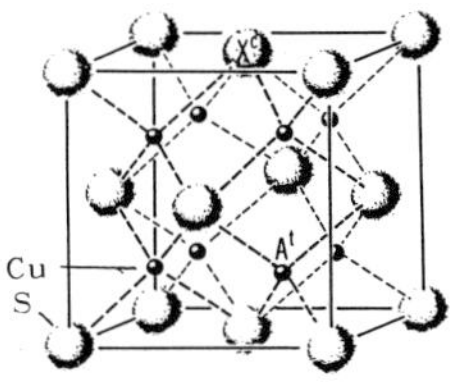

Fig. 1

Properties

Habit	Cleav.	Fract.	Twin.	Hardn.	Dens.	Colour	Transp.
octahedral, massive	(111)	conchoidal		2.5-3	5.546	blue to black	opaque

Refr. index/Reflect.	Birefr.	Luster	Streak	Melt.p.	CPI
18%					

Figures

Fig. 1. Drawing of the unit cell content of the antifluorite structure, which is practically equal to that of digenite (after Bloss, 1971).

Fig. 2. Condensed model of the antifluorite structure, which is practically equal to that of digenite. There are only some defects of occupation of the tetrahedral voids. The large open circles represent S atoms, and the small black circles correspond to Cu atoms in tetrahedral voids.

Description

Digenite is a defect antifluorite structure. The sulfur atoms form a cubic closest packing, where practically all the tetrahedral voids are occupied by copper atoms. According to Strunz (1982) the chemical formula is $Cu_9^t[S_5]^c$.

Bornite, $Cu_5^t Fe^t [S_4]^c$ is based on a a defect digenite structure. It is tetragonal, $P\bar{4}2_1c$, with a = 10.94 Å, c = 21.88 Å and Z = 16 (Strunz, 1982).

References

Kostov (1968) 150.
Palache et al. (1944) Vol.1, 180,181.
Pearson (1967) 65.
Bloss (1971) 251.
Strunz (1982) 109.

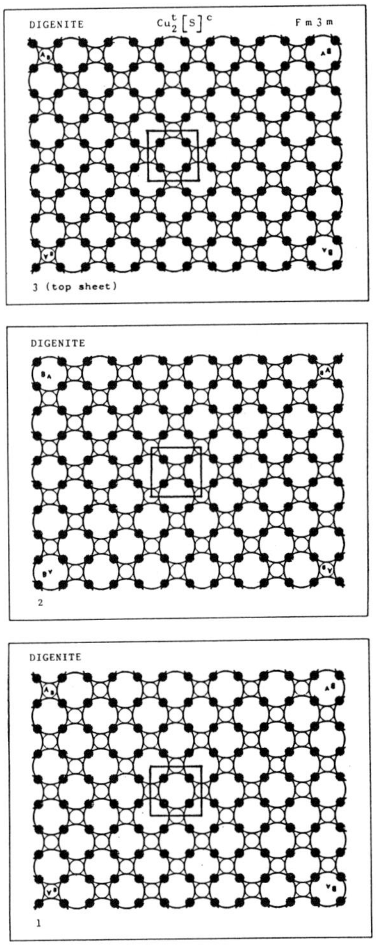

Fig. 2

BLENDE (Sphalerite)

$Zn^t[S]^c$

F $\bar{4}$ 3 m

a = 5.4093 Å (26°C) Zn(4a)

Z = 4 S (4c)

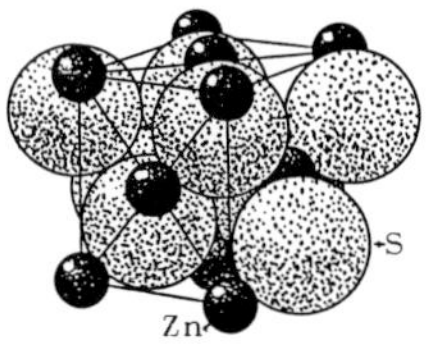

Fig. 1

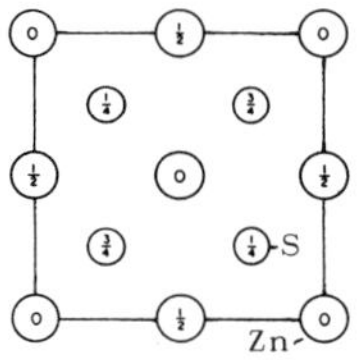

Fig. 2

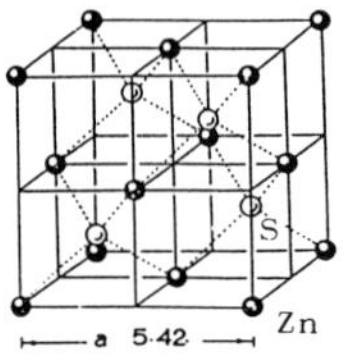

Fig. 3

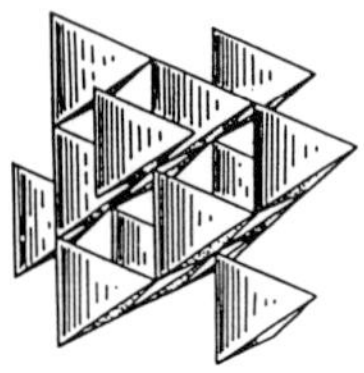

Fig. 4

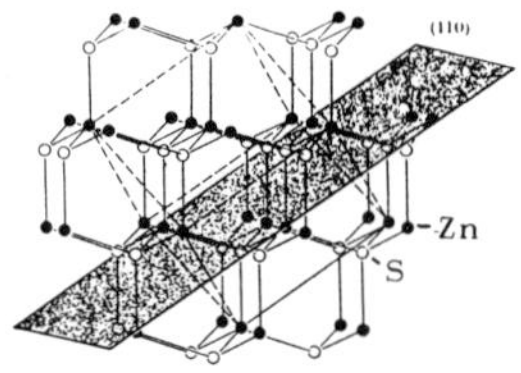

Fig. 5

Properties

Habit	Cleav.	Fract.	Twin.	Hardn.	Dens.	Colour	Transp.
tetrahedral, dodecahedral	perfect (001)	conchoidal	[111]	3.5 - 4	3.9-4.1	brown, black, variable	transparent to opaque

Refr. index/Reflect.		Birefr.	Luster	Streak	Melt.p.	CPI
n = 2.42	19%		resinous, adamantine	brownish, light yellow	1850°C	

Population

Metacinnabarite	$Hg^t[S]^c$
Hawleyite	$Cd^t[S]^c$
Stilleite	$Zn^t[Se]^c$
Tiemannite	$Hg^t[Se]^c$
Coloradoite	$Hg^t[Te]^c$
Nantokite	$Cu^t[Cl]^c$
Marshite	$Cu^t[I]^c$
Miersite	$Ag^t[I]^c$

Figures

Fig. 1. Packing drawing of the sphalerite structure (after Wyckoff, 1963, Vol. 1).

Fig. 2. Unit cell content of the sphalerite structure projected on (001).

Fig. 3. Ball and spoke model of the sphalerite structure (after Kostov, 1968).

Fig. 4. Polyhedral representation of the sphalerite structure (after Povarennykh, 1972).

Fig. 5. Ball and spoke model of the sphalerite structure with marked cleavage plane (110) (after Zemann, 1969).

Fig. 6. Condensed model of the sphalerite structure. The large open circles represent S atoms, and the small black circles Zn atoms.

References

Kostov (1968) 141, 142.

Wyckoff (1963) Vol. 1, 108-111.

Povarennykh (1972) 51, 211, 222, 637.

Pearson (1972) 366.

Zemann (1969) 108.

Palache et al. (1944) Vol. 1, 210,211.

Zoltai+Stout (1984) 382.

Description

The sphalerite structure is formed by a closest packing of S atoms, with Zn atoms occupying one half of the tetrahedral voids, with a square pattern.

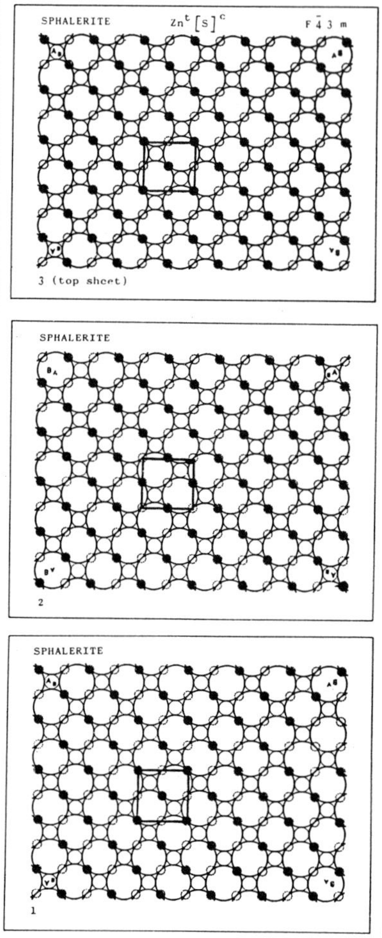

Fig. 6

CHALCOPYRITE

$Cu^{t}\ Fe^{t}\ [S_2]^{c}$

$I\bar{4}2d$

a = 5.24 Å	Cu(4a)
c = 10.30 Å	Fe(4b)
Z = 4	S (8d) x = 1/4

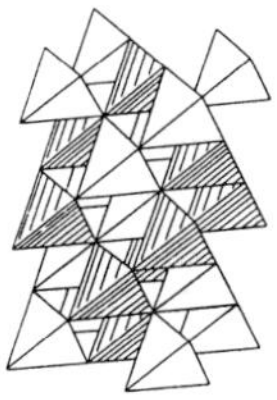

Fig. 1

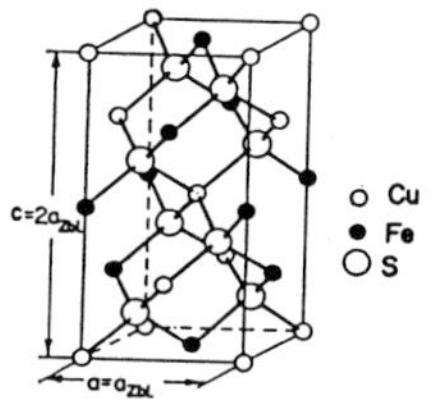

Fig. 2

Properties

Habit	Cleav.	Fract.	Twin.	Hardn.	Dens.	Colour	Transp.
tetrahedral bisphenoidal	poor (011)	uneven	(112)	3.5 - 4	4.1 - 4.3	brass-yellow	opaque

Refr. index/Reflect.	Birefr.	Luster	Streak	Melt.p.	CPI
40,5 %		metallic	greenish black		

Population

Gallite	$Cu^{t}\ Ga^{t}[S_2]^{c}$
Roquesite	$Cu^{t}\ In^{t}[S_2]^{c}$

Figures

Fig. 1. Polyhedral description of the chalcopyrite structure (after Kostov, 1968).

Fig. 2. Ball and spoke model of the chalcopyrite structure (after Parthé, 1964).

Fig. 3. Condensed model of the chalcopyrite structure. Large open circles represent S atoms, the small black circles Fe atoms, and the crossed small circles Cu atoms in tetrahedral voids.

Description

In the chalcopyrite structure the S atoms form a cubic closest packing, and iron and copper occupy one half of the tetrahedral voids. It is an interstitial substitution derivative of sphalerite.

References

Kostov (1968) 148.
Wyckoff (1964) Vol. 2, 336-339.
Parthé (1964) 37.
Palache et al. (1944) Vol. 1, 219,220.
Zoltai+Stout (1984) 382.
Strunz (1982) 115.

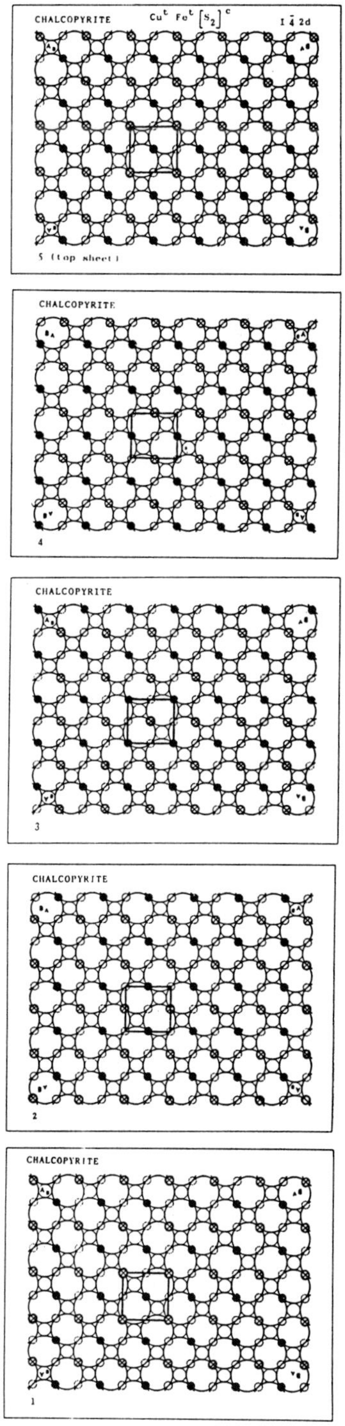

Fig. 3

FAMATINITE

$Cu_3^t\ Sb^t\ [S_4]^c$

$I\bar{4}2m$

a ≃ 5.46
c≃10.725
Z = 2

Cu_I (2b)
Cu_{II} (4d)
Sb (2a)
S (8i) u ≃ 1/4, v ≃ 1/8

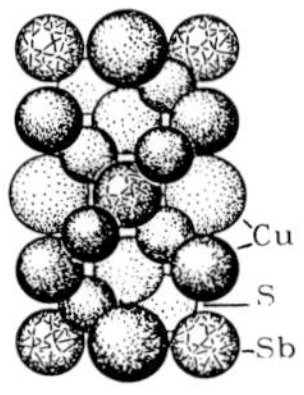

(a)

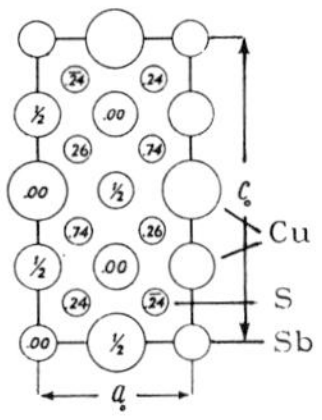

(b)

Fig. 1

Fig. 2

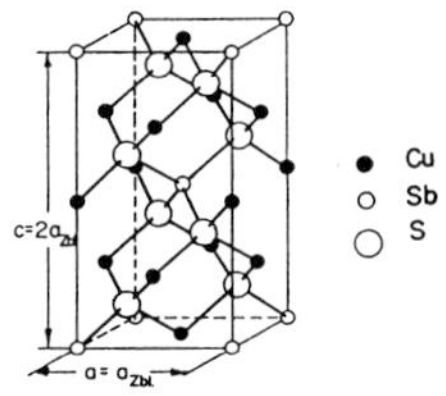

Fig. 3

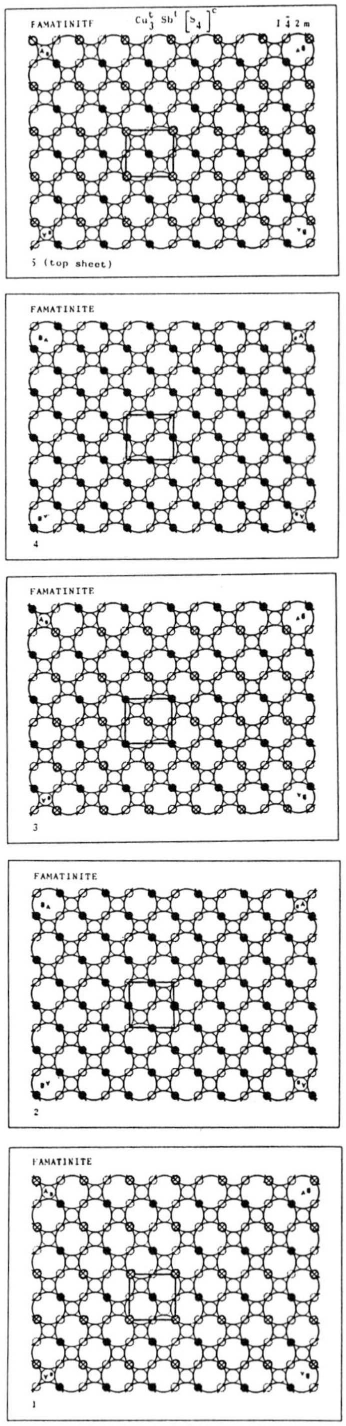

Fig. 4

Properties

Habit	Cleav.	Fract.	Twin.	Hardn.	Dens.	Colour	Transp.
massive, granular	good (101)	uneven		3.5	4.52	grey	opaque

Refr. index/Reflect.	Birefr.	Luster	Streak	Melt.p.	CPI
23.5 %		dull metallic	black		

Population

Luzonite $Cu_3^t\ (As,Sb)^t\ [S_4]^c$

Figures

Fig. 1. (a) Packing drawing of the famatinite structure, and (b) unit cell content projected along one of the a axis (adapted from Wyckoff, 1965, Vol. 3). Notice the inversion of the relative size of the atoms compared with Fig. 1 of sphalerite.

Fig. 2. Polyhedral description of the famatinite structure (adapted from Kostov, 1968).

Fig. 3. Ball and spoke model of the famatinite structure (after Parthé, 1964).

Fig. 4. Condensed model of the famatinite structure. The large open circles represent the S atoms, the small black circles correspond to Cu atoms and the crossed small circles to Sb atoms; Cu and Sb occupy one half of the tetrahedral voids.

Description

In the famatinite structure the S atom form a cubic closest packing with Cu and Sb atoms occupying one half of the tetrahedral voids. This structure is an interstitial substitution derivative of the sphalerite structure.

References

Kostov (1968) 169.
Wyckoff (1965) Vol. 3, 159,160.
Povarennykh (1972) 229,230.
Parthé (1964) 42.
Palache et al. (1944) Vol. 1, 387, 388.

STANNITE $Cu_2^t\ Fe^t\ Sn^t\ [S_4]^c$

$I\bar{4}2m$

a = 5.46 Å
c = 10.725 Å
Z = 2

Cu(4d)
Fe(2a)
Sn(2b)
S (8i) u = 0.245
v = 0.132

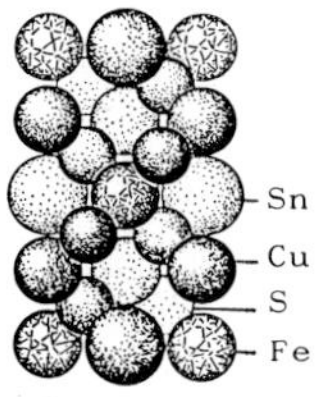

(a)

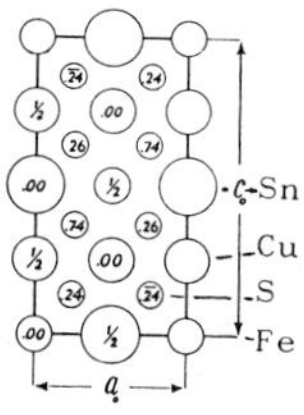

(b)

Fig. 1

Fig. 2

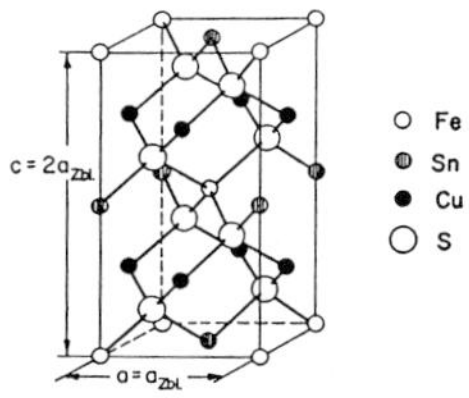

Fig. 3

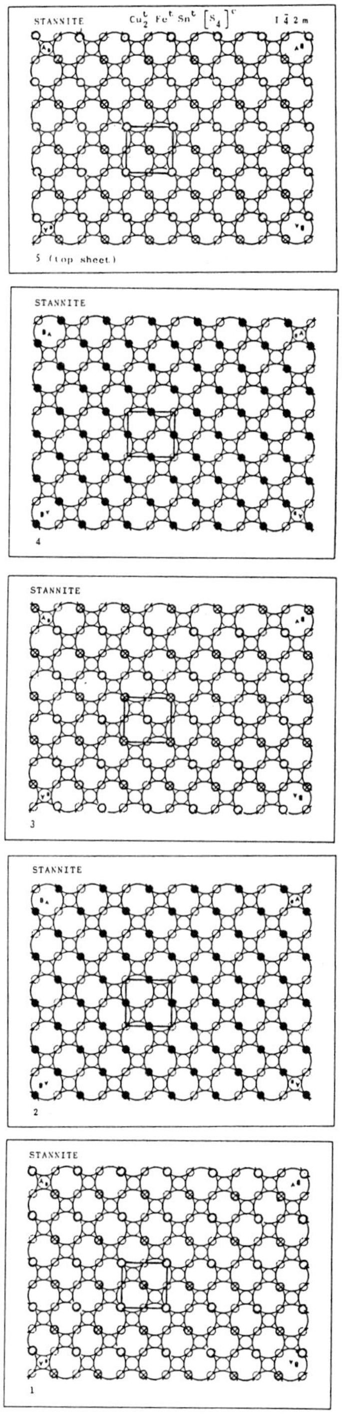

Fig. 4

Properties

Habit	Cleav.	Fract.	Twin.	Hardn.	Dens.	Colour	Transp.
pseudo-tetra-hedral	indistinct (110)	uneven	(102) (112)	4	4.3-4.5	steel--gray, iron--black	opaque

Refr. index/Reflect.	Birefr.	Luster	Streak	Melt.p.	CPI
21%		metallic	blackish		

Population

Kësterite	$Cu_2^t\ Zn^t\ Sn^t\ [S_4]^c$
Hocartite	$Ag_2^t\ Fe^t\ Sn^t\ [S_4]^c$
Briartite	$Cu_2^t\ (Fe,Zn)^t Ge^t\ [S_4]^c$

Figures

Fig. 1. (a) Packing drawing of the stannite structure, and (b) unit cell content projected on (010) (after Wyckoff, 1965, Vol. 3). Notice the inversion of the relative size of the atoms compared with the Fig. 1 of sphalerite.

Fig. 2. Polyhedral representation of the stannite structure (adapted from Kostov, 1968).

Fig. 3. Ball and spoke model of the stannite structure (after Parthé, 1964).

Fig. 4. Condensed model of the stannite structure. The large open circles represent S atoms forming the cubic closest packing. The small black circles correspond to Cu atoms, the crossed small circles to Fe atoms, and the small thick circles to Sn atoms, which are all in tetrahedral voids.

Description

The stannite structure is formed by the cubic closest packing of S atoms, with Fe, Cu and Sn in tetrahedral voids. It is a interstitial substitution derivative of sphalerite.

References

Kostov (1968) 146-148.
Wyckoff (1965) Vol. 3, 159,160.
Povarennykh (1972) 228.
Parthé (1964) 44.
Palache et al. (1944) Vol. 1, 224, 225.

WURTZITE

$Zn^{t}[S]^{h}$

P6_3mc

a=3.82Å
c=6.26Å
Z=2

Zn(2b) z=0
S (2b) z=0.371

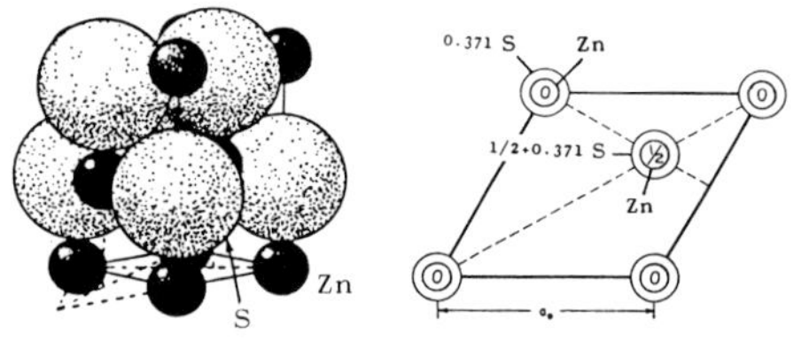

(a) (b)

Fig. 1

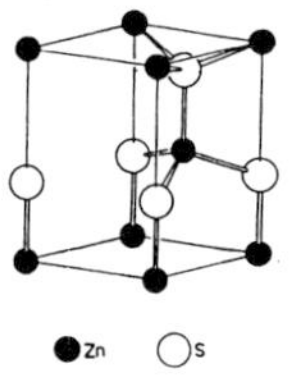

Fig. 2

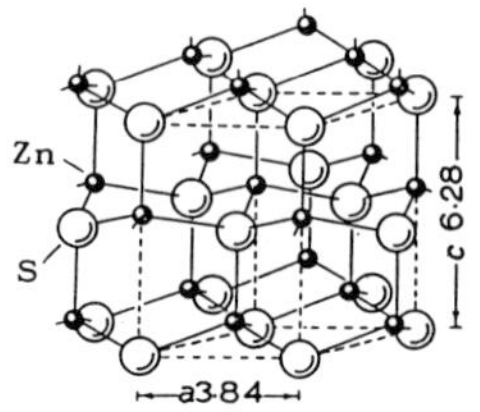

Fig. 3

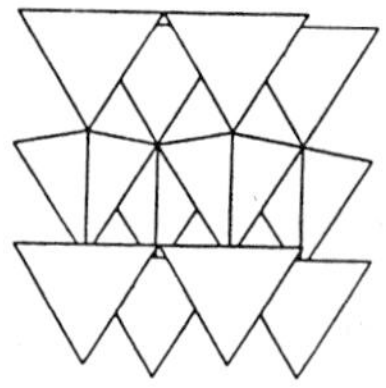

Fig. 4

Properties

Habit	Cleav.	Fract.	Twin.	Hardn.	Dens.	Colour	Transp.
pyramidal, short prismatic	good (11$\bar{2}$0)	conchoidal		3.5 - 4	3.98	brownish black	translucent

Refr. index/Reflect.		Birefr.	Luster	Streak	Melt.p.	CPI
n_ω = 2.356 n_ϵ = 2.378	19%	(+)	resinous	brown		

Population

Greenockite	$Cd^{t}[S]^{h}$
Cadmoselite	$Cd^{t}[Se]^{h}$
Zincite	$Zn^{t}[O]^{h}$
Bromellite	$Be^{t}[O]^{h}$
Iodargyrite	$Ag^{t}[I]^{h}$
Moissanite	$C^{t}[Si]^{h}$

Figures

Fig. 1. (a) Packing drawing of the wurtzite structure (after Wyckoff 1963, Vol. 1) and (b) unit cell content projected on (001).

Figs 2 and 3. Ball and spoke models of the wurtzite structure (after Wuensch, 1974, Kostov, 1968, respectively).

Fig. 4. Polyhedral representation of the wurtzite structure (after Zoltai, 1974).

Fig. 5. Condensed model of the wurtzite structure. The large open circles represent S atoms, and the small black circles correspond to the Zn atoms which occupy one half of the tetrahedral voids.

Description

The wurtzite structure is formed by a hexagonal closest packing of the S atoms, and Zn atoms occupying one half of the tetrahedral voids, with a triangular distribution pattern.

References

Kostov (1968) 144,145.
Povarennykh (1972) 199,200, 223, 272, 637.
Wyckoff (1963) Vol. 1, 111.
Zoltai (1974) 28.
Wuensch (1974) W-29.
Pearson (1967) 381.
Roberts et al. (1974) 676,677.

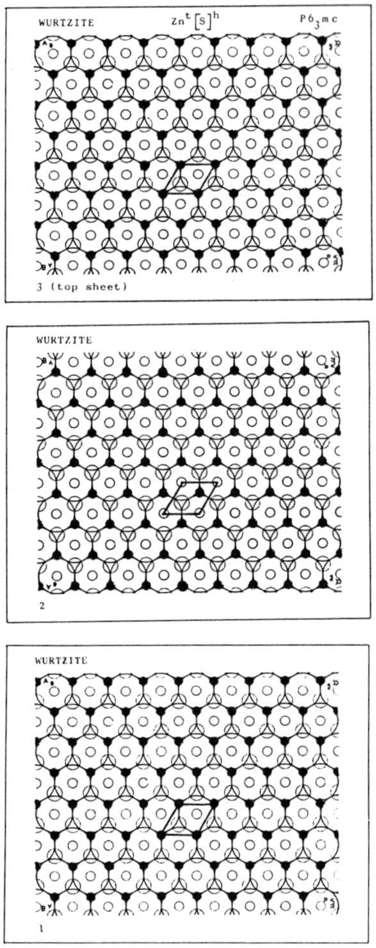

Fig. 5

CUBANITE

$Cu^{t} Fe^{t}_{2} [S_3]^{h}$

Pcmn

a = 6.46 Å
b = 11.117 Å
c = 6.233 Å
Z = 4

	x	y	z
Fe(8d)	0.0875	0.088	0.134
S_I(4c)	0.931	1/4	0.2625
S_{II}(8d)	0.413	0.0835	0.274
Cu(4c)	0.583	1/4	0.127

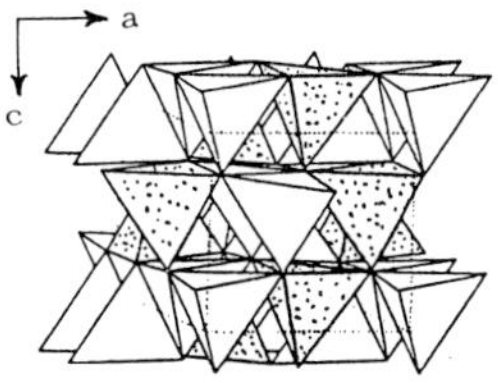

Fig. 1

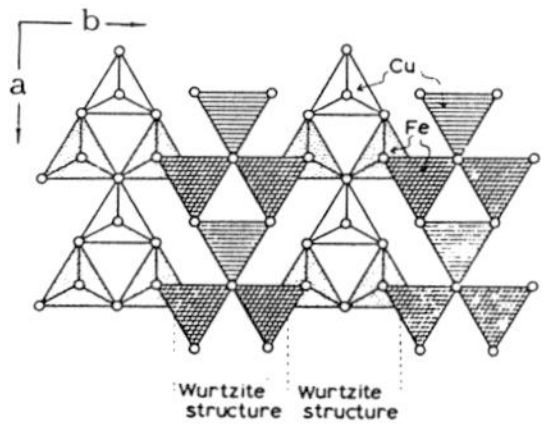

Fig. 2

Properties

Habit	Cleav.	Fract.	Twin.	Hardn.	Dens.	Colour	Transp.
tabular, massive	none	conchoidal	(110)	3.5	4.03–4.18	brass, bronze-yellow	opaque

Refr. index/Reflect.	Birefr.	Luster	Streak	Melt.p.	CPI
41%		metallic	grey		

Population

Argentopyrite $Ag^{t} Fe^{t}_{2} [S_3]^{h}$

Figures

Fig. 1. Polyhedral representation of the cubanite structure (after Zoltai, 1974).

Fig. 2. Polyhedral drawing of the cubanite structure showing that it may be imagined built of slices of the wurtzite arrangement (after Kostov, 1968).

Fig. 3. Condensed model of the cubanite structure. The large open circles represent the S atoms and the small black circles the iron atoms in tetrahedral voids. The dotted small circles correspond to the coppper atoms, also with tetrahedral coordination,

Description

The cubanite structure is based on an hexagonal closest packing of the S atoms, with copper and iron occupying one half of the tetrahedral voids. It can not be considered as an interstitial substitution derivative of wurtzite, but can be imagined formed by two dimensional slices of the wurtzite atomic arrangement, parallel to (001).

References

Kostov (1968) 147.
Wyckoff (1964) Vol. 2, 506,507.
Povarennykh (1972) 226,227.
Zoltai (1974) 29.

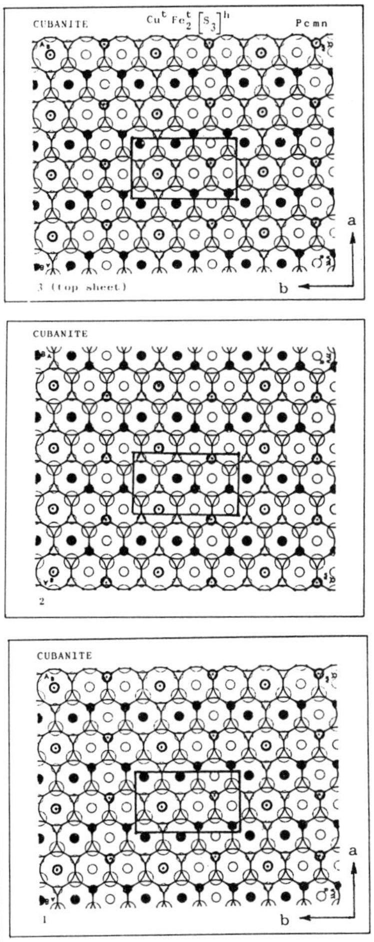

Fig. 3

ENARGITE

$Cu_3^t\ As^t\ [S_4]^h$

P n m 2_1

a = 6.46 Å
b = 7.43 Å
c = 6.18 Å
Z = 2

Cu_I	(2a)	x=0.165, y=0, z=0.500
Cu_{II}	(4b)	x=0.333, y=0.245, z=0.990
As	(2a)	x=0.82, y=0, z=0
S_I	(2a)	x=0.830, y=0, z=0.360
S_{II}	(2a)	x=0.140, y=0, z=0.875
S_{III}	(4b)	x=0.330, y=0.255, z=0.367

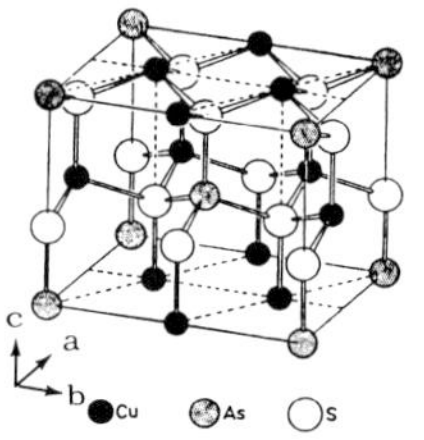

Fig. 1

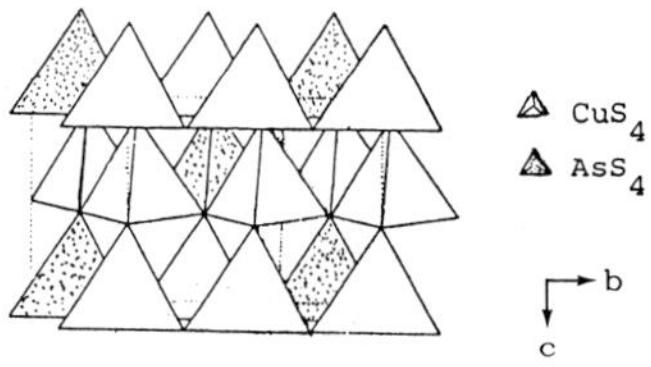

Fig. 2

Properties

Habit	Cleav.	Fract.	Twin.	Hardn.	Dens.	Colour	Transp.
tabular, prismatic, massive	perfect (110) good (100)	uneven	(320)	3	4.5	greyish black, iron-black	opaque

Refr. index/Reflect.	Birefr.	Luster	Streak	Melt.p.	CPI
21.5%		metallic	greyish black		

Figures

Fig. 1. Ball and spoke model of the enargite structure (after Wuensch, 1974).

Fig. 2. Polyhedral description of the enargite structure (after Zoltai, 1974).

Fig. 3. Condensed model of the enargite structure. The large circles represent the S atoms, the small black circles the Cu atoms, and the doted small circles the As atoms.

Description

The enargite structure is based on an hexagonal closest packing of the S atoms, with Cu and As occupying one half of the tetrahedral voids.

It is an interstitital substitution derivative of wurtzite.

References

Kostov (1969) 181.
Povarennykh (1972) 229.
Wyckoff (1965) Vol. 3, 157,158.
Zoltai (1974) 28.
Wuensch (1974) W-29.
Palache et al. (1944) Vol. 1, 390.
Roberts et al. (1974) 190.

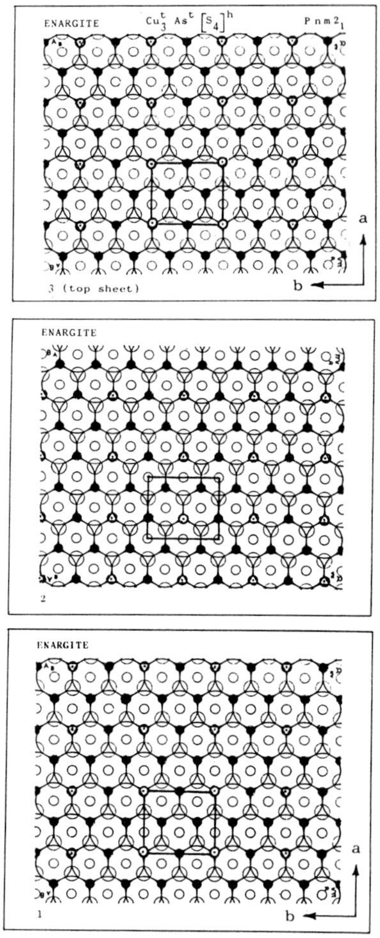

Fig. 3

LITHARGE $O^t[Pb]^c$

P 4/nmm

a = 3.975 Å
c = 5.023 Å
Z = 2

O (2a)
Pb(2c) u=0.2385

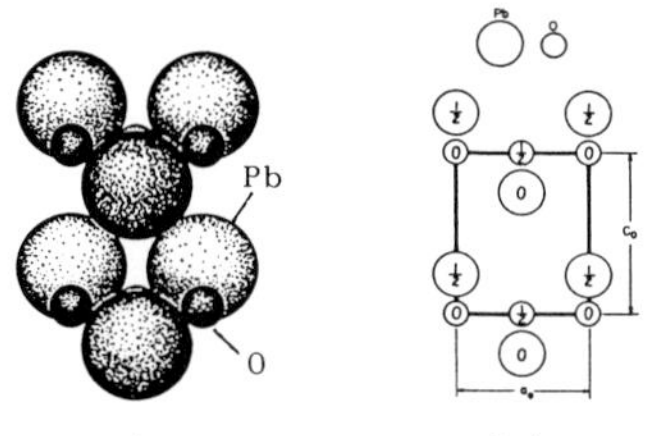

(a) (b)
Fig. 1

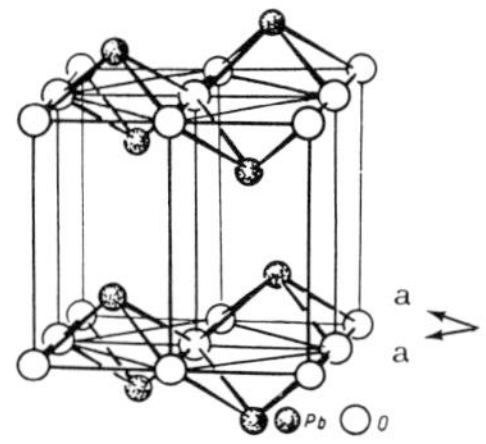

Fig. 2

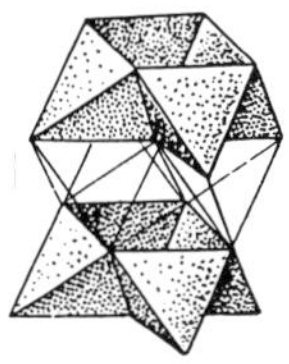

Fig. 3

Properties

Habit	Cleav.	Fract.	Twin.	Hardn.	Dens.	Colour	Transp.
tabular, crusts	(110)			2	9.14	red	transparent

Refr. index/Reflect.	Birefr.	Luster	Streak	Melt.p.	CPI
n_ω = 2.665 n_ϵ = 2.535	(-)	greasy to dull			

Population

Mackinawite $Fe^t[S]^c$

Distortion derivative

Massicot $O^t[Pb]^c$

Pcma

Figures

Fig. 1. (a) Packing representation of the litharge structure, and (b) unit cell projection on (100) (after Wyckoff (1963) Vol. 1).

Fig. 2. Ball and spoke model of the litharge structure (adapted from Povarennykh, 1972).

Fig. 3. Polyhedral description of the litharge structure (after Povarennykh, 1972).

Fig. 4. Condensed model of the litharge structure. The large open circles represent Pb atoms, and the small black circles oxygen atoms occupying tetrahedral voids. Notice that the structure is built of alternate layers, parallel to (001), one completely filled and the other completely empty in respect to the tetrahedral voids. To relate the unit cell marked on the condensed model with Fig. 1 and Fig. 3 look in the direction of the arrow, and to relate with Fig. 2 look in the direction of the double arrow.

Description

The litharge structure may be imagined formed by a cubic closest packing of Pb atoms with oxygen atoms occupying 1/2 of the tetrahedral voids. This interpretation is quite acceptable if we compare the c/a (real) = 1.26 with c/a (ideal) = 1.414 (c = 2x1.414R, a = 2R).

References

Kostov (1968) 259.
Povarennykh (1972) 310.
Wyckoff (1963) Vol. 1, 134-136.
Palache et al. (1944) Vol. 1, 514.

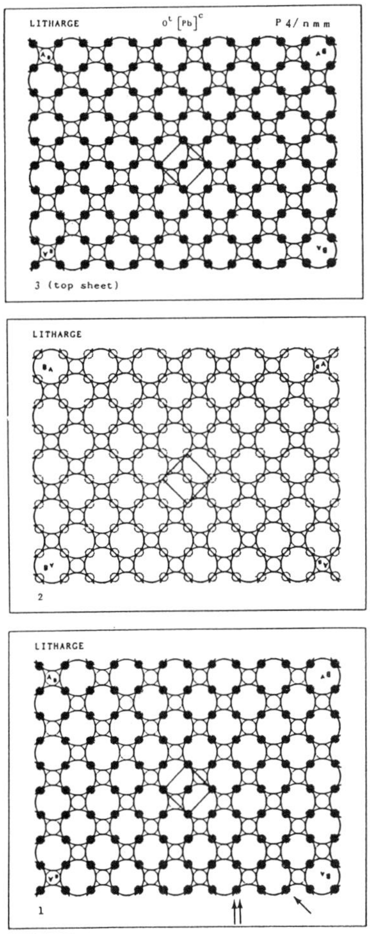

Fig. 4

BERTRANDITE

$Be^t_4\ Si^t_2\left[O_7(OH)_2\right]^h$

$Cmc2_1$

a=8.733 Å
b=15.31 Å
c=4.56 Å
Z=4

Be_I	(8b)	x=0.171	y=0.053	z=0.129
Be_{II}	(8b)	x=0.329	y=0.221	z=0.133
Si	(8b)	x=0.325	y=0.144	z=0.658
O_I	(8b)	x=0.288	y=0.125	z=0.000
O_{II}	(8b)	x=0.211	y=0.043	z=0.507
O_{III}	(8b)	x=0.292	y=0.209	z=0.469

...

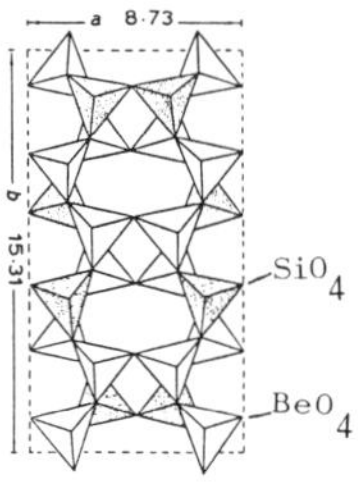

Fig. 1

Properties

Habit	Cleav.	Fract.	Twin.	Hardn.	Dens.	Colour	Transp.
tabular, prismatic	perfect (001)		(101)	6 - 7	2.60	colourless to pale yellow	transparent

Refr. index/Reflect.	Birefr.	Luster	Streak	Melt.p.	CPI
n_α = 1.591 n_β = 1.605 n_γ = 1.614	(-)	vitreous			

Figures

Fig. 1. Polyhedral representation of the bertrandite structure (after Soloveva + Belov, 1964, quoted by Kostov, 1968).

Fig. 2. Condensed model of the bertrandite structure. The large open circles represent oxygen atoms and the lined large circles (OH). The small black circles represent the Si atoms and the doted small circles the Be atoms, all in tetrahedral voids.

Description

The bertrandite structure is based on an hexagonal closest packing of oxygen and hydroxyls, with Be and Si atoms occupying 2/3 of the tetrahedral voids, forming together an honeycomb pattern. This structure can also be imagined formed by infinite distorted amphibole type chains (see Fig. 2), linked together forming a three dimensional network. However it should not be considered as a framework due to its closest packing characteristics.

Crystallographic data (continued)

O_{IV}	(4a)	x=0	y=-0.416	z=0.597
O_V	(4a)	x=0	y=-0.245	z=0.091
O_{VI}	(4a)	x=0	y=0.087	z=0.100

References

Kostov (1968) 275,276.
Povarennykh (1972) 401.
Wyckoff (1968) Vol. 4, 229-231.
Roberts et al. (1974) 63,64.
Soloveva + Belov (1964) 551.

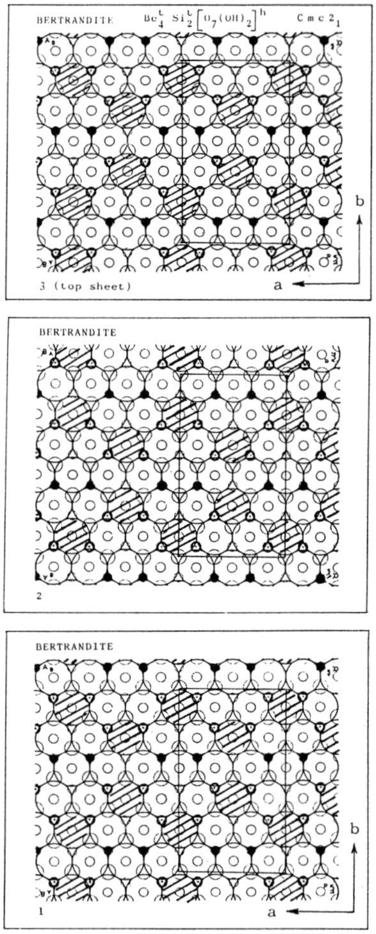

Fig. 2

BARYLITE

$Be_2^t\ Si_2^t\ [Ba\ O_7]^h$

$Pn2_1a$

a = 9.79 Å
b = 11.65 Å
c = 4.63 Å
Z = 4

		x	y	z
Si_I	(4a)	-0.096	0.88	0.75
Be_I	(4a)	0.175	0.001	0.15
Be_{II}	(4a)	-0.170	0.005	0.25
Si_{II}	(4a)	0.10	0.125	0.25
Ba	(4a)	0.142	0.75	0.25
O_I	(4a)	0.073	0.897	0.72
O_{II}	(4a)	-0.087	0.885	0.130

...

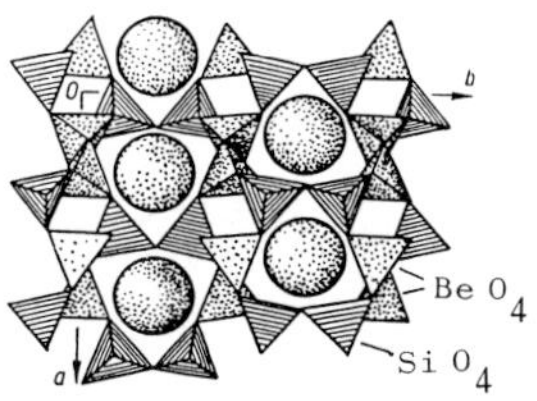

Fig. 1

Properties

Habit	Cleav.	Fract.	Twin.	Hardn.	Dens.	Colour	Transp.
tabular, prismatic	perfect (001) (100)			7	4.046	colourless, white, bluish	transparent

Refr. index/Reflect.	Birefr.	Luster	Streak	Melt.p.	CPI
n_α = 1.69 n_β = 1.70 n_γ = 1.70	(-) 2V = 40°27'	vitreous			

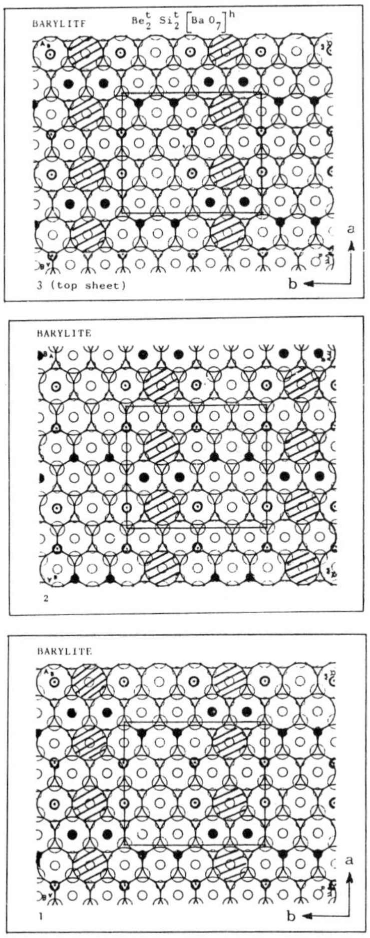

Fig. 2

Figures

Fig. 1. Polyhedral description of the barylite structure (after Povarennykh, 1972).

Fig. 2. Condensed model of the barylite structure. The large open circles represent the oxygen atoms, and the lined large circles the Ba atoms. The Si and Be atoms are represented by black and doted small circles, respectively, which are both in tetrahedral voids.

Description

The barylite structure is based on an hexagonal closest packing of oxygens, with Be and Si atoms in tetrahedral voids. This structure can also be imagined formed by $[Si_2\ O_7]$ groups and $[Be\ O_3]$ infinite chains linked together in a three dimensional network. However it should not be considered as a framework due to its closest packing characteristics.

References

Kostov (1968) 281.
Povarennykh (1972) 359.
Wyckoff (1968) Vol. 4, 222,223.
Roberts et al. (1974) 52.

Crystallographic data (continued)

		x	y	z
O_{III}	(4a)	0.181	0.03	0.452
O_{IV}	(4a)	-0.13	0.75	0.690
O_V	(4a)	-0.07	0.110	0.210
O_{VI}	(4a)	-0.185	-0.025	0.595
O_{VII}	(4a)	0.09	0.115	0.874

SPINEL $Al_2^o\ Mg^t\left[O_4\right]^c$

F d 3 m

a = 8.08Å
Z = 8

Al (16d)
Mg (8a)
O (32e) x = 0.387

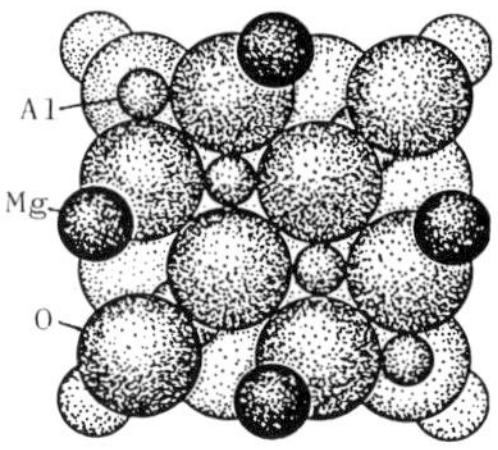

(a)

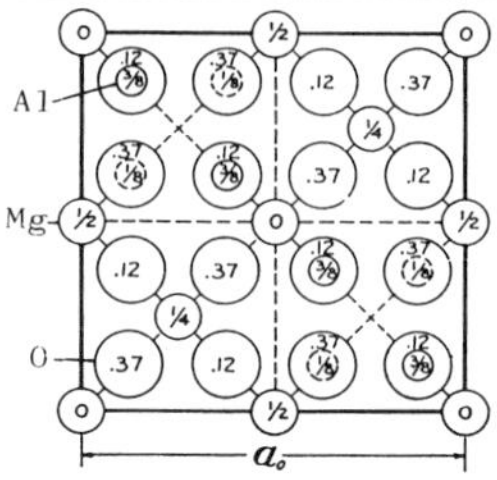

(b)

Fig. 1

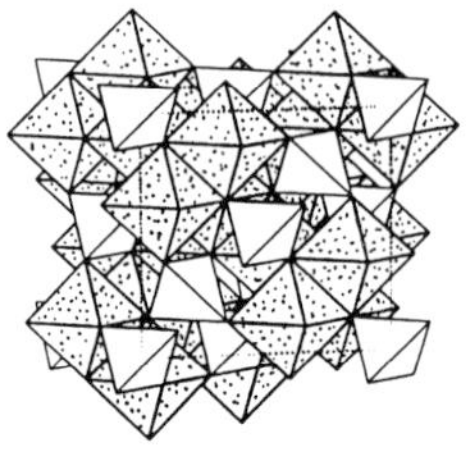

Fig. 2

Properties

Habit	Cleav.	Fract.	Twin.	Hardn.	Dens.	Colour	Transp.
octahedral, massive		conchoidal	(111)	7.5 - 8	3.55	variable	transparent to translucent

Refr. index/Reflect.	Birefr.	Luster	Streak	Melt.p.	CPI
n = 1.74		vitreous	white	2136° C	(SPI) 70

Population

Hercynite	$Al_2^o\ Fe^t\left[O_4\right]^c$
Gahnite	$Al_2^o\ Zn^t\left[O_4\right]^c$
Galaxite	$Al_2^o\ Mn^t\left[O_4\right]^c$
Franklinite	$Fe_2^o\ Zn^t\left[O_4\right]^c$
Magnesio-ferrite	$Fe_2^o\ Mg^t\left[O_4\right]^c$
Jacobsite	$Fe_2^o\ Mn^t\left[O_4\right]^c$
Magnetite	$Fe_2^o\ Fe^t\left[O_4\right]^c$
Trevorite	$Fe_2^o\ Ni^t\left[O_4\right]^c$
Ulvöspinel	$Fe_2^o\ Ti^t\left[O_4\right]^c$
Chromite	$Cr_2^o\ (Mg,Fe)^t\left[O_4\right]^c$
Coulsonite	$V_2^o\ Fe^t\left[O_4\right]^c$
Carrollite	$Co_2^o\ Cu^t\left[S_4\right]^c$
Violarite	$Ni_2^o\ Fe^t\left[S_4\right]^c$
Melnikovite	$Fe_2^o\ Fe^t\left[S_4\right]^c$
Indite	$In_2^o\ Fe^t\left[S_4\right]^c$
Polydymite	$(Ni,Co)_2^o(Co,Ni)^t\left[S_4\right]^c$
Linnaeite	$(Co,Ni)_2^o(Co,Ni)^t\left[S_4\right]^c$
Tyrrellite	$Co_2^o\ Cu^t\left[Se_4\right]^c$
Bornhardtite	$Co_2^o\ Co^t\left[Se_4\right]^c$
Trüstedtite	$Ni_2^o\ Ni^t\left[Se_4\right]^c$

Figures

Fig. 1. (a) Packing drawing of the spinel structure and (b) projection on a cube face; only the lower half of the unit cell is represented (after Wyckoff, 1965, Vol.3).

Fig. 2. Polyhedral representation of the spinel structure, showing the closest-packed layers (after Zoltai, 1977).

Fig. 3. Complete condensed model of the spinel structure based on square layers. Large open circles represent the oxygen atoms; the black circles correspond to Al atoms in octahedral voids, and the smaller black circles to Mg atoms in tetrahedral voids.

Description

The spinel structure is based on a cubic closest packing of oxygen atoms with a half of the octahedral voids occupied by Al atoms forming a T lattice complex, and 1/8 of the tetrahedral voids occupied by Mg atoms forming a F lattice complex. There are two kinds of spinels: normal and inverse. The structural formula for the normal spinel is $(A^{+3})_2^o\ (B^{+2})^t\left[X_4\right]^c$, and for the inverse spinel $(B^{+2})^o\ (A^{+3})^o\ (A^{+3})^t\left[X_4\right]^c$. Magnetite is an example of a inverse spinel and the formula is $(Fe^{+2})^o\ (Fe^{+3})^o\ (Fe^{+3})^t\left[O_4\right]^c$.

References

Kostov (1968) 123, 215.
Povarennykh (1972) 223, 283-286, 746.
Palache et al. (1944) Vol. 1, 690.
Wyckoff (1965) Vol. 3, 75, 76.
Zoltai + Stout (1984) 414.
Zoltai (1977) 6-40.
Ingerson (1955) 351.

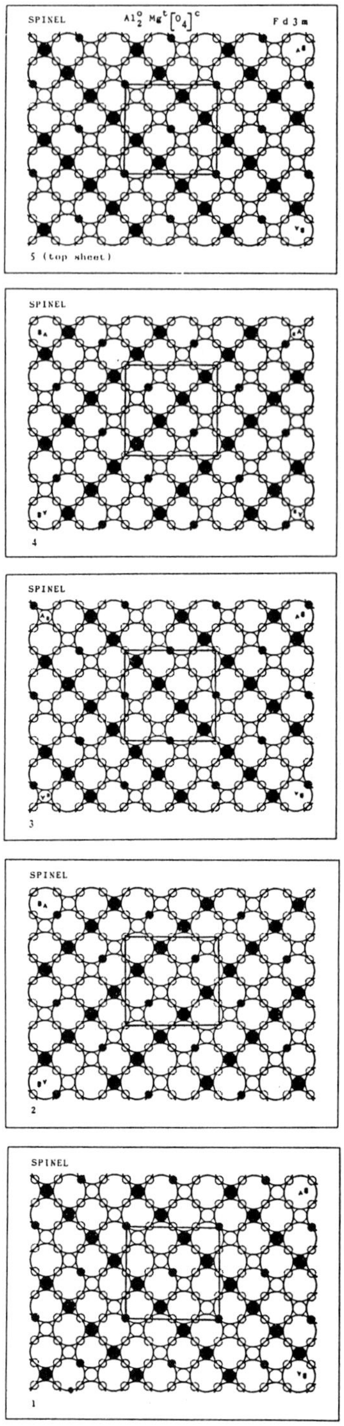

Fig. 3

HAUSMANNITE

$Mn_2^{II\,o}\ Mn^{IV\,t}\left[O_4\right]^c$

$I\,4_1/a\,m\,d$

a = 5.75 Å	$(Mn^{+2})^o$ (8d)	
c = 9.42 Å	$(Mn^{+4})^t$ (4a)	
Z = 4	O (16h)	u = 1/4, v = 3/8

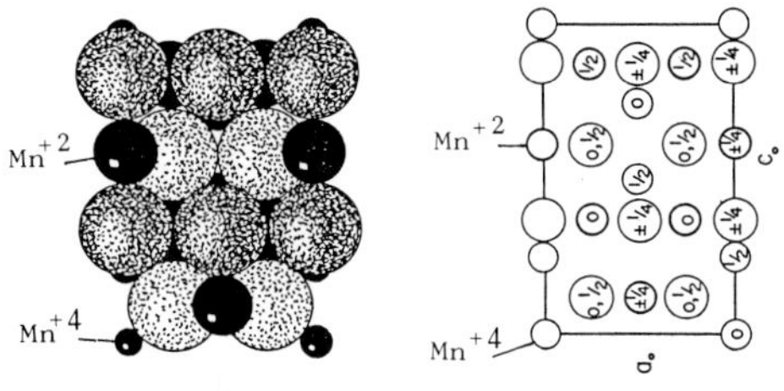

(a) (b)

Fig. 1

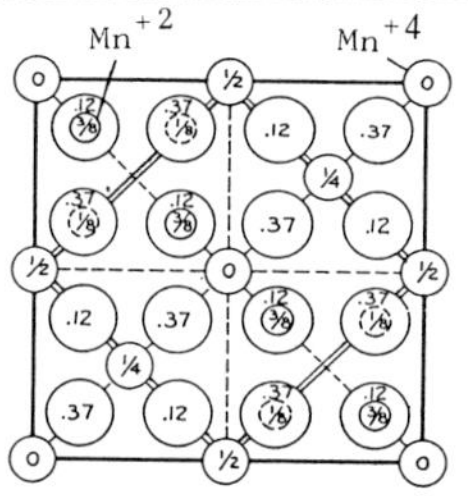

Fig. 2

Properties

Habit	Cleav.	Fract.	Twin.	Hardn.	Dens.	Colour	Transp.
pseudoc-tahedral	nearly perfect (001)	uneven	(112)	5 - 5.5	4.84	brownish black	transparent to opaque

Refr. index/Reflect.		Birefr.	Luster	Streak	Melt.p.	CPI
n_ω = 2.46, n_ϵ = 2.15	16,5%	(-)	subme-tallic	chestnut--brown		(SPI) 64

Figures

Fig. 1. (a) Packing drawing of the hausmannite structure, and (b) projection of the hausmannite structure along the b axis (after Wyckoff, 1965, Vol. 3).

Fig. 2. Unit cell (only half of the lower part) of spinel in order to show the relation with the hausmannite unit cell (marked with double line) (adapted from Wyckoff, 1965, Vol. 3).

Fig. 3. Condensed model of the hausmannite structure. Large open circles represent the oxygen atoms, small black circles the Mn^{+2} ions in octahedral voids, and the smaller black circles the Mn^{+4} ions in tetrahedral voids. In order to compare with Fig. 1 the unit cell should be looked in the direction of the arrow. The double line correspond to the hausmannite unit cell and the single line to the spinel unit cell.

Description

The hausmannite structure is a distortion derivative of spinel. It is based on a cubic closest packing of oxygens, with Mn^{+2} in octahedral voids, and Mn^{+4} in tetrahedral voids.

References

Kostov (1968) 236.
Povarennykh (1972) 287.
Wyckoff (1965) Vol. 3, 76, 84.
Zoltai + Stout (1984) 416.
Palache et al. (1944) Vol. 1, 713, 714.

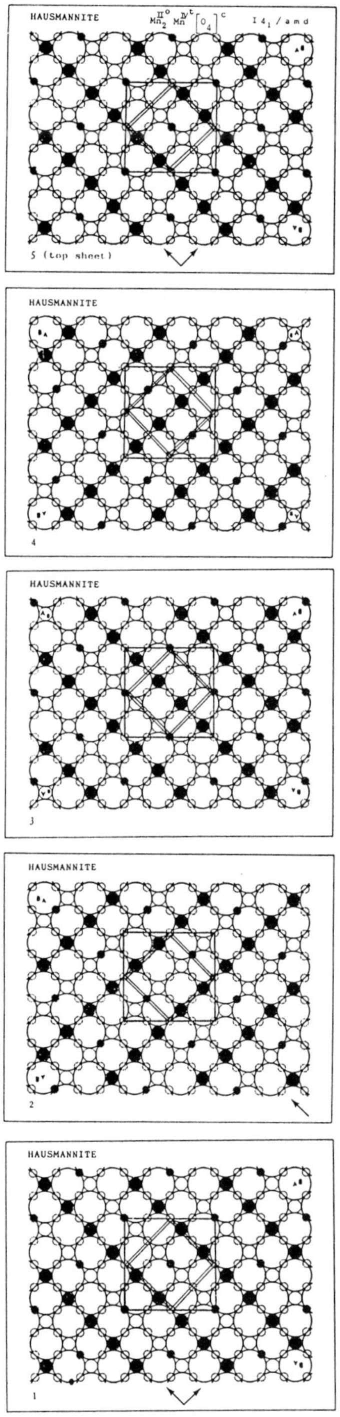

Fig. 3

OLIVINE

[Forsterite (Mg) - fayalite (Fe)]

$(Mg,Fe)_2^o\ Si^t\ [O_4]^h$ P n m a

(for forsterite)
a = 10.26 Å
b = 6.00 Å
c = 4.77 Å
Z = 4

$(Mg,Fe)_I$	(4a)	
$(Mg,Fe)_{II}$	(4c)	x=0.2775, y=1/4, z=-0.010
Si	(4c)	x=0.0945, y=1/4, z=0.426
O_I	(4c)	x=0.092, y=1/4, z=0.767
O_{II}	(4c)	x=0.449, y=1/4, z=0.219
O_{III}	(8d)	x=0.163, y=0.0365, z=0.277

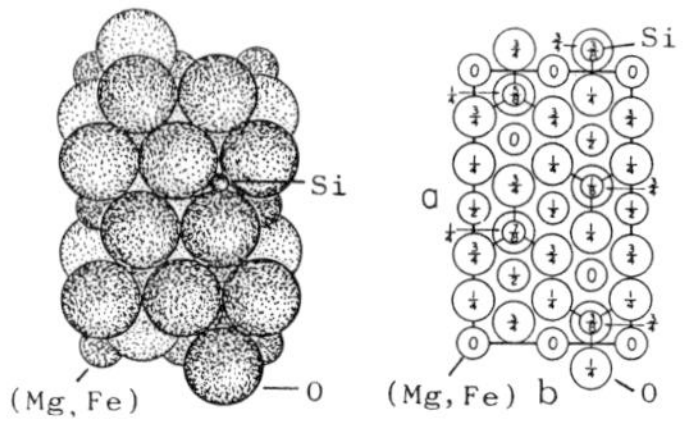

(a) (b)
Fig. 1

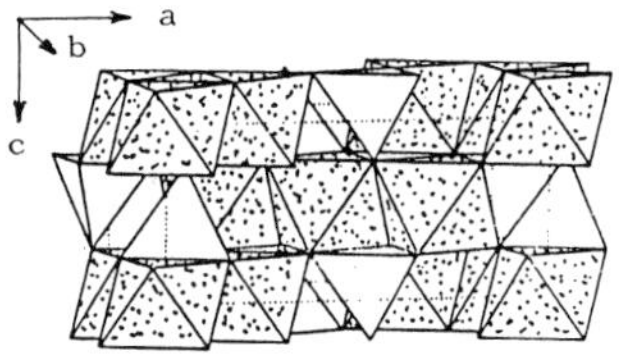

Fig. 2

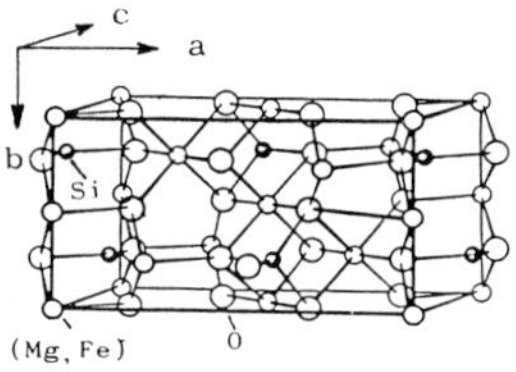

Fig. 3

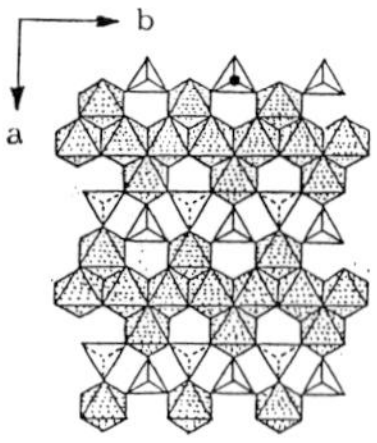

Fig. 4

Properties

Habit	Cleav.	Fract.	Twin.	Hardn.	Dens.	Colour	Transp.
tabular prismatic	good (100)	conchoi-dal	(001) (110) (120)	6.5	3.2	colourless, green	transparent to trans-lucent

Refr. index/Reflect.	Birefr.	Luster	Streak	Melt.p.	CPI
n_α = 1.635 n_β = 1.651 n_γ = 1.670	(+) 2V= 85°-90°	vitreous	white	1890°C (forsterite) 1205°C (fayalite)	(SPI) 65

Population

Knebelite	$(Mn,Fe)_2^o\ Si^t\ [O_4]^h$
Chrysoberyl	$Al_2^o\ Be^t\ [O_4]^h$
Sinhalite	$Al^o\ Mg^o\ B^t\ [O_4]^h$

Description

The olivine structure is based on a nearly perfect hexagonal closest packing of oxygen atoms with Mg and Fe atoms occupying 1/2 of octahedral voids forming a zig-zag pattern, and Si atoms occupying 1/8 of the tetrahedral voids forming also a zig-zag pattern.

References

Kostov (1968) 291-294.
Povarennykh (1972) 283, 384,385, 465.
Wyckoff (1965) Vol. 3, 91-93.
Wyckoff (1968) Vol. 4, 159-161.
Zoltai + Stout (1984) 359.
Ingerson (1955) 351.
Zoltai (1975) II-3.
Zoltai (1977) 6-40.

Figures

Fig. 1. Packing drawing of the olivine structure (a) and corresponding unit-cell projection on (010) plane (b) (after Wyckoff, 1965, Vol. 3).

Fig. 2. Polyhedral representation of the olivine structure (adapted from Zoltai, 1977).

Fig. 3. Ball and spoke model of the olivine structure (adapted from Povarennykh, 1972).

Fig. 4. Mixed octahedral-tetrahedral layer in the (010) plane of olivine (adapted from Zoltai, 1975).

Fig. 5. Condensed model of the olivine structure. The large open circles represent oxygen atoms. The small black circles (Mg, Fe) atoms in octahedral voids, and the smaller black circles Si atoms in tetrahedral voids. Notice the zig-zag pattern of both occupied octahedral and tetrahedral voids.

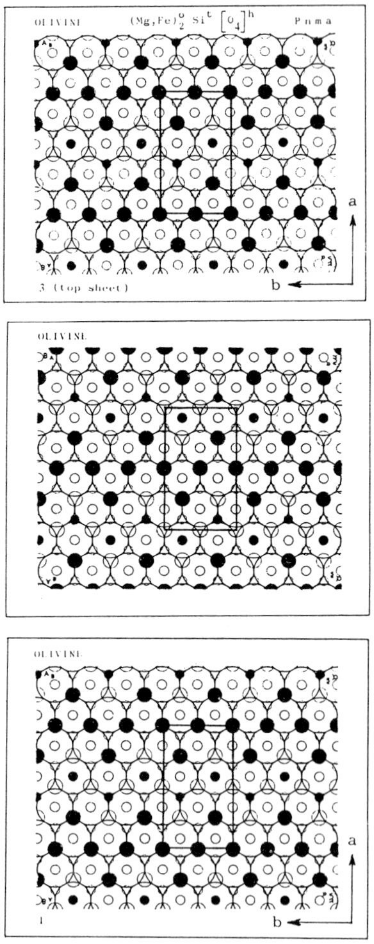

Fig. 5

NORBERGITE $Mg^o_3\ Si^t\left[O_4\ (OH,F)_2\right]^h$

P b n m

a=4.71Å
b=10.2Å
c=8.74Å
Z=4

Mg_I	(8d)	x=0.00	y=-0.36	z=0.08
Mg_{II}	(4c)	x=0.01	y=-0.11	z=1/4
Si	(4c)	x=0.39	y=-0.28	z=1/4
O_I	(8d)	x=0.25	y=0.29	z=0.08
O_{II}	(4c)	x=0.25	y=0.04	z=1/4
O_{III}	(4c)	x=0.75	y=0.21	z=1/4
...				

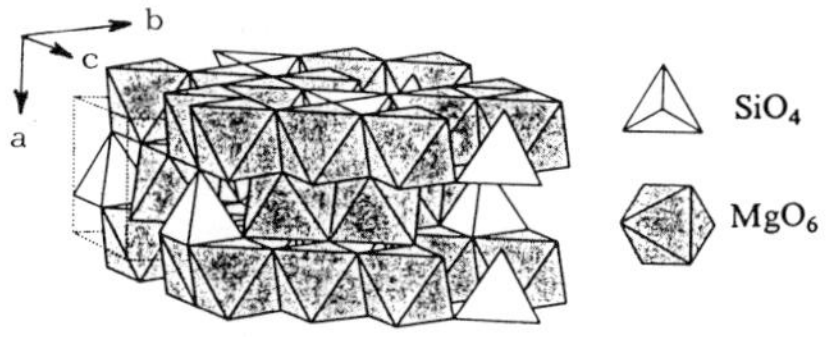

Fig. 1

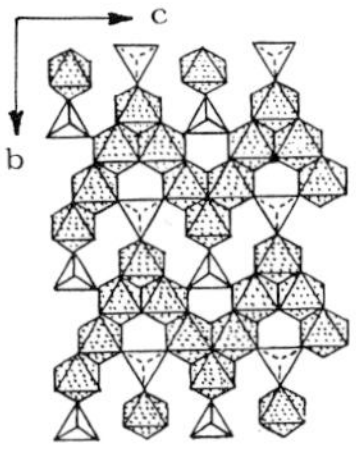

Fig. 2

Properties

Habit	Cleav.	Fract.	Twin.	Hardn.	Dens.	Colour	Transp.
tabular		subcon-choidal	(001)	6.5	3.16	white, yellow	transparent to trans-lucent

Refr. index/Reflect.	Birefr.	Luster	Streak	Melt.p.	CPI
n_α = 1.561 n_β = 1.570 n_γ = 1.587	(+) 2V= 44° - 50°	vitreous	white		(SPI) 64

Figures

Fig. 1. Polyhedral drawing of the norbergite structure (after Zoltai + Stout. 1984).

Fig. 2. Mixed octahedral-tetrahedral layer in the (100) plane of norbergite (after Zoltai, 1975).

Fig. 3. Condensed model of the norbergite structure. Large open circles represent oxygen atoms, and large lined circles (OH) or F atoms. The black circles correspond to Mg atoms in octahedral voids, and the smaller black circles to Si atoms in tetrahedral voids.

Description

The norbergite structure is based on an hexagonal closest packing of O, F and (OH), with Mg atoms in octahedral voids forming a zig-zag pattern, and Si atoms in tetrahedral voids. It can be considered derived from the olivine structure by adding strips of $Mg(OH,F)_2$ parallel to the (001) plane that is, Mg_2SiO_4 (olivine) + $Mg(OH,F)_2$ (brucite like structure).

References

Kostov (1968) 294.
Povarennykh (1972) 391,392.
Wyckoff (1968) Vol. 4, 175-177.
Zoltai + Stout (1984) 360,361.
Zoltai (1975) I-15.

Crystallographic data (continued)

(OH,F)	(8d)	x=0.75	y=-0.04	z=0.08

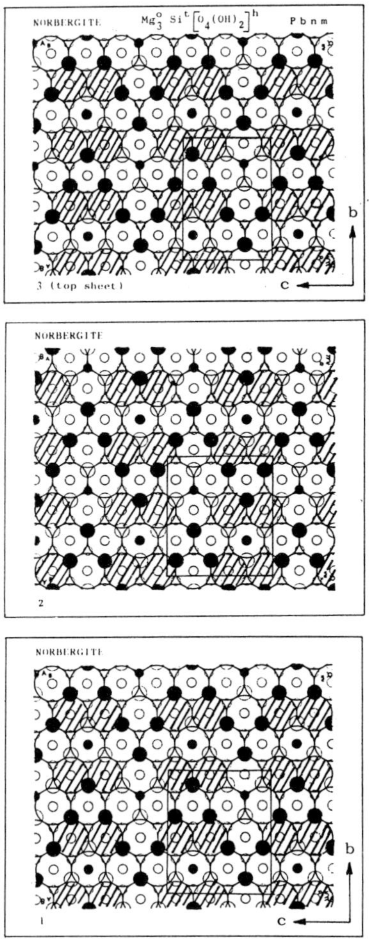

Fig. 3

CHONDRODITE $Mg_5^o\ Si_2^t\left[O_8(OH,F)_2\right]^h$

P $2_1/a$

a=10.29 Å
b=4.742 Å
c=7.87 Å
β=109°2'
Z=2

Mg_I	(4c)	x=0.17	y=0	z=0.30
Mg_{II}	(2d)	x=0	y=1/2	z=1/2
Mg_{III}	(4e)	x=-0.10	y=0.50	z=0.10
Si	(4e)	x=0.15	y=0.10	z=-0.30
O_I	(4e)	x=0.03	y=0.75	z=0.30
O_{II}	(4e)	x=0.23	y=0.75	z=0.10

...

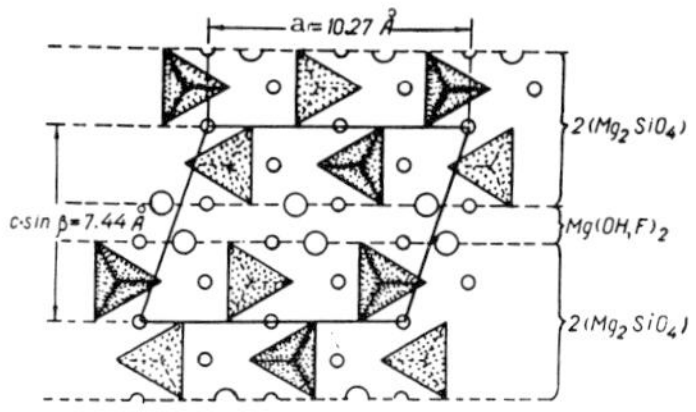

Fig. 1

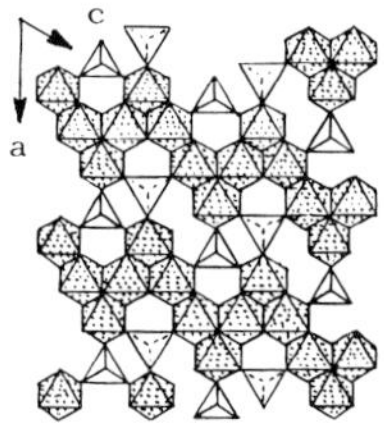

Fig. 2

Properties

Habit	Cleav.	Fract.	Twin.	Hardn.	Dens.	Colour	Transp.
equant	poor (010)	subconchoidal	(001)	6.5	3.16-3.26	white, yellow	translucent

Refr. index/Reflect.	Birefr.	Luster	Streak	Melt.p.	CPI
n_α = 1.60 n_β = 1.62 n_γ = 1.63	(+) 2V = 60° - 90°	vitreous	white		(SPI) 62

Population

Alleghanyite $Mn_5^oSi_2^t\left[O_8(OH)_2\right]^h$

Figures

Fig. 1. Projection of the chondrodite structure along the b axis emphasizing its relation with olivine and brucite (after Povarennykh, 1972).

Fig. 2. Mixed octahedral-tetrahedral layer in the (010) plane of chondrodite (after Zoltai, 1975).

Fig. 3. Condensed model of the chondrodite structure. The large open circles represent oxygen atoms, and large lined circles OH or F atoms. The black circles correspond to Mg atoms in octahedral voids, and the smaller black circles to Si atoms in tetrahedral voids.

Description

The chondrodite structure is based on an hexagonal closest packing of O, OH and F, with Mg atoms in octahedral voids forming a zig-zag pattern, and Si in tetrahedral voids. It can be considered derived from the olivine structure by adding strips of brucite, $Mg(OH,F)_2$, in a proportion of 2 olivine plus 1 brucite, $2Mg_2SiO_4 + Mg(OH,F)_2$.

References

Kostov (1968) 294.
Povarennykh (1972) 391,392.
Wyckoff (1968) Vol. 4, 178-180.
Zoltai + Stout (1984) 361.
Zoltai (1975) 1-I5.

Crystallographic data (continued)

O_{III}	(4e)	x=0.17	y=0.25	z=0.50
O_{IV}	(4e)	x=0.13	y=0.25	z=0.30
(OH,F)	(4e)	x=0.07	y=0.25	z=0.10

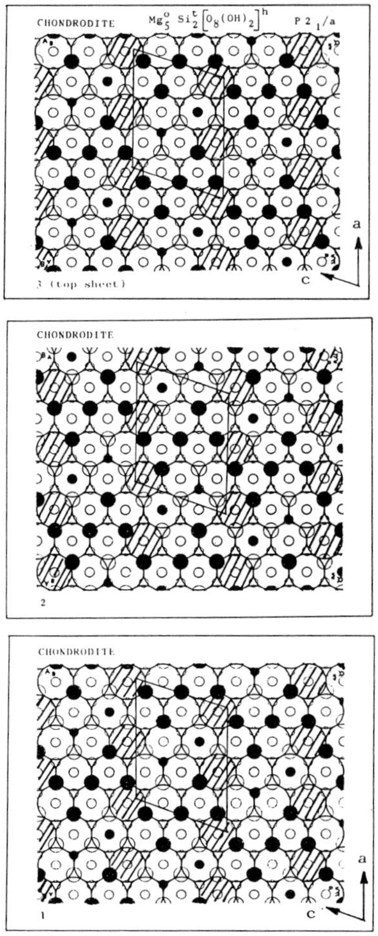

Fig. 3

HUMITE

Pbnm

$Mg^{o}_{7}\ Si^{t}_{3}\ [O_{12}(OH,F)_{2}]^{h}$

a=4.748Å
b=10.25Å
c=20.90Å
Z=4

Mg_I	(4c)	x=0.50 y=0.15 z=1/4
Mg_{II}	(8d)	x=0.00 y=0.09 z=0.11
Mg_{III}	(8d)	x= 0.50 y=-0.12 z= 0.04
Mg_{IV}	(8d)	x=0.00 y=0.37 z=0.18
Si_I	(4c)	x=0.10 y=-0.02 z=1/4
Si_{II}	(8d)	x=0.60 y=0.27 z=0.11
O_I	(8d)	x=0.75 y=0.21 z=0.18
O_{II}	(8d)	x=0.25 y=0.04 z=0.18
O_{III}	(8d)	x=0.75 y=0.21 z=0.04
...		

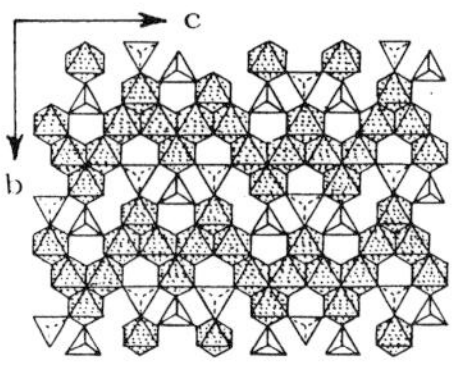

Fig. 1

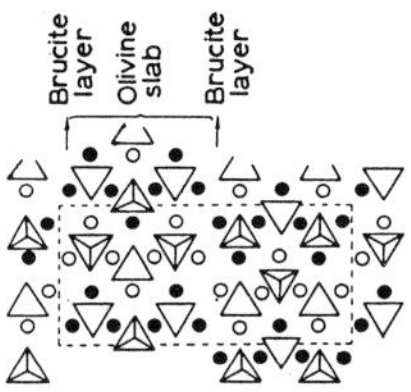

Fig. 2

Properties

Habit	Cleav.	Fract.	Twin.	Hardn.	Dens.	Colour	Transp.
varied	distinct (001)	subconchoidal to uneven		6	3.20-3.32	white, yellowish	transparent to translucent

Refr. index/Reflect.	Birefr.	Luster	Streak	Melt.p.	CPI
n_α = 1.607-1.643 n_β = 1.619-1.653 n_γ = 1.639-1.675	(+) 2V= 65°-84°	vitreous			

Population

Manganhumite $Mn^{o}_{7}\ Si^{t}_{3}\ [O_{12}(OH)_{2}]^{h}$

Figures

Fig. 1. Mixed octahedral-tetrahedral layer in the (100) plane of humite (after Zoltai, 1975).

Fig. 2. Structural scheme of humite emphasizing its relation with olivine and brucite (after Kostov, 1968).

Fig. 3. Condensed model of the humite structure. Large open circles represent oxygen atoms, and large lined circles OH or F atoms. The black circles correspond to Mg atoms in octahedral voids, and the smaller black circles to Si atoms in tetradral voids.

Description

The humite structure is based on an hexagonal closest packing of O, (OH) and F, with Mg atoms in octahedral voids forming a zig-zag pattern, and Si atoms in tetrahedral voids. It can be considered derived from olivine and brucite.
3 Mg_2SiO_4 (olivine) + $Mg(OH,F)_2$ (brucite like structure).

References

Kostov (1968) 292-294.
Povarennykh (1972) 391,392.
Wyckoff (1968) Vol. 4, 175-178.
Zoltai (1975) I-14.
Roberts et al. (1974) 284.

Crystallographic data (continued)

O_{IV}	(8d)	x=0.25 y=0.29 z=0.11
O_V	(4c)	x=0.25 y=0.29 z=1/4
O_{VI}	(4c)	x=0.75 y=-0.04 z=1/4
O_{VII}	(8d)	x=0.75 y=-0.04 z=0.11
(OH,F)	(8d)	x=0.25 y=0.04 z=0.04

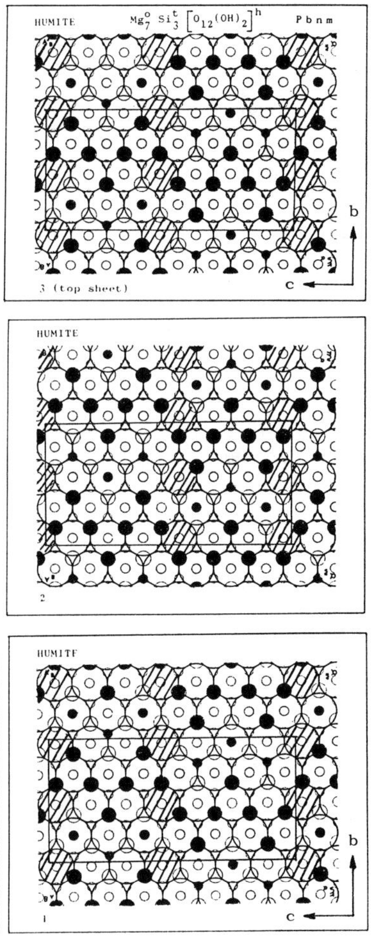

Fig. 3

CLINOHUMITE

$P\,2_1/a$

$Mg^{o}_{9}\,Si^{t}_{4}\,[O_{16}\,(OH,F)_2]^{h}$

a=10.29 Å	Mg_I (4e) x=0.14 y=0.00 z=0.17	Mg_{IV}(2d) x=0 y=1/2 z=1/2	Si_{II}(4e) x=0.07 y=0.10 z=0.39
b=4.754 Å			
c=13.71 Å	Mg_{II} (4e) x=-0.056 y=0.50 z=0.28	Mg_V (4e) x=0.25 y=0.50 z=0.39	O_I (4e) x=0.00 y=0.75 z=0.17
β=100°50'			
Z=2	Mg_{III}(4e) x=-0.11 y=0.50 z=0.056	Si_I (4e) x=0.18 y=0.10 z=-0.17	O_{II} (4e) x=0.22 y=0.75 z=0.056 ...

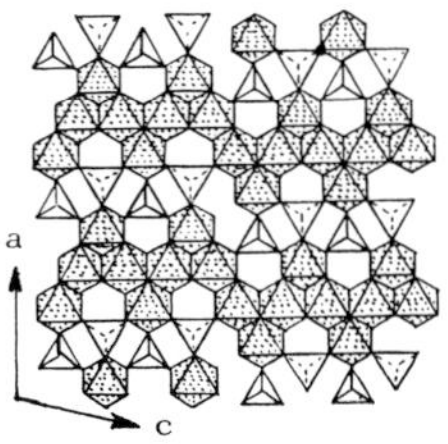

Fig. 1

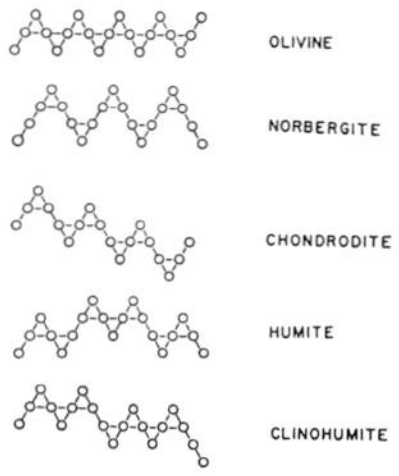

Fig. 2

Properties

Habit	Cleav.	Fract.	Twin.	Hardn.	Dens.	Colour	Transp.
equant	poor (010)	subconchoidal	(001)	6	3.21-3.35	white, yellow	translucent

Refr. index/Reflect.	Birefr.	Luster	Streak	Melt.p.	CPI
n_α = 1.63 n_β = 1.64 n_γ = 1.59	(+) 2V = 73° - 76°	vitreous	white		(SPI) 64

Population

Sonolite $Mn^{o}_{9}\,Si^{t}_{4}\,[O_{16}(OH)_2]^{h}$

Description

The clinohumite structure is based on an hexagonal closest packing of O, (OH) and F, with Mg atoms in octahedral voids forming a zig-zag pattern, and Si atoms in tetrahedral voids. It can be imagined derived from olivine and brucite $4\,Mg_2\,Si\,O_4$ (olivine) + $Mg\,(OH,F)_2$ (brucite like structure).

Figures

Fig. 1. Mixed octahedral-tetrahedral layer in the (010) plane of clinohumite (after Zoltai, 1975).

Fig. 2. Schematic representation of the octahedral zig-zag pattern of the structures of olivine, norbergite, chondrodite, humite and clinohumite (after Ribbe et al., 1968).

Fig. 3. Condensed model of the clinohumite structure. Large open circles represent oxygen atoms, and large lined circles OH and F atoms. The black circles correspond to Mg atoms in octahedral voids, and the smaller black circles to Si atoms in tetrahedral voids.

References

Kostov (1968) 294.
Wyckoff (1968) Vol. 4, 178-181.
Zoltai + Stout (1984) 361.
Zoltai (1975) I-13.
Ribbe et al. (1968) 970.

Crystallographic data (continued)

O_{III}(4e) x=0.11 y=0.25 z=0.28	O_{VI} (4e) x=0.39 y=0.25 z=0.39
O_{IV} (4e) x=0.28 y=0.75 z=0.28	O_{VII} (4e) x=0.056 y=0.75 z=0.39
O_V (4e) x=0.33 y=0.25 z=0.17	O_{VIII}(4e) x=0.17 y=0.25 z=0.50
	(OH,F)(4e) x=0.056 y=0.25 z=0.056

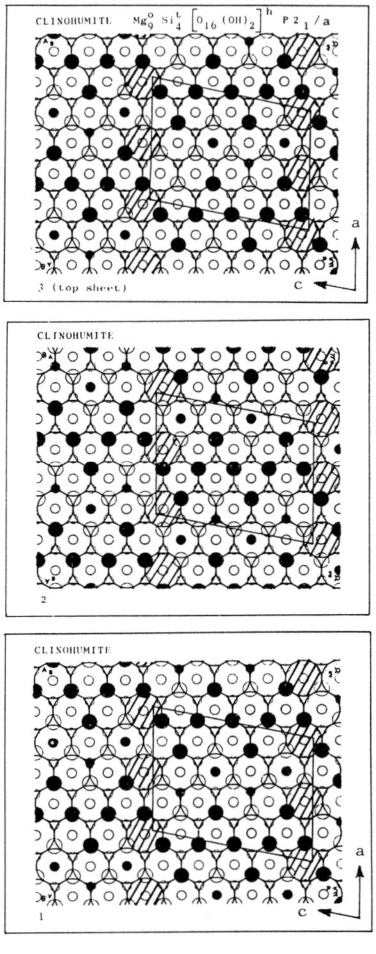

Fig. 3

KYANITE (Cyanite) (Disthene)	$Al_2^o\ Si^t\ [O_5]\ c$ P 1	a = 7.1192 Å b = 7.8473 Å c = 5.5724 Å α = 89.977° β = 101.121° γ = 106.006° Z = 4	Al_I (2i) x=0.17504 y=-0.20384 z=0.04243 Al_{II} (2i) x=0.20222 y=-0.19893 z=-0.45032 Al_{III} (2i) x=0.39951 y=0.11370 z=-0.14061 ...

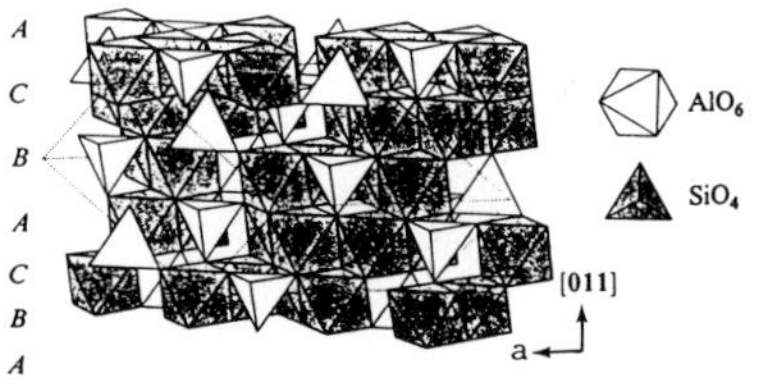

Fig. 1

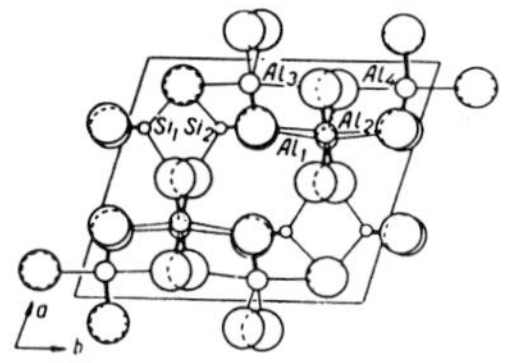

Fig. 2

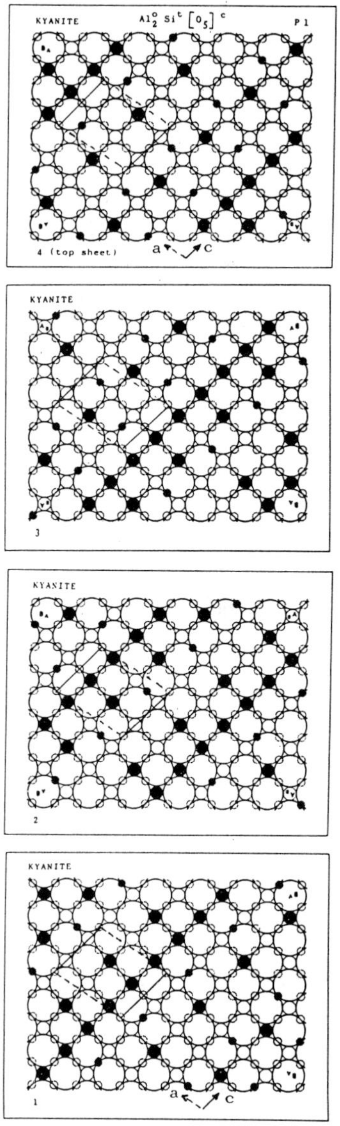

Fig. 3

Properties

Habit	Cleav.	Fract.	Twin.	Hardn.	Dens.	Colour	Transp.
blady, tabular	perfect (100) good (010)	uneven	(100)	5 - 7	3.60	blue, white	transparent to translucent

Refr. index/Reflect.	Birefr.	Luster	Streak	Melt.p.	CPI
nα = 1.712 nβ = 1.720 nγ = 1.728	(-) 2V= 82° - 83°	vitreous, pearly	white		(SPI) 68

Figures

Fig. 1. Polyhedral description of the kyanite structure (after Zoltai + Stout, 1984).

Fig. 2. Projection of the kyanite structure on the (001) plane (after Povarennykh, 1972).

Fig. 3. Condensed model of kyanite. Large open circles represent O atoms, and the small black circles Al in octahedral voids. The still smaller black circles correspond to Si atoms in tetrahedral voids (after Figueiredo, 1976).

Description

The kyanite structure is based on a cubic closest packing of the O atoms, with Al in octahedral voids, and Si occupying tetrahedral voids.

References

Kostov (1968) 283, 287.
Wyckoff (1968) Vol. 4, 180-183.
Povarennykh (1972) 386.
Zoltai + Stout (1984) 363,364.
Figueiredo (1976).

Crystallographic data (continued)

Atom	Position	x	y	z	Atom	Position	x	y	z
Al_{IV}	(2i)	0.38773	-0.41754	0.33561	O_V	(2i)	0.39323	0.34866	-0.16572
Si_I	(2i)	0.20806	0.16815	0.31076	O_{VI}	(2i)	0.38788	0.35309	0.37156
Si_{II}	(2i)	0.20327	0.43503	-0.20644	O_{VII}	(2i)	0.20915	-0.43554	-0.43545
O_I	(2i)	-0.00220	0.26859	-0.25456	O_{VIII}	(2i)	0.22172	0.05444	0.07146
O_{II}	(2i)	-0.00080	0.22481	0.25581	O_{IX}	(2i)	0.37763	-0.13082	-0.14053
O_{III}	(2i)	0.21569	-0.44677	0.03635	O_X	(2i)	0.37830	-0.18550	0.31866
O_{IV}	(2i)	0.22259	0.04515	-0.45589					

TOPAZ

P b n m

$Al_2^{o}\ Si^{t}\left[O_4\,(OH,F)_2\right]^{ch}$

a=4.6499 Å

b=8.7968 Å

c=8.3909 Å

Z=4

Atom	Position	x	y	z
Al	(8d)	0.90354	0.13102	0.08236
Si	(4c)	0.39784	0.94049	1/4
O_I	(4c)	0.79608	0.53208	1/4
O_{II}	(4c)	0.45691	0.75625	1/4
O_{III}	(8d)	0.78931	0.01058	0.90788
(OH,F)	(8d)	0.90176	0.75258	0.05720

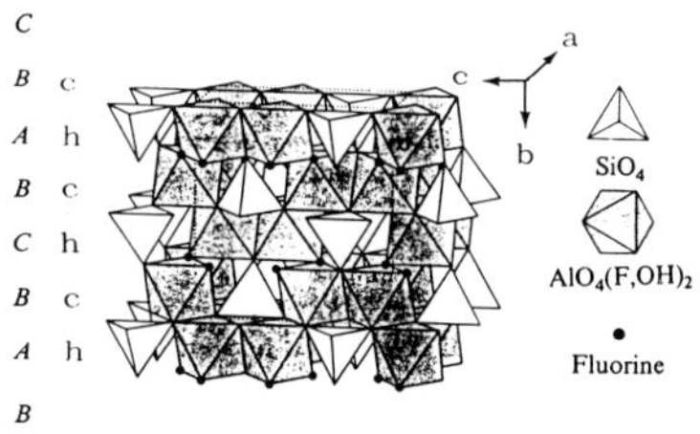

Fig. 1

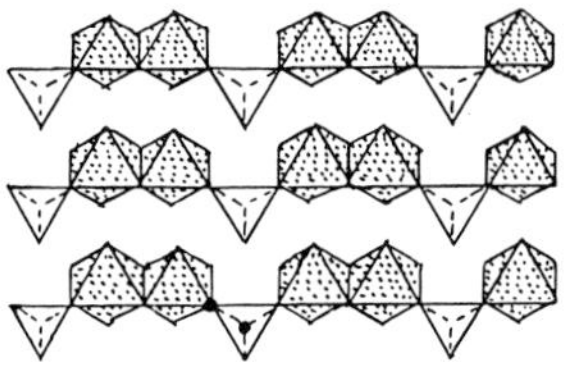

Fig. 2

Properties

Habit	Cleav.	Fract.	Twin.	Hardn.	Dens.	Colour	Transp.
prismatic	perfect (001)	subcon-choidal		8	3.5-3.6	colourless, variable	transparent to trans-lucent

Refr. index/Reflect.	Birefr.	Luster	Streak	Melt.p.	CPI
n_α = 1.61 n_β = 1.61 n_γ = 1.62	(+) 2V = 48-65°	vitreous	white		(SPI) 66

Figures

Fig. 1. Polyhedral representation of the topaz structure normal to the closest packed layers (after Zoltai + Stout, 1984).

Fig. 2. Mixed octahedral-tetrahedral distribution of the topaz structure parallel to the (010) plane (after Zoltai, 1975).

Fig. 3. Condensed model of the topaz structure. Large open circles represent oxygen atoms, and large lined circles OH or F atoms. Small black circles correspond to Al atoms in octahedral voids, and smaller black circles to Si atoms in tetrahedral voids.

Description

The topaz structure is based on a ch closest packing of OH and the oxygen atoms. The Al atoms occupy octahedral voids, and Si atoms tetrahedral voids.

References

Kostov (1968) 288.

Wyckoff (1968) Vol. 4, 174,175.

Zoltai + Stout (1984) 368,369.

Zoltai (1975) I-13.

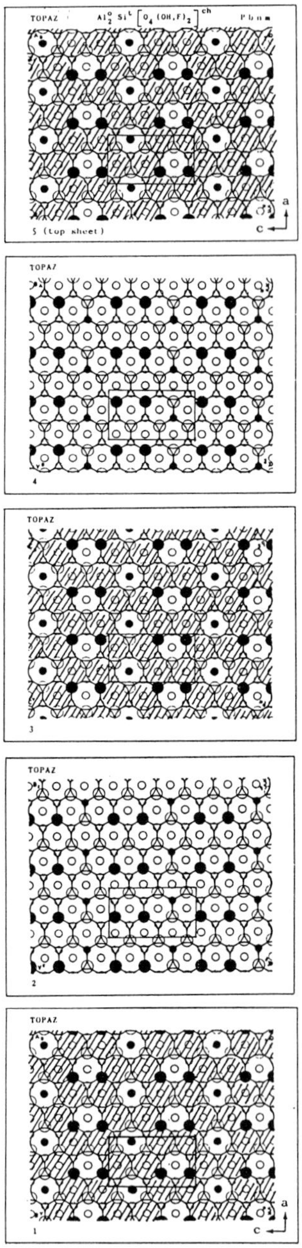

Fig. 3

ANTLERITE

$Cu_3^{o}\ S^{t}\ [O_4(OH)_4]^{c}$

P n m a

a = 8.226 Å	Cu_I (4c) x=0.0052 y=1/4 z=0.0015	O_I (4c) x=0.262 y=1/4 z=0.283
b = 6.046 Å	Cu_{II} (8d) x=0.2897 y=0.0032 z=0.1259	O_{II} (4c) x=0.195 y=1/4 z=0.476
c = 11.978 Å	S (4c) x=0.1304 y=1/4 z=0.3641	O_{III} (8d) x=0.033 y=0.047 z=0.348
Z = 4		...

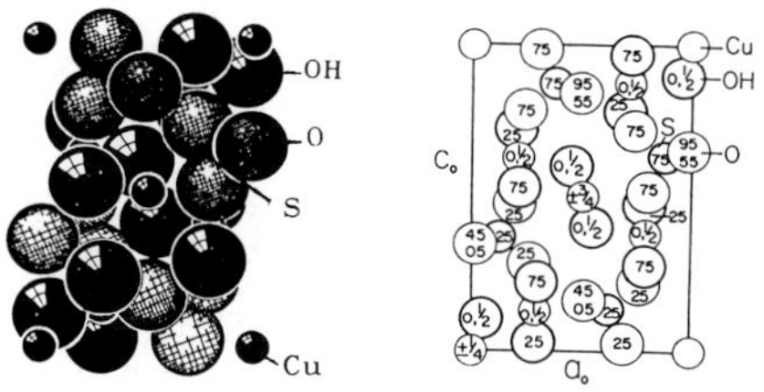

(a) Fig. 1 (b)

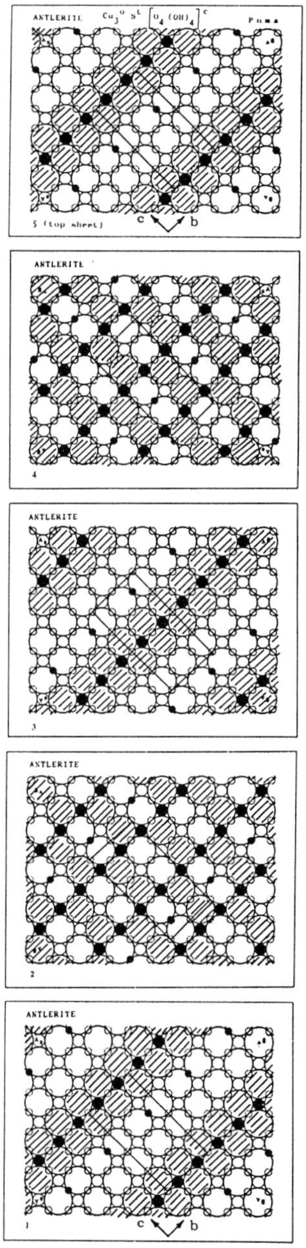

Fig. 2

Properties

Habit	Cleav.	Fract.	Twin.	Hardn.	Dens.	Colour	Transp.
prismatic, tabular	perfect (001)	uneven		3.5-4	3.9	white grey	transparent to trans-lucent

Refr. index/Reflect.	Birefr.	Luster	Streak	Melt.p.	CPI
n_α = 1.726 n_β = 1.738 n_γ = 1.789	(+) 2V = 53°	vitreous	green grey		(SPI) 54

Figures

Fig. 1. (a) Packing drawing of the antlerite structure, and (b) unit cell content projected along the b axis (after Wyckoff, 1965, Vol. 3).

Fig. 2. Condensed model of the antlerite structure. The large open circles represent oxygen atoms and the large lined circles correspond to OH. The small black circles represent Cu atoms in octahedral voids forming a row pattern, and the smaller black circles correspond to S atoms in tetrahedral voids(after Figueiredo, 1976).

Description

The antlerite structure is based on a cubic closest packing of OH and O atoms. The Cu atoms occupy 3/8 of the octahedral voids, and the S atoms 1/16 of the tetrahedral voids.

References

Kostov (1968) 512-513.
Wyckoff (1965) Vol. 3, 205,206.
Zoltai + Stout (1984) 441.
Figueiredo (1976).

Crystallographic data (continued)

$(OH)_I$ (4c) x=0.285 y=1/4 z=0.026

$(OH)_{II}$ (4c) x=0.701 y=1/4 z=0.779

$(OH)_{III}$ (8d) x=0.046 y=0.503 z=0.102

STAUROLITE

C 2/ m

$Al_9^o Fe_2^t Si_4^t [O_{23}(OH)]^c$

a = 7.82 Å	Al_I (2a)	Si (8j) x = 0.125, y = 0.167, z = 0.270
b = 16.52Å	Al_{II} (4h) u = 0.322	O_I (4i) x = 0.222, y = 0, z = 0.980
c = 5.63 Å	Al_{III}(4g) u = 0.322	
β = 90°		
Z = 2	Al_{IV} (8j) x = 0.250, y = 0.408, z = 0.250	O_{II}(4i) x = 0.222, y = 0, z = 0.520
Fe(4i) u=0.389, v=0.250		...

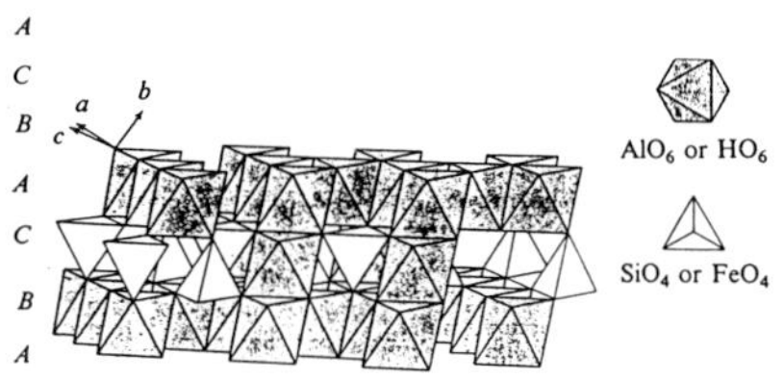

Fig. 1

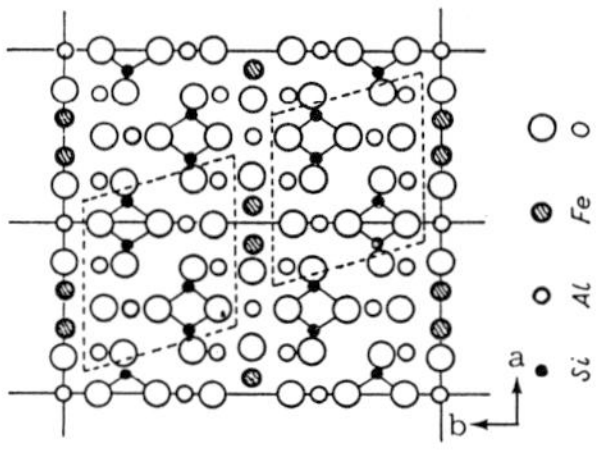

Fig. 2

Properties

Habit	Cleav.	Fract.	Twin.	Hardn.	Dens.	Colour	Transp.
prismatic, tabular	perfect (011)	subcon-choidal	(031)	7-7.5	3.75	black, brown	tanslucent

Refr. index/Reflect.	Birefr.	Luster	Streak	Melt.p.	CPI
n_α = 1.740	(+)	resinous,	white,		(SPI)
n_β = 1.744		vitreous	grey		67
n_γ = 1.753	2V = 80°-88°				

Figures

Fig. 1. Polyhedral drawing of the staurolite structure (after Zoltai + Stout, 1984).

Fig. 2. Relation between the structure of staurolite and that of kyanite.The unit cell of kyanite is marked with dashed lines (after Wells, 1962).

Fig. 3. Condensed model of the staurolite structure. Large open circles represent OH and O atoms; small black circles correspond to Al atoms in octadral voids, and the smaller black circles to Si atoms in tetrahedral voids. The small crossed circles represent Fe atoms in tetrahedral voids.

Description

The staurolite structure is based on a cubic closest packing of OH and O atoms, with Al atoms in octahedral voids, and Fe an Si atoms in tetrahedral voids. H atoms are assumed to be located at octahedral voids.

References

Kostov (1968) 289,290.

Wyckoff (1968) Vol. 4, 184-186.

Zoltai + Stout (1984) 365,366.

Wells (1962) 796.

Crystallographic data (continued)

O_{III}(8j) x = 0.237, y = 0.158, z = 0.020	O_{VI} (8j) x = 0.000, y = 0.250, z = 0.230
O_{IV} (8j) x = 0.763, y = 0.158, z = 0.520	O_{VII}(8j) x = 0.000, y = 0.394, z = 0.230
O_V (8j) x = 0.000, y = 0.094, z = 0.270	H (2c)

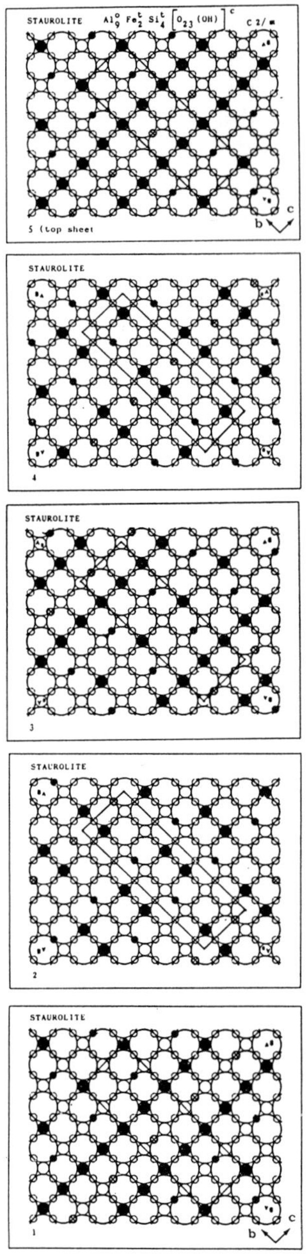

Fig. 3

PENTLANDITE

F m 3 m

$(Ni,Fe)^{o}\ (Ni,Fe)^{t}_{8}\ [S_8]^{c}$

a = 10.044 Å
Z = 4

$(Ni,Fe)^{o}$ (4b)
$(Ni,Fe)^{t}$ (32f) x = 0.1261
S_I (8c)
S_{II} (24e) x = 0.2629

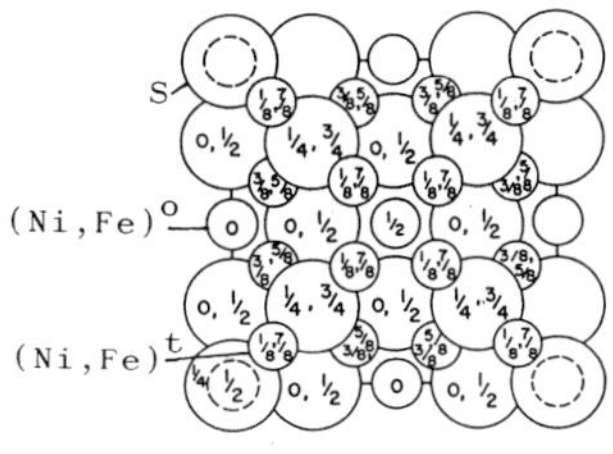

Fig. 1

Fig. 2

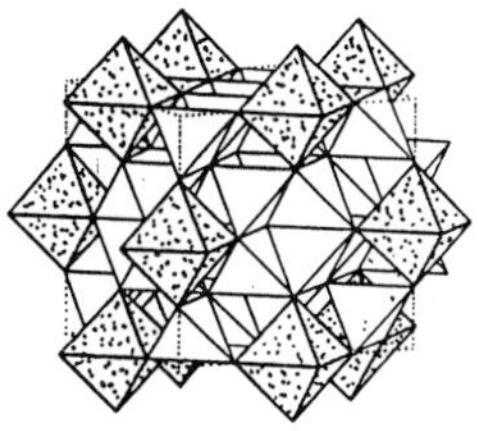

Fig. 3

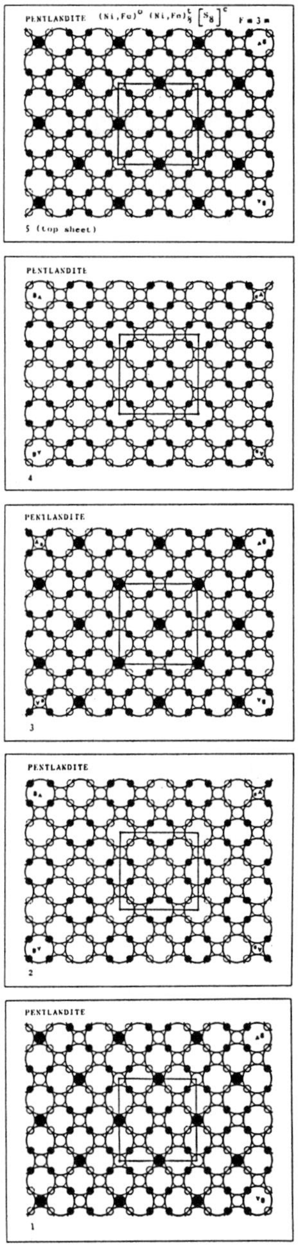

Fig. 4

Properties

Habit	Cleav.	Fract.	Twin.	Hardn.	Dens.	Colour	Transp.
massive		uneven		3.5-4	5.0	light bronze	opaque

Refr. index/Reflect.	Birefr.	Luster	Streak	Melt.p.	CPI
51%		metallic	light bronze	878°C	

Figures

Fig. 1. Projection of the pentlandite structure on a cubic face (after Wyckoff, 1964, Vol. 2).

Fig. 2. Polyhedral drawing of the pentlandite structure (after Povarennykh, 1972).

Fig. 3. Another polyhedral drawing of the pentlandite structure (after Zoltai, 1977).

Fig. 4. Condensed model of the pentlandite structure. The large open circles represent the S atoms. The small black circles (Ni, Fe) in octahedral voids, with a square pattern, and the smaller black circles the (Ni, Fe) in tetrahedral voids.

Description

The pentlandite structure is based on a cubic closest packing of the S atoms, with (Ni,Fe) occupying 1/8 of the octahedral voids, forming a square pattern, and also occupying 1/2 of the tetrahedral voids. It can be imagined derived from the $Na^{o}[Cl]^{c}$ by replacing part of the Na by a cluster of 8 $(Ni,Fe)S_4$ tetrahedra.

References

Kostov (1968) 118.
Povarennykh (1972) 226.
Wyckoff (1964) Vol. 2, 222-224.
Zoltai + Stout (1984) 388.
Palache et al. (1944) Vol. 1, 242.
Zoltai (1977) 6-40.

GERSTMANNITE

Bbcm

$(Mn,Mg)^{o} Mg^{o} Zn^{t} Si^{t} \left[O_4 (OH)_2\right]^{c}$

a=8.185 Å
b=18.65 Å
c=6.256 Å
Z=8

(Mn,Mg)(8f)	x=0.00721, y=0.39594, z=1/2	Si(8f)	x=0.14422, y=0.34580, z=0
Mg (8c)	x=1/4, y=1/2, z=1/4	O_I(8f)	x=0.2484, y=0.4205, z=0
Zn (8e)	x=0.40033, y=1/4, z=1/4	O_{II}(8f)	x=0.2775, y=0.2793, z=0
		...	

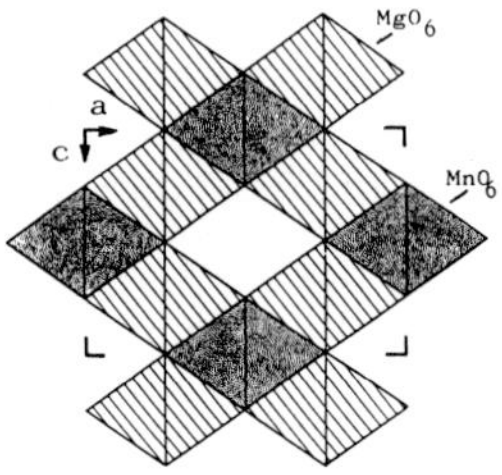

Fig. 1

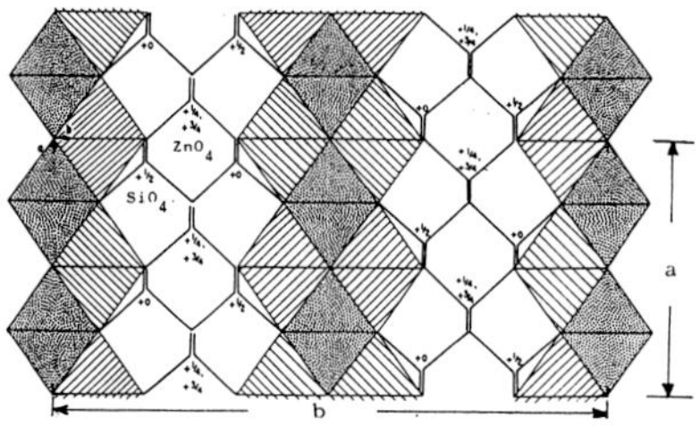

Fig. 2

Properties

Habit	Cleav.	Fract.	Twin.	Hardn.	Dens.	Colour	Transp.
prismatic	good (010)			4.5	3.68	white pale-pink	translucent to opaque

Refr. index/Reflect.	Birefr.	Luster	Streak	Melt.p.	CPI
n_α = 1.665 n_β = 1.675 n_γ = 1.678	(-) 2V ≃ 50°- 60°	vitreous	white		

Figures

Fig. 1. Idealized octahedral sheet of the gerstmannite structure down [010] (after Moore + Araki, 1977).

Fig. 2. Idealized polyhedral and spoke diagram of the gerstmannite structure down [001] (after Moore + Araki, 1977).

Fig. 3. Condensed model of the gerstmannite structure. Large open circles represent O atoms, and large lined circles OH. The medium black circles correspond to certain Mg atoms which are in octahedral voids. The medium lined circles represent Mn and the other Mg atoms, also in octahedral voids. The smaller black circles correspond to Si atoms and the smaller crossed circles to Zn, all in tetrahedral voids.

Description

The gerstmannite structure is based on a cubic closest packing of OH and O atoms. Mn and Mg occupy octahedral voids, and Zn and Si tetrahedral voids.

References

Moore + Araki (1977) 51-59.

O_{III}(16g)	x=0.294, y=0.3373, z=0.2112
$(OH)_I$(8d)	x=0, y=1/2, z=0.2822
$(OH)_{II}$(8f)	x=0.2501, y=0.4306, z=1/2

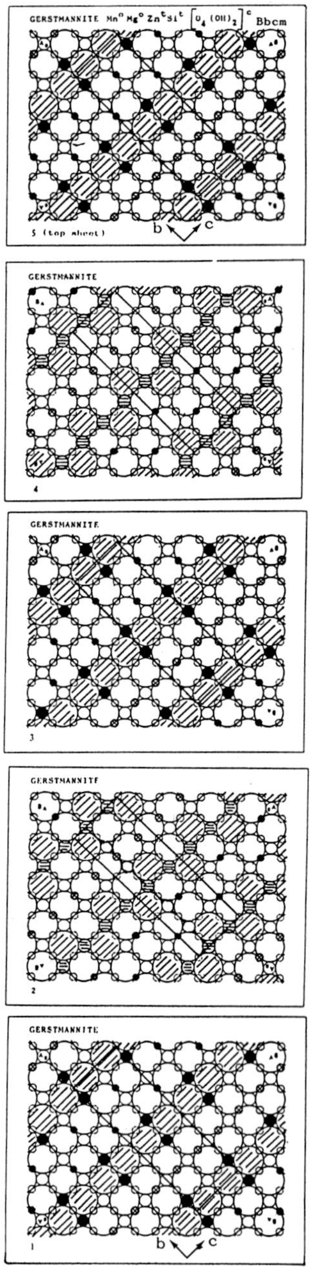

Fig. 3

SWEDENBORGITE

$Sb^{o}\ Be^{t}_{4}\ [Na\ O_7]^{ch}$

$P\ 6_3\ m\ c$

a=5.43 Å
c=8.82 Å
Z=2

Na	(2b)	x=1/3, y=2/3, z=0.75
Sb	(2b)	x=1/3, y=2/3, z=0.12
Be_I	(2a)	x=0, y=0, z=0.19
Be_{II}	(6c)	x=0.17, y=-0.17, z=0.44
O_I	(2a)	x=0, y=0, z=0.00
O_{II}	(6c)	x=0.50, y=0.50, z=0.00
O_{III}	(6c)	x=0.17, y=-0.17, z=0.25

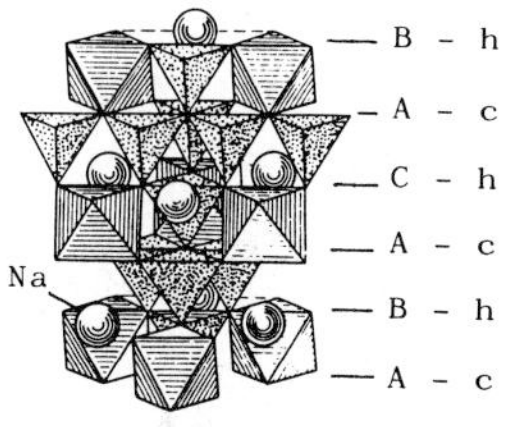

Fig. 1

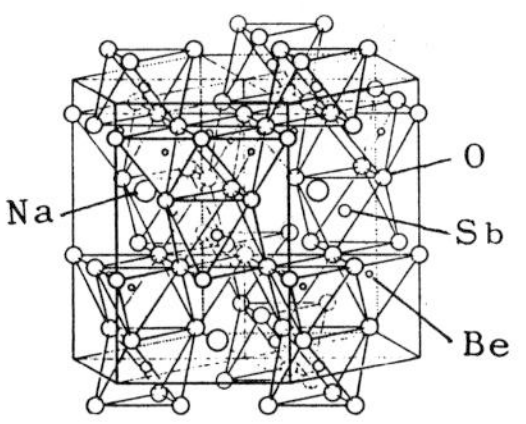

Fig. 2

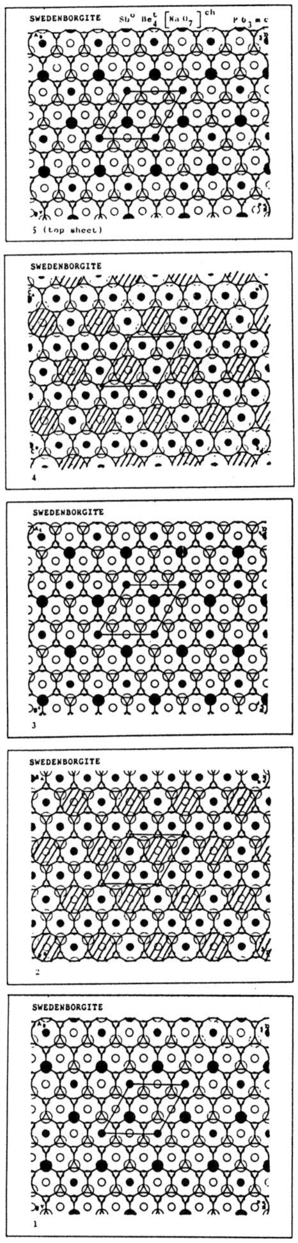

Fig. 3

Properties

Habit	Cleav.	Fract.	Twin.	Hardn.	Dens.	Colour	Transp.
short prismatic	distinct (0001)	subcon-choidal		~8	4.285	colourless to wine--yellow	transparent

Refr. index/Reflect.	Birefr.	Luster	Streak	Melt.p.	CPI
n_{ω} = 1.7724 n_{ϵ} = 1.7700	(-)	vitreous			

Figures

Fig. 1. Polyhedral description of the swedenborgite structure (after Povarennykh, 1972)

Fig. 2. Another polyhedral drawing of the swedenborgite structure (quoted in Strukturbericht, 1937, Vol. 3).

Fig. 3. Condensed model of the swedenborgite structure. The large open circles represent oxygen atoms, and the large lined circles Na atoms. The small black circles correspond to Sb atoms in octahedral voids, forming a triangular pattern, and the smaller black circles represent Be atoms in tetrahedral voids with two kinds of alternate layers, one with a triangular pattern and another with a Kagomé pattern.

Description

The swedenborgite structure is based on a <u>ch</u> closest packing of oxygens and Na atoms. The Sb atoms occupy 1/8 of the octahedral voids, and the Be atoms 1/4 of the tetrahedral voids. The structure is built by two alternate kind of layers.

References

Kostov (1968) 213.
Povarennykh (1972) 279.
Wyckoff (1968) Vol. 4, 72,73.
Strukturbericht (1937) Vol. 3, 69-71.
Palache et al. (1951) Vol. 2, 1027,1028.
Roberts et al. (1974) 594.

HODGKINSONITE

$P\,2_1/a$

$Mn^{o}\,Zn_2^{t}\,Si^{t}\,[O_4\,(OH)_2]^{h}$

a = 8.170 Å
b = 5.316 Å
c = 11.761 Å
β = 95° 15'
Z = 4

Atom		
Mn (4e)	x = 0.1063	y = 0.2420	z = 0.5482
Zn_I (4e)	x = 0.2732	y = 0.0657	z = 0.0788
Zn_{II} (4e)	x = 0.6084	y = 0.0623	z = 0.2491

...

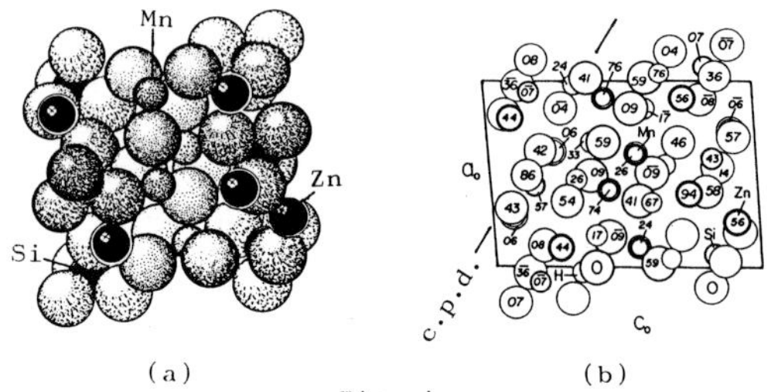

Fig. 1

Properties

Habit	Cleav.	Fract.	Twin.	Hardn.	Dens.	Colour	Transp.
pyramidal, prismatic, massive	perfect (100)			4.5-5	3.91-3.99	bright pink, reddish brown	transparent to translucent

Refr. index/Reflect.	Birefr.	Luster	Streak	Melt.p.	CPI
n_α = 1.720 n_β = 1.741 n_γ = 1.746	(-) 2V= 52°	vitreous			

Figures

Fig. 1. (a) Packing drawing of the hodgkinsonite structure, and (b) projection along the b axis (after Wyckoff, 1968, Vol. 4). The close packing direction is marked c.p.d.

Fig. 2. Condensed model of the hodgkinsonite structure. Large open circles represent oxygen atoms, and large lined circles correspond to OH. The medium black circles represent Mn atoms, in octahedral voids; the smaller black circles represent Si atoms in tetrahedral voids. The smaller dotted circles correspond to Zn atoms also in tetrahedral voids. The unit cell is inclined in relation to the closest packed layers (after Figueiredo, 1976 b).

Description

The hodgkinsonite structure is based on a hexagonal closest packing of OH and O atoms. The Mn atoms occupy octahedral voids, and Si and Zn tetrahedral voids.

References

Kostov (1968) 324.
Wyckoff (1968) Vol. 4, 204-206.
Roberts et al. (1974) 276.
Figueiredo (1976 b).

Crystallographic data (continued)

Atom	x	y	z
Si (4e)	0.0673	0.0666	0.8292
O_I (4e)	0.1067	0.0810	0.1911
O_{II} (4e)	0.1657	0.0405	0.7151
O_{III} (4e)	0.8193	0.0702	0.0660
O_{IV} (4e)	0.5384	0.1441	0.8621
O_V (4e)	0.4902	0.0892	0.3836
O_{VI} (4e)	0.8491	0.0882	0.5560
H_I (4e)	0.466	0.260	0.333
H_{II} (4e)	0.842	-0.83	0.600

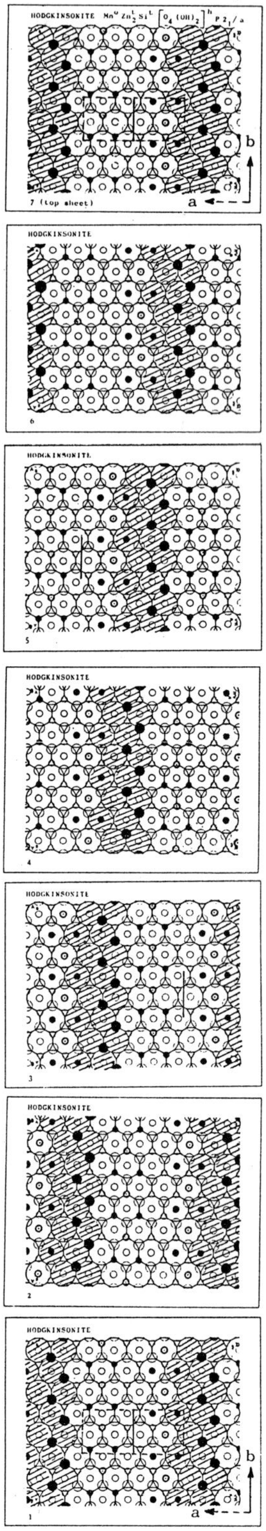

Fig. 2

IRON

$[Fe]^{Bb}$ a=2.87 Å   Fe(2a)

Z=2

I m 3 m

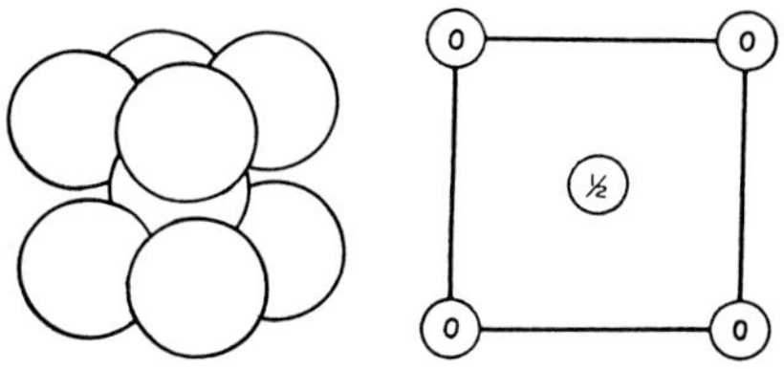

(a) Fig. 1 (b)

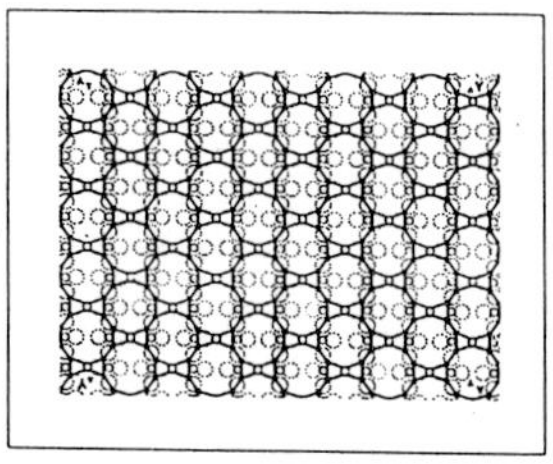

Fig. 2

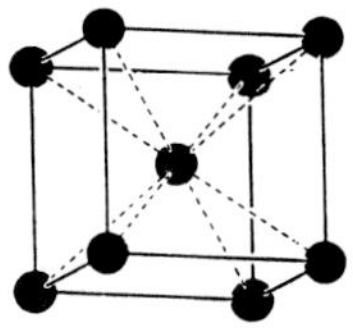

Fig. 3

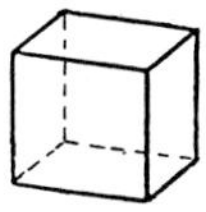

Fig. 4

Properties

Habit	Cleav.	Fract.	Twin.	Hardn.	Dens.	Colour	Transp.
irregular grains	perfect (100)	hackly	(111)	4.5	7.3-7.8	steel- -grey	opaque

Refr. index	Reflect.	Birefr.	Luster	Streak	Melt.p.	CPI
n = 2.36	56%		metallic		1535°C	(SPI) 68

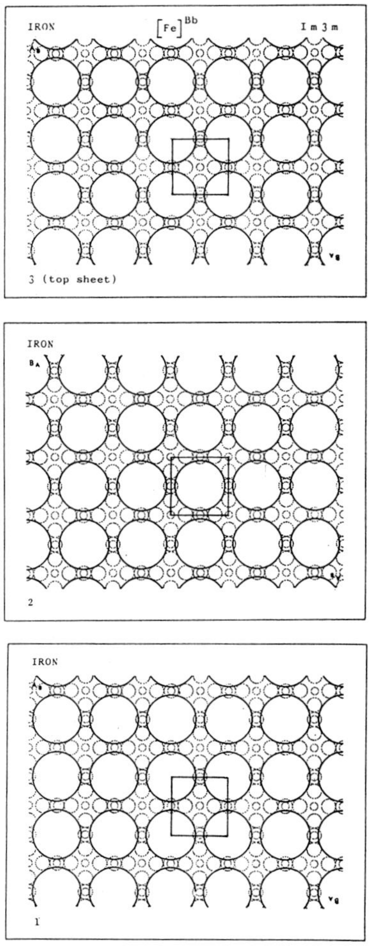

Fig. 5

Figures

Fig. 1. (a) Packing drawing of the iron structure and (b) projection on a cubic face.

Fig. 2. Condensed model sheet for the representation of body centered cubic packing, where the centres of the large circles, representing the packing atoms, form a tesselation of lozenges with 70°32' as smallest angle. This kind of packing layer is called B, and is an intermediate between the T (closest) and the Q layers. It should be stacked over other sheets on the sequence b, over valleys.

Fig. 3. Ball and spoke model of the iron structure.

Fig. 4. Cubic coordination of the packing atoms in the body centered cubic packing.

Fig. 5. Condensed model of the iron structure. The large open circles correspond to Fe atoms.

The layers are of ≪Q≫ type and their stacking is f. The packing ≪Q≫ f is equal to the packing Bb.

Description

The iron structure is formed by the body-centered cubic packing of the iron atoms.

References

Kostov (1968) 90,91.

Wyckoff (1963) Vol. 1, 15,16.

Zoltai + Stout (1984) 375.

Ingerson (1955). 350.

WAIRAUITE

$[\mathrm{Fe\,Co}]^{Bb}$

P m 3 m

a = 2.856 Å Co (1a)

Z = 1 Fe (1b)

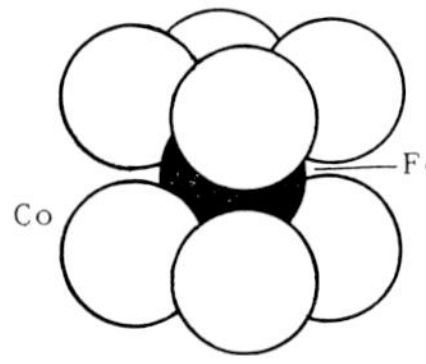

(a)

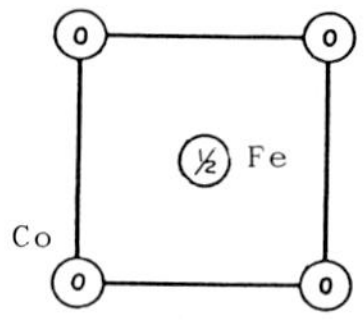

(b)

Fig. 1

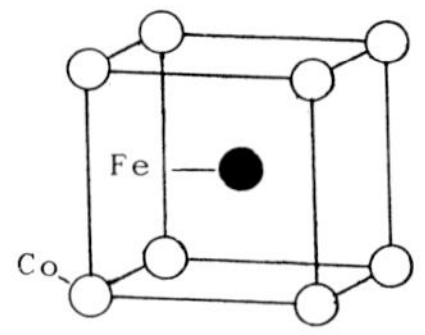

Fig. 2

Properties

Habit	Cleav.	Fract.	Twin.	Hardn.	Dens.	Colour	Transp.
cubic, octahedral				5	8.23	steel grey	opaque

Refr. index/Reflect.	Birefr.	Luster	Streak	Melt.p.	CPI
		metallic			

Figures

Fig. 1. (a) Packing representation of the wairauite structure, and (b) projection on a cube face.

Fig. 2. Ball and spoke model of the wairauite structure.

Fig. 3. Condensed model of the wairauite structure. Large open circles represent Co atoms and large lined circles Fe atoms.

Description

The wairauite structure has been considered as isotypic with the cesium chloride strcture that is $\mathrm{Fe}^{cb}[\mathrm{Co}]^{Qs}$. However, as the Co and Fe atomic radii are practically the same (R_{Fe}= 1.26 Å and R_{Co}= 1.25 Å, after the periodic table of Sargent & Co.) a more correct interpretation is to consider that wairauite is a substitution derivative of the iron structure which forms a body-centred cubic packing. Therefore its formula should be $[\mathrm{Fe\,Co}]^{Bb}$, or $[\mathrm{FeCo}]^{\ll Q \gg f}$.

References

Kostov (1968) 190.

Wyckoff (1963) Vol. 1, 15.

Roberts et al. (1974) 660.

Povarennykh (1972) 193.

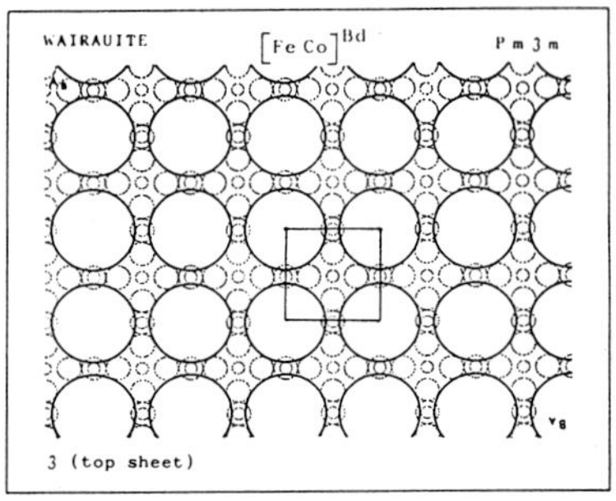

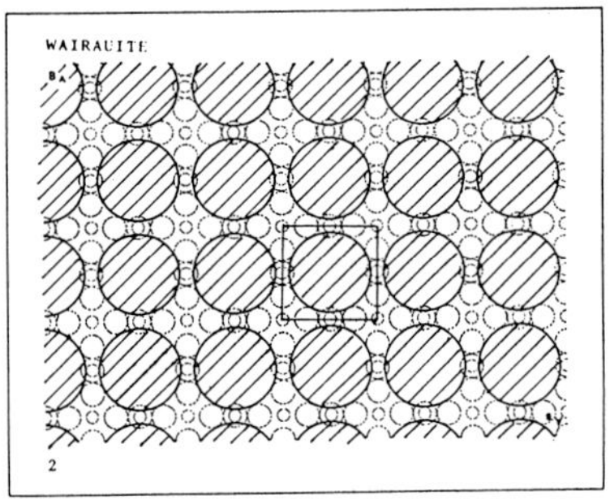

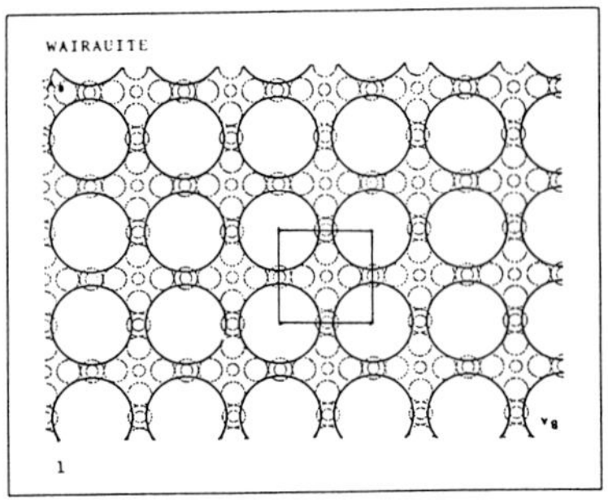

Fig. 3

CAROBBIITE

$[K\,F]^{Qs}$ a = 5.347 Å F (4a)

Z = 4 K (4b)

F m 3 m

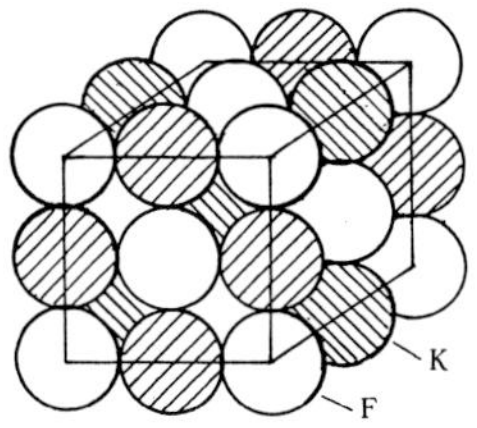

Fig. 1

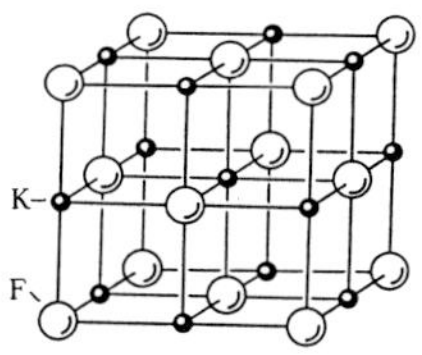

Fig. 2

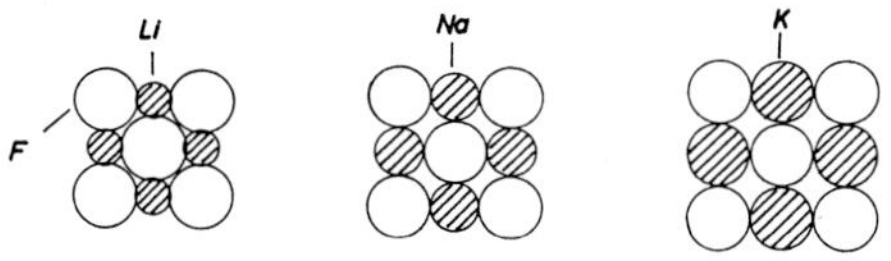

Fig. 3

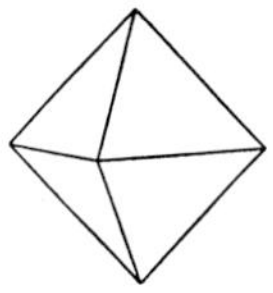

Fig. 4

Properties

Habit	Cleav.	Fract.	Twin.	Hardn.	Dens.	Colour	Transp.
cubic	(100)				2.524	colourless	

Refr. index/Reflect.	Birefr.	Luster	Streak	Melt.p.	CPI
n = 1.362					

Figures

Fig. 1. Packing drawing of the carobbiite structure.

Fig. 2. Ball and spoke model of the carobbiite structure (after Kostov, 1968).

Fig. 3. Different radius ratios attributed to the Na Cl type of structure (adapted from Krebs, 1968).

Fig. 4. Octahedral coordination of the packing atoms in the simple cubic packing (Qs) and not cuboctahedral like in cubic closest packing.

Fig. 5. Condensed model of the carobbiite structure. The large open circles represent F atoms and the large lined circles K atoms.

Description

The carobbiite structure is geometrically equal to the halite structure, $Na^{o}[Cl]^{c}$. However, as the sizes of F^- and K^+ ions are practically equal, the correct description is that carobbiite is based on a simple cubic packing. It is what has been called (Lima-de-Faria + Figueiredo, 1969) a "limiting structure" of the halite structure.

References

Kostov (1968) 155, 189.

Wyckoff (1963) Vol. 1, 85-87.

Krebs (1968) 175.

Roberts et al. (1974) 109.

Lima-de-Faria + Figueiredo (1969) 47.

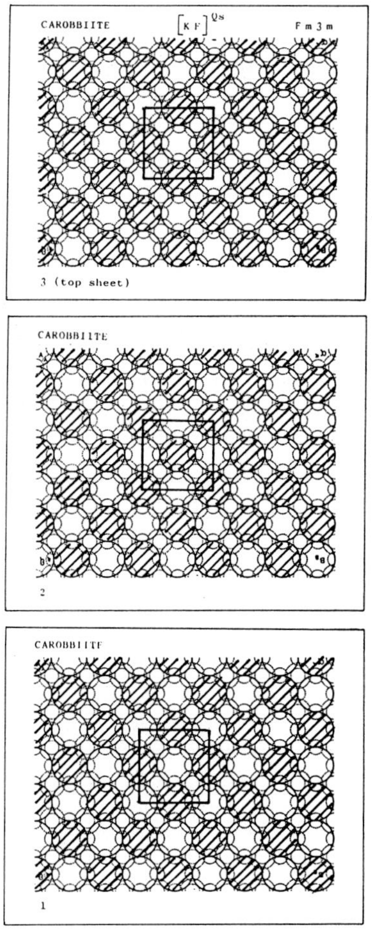

Fig. 5

FLUORITE $Ca^{cb}[F_2]^{Qs}$

$a = 5.46295 Å$ (at 28°C) Ca(4a)

Z = 4 F (8c)

F m $\bar{3}$ m

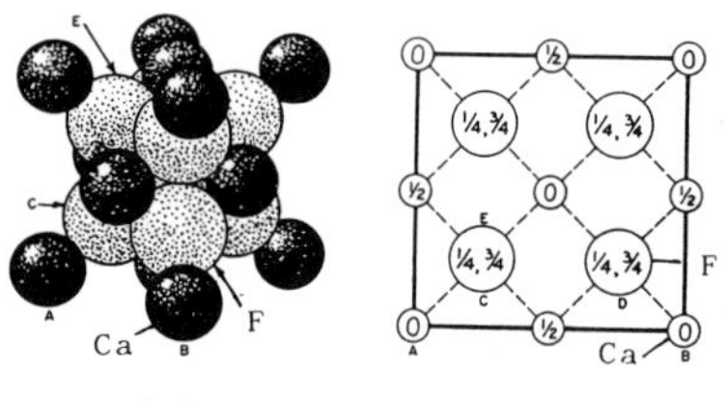

Fig. 1

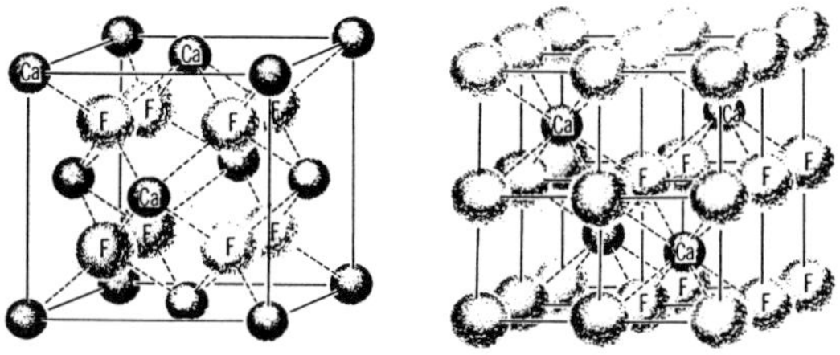

Fig. 2

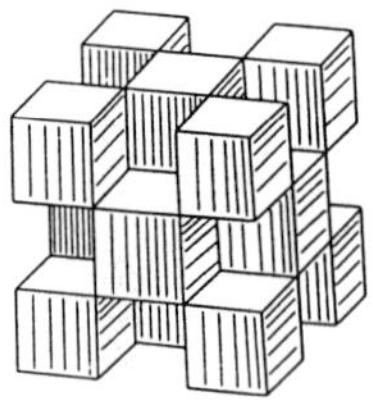

Fig. 3

Properties

Habit	Cleav.	Fract.	Twin.	Hardn.	Dens.	Colour	Transp.
cubic, octahedral	perfect (111)	conchoidal, splintery	(111)	4	3.18	colourless, variable	transparent

Refr. index/Reflect.	Birefr.	Luster	Streak	Melt.p.	CPI
n = 1.434		vitreous	white	1378°C	(SPI) 60

Population

Thorianite	$Th^{cb}[O_2]^{Qs}$
Cerianite	$Ce^{cb}[O_2]^{Qs}$
Uraninite	$U^{cb}[O_2]^{Qs}$ (approx.)

Description

The fluorite structure is based on a simple cubic packing of the F atoms, with Ca occupying 1/2 of the cubic voids forming a square pattern. Another interpretation of this structure is that Ca form a cubic closest packing with F in tetrahedral voids. This is a wrong interpretation because the F atoms are larger than the Ca atoms, and consequently Ca atoms can not form a close packing in the fluorite structure.

Figures

Fig. 1. (a) Packing drawing of the fluorite structure, and (b) unit cell content projected on a cube face (after Wyckoff, 1963, Vol. 1).

Fig. 2. Two ball and spoke models of the fluorite structure corresponding to different choices for the origin of the unit cell, in order to emphasize the simple cubic packing of the F^- ions (after Bloss, 1971).

Fig. 3. Polyhedral representation of the fluorite structure (after Kostov, 1969) showing the occupied cubic voids by Ca atoms.

Fig. 4. Condensed model of the fluorite structure. The large open circles represent F atoms. The smaller black circles correspond to Ca atoms in cubic (cb) voids.

References

Kostov (1968) 190-192

Povarennykh (1972) 269, 273,274, 661.

Wyckoff (1963) Vol. 1, 239-244.

Zoltai + Stout (1984) 403.

Bloss (1971) 248.

Ingerson (1955) 351.

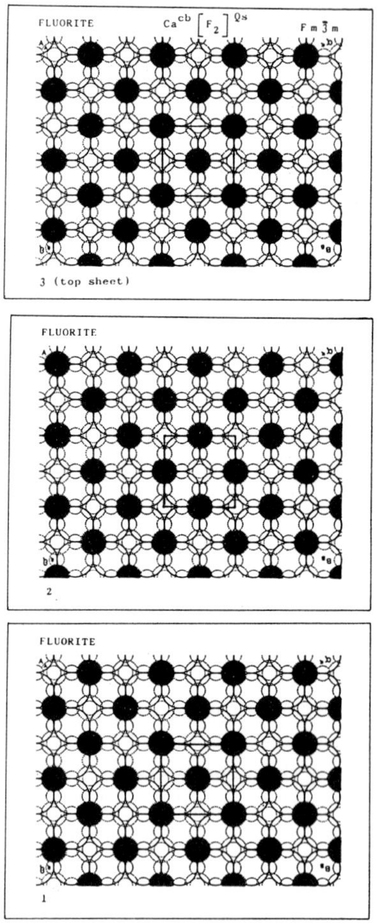

Fig. 4

BIXBYITE $(Fe, Mn)^{6}_{2}\left[O_{3}\square\right]^{Qs}$

a = 9.365 Å

Z = 16

I a 3

$(Fe, Mn)_{I}$ (8b)

$(Fe, Mn)_{II}$ (24d) u = -0.034

O (48e) x = 0.375, y = 0.162, z = 0.400

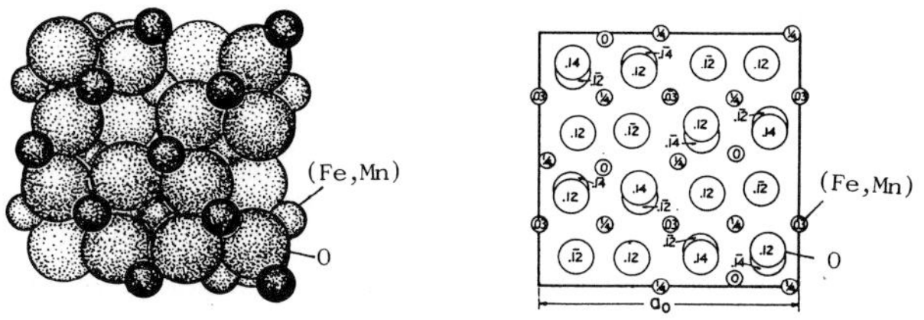

(a) Fig. 1 (b)

Properties

Habit	Cleav.	Fract.	Twin.	Hardn.	Dens.	Colour	Transp.
cubic	(111)	irregular	(111)	6-6.5	4.945	black	opaque

Refr. index/Reflect.	Birefr.	Luster	Streak	Melt.p.	CPI
		submetallic	black		

Population

Avicennite $Te^{6}_{2}\left[O_{3}\square\right]^{Qs}$

Description

The bixbyite structure is based on a defect simple cubic packing of oxygens, with 1/2 of the cubic voids occupied by Fe and Mn atoms. It may be considered as a defect packing derivative of fluorite.

Figures

Fig. 1. (a) Packing drawing of the bixbyite structure, and (b) unit cell content projected on a cube face (after Wyckoff, 1964, Vol. 2).

Fig. 2. Condensed model of the bixbyite structure. Large open circles represent oxygen atoms in simple cubic packing, and the crosses on these mean vacancies. The black circles correspond to Fe or Mn atoms in cubic voids, forming a square pattern.

References

Kostov (1968) 231,232, 253.

Povarennykh (1972) 271.

Wyckoff (1968) Vol. 2, 2-4.

Palache et al. (1944) Vol. 1, 550.

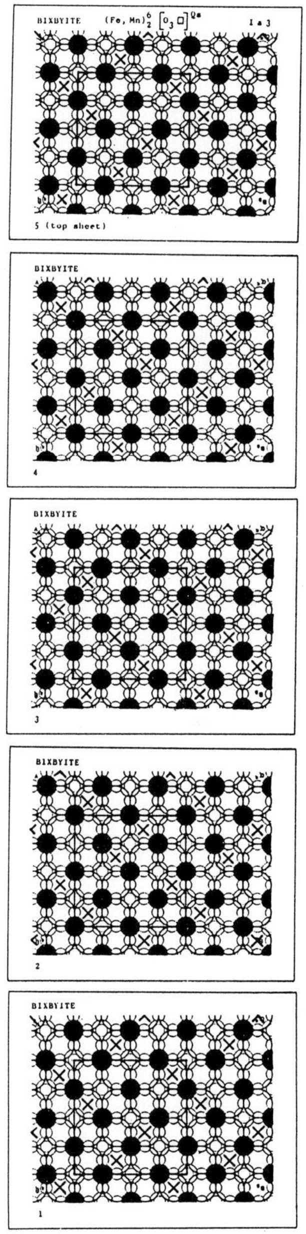

Fig. 2

PYROCHLORE

F d 3 m

$(Ca, Na)_2^{cb}\ Nb_2^6\ [O_6\ (OH)\ \square]^{Qs}$

a = 10.397 Å

Z = 8

(Ca, Na) (16c)

Nb (16d)

$(O, OH)_I$ (8b)

$(O,OH)_{II}$ (48f) u ≈ 0.19

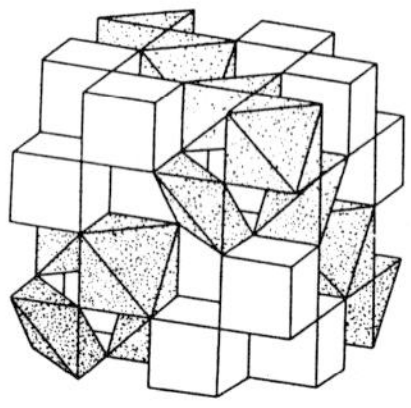

Fig. 1

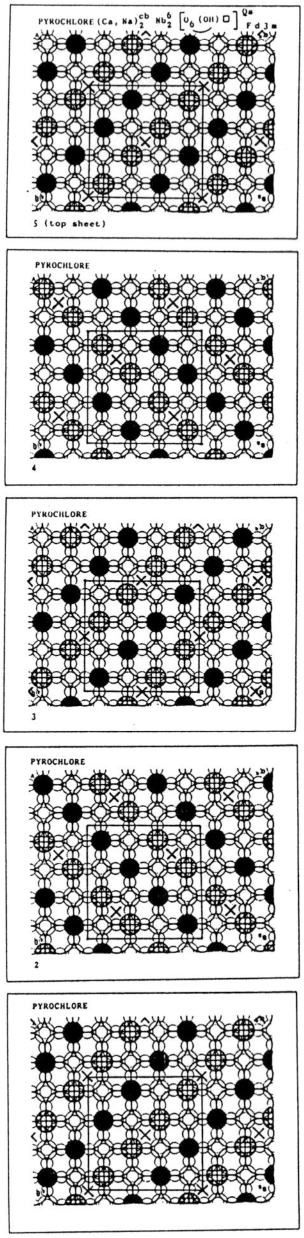

Fig. 2

Properties

Habit	Cleav.	Fract.	Twin.	Hardn.	Dens.	Colour	Transp.
octahedral, in grains	distinct (111)	subcon- choidal	rare (111)	5 - 5.5	4.2	brown, black, reddish	translucent to opaque

Refr. index/Reflect.	Birefr.	Luster	Streak	Melt.p.	CPI
n = 1.96-2.01		vitreous resinous	light brown		

Population

Microlite	$(Ca,Na)_2^{cb}\ Ta_2^6[O_6(OH,F)\square]^{Qs}$
Betafite	$(Ca,U)_2^{cb}\ (Nb,Ti)_2^6\ [O_6\ (OH,F)\square]^{Qs}$
Pandaite	$Ba_{2-x}^{cb}\ Nb_2^6\ [O_{7-x}(H_2O)_x\square]^{Qs}$
Rijkeboe- rite	$Ba_{2-x}^{cb}\ Ta_2^6[O_{7-x}(H_2O)_x\square]^{Qs}$
Sukulaite	$Sn_2^{cb}\ Ta_2^6\ [O_7\square]^{Qs}$
Romeite	$Ca_2^{cb}\ Sb_2^6\ [O_7\square]^{Qs}$
Bindhei- mite	$Pb_2^{cb}\ Sb_2^6\ [O_7\square]^{Qs}$
Partzite	$Cu_2^{cb}\ Sb_2^6\ [O_6\ (H_2O)\square]^{Qs}$
Stetefeld- tite	$Ag_2^{cb}\ Sb_2^6\ [O_6\ (H_2O)\square]^{Qs}$

Figures

Fig. 1. Polyhedral drawing of the pyrochlore structure (after Belov, quoted by Kostov, 1968).

Fig. 2. Condensed model of the pyrochlore structure. Large open circles represent OH and O atoms. The black circles correspond to Ca or Na atoms with cubic coordination, and the crossed lined circles to Nb with coordination six. The crosses represent vacancies in the simple cubic packing.

Description

The pyrochlore structure is based on a defect simple cubic packing of the OH and O atoms, with Na, Ca and Nb occupying 1/2 of the cubic voids. The large cations Na and Ca have cubic coordination. and Nb atoms 6 coordination, due to the vacancies (□). Both have a row pattern.

References

Kostov (1968) 253.
Wyckoff (1965) Vol. 3, 439,440.
Povarennykh (1972) 276,277.
Palache et al. (1944) Vol. 1, 748,749.
Roberts et al. (1974) 499.

BRAUNITE

$I\,4_1/acd$

$Mn^{cb}\,Mn_6^{[6]}\,Si^{t}\,[O_{12}\,\square_4]$ Qs

a = 9.408 Å
c = 18.668 Å
Z = 8

		x	y	z
Mn^{cb}	(8b)			
Mn_I^6	(16c)			
Mn_{II}^6	(16e)	1/4	0.2157	0
Mn_{III}	(16f)	0.2318	1/4+x	1/8
Si	(8a)			
O_I	(32g)	0.1487	0.8537	0.9453
O_{II}	(32g)	0.1457	0.0734	0.0569
O_{III}	(32g)	0.0787	0.1347	0.9250

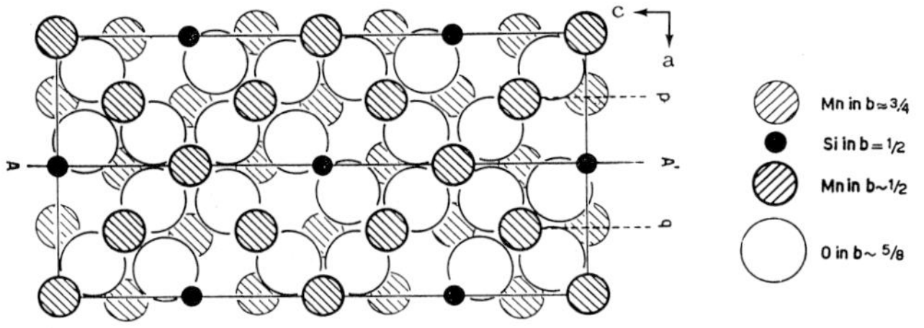

Fig. 1

Properties

Habit	Cleav.	Fract.	Twin.	Hardn.	Dens.	Colour	Transp.
pyramidal	perfect (112)		(112)	6-6.5	4.722-4.83	brownish black to steel- grey	opaque

Refr. index/Reflect.	Birefr.	Luster	Streak	Melt.p.	CPI
		submetallic	brawnish black		

Figures

Fig. 1. Projection of the braunite structure on an a axis, (after Byström and Mason, 1943, quoted in Structure Reports, 1955, Vol. 9).

Fig. 2. Condensed model of the braunite structure. The large open circles represent oxygen atoms in a simple cubic packing, and the crosses on these mean vacancies. Some of the manganese atoms (double circles) have cubic coordination, others 6 coordination (lined circles) and the silicon atoms have tetrahedral coordination (black circles), as a consequence of the packing vacancies. To compare Fig. 2 with Fig. 1 look in the direction of the arrow.

Description

The braunite structure is based on a defect simple cubic packing of oxygen atoms with Mn atoms with two different coordinatios, cubic and [6], and Si atoms with tetrahedral coordination. We may imagine Mn and Si all in cubic voids, the corresponding coordination [6] and t being the result of the packing vacancies. This structure can be considered as an interstitial substitution and defect derivative of fluorite.

References

Kostov (1968) 233-236.
Moore + Araki (1976) 1226-1240.
Wyckoff (1968) Vol. 4, 203,204.
Structure Reports (1955) Vol. 9, 253.
Povarennykh (1972) 287,288.

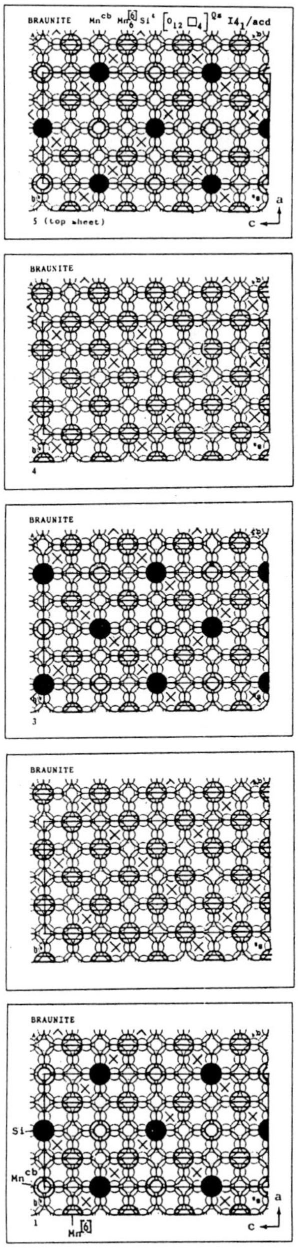

Fig. 2

COOPERITE $Pt^{sq}[S]^{Qs}$

P 4_2/m m c

a = 3.47 Å		Pt (2c)
c = 6.10 Å		S (2e)
Z = 2		

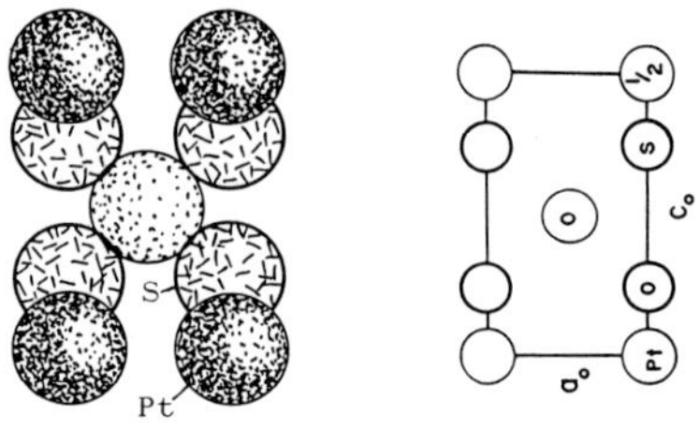

(a) Fig. 1 (b)

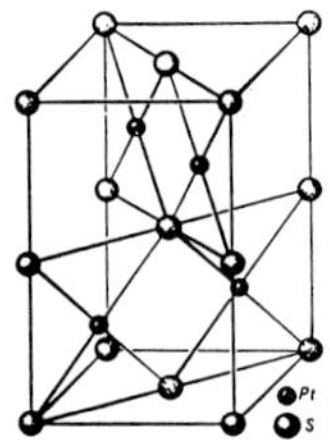

Fig. 2

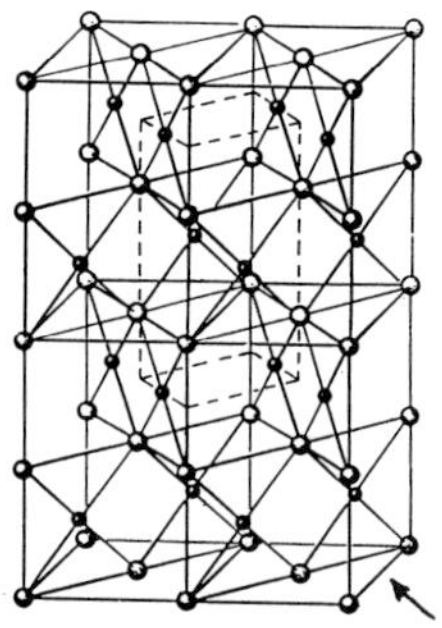

Fig. 3

Properties

Habit	Cleav.	Fract.	Twin.	Hardn.	Dens.	Colour	Transp.
irregular grains	(011)	conchoi-dal		4-5	9.5	steel--grey	opaque
Refr. index/Reflect.		**Birefr.**		**Luster**	**Streak**	**Melt.p.**	**CPI**
				metallic			

Figures

Fig. 1. (a) Packing drawing of the cooperite structure, and (b) unit cell content projected on (010) plane (after Wyckoff, 1963, Vol. 1).

Fig. 2. Ball and spoke model of the cooperite structure (after Povarennykh, 1972).

Fig. 3. More complete ball and spoke model of the cooperite structure in order to mark the unit cell (dashed line) used by Wyckoff, Fig. 1 (adapted from Povarennykh, 1972). Look at the direction of the arrow in order to relate it to Fig. 1.

Fig. 4. Condensed model of the cooperite structure. Large open circles represent S atoms, which form a simple cubic packing, with Pt atoms in square voids, forming a square pattern. When using the condensed model look in the direction of the arrow to compare with Fig. 1, and consider the unit cell marked with dashed lines to understand Fig. 2.

Description

The cooperite structure is based on a simple cubic packing of S atoms with Pt atoms occupying 1/2 of the square voids, forming a square pattern.

References

Kostov (1968) 115.
Povarennykh (1972) 221,222.
Wyckoff (1963) Vol. 1, 139,140.
Palache et al. (1944) Vol. I, 258.

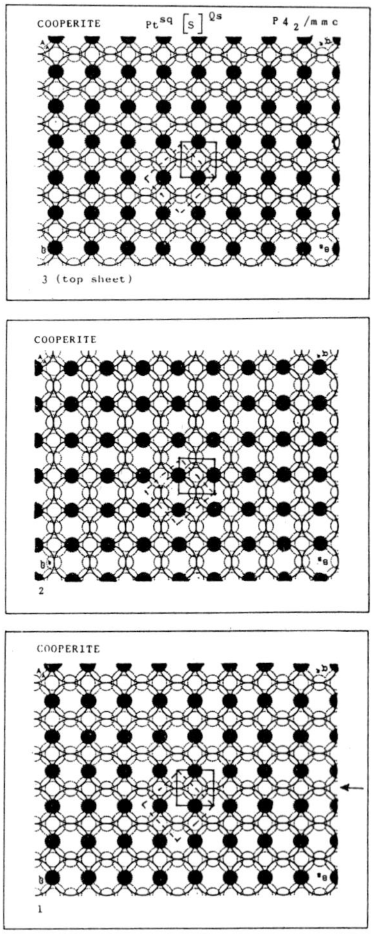

Fig. 4

TENORITE

$Cu^{sq}[O]^{Qs}$

C 2/c

a=4.653 Å
b=3.410 Å
c=5.108 Å
β=99°29'
Z=4

Cu(4c)
O(4e) u = -0.584

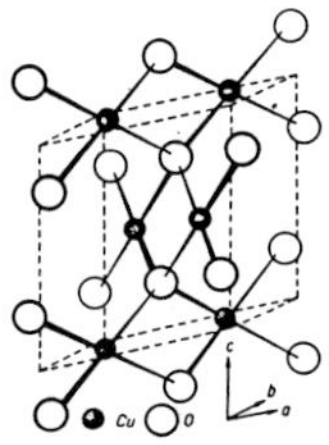

Fig. 1

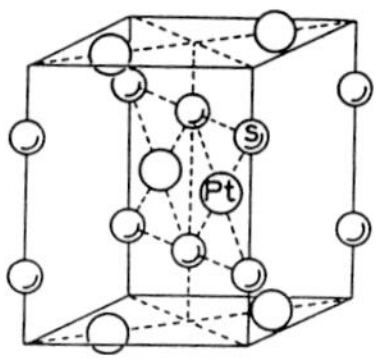

Fig. 2

Properties

Habit	Cleav.	Fract.	Twin.	Hardn.	Dens.	Colour	Transp.
twinned aggregates	[011] [0$\bar{1}$1]	conchoidal to uneven	(011)	3.5	5.8-6.4	steel or iron-grey to black	opaque

Refr. index/Reflect.	Birefr.	Luster	Streak	Melt.p.	CPI
n = 2.63		metallic			

Figures

Fig. 1. Ball and spoke model of the tenorite structure (after Povarennykh, 1972).

Fig. 2. Ball and spoke model of the cooperite structure, $Pt^{sq}[S]^{Qs}$ (after Kostov, 1968). which enables a easier comparison between tenorite and cooperite.

Fig. 3. Condensed model of the tenorite structure. The large open circles represent oxygen atoms in a simple cubic packing, and the small black circles correspond to Cu atoms in square voids, showing a square pattern. One should look in the direction of the arrow, and consider the unit cell marked with dashed lines,in order to compare with Fig. 2.

Description

The tenorite structure is a distortion derivative of cooperite. It is based on a simple cubic packing of the O atoms, with Cu atoms occupying 1/2 of the square voids.

References

Kostov (1968) 115, 257.
Povarennykh (1972) 272.
Wyckoff (1963) Vol. 1, 140,141.
Palache et al. (1944) Vol. 1, 507,508.

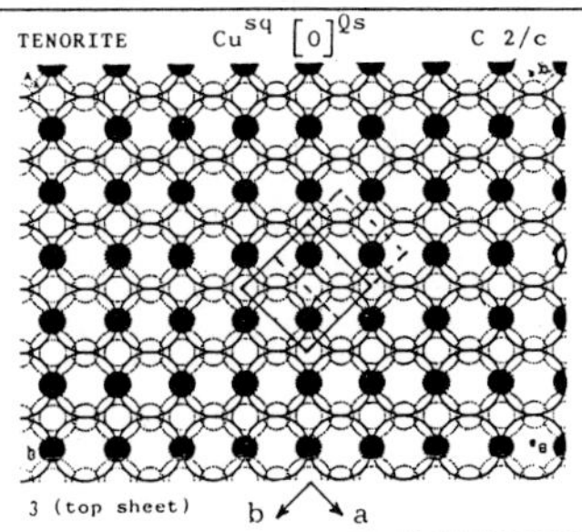

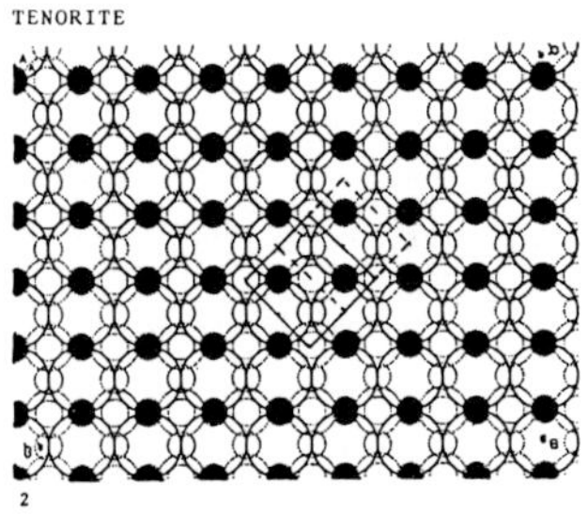

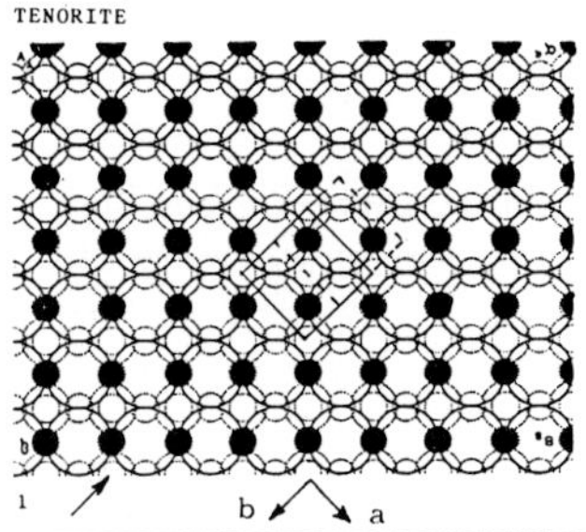

Fig. 3

ANDALUSITE

Pnnm

$Al^{o}\ Al^{[5]}\ Si^{t}\ [O_5 \square]^{R^{21}_{by}}$

a = 7.7942 Å
b = 7.8985 Å
c = 5.559 Å
Z = 4

Atom	x	y	z
Al^{o} (4e)	0	0	0.2422
Al^{t} (4g)	-0.1292	0.3613	0
Si (4g)	0.2462	0.2529	0
O_I (4g)	0.0768	-0.1373	0
O_{II} (4g)	0.4240	0.3629	0
O_{III} (4g)	0.1038	0.4013	0
O_{IV} (8h)	0.2303	0.1343	0.2390

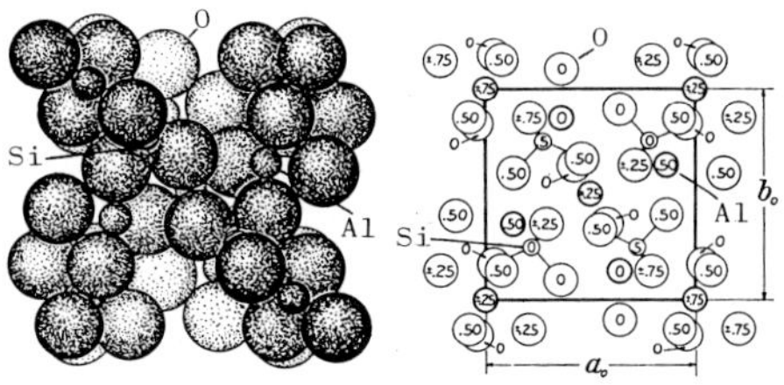

(a) (b)

Fig. 1

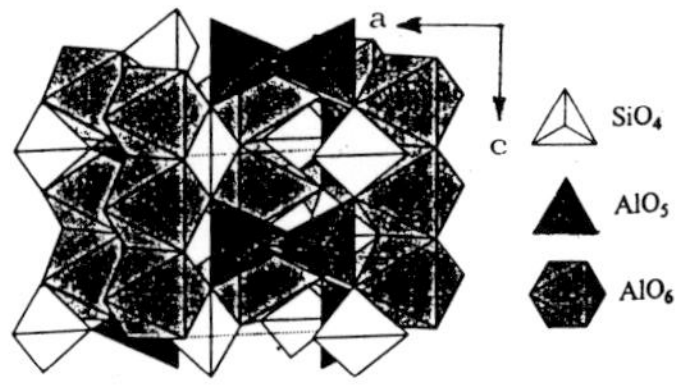

Fig. 2

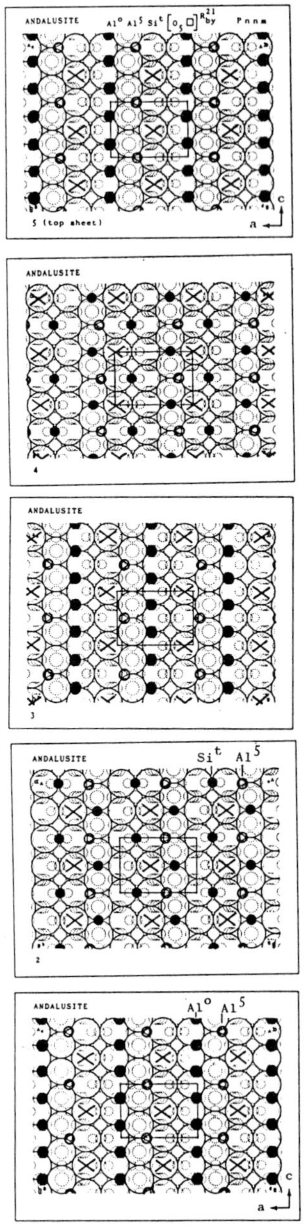

Fig. 3

Properties

Habit	Cleav.	Fract.	Twin.	Hardn.	Dens.	Colour	Transp.
prismatic, massive	good (110)	subconchoidal	rare (011)	7.5	3.18	brown, red	transparent to translucent

Refr. index/Reflect.	Birefr.	Luster	Streak	Melt.p.	CPI
n_α = 1.632 n_β = 1.640 n_γ = 1.442	(-) 2V = 75-85°	vitreous	white		(SPI) 58

Population

Olivenite	$Cu^{o}Cu^{[5]}As^{t}[O_4(OH)\square]^{R^{21}_{by}}$
Adamite	$Zn^{o}Zn^{[5]}As^{t}[O_4(OH)\square]^{R^{21}_{by}}$
Eveite	$Mn^{o}Mn^{[5]}As^{t}[O_4(OH)\square]^{R^{21}_{by}}$
Libethenite	$Cu^{o}Cu^{[5]}P^{t}[O_4(OH)\square]^{R^{21}_{by}}$

Figures

Fig. 1. (a) Packing drawing of the andalusite structure, and (b) unit cell content projected along the c axis (after Wyckoff, 1968, Vol. 4).

Fig. 2. Polyhedral representation of the andalusite structure (after Zoltai + Stout, 1984).

Fig. 3. Condensed model of the andalusite structure. The large open circles represent oxygen atoms forming R^{21} layers. The packing vacancies are marked with a cross. The medium black circles correspond to certain Al atoms in octahedral voids, the smaller black circles to Si atoms in tetrahedral voids, and the small double circles to other Al atoms with coordination [5] (after Figueiredo, 1976).

Description

The andalusite structure is based on defect R^{21} packing layers of oxygen atoms, with certain Al atoms in octahedral voids, other Al atoms with coordination [5] , and Si atoms in tetrahedral voids. The R^{21} packing layers are formed by two rows forming a closest packing (two triangular rows) and one row with squares.

References

Kostov (1968) 282-285.

Povarennykh (1972) 386,387, 510, 540.

Wyckoff (1968) Vol. 4, 188,189.

Zoltai + Stout (1984) 362,363.

Figueiredo (1976).

PHENAKITE $R\bar{3}$

$Be^{t}_{4}\ Si^{t}_{2}\ [O_8\square]^{(3R^{21}_{o})by}$

a_R = 7.70	Be_I (18f) x=-0.21 y=-0.01 z=0.58	O_I (18f) x=-0.092 y=0.120 z=0.250
α = 108°1'		
Z_R = 6	Be_{II} (18f) x=0.13 y=-0.38 z=0.58	O_{II} (18f) x=-0.32 y=0.000 z=0.250
a_H = 12.42 Å		
c = 8.24 Å	Si (18f) x=-0.211 y=-0.011 z=0.250	O_{III} (18f) x=-0.205 y=-0.074 z=0.083
Z = 18		...

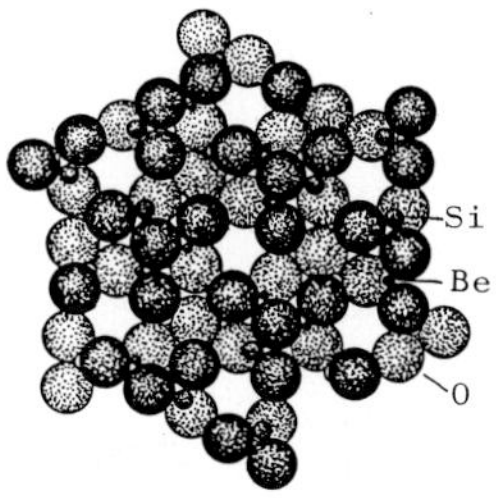

Fig. 1

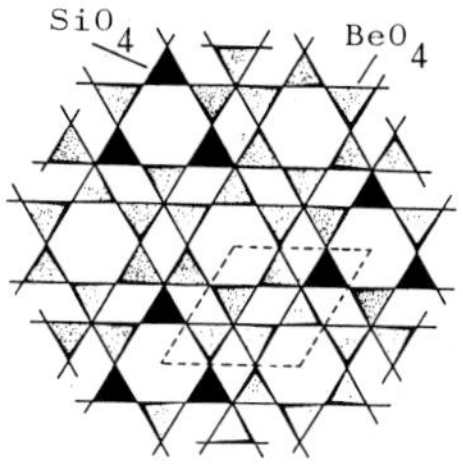

Fig. 2

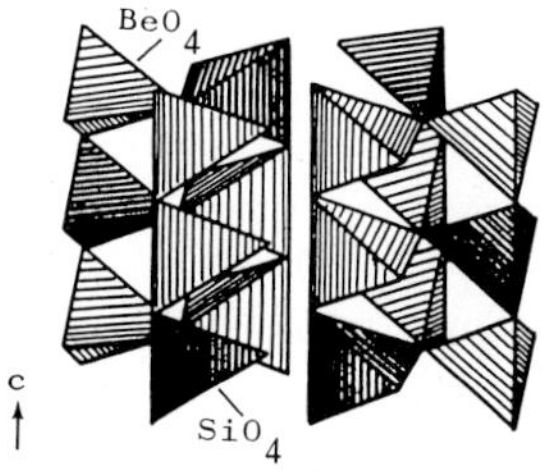

(a)

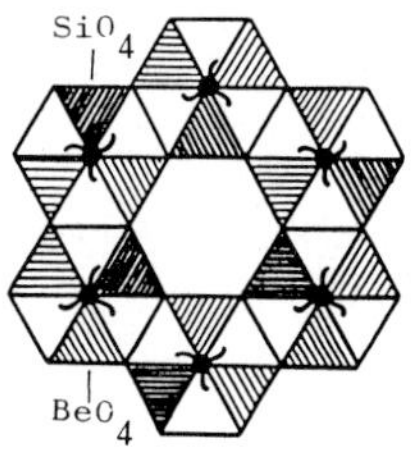

Fig. 3 (b)

Properties

Habit	Cleav.	Fract.	Twin.	Hardn.	Dens.	Colour	Transp.
rhombo-hedral prismatic	distinct (11$\bar{2}$0)	conchoi-dal	(10$\bar{1}$0)	7.5--8	2.93-3.0	colourless wine--yellow	transpa-rent

Refr. index	/Reflect.	Birefr.	Luster	Streak	Melt.p.	CPI
n_ω = 1.654 n_ϵ = 1.670		(+)	vitreous			

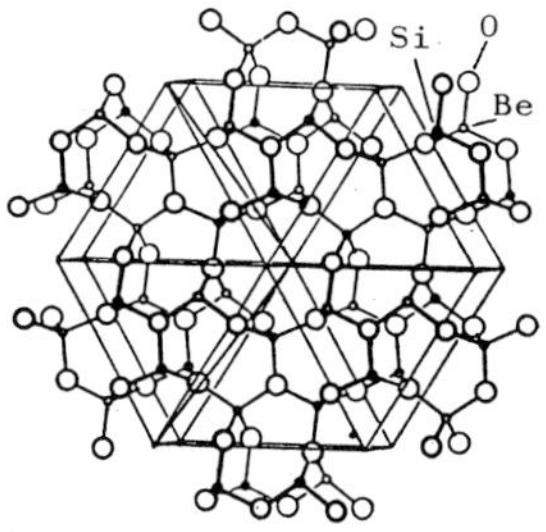

Fig. 4

Population

Willemite $Zn^{t}_{4}\ Si^{t}_{2}[O_8\square]^{(3R^{21}_{o})by}$

Figures

Fig. 1. Packing drawing of the phenakite structure projected along the c axis (after Wyckoff, 1968, Vol. 4).

Fig. 2. Structure of phenakite viewed along the c axis (after Kostov, 1968).

Fig. 3. Polyhedral representation of the phenakite structure (a) left and right triple strips of BeO_4 and SiO_4 tetrahedra, and (b) general view in projection on the basal plane (after Povarennykh, 1972).

Fig. 4. Phenakite structure (after Strukturbericht, 1931, Vol. 1).

Description

The structure of phenakite is based on a defect packing of R^{21} layers, with Be and Si atoms occupying tetrahedral voids.

Crystallographic data (continued)

O_{IV} (18f) x=-0.205 y=-0.074 z=0.417

References

Kostov (1968) 276, 324.
Povarennykh (1972) 383,384.
Wyckoff (1968) Vol. 4, 283.
Roberts et al. (1974) 475.
Strukturbericht (1931) Vol. 1, 356.

LEPIDOCROCITE

$Fe^{o}\left[O\,(OH)\right]\,R^{31}_{my}$

Bbmn

a=12.4 Å
b=3.87 Å
c=3.06 Å
Z=4

Fe(4c) u=-0.178
O(4c) u=0.21
OH(4c) u=0.425

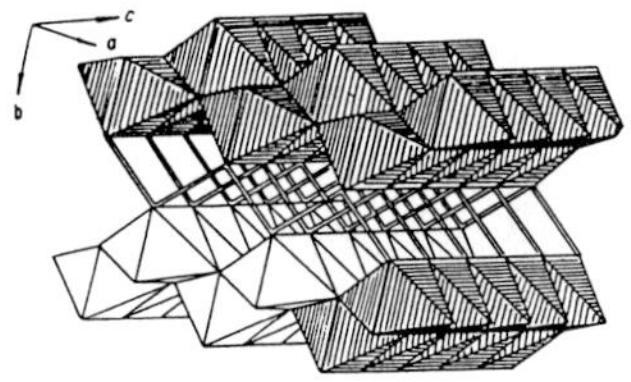

Fig. 1

Properties

Habit	Cleav.	Fract.	Twin.	Hardn.	Dens.	Colour	Transp.
bladed, tabular	perfect (100) good (001)			5	4.0	red brown	trans- lucent

Refr. index /Reflect.	Birefr.	Luster	Streak	Melt.p.	CPI
n_α = 2.20, n_β = 1.94, n_γ = 2.51	(-) 2V= 83°	subme- tallic	orange		(SPI) 52

Population

Boehmite $Al^{o}\left[O\,(OH)\right]\,R^{31}_{my}$

Figures

Fig. 1. Polyhedral representation of the lepidocrocite structure (adapted from Povarennykh, 1972).

Description

The lepidocrocite structure is based on R^{31} packing layers of OH and O atoms. The Fe atoms are located in octahedral voids, forming a double row pattern.

References

Kostov (1968) 231.
Povarennykh (1972) 146, 322,323.
Wyckoff (1963) Vol. 1, 293.
Zoltai + Stout (1984) 423,424.

VYSOTSKITE

$Pd^{sq}[S]^{21}_{Ns}$

$P4_2/m$

$a = 6.4287$ Å	Pd_I (2e)		S(8k) $x = 0.19$, $y = 0.32$, $z = 0.23$
$c = 6.6082$ Å	Pd_{II} (2c)		
$Z = 8$	Pd_{III}(4j)	$u = 0.48$, $v = 0.25$	

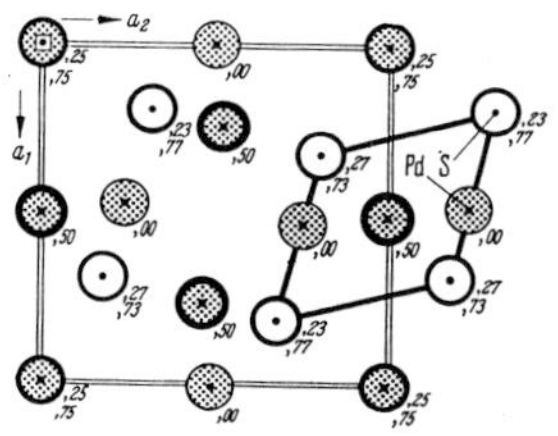

Fig. 1

Properties

Habit	Cleav.	Fract.	Twin.	Hardn.	Dens.	Colour	Transp.
irregular masses				5	6.69	silvery	opaque

Refr. index/Reflect.	Birefr.	Luster	Streak	Melt.p.	CPI
		metallic			

Figures

Fig. 1. Projection of the vysotskite structure along the c axis (after Schubert, 1964).

Fig. 2. Condensed model of the structure of vysotskite. The large open circles represented S atoms and the black circles correspond to Pd atoms in square voids.

Description

The vysotskite structure is based on the Ns^{21} packing of S atoms, where N^{21} means a layer formed by triangles and squares in a proportion of 2 to 1 with a tetragonal symmetry, superimposed over one another (s). The Pd atoms occupy square voids in the Ns^{21} packing.

References

Kostov (1968) 115

Wyckoff (1963) Vol. 1, 142,143.

Schubert (1964) 195.

Roberts et al. (1974) 658

Povarennykh (1972) 221.

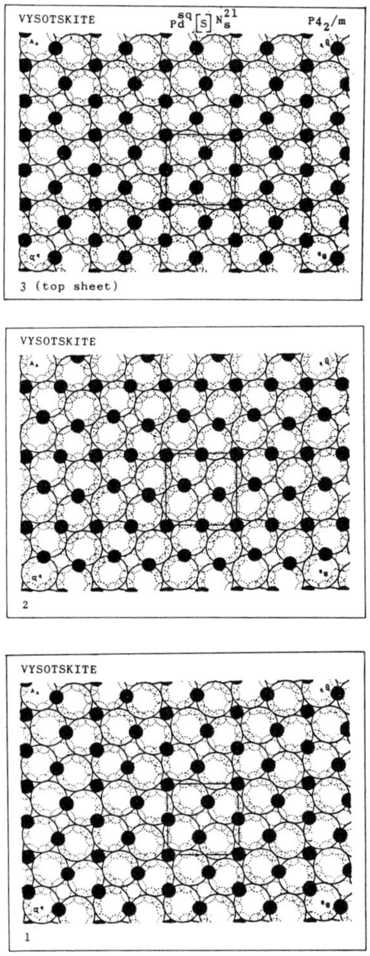

Fig. 2

GARNET $Ca_3^{do}\ Al_2^{o}\ Si_3^{t}\left[O_{12}\right]$ *
(Grossular)

I a 3 d

a = 11.855 Å

Z = 8

Ca(24c)

Al(16a)

Si(24d)

O (96h) x = −0.0382 y = 0.0457 z = 0.1512

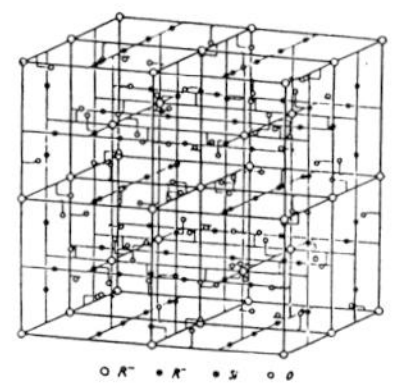

Fig. 1

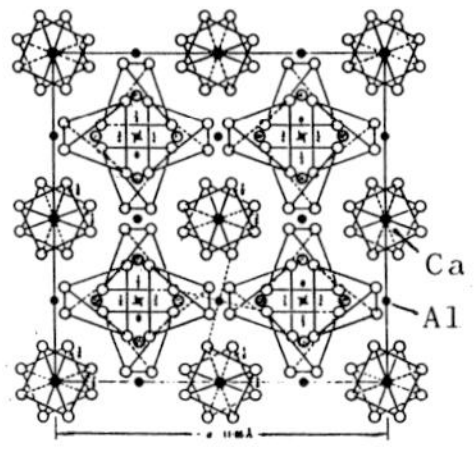

Fig. 2

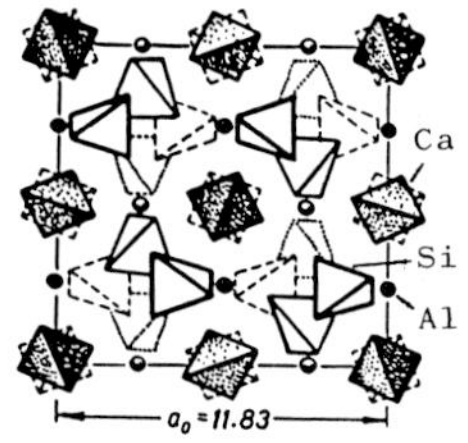

Fig. 3

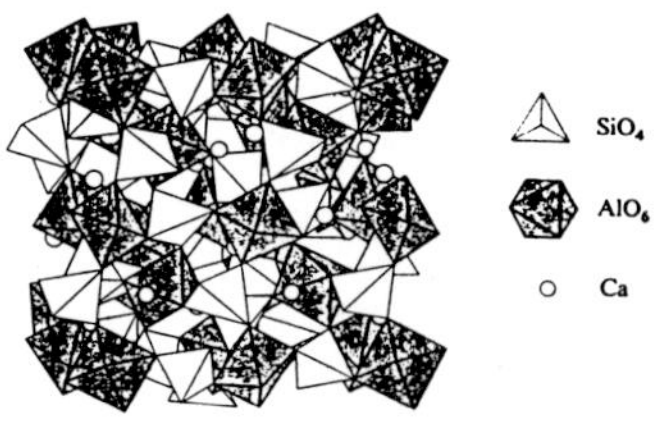

Fig. 4

Fig. 5

Properties

Habit	Cleav.	Fract.	Twin.	Hardn.	Dens.	Colour	Transp.
dodecahedral		subconchoidal		6.5 − 7.5	3.56	colourless, red, yellow	transparent

Refr. index/Reflect.	Birefr.	Luster	Streak	Melt.p.	CPI
n = 1.75		resinous	white		(SPI) 59

Population

Uvarovite	$Ca_3^{do}\ Cr_2^{o}\ Si_3^{t}\left[O_{12}\right]$ *
Pyrope	$(Mg,Fe)_3^{do}\ Al_2^{o}\ Si_3^{t}\left[O_{12}\right]$ *
Spessartine	$(Mn,Fe)_3^{do}\ Al_2^{o}\ Si_3^{t}\left[O_{12}\right]$ *
Calderite	$Mn_3^{do}\ Fe_2^{o}\ Si_3^{t}\left[O_{12}\right]$ *
Goldmanite	$Ca_3^{do}\ V_2^{o}\ Si_3^{t}\left[O_{12}\right]$ *
Kimzeyite	$Ca_3^{do}\ Zr_2^{o}\ (Si,Al)_3^{t}\left[O_{12}\right]$ *
Almandine	$Fe_3^{do}\ Al_2^{o}\ Si_3^{t}\left[O_{12}\right]$ *
Andradite	$Ca_3^{do}\ Fe_2^{o}\ Si_3^{t}\left[O_{12}\right]$ *

Figures

Fig. 1. Unit cell of the garnet structure in perspective (after Menzer, 1928, quoted by Bragg, 1929).

Fig. 2. Projection on (001) of the grossular structure (after Strunz, 1957, quoted by Deer Howie and Zussman, Vol. 1, 1962).

Fig. 3. Similar polyhedral description of the garnet structure on the same orientation as Fig. 1. Partial representation (after Povarennykh, 1972).

Fig. 4. Polyhedral description of the garnet structure (after Zoltai + Stout, 1984). Notice that the polyhedra around Ca are not represented.

Fig. 5. Partial description of the garnet structure in terms of the cubic body centered packing of rods of Al octahedra and empty triangular prisms (after Andersson + O'Keeffe, 1977).

Description

Garnet is a non-layered structure based on a three dimensional close packing of the oxygen atoms, with Al atoms in octahedral voids, Si atoms in tetrahedral voids, and Ca with a special [8] coordination forming a dodecahedron. The (SPI) = 59 confirms the close packing character of this structure, and is against of a framework classification. Fig. 5 also explains the 3D architecture, and the non-layered character (*) of the garnet structure.

References

Kostov (1968) 315-318.

Povarennykh (1972) 382,383.

Wyckoff (1968) Vol. 4, 159.

Zoltai + Stout (1984) 355-358.

Bragg (1929) 306.

Anderson + O'Keeffe (1977) 605.

Deer et al. (1962) Vol. 1, 78.

ZIRCON

$Zr^{do}\ Si^{t}\left[O_4\right]^{*}$

$I\,4_1/a\,m\,d$

$a = 6.6164$ Å Zr(4a)
$c = 6.0150$ Å Si(4b)
Z = 4

O(16h) u = 0.067
  v = 0.198

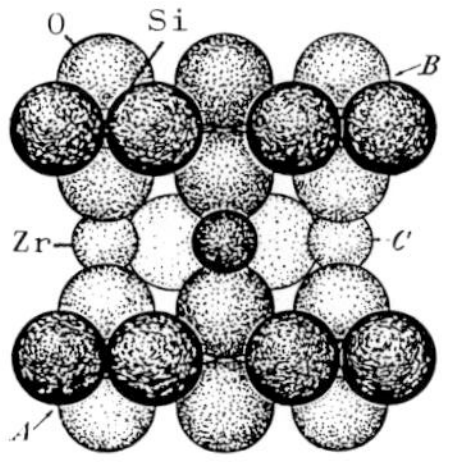

Fig. 1

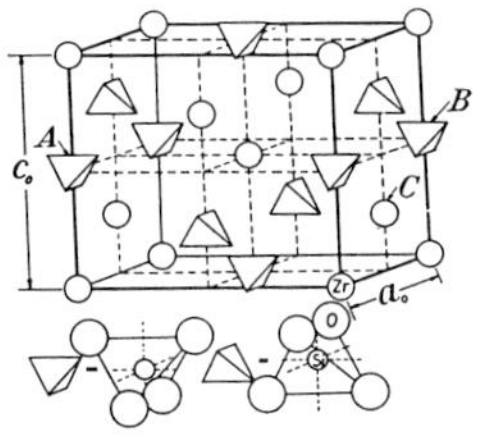

Fig. 2

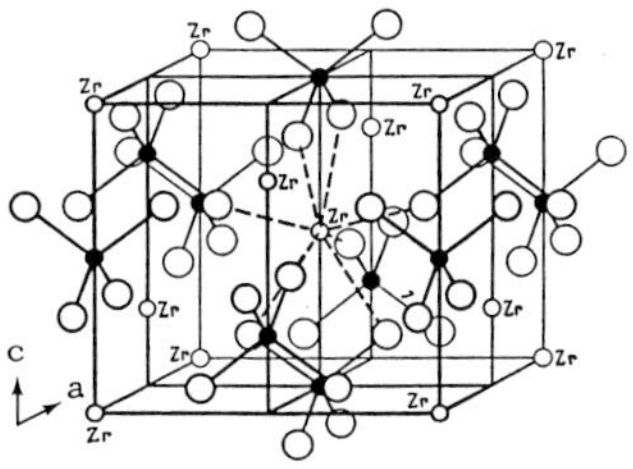

Fig. 3

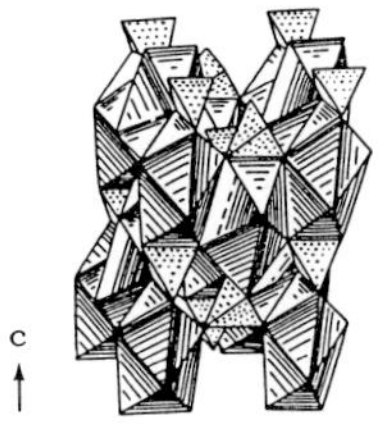

Fig. 4

Properties

Habit	Cleav.	Fract.	Twin.	Hardn.	Dens.	Colour	Transp.
prismatic	poor (100)	conchoidal	(111)	7.5	4.68	brown, green	transparent to translucent

Refr. index /Reflect.	Birefr.	Luster	Streak	Melt.p.	CPI
$n_\omega = 1.99$ $n_\epsilon = 1.93$	(+)	adamantine	white	2430° C	(SPI) 63

Population

Coffinite	$U^{do}\ Si^{t}\left[O_{4-x}(OH)_x\right]^{*}$
Thorite	$Th^{do}Si^{t}\left[O_4\right]^{*}$
Behierite	$Ta^{do}\ B^{t}\left[O_4\right]^{*}$
Wakefieldite	$Y^{do}\ V^{t}\left[O_4\right]^{*}$
Chernovite	$Y^{do}\ As^{t}\left[O_4\right]^{*}$
Xenotime	$Y^{do}\ P^{t}\left[O_4\right]^{*}$
Chromatite	$Ca^{do}\ Cr^{t}\left[O_4\right]^{*}$

Description

The zircon structure should be consider as a close packed structure, and not a framework, because of the high "close packing index" (SPI 63). However this close packing do not fit any layered close packing, due to the existance of dodecahedra in the structure. The O atoms form a special three dimensional close packing, symbolised by * , where the Zr atoms have dodecahedral coordination, and the Si atoms tetrahedral coordination.

Figures

Fig. 1. Packing drawing of the zircon structure projected on (001) and passing through centers of the tetrahedra A and B of Fig. 2 (after Wyckoff. 1965, Vol. 3).

Fig. 2. Perspective drawing of the zircon unit cell (after Wyckoff, 1965, Vol. 3).

Fig. 3. Ball and spoke model of the zircon structure (after Bragg + Claringbull, 1965).

Fig. 4. Another polyhedral representation of the zircon structure (after Povarennykh, 1972).

References

Kostov (1968) 294-296.
Povarennykh (1972) 381,382, 465. 493, 508,509, 536, 575.
Wyckoff (1965) Vol. 3, 16.
Wyckoff (1968) Vol. 4, 158.
Smith (1982) 80, 335.
Zoltai + Stout (1984) 369.
Bragg + Claringbull (1965) 185.
Ingerson (1955) 352.

SCHEELITE $Ca^{do} W^{t} [O_4]^{*}$

$I\,4_1/a$

a = 5.24 Å
c = 11.38 Å
Z = 4

Ca(4b)
W (4a)
O (16f) x = 0.25, y = 0.15, z = 0.075

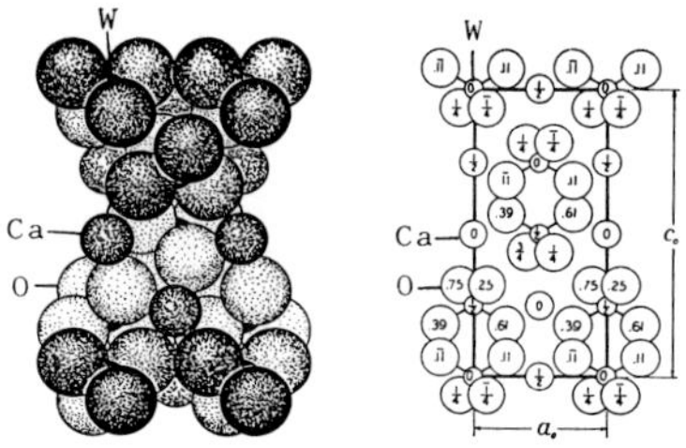

(a) Fig. 1 (b)

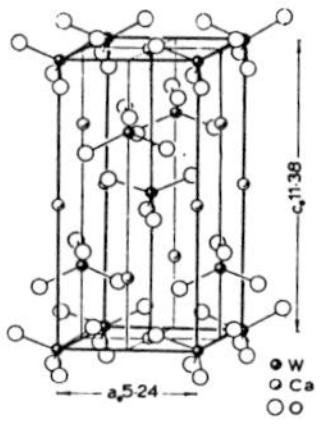

Fig. 2

Properties

Habit	Cleav.	Fract.	Twin.	Hardn.	Dens.	Colour	Transp.
bipyrami-dal, tabular	good (101)	uneven	(110)	4.5-5	6.11	white, yellow, brown, variable	transparent to translu-cent

Refr. index/Reflect.	Birefr.	Luster	Streak	Melt.p.	CPI
n_ω = 1.920, n_ϵ = 1.934	(+)	subadaman-tine	white		(SPI) 69

Population

Wulfenite	$Pb^{do} Mo^{t} [O_4]^{*}$
Powellite	$Cd^{do} Mo^{t} [O_4]^{*}$
Stolzite	$Pb^{do} W^{t} [O_4]^{*}$

Figures

Fig. 1. (a) Packing drawing of the scheelite structure, and (b) unit cell content projected along the b axis (after Wyckoff, 1965, Vol. 3).

Fig. 2. Ball and spoke model of the scheelite structure (after Kostov, 1968).

Description

The scheelite structure is a non layered structure, it is formed by isolated WO_4 tetrahedra separated by Ca atoms, each of which is bonded to eight oxygen atoms forming a dodecahedron (Fig. 3).

It is a distortion derivative of the zircon structure.

Notice the high value 69 of (SPI), which confirms the classification of scheelite as a three dimensional close packed structure.

References

Kostov (1968) 484,485.
Povarennykh (1972) 569,570.
Wyckoff (1965) Vol. 3, 19-23.
Zoltai + Stout (1984) 445-447.

ANHYDRITE

$Ca^{8} S^{t} [O_4]$ *

B b m m

a = 6.238 Å
b = 6.991 Å
c = 6.996 Å
Z = 4

Atom	x	y	z
Ca(4c)	0.654	1/4	0
S (4c)	0.155	1/4	0
O_I(8f)	0.298	0.08	0
O_{II}(8g)	0.015	1/4	0.171

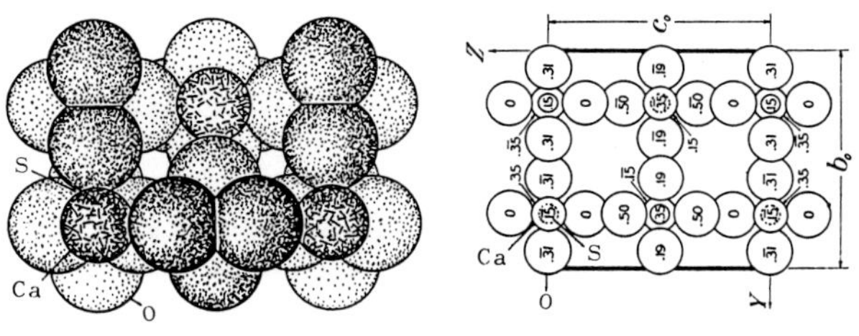

(a) Fig. 1 (b)

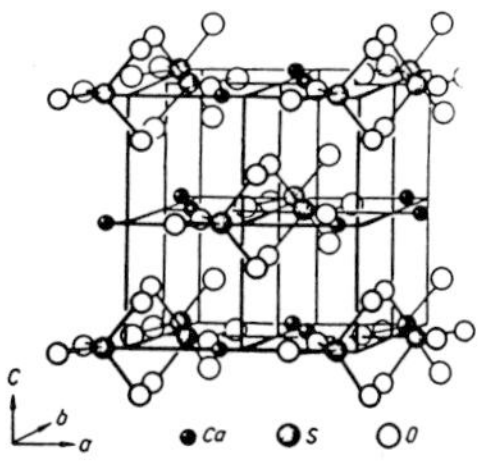

Fig. 2

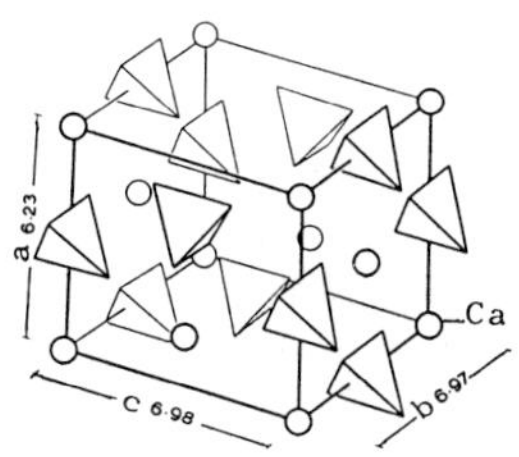

Fig. 3

Properties

Habit	Cleav.	Fract.	Twin.	Hardn.	Dens.	Colour	Transp.
tabular, lamellar, fibrous	good (001) (010)	uneven, splintery		3.5	2.98	colourless, white, bluish	transparent

Refr. index/Reflect.	Birefr.	Luster	Streak	Melt.p.	CPI
n_α = 1.570 n_β = 1.575 n_γ = 1.614	(+) 2V = 44°	vitreous, pearly	white		(SPI) 70

Figures

Fig. 1. (a) Packing drawing of the $NaClO_4$ structure, which is isotypic with the anhydrite structure, and (b) unit cell content projected along the a axis (after Wyckoff, 1965, Vol. 3).

Fig. 2. Ball and spoke model of the anhydrite structure (after Povarennykh, 1972).

Fig. 3. Structural scheme of the anhydrite structure (after Kostov, 1968).

Description

Anhydrite is a distortion derivative of zircon. It is formed by SO_4 tetrahedra linked to CaO_8 dodecahedra.

Notice the high value 70 of (SPI), which confirms the classification of scheelite as a three dimensional close packed structure.

References

Kostov (1968) 505,506.
Povarennykh (1972) 581.
Wyckoff (1965) Vol. 3, 18,19.
Zoltai + Stout (1984) 439.
Ingerson (1955) 351.

HETEROGENITE $Co^{o}\left[O(OH)\right]^{(2Ts)c}$

$R\bar{3}m$

$A_R = 4.676$ Å
$\alpha = 35^{\circ}28'$
$Z_R = 1$
$a_H = 2.894$ Å
$c = 13.130$ Å
$Z_H = 3$

(rhombohedral description)

Co (1a) x = 0, y = 0, z = 0

O,OH(2c) x = 0.41

Fig. 1

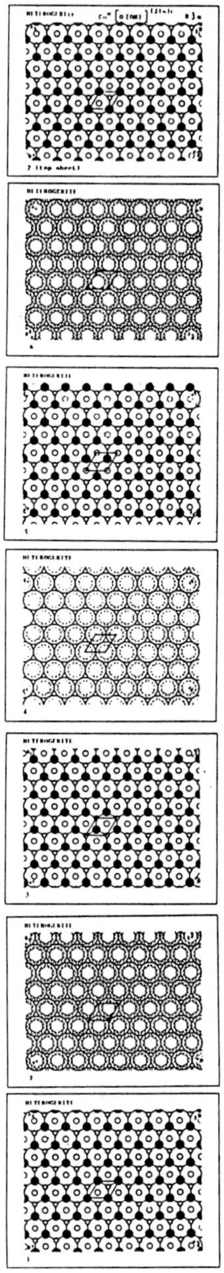

Fig. 2

Properties

Habit	Cleav.	Fract.	Twin.	Hardn.	Dens.	Colour	Transp.
needle like, massive	perfect (001)	conchoidal		4.5	4.13-4.47	black, steel grey	opaque

Refr. index/Reflect.	Birefr.	Luster	Streak	Melt.p.	CPI
		metallic, dull	black, dark brown		

Figures

Fig. 1. Polyhedral drawing of $NaHF_2$ which is isotypic with heterogenite (after Povarennykh, 1972).

Fig. 2. Condensed model of the heterogenite structure. Large open circles represent oxygen, and OH. The small black circles represent Co atoms in octahedral voids.

Description

The heterogenite structure is based on a heterogeneous packing (2Ts)c of OH and O atoms. Two layers O + OH are superimposed in a simple hexagonal packing (Ts) and these double layers are stacked in a closest packing way with a cubic sequence (c), giving rise to empty trigonal prismatic voids. The Co atoms fill up completely the octahedral voids of the double layers O(OH) stacked in a closest way.

References

Kostov (1968) 228.
Kondrasev + Fedorova (1954) 229-231.
Structure Reports (1961) Vol. 18, 515,516.
Wyckoff (1964) Vol. 2, 294.
Povarennykh (1972) 144,145, 322.
Roberts et al. (1974) 271.

LITHIOPHORITE

C 2 / m

$(Al, Li)^{o}\ Mn^{o}\ \left[O_2\ (OH)_2\right]^{(2Ts)c}$

a = 5.06 Å
b = 2.91 Å
c = 9.55 Å
β = 100° 30'
Z = 2

(Al,Li) (2d)
Mn (2a)
O (4i) x = 0.696, z = 0.103
(OH)(4i) x = 0.794, z = 0.397

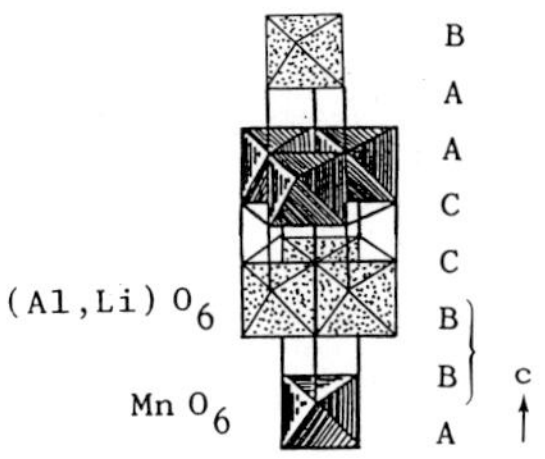

Fig. 1

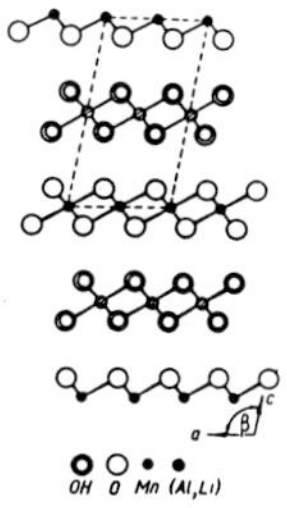

Fig. 2

Properties

Habit	Cleav.	Fract.	Twin.	Hardn.	Dens.	Colour	Transp.
massive, compact, botryoidal	perfect (001)			3	3.14 - 3.4	bluish black	opaque

Refr. index/Reflect.	Birefr.	Luster	Streak	Melt.p.	CPI
		dull, metallic	blackish grey		

Figures

Fig. 1. Polyhedral drawing of the lithiophorite structure (adapted from Povarennykh, 1972).

Fig. 2. Layers of the lithiophorite structure projected on the b axis. The unit cell is shown by broken lines (after Wadsley, 1952, quoted in Structure Reports, 1959, Vol. 16).

Fig. 3. Condensed model of the lithiophorite structure. The large open circles represent oxygens atoms, and the large lined circles OH. The small black circles correspond to Mn atoms in octahedral voids and the small crossed circles to Al and Li atoms, also in octahedral voids.

Description

The lithiophorite structure is based on a heterogeneous packing (2Ts)c, of OH and O atoms. Two layers of OH alternate with two layers of O atoms. These double packing layers superimpose on each other giving rise to a simple hexagonal packing (Ts) in between them, with empty trigonal prismatic voids. The double layers are stacked in a closest packing way with cubic sequence (c). The Mn atoms and the Al and Li atoms fill completely, and alternately the octahedral voids within the double layers.

Lithiophorite is an interstitial substitution and distortion derivative of heterogenite.

References

Kostov (1968) 236.
Wadsley (1952) 676-688.
Structure Reports (1959) Vol. 16, 266-268.
Povarennykh (1972) 144,145, 328,329.
Roberts et al. (1974) 361.

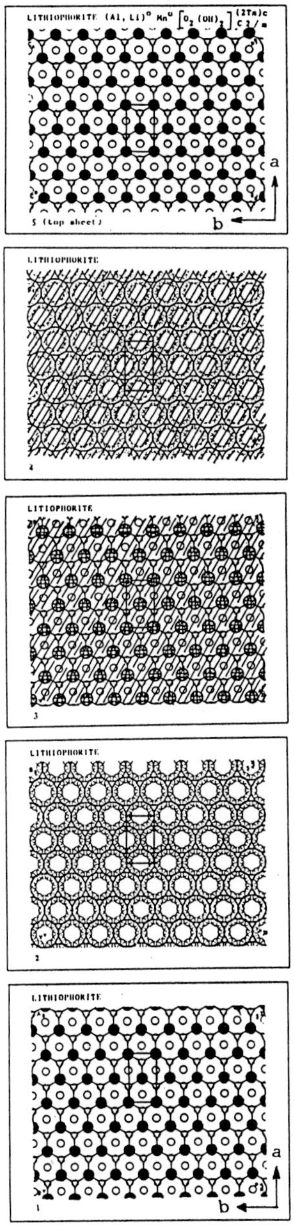

Fig. 3

MOLYBDENITE

$Mo^{p}\left[S_2\right]^{(2Ts)h}$

$P\,6_3/mmc$

a = 3.1604 Å
c = 12.295 Å
Z = 2

Mo(2c)
S (4f) u = 0.629

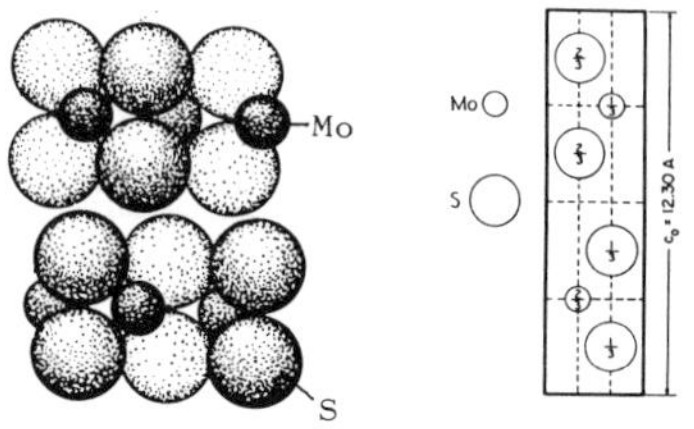

(a) Fig. 1 (b)

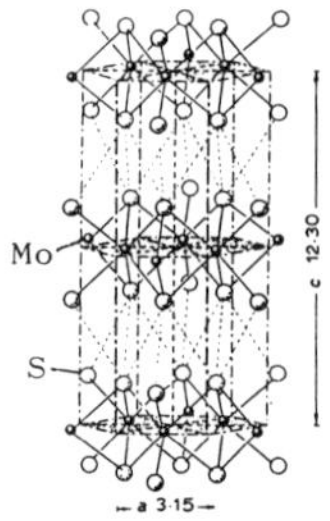

Fig. 2

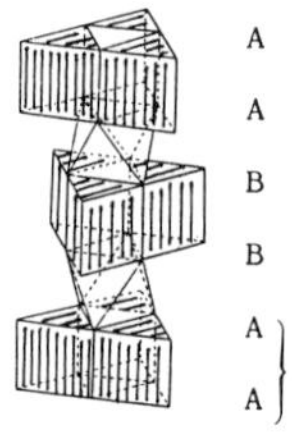

Fig. 3

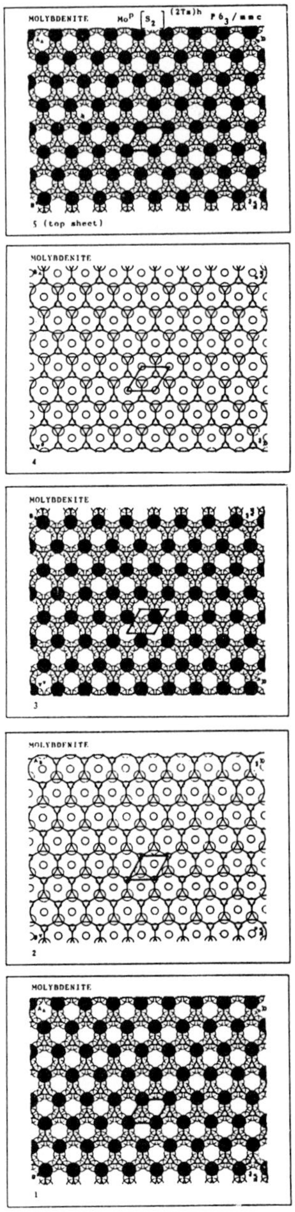

Fig. 4

Properties

Habit	Cleav.	Fract.	Twin.	Hardn.	Dens.	Colour	Transp.
hexagonal, tabular, flexible, massive	perfect (0001)		(0001)	1 - 1.5	4.62-4.73	lead-grey	opaque

Refr. index	/Reflect.	Birefr.	Luster	Streak	Melt.p.	CPI
n_ω = 4.336 n_ϵ = 2.035	31,5 %	(-)	metallic	greenish	1185°C	

Population

Tungstenite $W^{p}\left[S_2\right]^{(2Ts)h}$

Figures

Fig. 1. (a) Packing drawing of the molybdenite structure, and (b) unit cell content projected on a side face of the hexagonal cell (after Wyckoff, 1963, Vol. 1).

Fig. 2. Ball and spoke model of the molybdenite structure (after Kostov, 1968).

Fig. 3. Polyhedral description of the molybdenite structure (adapted from Kostov, 1968).

Fig. 4. Condensed model of the molybdenite structure. The large open circles represent S atoms, and the small black circles to Mo atoms in prismatic voids.

Description

The molybdenite structure is based on a heterogeneous packing. Double simple hexagonal layers are stacked in a hexagonal closest way with an hexagonal sequ ence (h). The Mo atoms occupy 1/2 of the trigonal prismatic voids, and the octahedral voids within the closest packing layers ara all empty.

References

Kostov (1968) 137-139.
Wyckoff (1963) Vol. 1, 280-282.
Povarennykh (1972) 259,260.
Palache et al. (1944) Vol. 1, 328,329.
Roberts et al. (1974) 412.

GIBBSITE $Al^{o}\left[(OH)_3\right]^{(2Ts)h}$

(Hydrargillite)

$P\,2_1/n$

a = 8.6236 Å
b = 5.0602 Å
c = 9.699 Å
β = 85° 26'
Z = 8

		x	y	z
Al_I	(4e)	0.176	0.520	0.005
Al_{II}	(4e)	0.333	0.020	0.005
O_I	(4e)	0.181	0.205	0.110
O_{II}	(4e)	0.681	0.671	0.110
O_{III}	(4e)	0.515	0.131	0.110
O_{IV}	(4e)	-0.015	0.631	0.110
O_V	(4e)	0.298	0.701	0.100
O_{VI}	(4e)	0.838	0.171	0.100

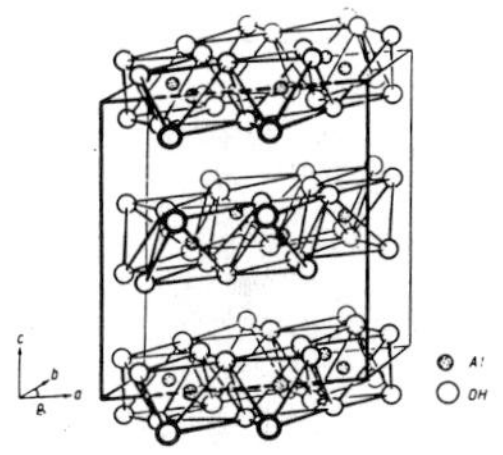

Fig. 1

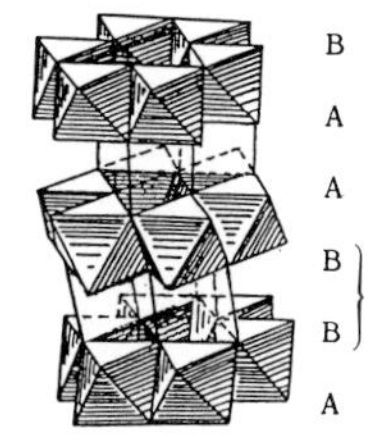

Fig. 2

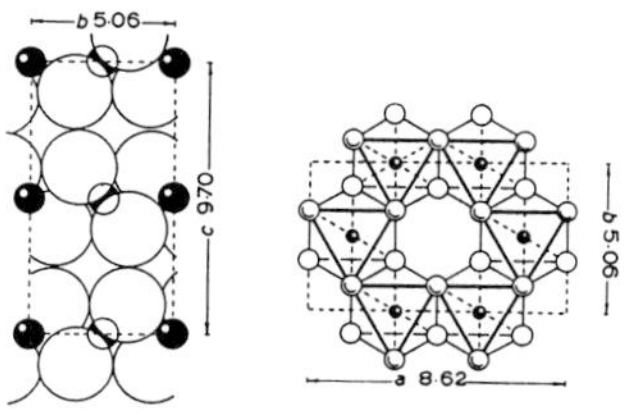

(a) Fig. 3 (b)

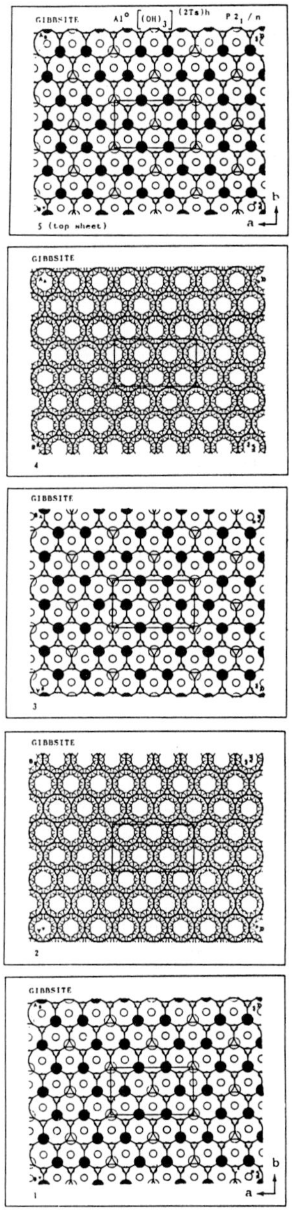

Fig. 4

Properties

Habit	Cleav.	Fract.	Twin.	Hardn.	Dens.	Colour	Transp.
tabular, foliated	perfect (001)	uneven	(310) (001)	2.5-3.5	2.40	white, grey	transparent to translucent

Refr. index/Reflect.	Birefr.	Luster	Streak	Melt.p.	CPI
n_α = 1.57 n_β = 1.57 n_γ = 1.59	(+) 2V= 0°-40°	pearly, vitreous	white		(SPI) 39

Distortion derivative

Bayerite $Al^{o}\left[(OH)_3\right]^{(2Ts)h}$ $P\,2_1/a$

Description

The gibbsite structure is based on a heterogeneous packing, with alternate layers of simple hexagonal and closest packed layers. The Al atoms occupy 2/3 of the voids within the closest packed layers, forming a honeycomb pattern. The double simple hexagonal layers have all the prismatic voids empty.

Figures

Fig. 1. Ball and spoke model of the gibbsite structure (after Povarennykh, 1972).

Fig. 2. Polyhedral description of the gibbsite structure (after Povarennykh, 1972).

Fig. 3. Structure of gibbsite: (a) projected on (100), (b) seen along the pseudohexagonal axis (after Kostov, 1968).

Fig. 4. Condensed model of the gibbsite structure. The large open circles represent OH, and the small black circles Al atoms in 2/3 of the octahedral voids, forming a honeycomb pattern.

References

Kostov (1968) 219-221.
Povarennykh (1972) 323,324.
Wyckoff (1964) Vol. 2, 78-80.
Zoltai + Stout (1984) 420,421.

COVELLINE (Covellite)	$Cu^{t}_{2}\,Cu^{tr}[S_3]$ (3Th)s $P\,6_3/m\,m\,c$	a = 3.7938 Å c = 16.341 Å Z = 2	Cu^{tr} (2d) Cu^{t} (4f) – z = 0.10733 S_{I} (2c) S_{II} (4e) – z = 0.06337

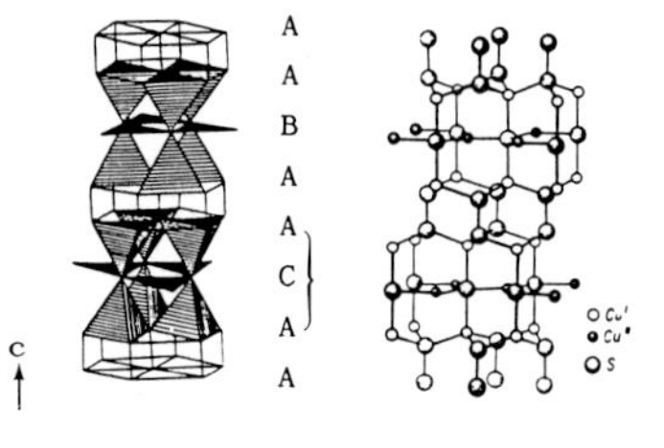

(a) Fig. 1 (b)

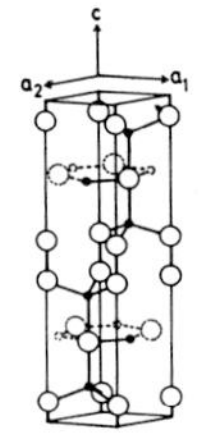

Fig. 2

Properties

Habit	Cleav.	Fract.	Twin.	Hardn.	Dens.	Colour	Transp.
platy, massive, spheroidal	perfect (0001)	conchoidal		1.5 - 2	4.6-4.76	indigo-blue	opaque

Refr. index	/Reflect.	Birefr.	Luster	Streak	Melt.p.	CPI
n_{ω} = 1.60 n_{ϵ} = 1.45	ω = 15% ϵ = 24%	(+)	submetallic	lead-grey		

Figures

Fig. 1. (a) Polyhedral description of the covelline structure and (b) corresponding ball and spoke model (after Povarennykh, 1972).

Fig. 2. Perspective of the unit cell of covelline structure (after Goble, 1985).

Fig. 3. Condensed model of the covelline structure. The large open circles represent S atoms. The small black circles correspond to Cu atoms in tetrahedral voids, and the still smaller black circles to other Cu atoms in triangular voids.

Description

The covelline structure is based on a heterogeneous packing of S atoms. The large sequence is ACAABAACA .... =(3Th)s.

The Cu atoms are of two categories: one occupy tetrahedral voids, and the other triangular voids.

References

Kostov (1968) 152.
Povarennykh (1972) 265,266.
Evans + Konnert (1976) 996-1000.
Zoltai + Stout (1984) 391-393.
Goble (1985) 62.

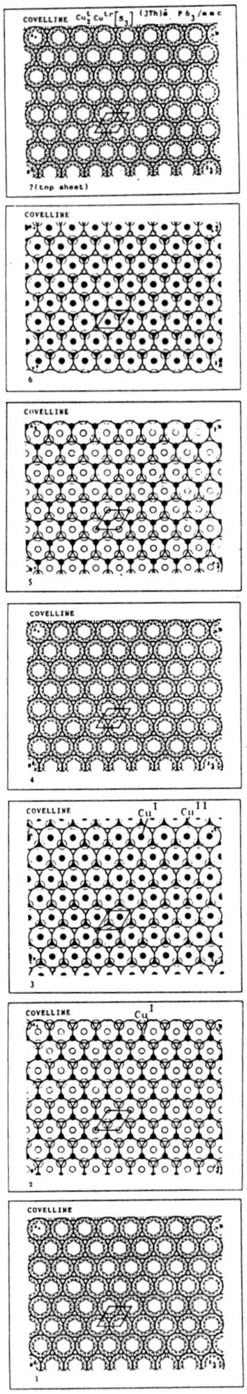

Fig. 3

MATLOCKITE

P 4/n m m

a=4.106 Å
c=7.23 Å
Z=2

Pb(2c) u=0.20
F (2a)
Cl(2c) u=0.65

$Pb^{[9]}\left[[F]\,[Cl]\right]\,(Q^2, 2Q^1f)f$

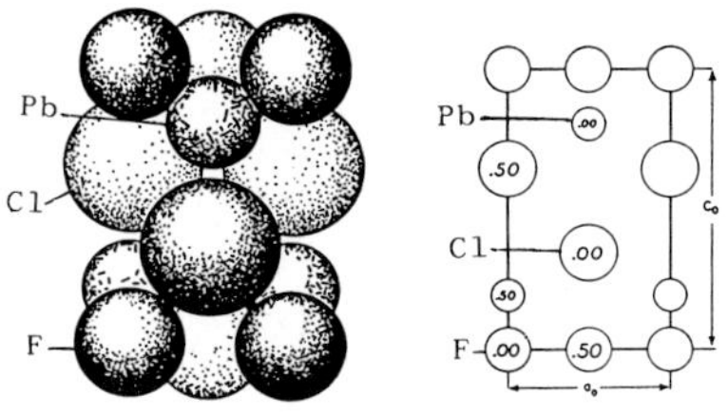

(a) (b)

Fig. 1

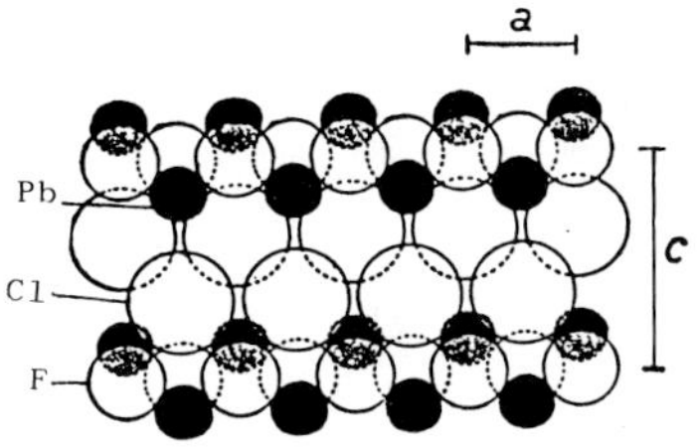

Fig. 2

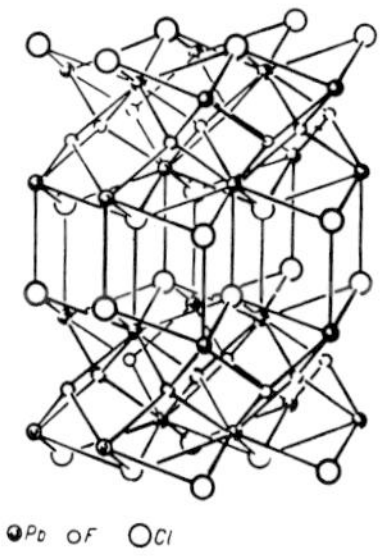

Fig. 3

Properties

Habit	Cleav.	Fract.	Twin.	Hardn.	Dens.	Colour	Transp.
tabular pyramidal, massive	perfect (001)	uneven to sub-conchoidal		2.5-3	7.12	colourless, yellow, greenish	transparent

Refr. index/Reflect.	Birefr.	Luster	Streak	Melt.p.	CPI
n_ω = 2.145 n_ϵ = 2.006	(-)	adamantine to pearly		601° C	

Population

Bismoclite	$Bi^{[9]}\{[O]\,[Cl]\}^{(Q^2, 2Q^1f)f}$
Zavaritskite	$Bi^{[9]}\{[O]\,[F]\}^{(Q^2, 2Q^1f)f}$

Figures

Fig. 1. (a) Packing drawing of the matlockite structure, and (b) unit cell content projected along one of the a axis (after Wyckoff, 1963, Vol. 1).

Fig. 2. Another packing drawing of the matlockite structure (after Wells, 1962).

Fig. 3. Ball and spoke model of the matlockite structure (after Povarennykh, 1972).

Fig. 4. Condensed model of the matlockite structure. The larger open circles represent Cl atoms and the other open circles (smaller) correspond to F atoms. The small black circles represent Pb with a coordination [9], four F, four Cl atoms, and more a Cl atom from the third layer.

Description

The matlockite structure is based on an heterogeneous packing, where the packing atoms are not all alike but of two categories: Cl (larger) and F. All form square layers (Q) which stack in the closest way. The relation between Cl and F square layers is 1F, (Q^2), to 2 Cl, ($2Q^1$). The sequence is 2 square Cl layers stacked in the closest way (f), therefore $2Q^1f$, and a Q^2 layer of F atoms also stacked in the closest way in relation to Q^1. The complete stacking sequence is (Q^2, $2Q^1f$)f. The Pb atoms are located between Q^2 and Q^1 layers, and are [9] coordinated (9 = 8+1).

References

Kostov (1968) 203,204.
Povarennykh (1972) 644,645, 653,654.
Wyckoff (1963) Vol. 1, 295,296.
Wells (1962) 674.
Palache et al. (1951) Vol. 2, 59.
Roberts et al. (1974) 386.

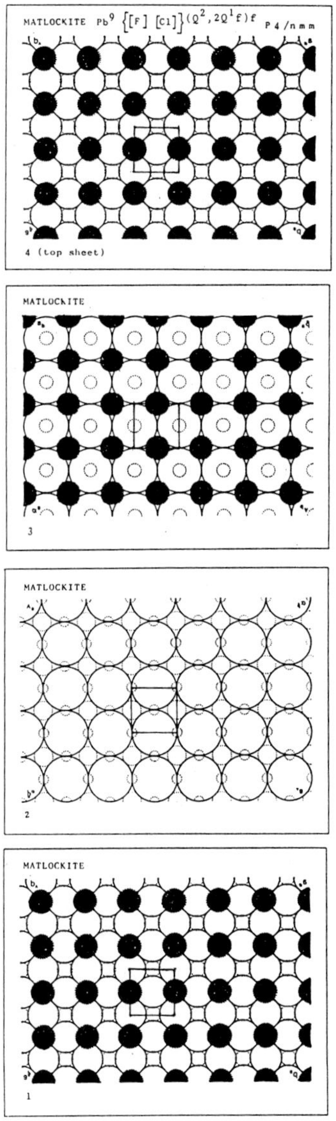

Fig. 4

DIABOLEITE

P 4 m m

$Pb_2^{[8]}\ Cu^{sq}\left[\left[(OH)\right]_4\left[Cl\right]_2\right]\ (Q^2Q^1)\ f$

a = 5.870 Å
c = 5.494 Å
Z = 1
(OH) (4d) u = 0.250, v = 0.278

Pb (2c) u = 0.00
Cu (1a) u = 0.278
Cl_I(1a) u = -0.222
Cl_{II}(1b) u = -0.125

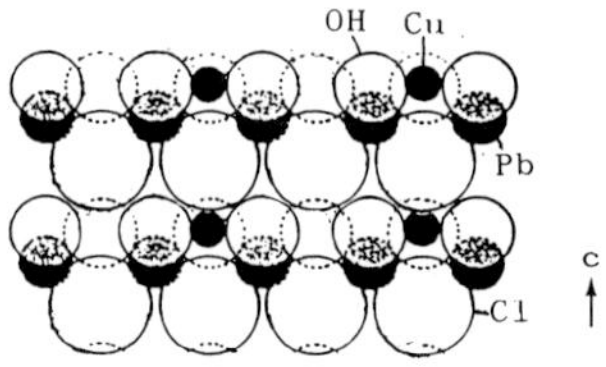

Fig. 1

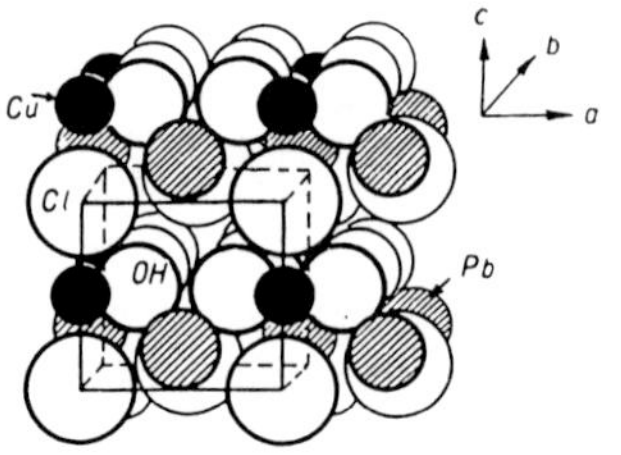

Fig. 2

Properties

Habit	Cleav.	Fract.	Twin.	Hardn.	Dens.	Colour	Transp.
tabular, thin plates	perfect (001)	conchoi- dal		2.5	5.42	deep blue	transparent

Refr. index/Reflect.	Birefr.	Luster	Streak	Melt.p.	CPI
n_ω = 1.98 n_ϵ = 1.85	(-)	vitreous	pale blue		

Figures

Fig. 1. Packing drawing of the diaboleite structure (adapted from Wells, 1962).

Fig. 2. Packing drawing in a three dimentional perspective of the diaboleite structure (after Povarennykh, 1972).

Fig. 3. Condensed model of the diaboleite structure. Large open circles represent Cl atoms, and smaller open circles correspond to OH. On the Cl square layer the crossed circles represent Pb atoms with coordination [8] . On the layers formed by OH the small black circles correspond to Cu atoms in square voids.

Description

The diaboleite structure is based on an heterogeneous packing of Cl, (Q^1), and of OH, (Q^2). The square packing layers of Cl and of OH, alternate and stack in the closest way; the stacking symbol for such kind of sequences is ($Q^2\ Q^1$)f. The Cu atoms occupy 1/4 of the square voids of the Q^2 layers of OH, and the Pb atoms all the anti-cubic voids of the Q^1 layers of Cl atoms.

References

Kostov (1968) 203.
Povarennykh (1972) 656,657.
Wyckoff (1968) Vol. 4, 98,99.
Wells (1962), 674.
Palache et al. (1951) Vol. 2, 82,83.
Roberts et al. (1974) 171.

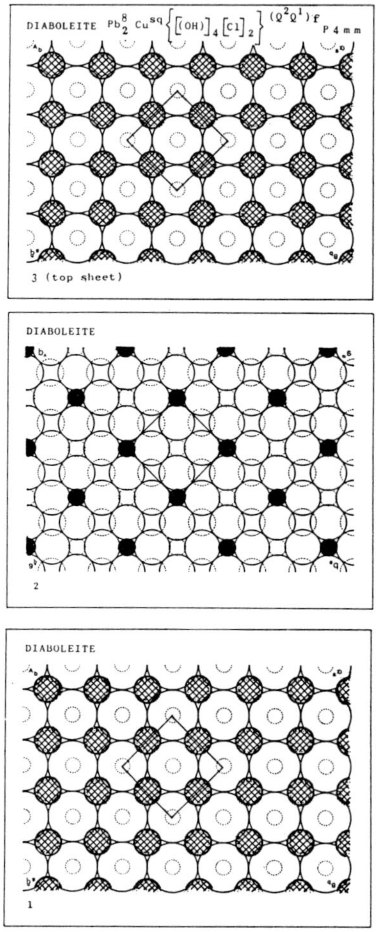

Fig. 3

PERITE

B m m b

$Pb^{[8]} Bi^{[8]} \left[[O]_2 [Cl] \right]^{(Q^2Q^1)f}$

a = 5.627 Å	Pb(4c) u = 0.385
b = 5.575 Å	Bi(4c) u = 0.090
c = 12.425 Å	O (8e) u = 0.25
Z = 4	Cl(4c) u = 0.75

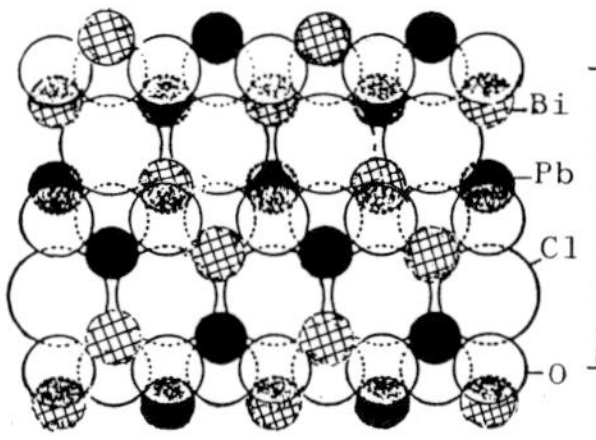

Fig. 1

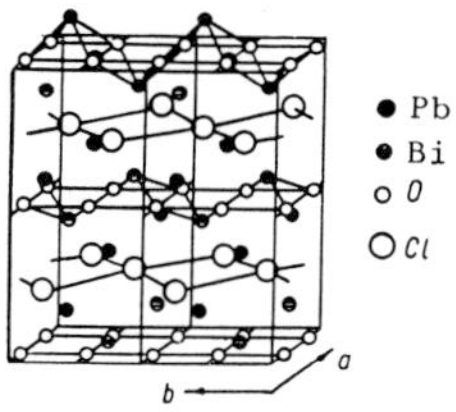

Fig. 2

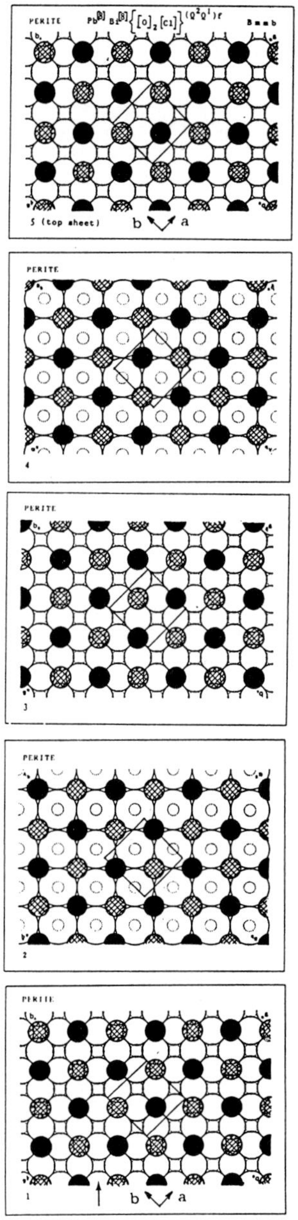

Fig. 3

Properties

Habit	Cleav.	Fract.	Twin.	Hardn.	Dens.	Colour	Transp.
tabular	distinct (001)			3	8.16	sulphur yellow	

Refr. index /Reflect.	Birefr.	Luster	Streak	Melt.p.	CPI
n = 2 - 2.4	(-)	adamantine			

Figures

Fig. 1. Packing drawing projected along the direction of the arrow Fig. 3 (adapted from Wells, 1962).

Fig. 2. Ball and spoke representation of the perite structure (adapted from Povarennykh, 1972).

Fig. 3. Condensed model of the perite structure. Large open circles represent Cl atoms, and smaller open circles O atoms. The small black circles correspond to Pb atoms with coordination [8] , and the crossed circles to Bi atoms.

Description

The perite structure is based on an heterogeneous packing of Cl, Q^2, and of O atoms, Q^1. The square packing layers of Cl and of O atoms alternate and stack in the closest way; the stacking symbol for such kind of sequence is $(Q^2 Q^1)f$. The Pb and Bi atoms occupy 1/2 of the [8] coordinated voids of this packing.

References

Kostov (1968) 204.
Povarennykh (1972) 655,656.
Wyckoff (1968) Vol. 4, 96,97.
Wells (1962) 674.
Roberts et al. (1974) 472.

STIBNITE (Antimonite)	$[Sb_2S_3]$ c/h Pbnm	a = 11.299Å, b = 11.310Å, c = 3.8389, Z = 4	Sb_I (4c) u = 0.326, v = 0.030; Sb_{II}(4c) u = 0.536, v = 0.351; S_I (4c) u = 0.880, v = 0.055	S_{II} (4c) u = 0.559, v = 0.869; S_{III}(4c) u = 0.189, v = 0.214

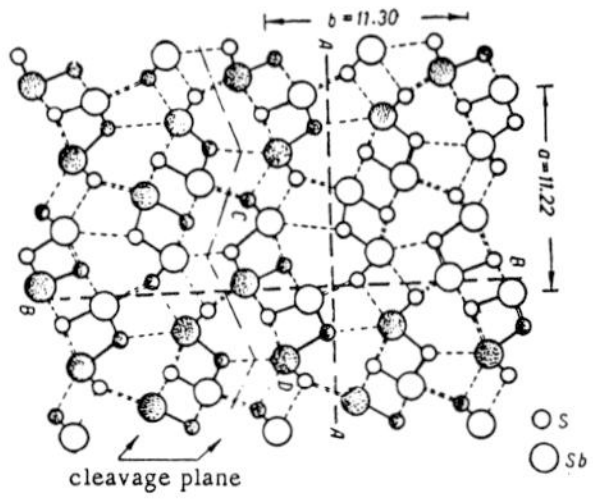

(a) Fig. 1 (b)

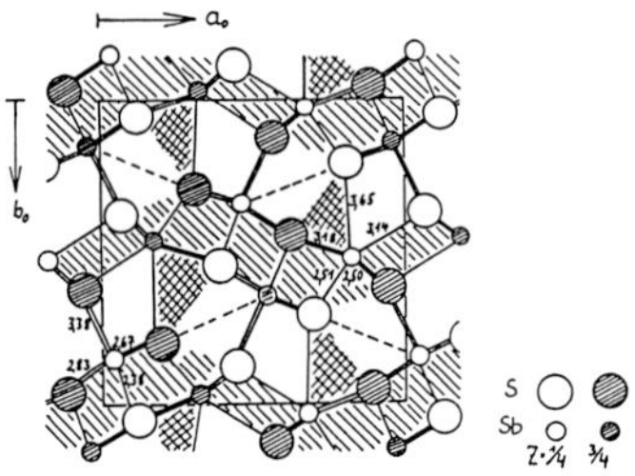

Fig. 2

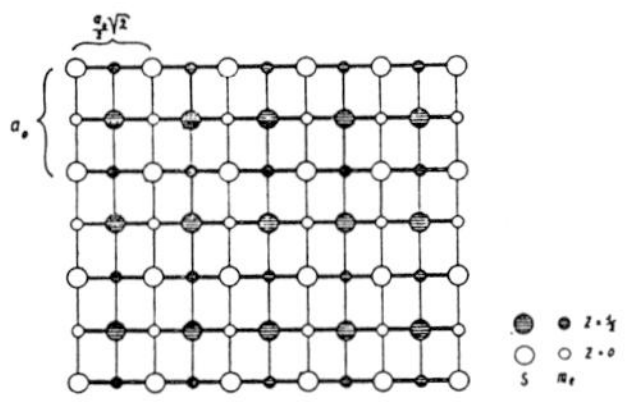

Fig. 3

Properties

Habit	Cleav.	Fract.	Twin.	Hardn.	Dens.	Colour	Transp.
prismatic acicular, massive	perfect (010)	subcon-choidal	(130)	2	4.63	lead--grey	opaque

Refr. index	/Reflect.	Birefr.	Luster	Streak	Melt.p.	CPI
n_α = 3.194, n_β = 4.046, n_γ = 4.303	34,2 %	(-) 2V = 25° 45'	metallic	lead--grey	546°C	

Figures

Fig. 1. (a) Representation of the stibnite structure: (a) projected along the c axis, and (b) showing the way the Sb_2S_3 chains are linked together (after Povarennykh, 1972).

Fig. 2. The structure of stibnite projected along the c axis (after Hellner, 1958).

Fig. 3. The structure of galena, PbS, projected on (110) (after Hellner, 1958).

Description

The stibnite structure is a recombination structure based on a heterogeneous packing formed by interpenetrated slabs of closest packings (c/h) of deformed galena type structure parallel to (110).

The stibnite structure may also be considered based on Sb_2S_3 chains linked together along the a axis forming bands, what is in agreement with the perfect (010) cleavage.

References

Kostov (1968) 164,165.
Povarennykh (1972) 128, 246.
Wyckoff (1964) Vol. 2, 27,28.
Hellner (1958) 504, 519.
Palache et al. (1944) Vol. 1, 271-273.
Ingerson (1955) 350.

BOURNONITE

$Pn2_1m$

$\left[Pb\ Cu\ Sb\ S_3\right]^{c/h}$

a = 8.162 Å
b = 8.7105 Å
c = 7.8105 Å
Z = 4

Atom	Position	x	y	z
Cu	(4b)	0.269	0.440	0.249
Pb_I	(2a)	0.079	0	0
Pb_{II}	(2a)	0.440	0.670	0
Sb_I	(2a)	-0.073	0.550	0
Sb_{II}	(2a)	0.515	0.125	0
S_I	(2a)	0.225	0.310	0

...

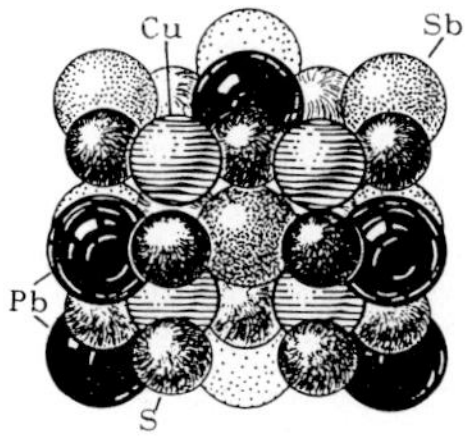

(a)

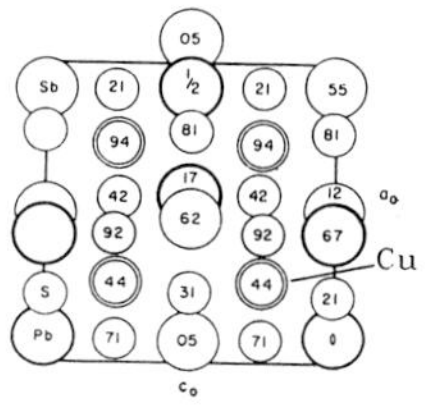

(b)

Fig. 1

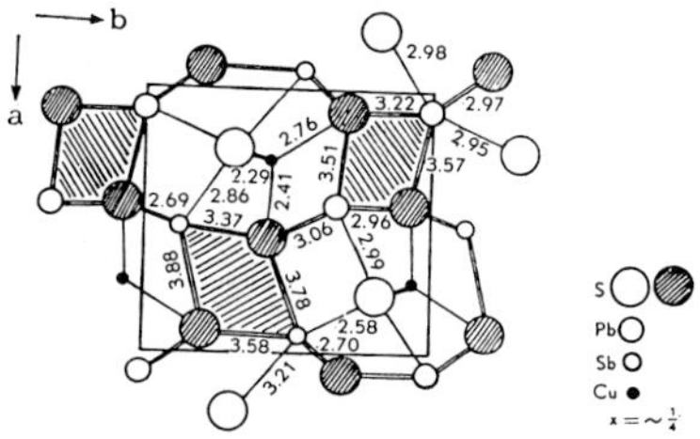

Fig. 2

Properties

Habit	Cleav.	Fract.	Twin.	Hardn.	Dens.	Colour	Transp.
prismatic, tabular, massive	imperfect (010) (100)	subconchoidal	(110)	2.5 - 3	5.83	steel-grey	opaque

Refr. index/Reflect.	Birefr.	Luster	Streak	Melt.p.	CPI
30 %		metallic	steel-grey		

Population

Seligmannite

$\left[Pb\ Cu\ As\ S_3\right]^{c/h}$

Description

Bournonite is a recombination structure based on a heterogeneous close packing formed by interpenetrated slabs of closest packings (c/h) of deformed galena-type structure.

Figures

Fig. 1. (a) Packing drawing of the bournonite structure, and (b) corresponding unit cell content projected along the b axis (after Wyckoff, 1964, Vol. 2).

Fig. 2. Ball and spoke description of the bournonite structure (after Hellner + Leineweber, 1956, quoted in Structure Reports, 1963, Vol. 20).

References

Kostov (1968) 172.

Povarennykh (1972) 237.

Wyckoff (1964) Vol. 2, 508,509

Structure Reports (1963) Vol. 20, 30-32.

Hellner + Leineweber (1956) 150-154.

Palache et al. (1944) Vol. 1, 407.

Crystallographic data (continued)

Atom	Position	x	y	z
S_{II}	(2a)	-0.225	0.810	0
S_{III}	(4b)	0.084	0.706	0.248
S_{IV}	(4b)	0.564	0.425	0.267

COSALITE

Pbnm

$\left[Pb_2 Bi_2 S_5\right]^{c/h}$

a = 19.101Å
b = 23.913Å
c = 4.061Å
Z = 8

Atom		u	v
$(Pb,Bi)_I$	(4c)	-0.018	0.155
$(Pb,Bi)_{II}$	(4c)	0.093	0.302
$(Pb,Bi)_{III}$	(4c)	0.274	0.040
$(Pb,Bi)_{IV}$	(4c)	-0.035	-0.433
$(Pb,Bi)_V$	(4c)	-0.091	-0.011
$(Pb,Bi)_{VI}$	(4c)	-0.186	-0.171

...

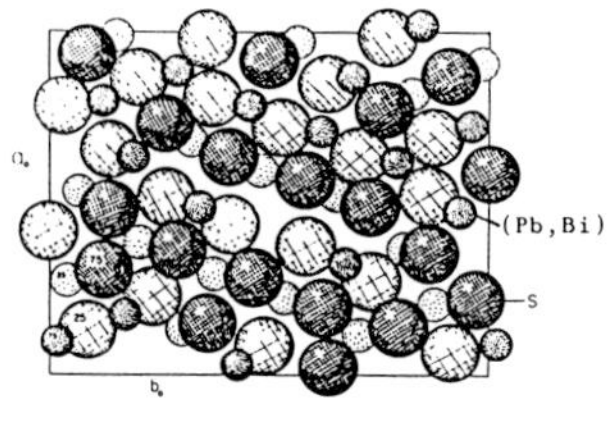

Fig. 1

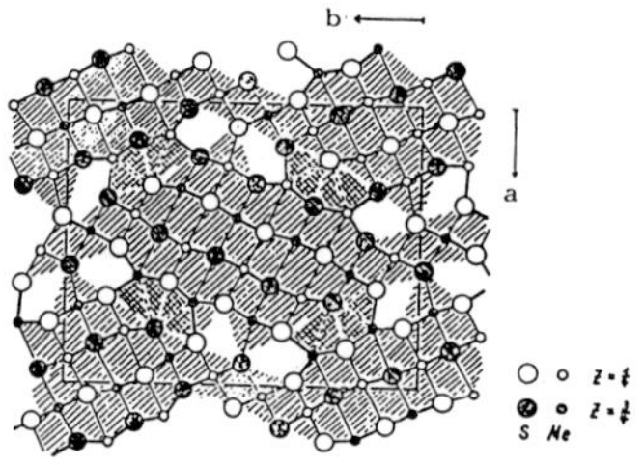

Fig. 2

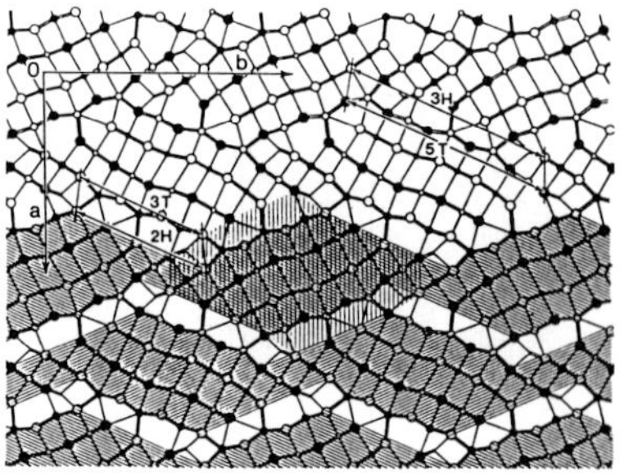

Fig. 3

Properties

Habit	Cleav.	Fract.	Twin.	Hardn.	Dens.	Colour	Transp.
prismatic, acicular, fibrous, massive	perfect prismatic (001)	uneven		2.5 - 3	6.76	lead--grey, steel--grey	opaque

Refr. index/Reflect.	Birefr.	Luster	Streak	Melt.p.	CPI
		metallic	black		

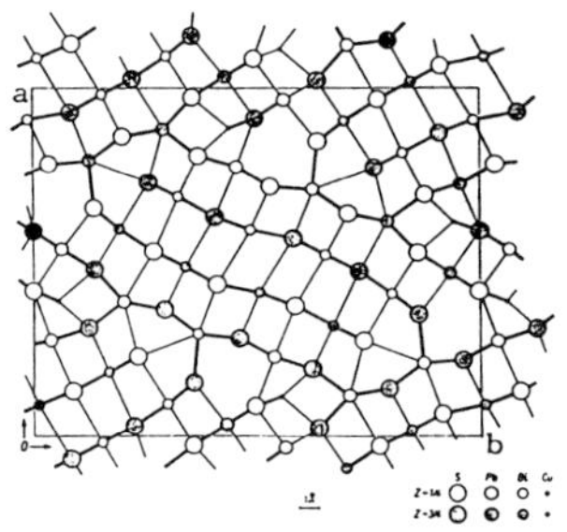

Fig. 4

Figures

Fig. 1. Projection of the cosalite structure along the c axis (after Wyckoff, 1965, Vol. 3).

Fig. 2. Structure of cosalite projected along the c axis, showing the different kinds of slabs of deformed galena-type structure (after Hellner, 1958).

Fig. 3. The structure of cosalite. Sulphur atoms are denoted by large circles, metal atoms (Pb, Bi) by small circles. Atoms at Z = 0 and 1/2 are indicated by empty and filled circles respectively. The zig--zag octahedral layers are hatched; interlayer spaces with complicated Pb-S bond geometry are left blank. Aproximate H: T subcell matches are indicated (after Makovicky, 1985).

Fig. 4. Structure of cosalite projected along the c axis (after Sriksishnan + Nowacki, 1974).

Description

The cosalite structure is a recombination structure based on a heterogeneous packing formed by interpenetrated slabs of closest packings (c/h) of deformed galena--type structure (77%), (after Hellner, 1958) some parallel to (110) and others to (111).

References

Kostov (1968) 181.
Povarennykh (1972) 252.
Wyckoff (1965) Vol. 3, 317-319.
Srikrishnan + Nowacki (1974) 134.
Hellner (1958) 503-525.
Makovicky (1985) 15.
Palache et al. (1944) Vol. 1, 445.

Crystallographic data (continued)

Atom		u	v
$(Pb,Bi)_{VII}$	(4c)	0.383	0.206
$(Pb,Bi)_{VIII}$	(4c)	0.293	0.374
S_I	(4c)	-0.476	-0.138
S_{II}	(4c)	-0.029	-0.230
S_{III}	(4c)	0.413	-0.002
S_{IV}	(4c)	-0.152	-0.361
S_V	(4c)	-0.285	-0.465
S_{VI}	(4c)	0.228	0.235
S_{VII}	(4c)	0.131	0.088
S_{VIII}	(4c)	-0.306	-0.125
S_{IX}	(4c)	-0.360	-0.296
S_X	(4c)	0.434	0.418

GALENOBISMUTITE

$Pb^{7}\left[Bi_2^{6/7}\ S_4\right]^{c/h}$

Pnma

a = 11.79Å
b = 4.10Å
c = 14.59Å
Z = 4

Atom	Position	u	v
Pb	(4c)	0.2479	0.6513
Bi_I	(4c)	0.0675	0.3901
Bi_{II}	(4c)	0.1043	0.9056
S_I	(4c)	0.3307	0.0141
S_{II}	(4c)	0.2609	0.2997
S_{III}	(4c)	0.0550	0.0927

...

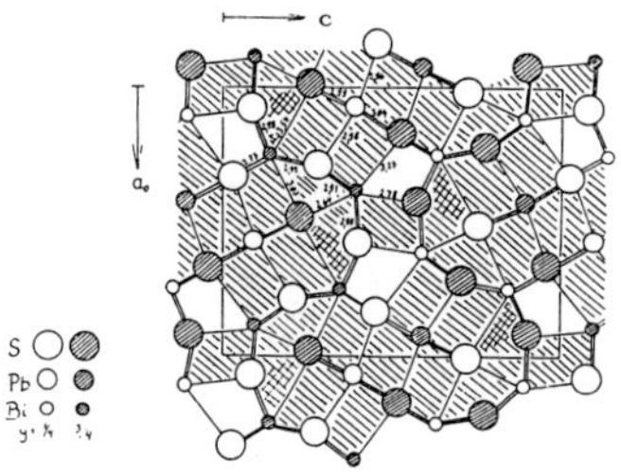

Fig. 1

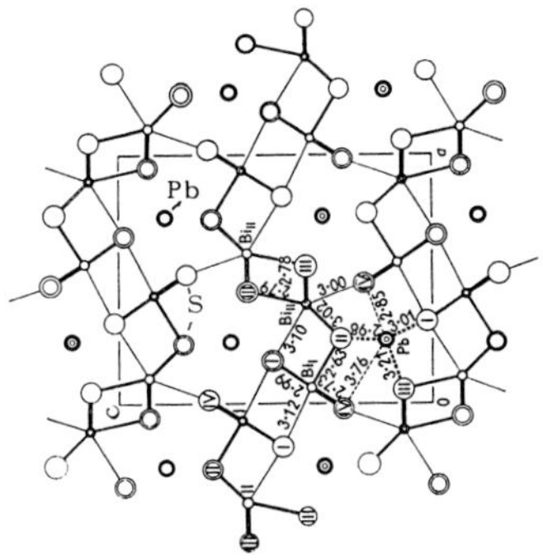

Fig. 2

Properties

Habit	Cleav.	Fract.	Twin.	Hardn.	Dens.	Colour	Transp.
lathlike, needles, massive	good (101)			2.5 - 3.5	7.04	light grey, tin-white	opaque

Refr. index/Reflect.	Birefr.	Luster	Streak	Melt.p.	CPI
		metallic	black		

Figures

Fig. 1. Projection of the galenobismutite structure along the b axis showing the different slabs from which can be derived. It contains 82% of the distorted galena-type arrangement (after Hellner, 1958).

Fig. 2. Projection of the structure of galenobismutite along the b axis (after Iitaka + Nowacki, 1962).

Description

The galenobismutite is a recombination structure based on a heterogeneous packing formed by interpenetrated slabs of closest packings (c/h) of deformed galena-type structure (82%), some parallel to (110) and others to (111).

Crystallographic data (continued)

S_{IV} (4c) u = 0.0181
v = 0.7120

References

Kostov (1968) 181.
Povarennykh (1972) 250,251.
Wyckoff (1965) Vol. 3, 134-136.
Hellner (1958) 504, 514, 517.
Iitaka + Nowacki (1962) 696.
Palache et al. (1944) Vol. 1, 471,472.

BOULANGERITE

P b n m

$Pb_5\left[Sb_4\,S_{11}\right]$ c/h

a = 21.14 Å
b = 23.46 Å
c = 4.035 Å
Z = 4

Atom	Position	x	y	z
Pb_I	(4c)	0.307	0.160	1/4
Pb_{II}	(4c)	0.117	0.498	1/4
Pb_{III}	(4c)	0.206	0.323	1/4
$(Pb,Sb)_I$	(4c)	0.458	0.433	3/4
$(Pb,Sb)_{II}$	(4c)	0.132	0.098	3/4
$(Pb,Sb)_{III}$	(4c)	0.486	0.129	3/4
...				

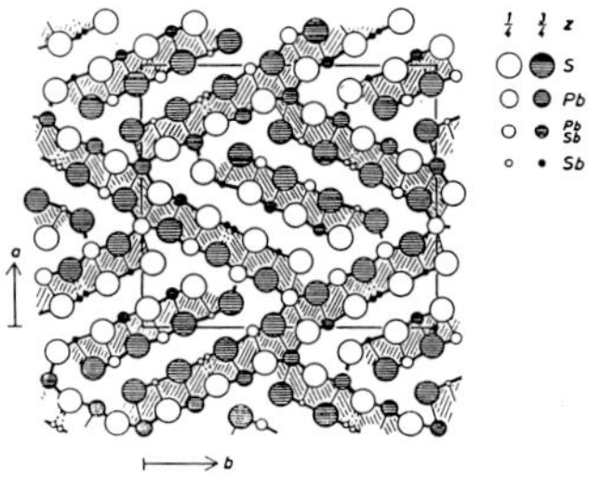

Fig. 1

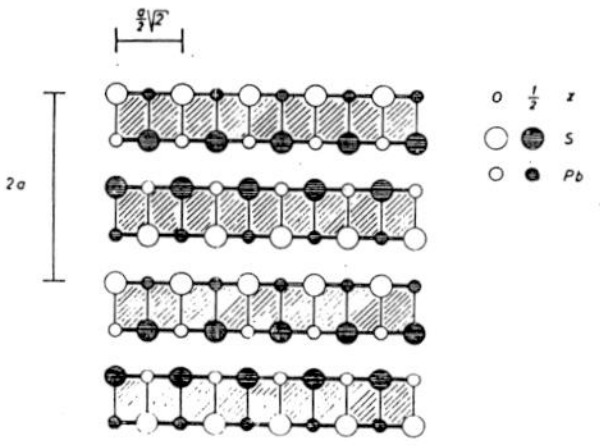

Fig. 2

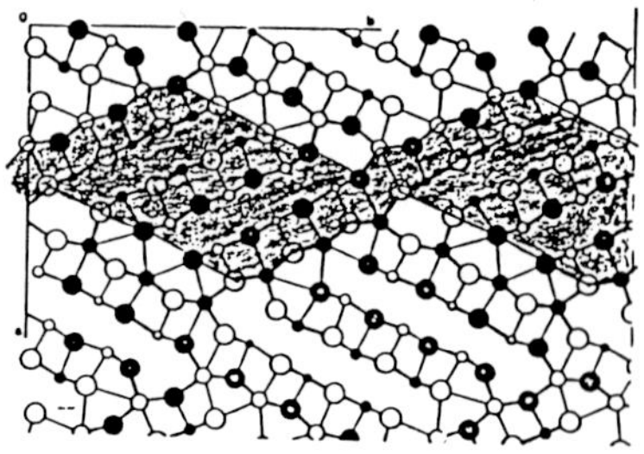

Fig. 3

Properties

Habit	Cleav.	Fract.	Twin.	Hardn.	Dens.	Colour	Transp.
prismatic, acicular, fibrous masses	good (100)			2.5 - 3	6.23	bluish lead-grey	opaque

Refr. index/Reflect.	Birefr.	Luster	Streak	Melt.p.	CPI
34,5%		metallic	brownish grey		

Figures

Fig. 1. Structure of boulangerite for the sub-cell, projected on (001); the hatched parts indicate the chains in the c-direction (after Born + Hellner, 1960).

Fig. 2. Projection of the galena type on (110), but half of the z parameters of the sulphur and metal atoms are interchanged in such a way that the 3 - dimensional network of the galena type changes into 2 - dimensional infinite nets (after Born + Hellner, 1960).

Fig. 3. Boulangerite structure projected on (001) (after Petrova et al., 1978, and quoted by Makovicky 1983).

Description

Boulangerite is a recombination structure based on a heterogeneous packing formed by interpenetrated slabs of closest packings (c/h) of deformed galena-type structure, most of them parallel to (110).

Also described as $P2_1/a$ with a = 21.56, b = 23.51, c = 8.09, $\beta = 100^\circ 48'$, Z = 8 (quoted in Strunz 1982).

References

Kostov (1968) 179.
Born + Hellner (1960) 1266-1271.
Makovicky (1983).
Strunz (1982) 149.

Crystallographic data (continued)

Atom	Position	x	y	z
Sb_I	(4c)	0.287	0.462	3/4
Sb_{II}	(4c)	0.046	0.232	1/4
1/2 Sb_I	(4c)	0.388	0.283	3/4
1/2 Sb_{II}	(4c)	0.372	0.307	3/4
S_I	(4c)	0.069	0.017	1/4
S_{II}	(4c)	0.247	0.028	1/4
S_{III}	(4c)	0.156	0.175	1/4
S_{IV}	(4c)	0.420	0.219	1/4
S_V	(4c)	0.330	0.375	1/4
S_{VI}	(4c)	0.372	0.088	3/4
S_{VII}	(4c)	0.014	0.138	3/4
S_{VIII}	(4c)	0.278	0.251	3/4
S_{IX}	(4c)	0.096	0.295	3/4
S_X	(4c)	0.186	0.415	3/4
S_{XI}	(4c)	0.013	0.441	3/4

APATITE

$Ca^{P}_{5}\ P^{t}_{3}\ [O_{12}\ (OH,F)]^{Ts/h}$

$P\ 6_3/m$

(for fluorapatite)
a = 9.3684 Å
c = 6.8841 Å
Z = 2

Ca_I	(4f)	u = 0.001
Ca_{II}	(6h)	u = 0.246, v = -0.007
F	(2a)	
P	(6h)	u = 0.400, v = 0.369
O_I	(6h)	u = 0.329, v = 0.484
O_{II}	(6h)	u = 0.589, v = 0.466
O_{III}	(12i)	x = 0.348, y = 0.259, z = 0.073

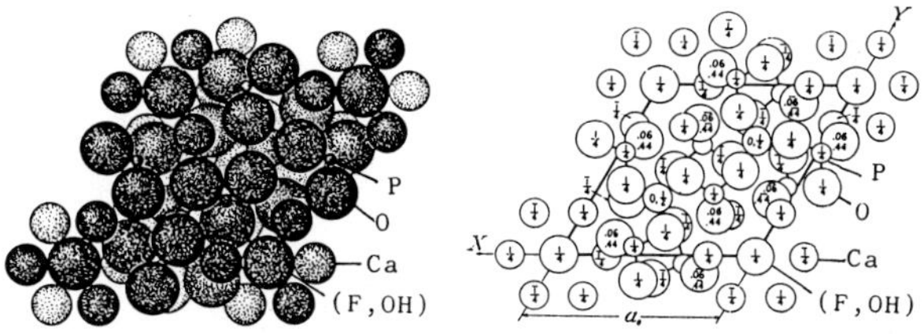

(a) Fig. 1 (b)

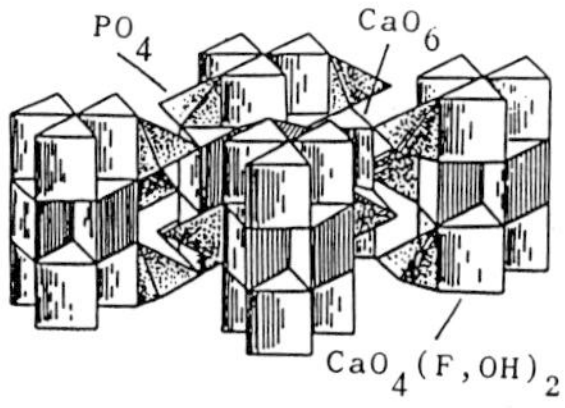

Fig. 2

Fig. 3

Properties

Habit	Cleav.	Fract.	Twin.	Hardn.	Dens.	Colour	Transp.
prismatic, tabular, fibrous	good (0001)	conchoidal		5	3.2	green, yellow, variable	transparent to transcent

Refr. index/Reflect.	Birefr.	Luster	Streak	Melt.p.	CPI
n_ω = 1.633 n_ϵ = 1.630	(-)	subresinous	white	1270° C	(SPI) 50

Population

Pyromorphite	$Pb^{P}_{5}\ P^{t}_{3}\ [O_{12}Cl]^{Ts/h}$
Vanadinite	$Pb^{P}_{5}\ V^{t}_{3}\ [O_{12}Cl]^{Ts/h}$
Mimetite	$Pb^{P}_{5}\ As^{t}_{3}[O_{12}Cl]^{Ts/h}$

Figures

Fig. 1. (a) Packing drawing of the apatite structure, and (b) unit cell content projected along the c axis (after Wyckoff, 1965, Vol. 3).

Fig. 2. Polyhedral description of the apatite structure. The prisms correspond to the triangular prismatic coordination of calcium, and the tetrahedra to PO_4 (after Povarennykh, 1972).

Fig. 3. Projection on (0001) of the polyhedral representation of apatite of Fig.1 (after Povarennykh, 1972). This aspect may be obtained with the superimposition of the first two sheets of the condensed model (Fig. 4).

Fig. 4. Condensed model of the apatite structure. The large open circles represent O atoms, and the large lined circles OH or F atoms. The black circles and the crossed circles correspond to Ca atoms in trigonal prismatic voids; the smaller black circles represent the P atoms with tetrahedral coordination.

Description

The apatite structure is based on a puckered simple hexagonal packing (∼Ts), or on the interpenetration of slabs of simple hexagonal and closest packings (Ts/h). If we imagine the structure built of simple hexagonal packing sheets (Fig. 4, and heavily outlined the packing atoms that are displaced 1/2 in relation to the other packing atoms, we obtain the analogue of the apatite structure. Ca atoms occupy the trigonal prismatic voids and the P atoms, which have tetrahedral coordination, are located on the fronteirs of these displaced parts of simple hexagonal packing, where the contact packing is closest.

References

Kostov (1968) 458-460, 468.
Povarennykh (1972) 494, 512, 541,542.
Wyckoff (1965) Vol. 3, 228-234.
Zoltai + Stout (1984) 449.
Ingerson (1955) 351.

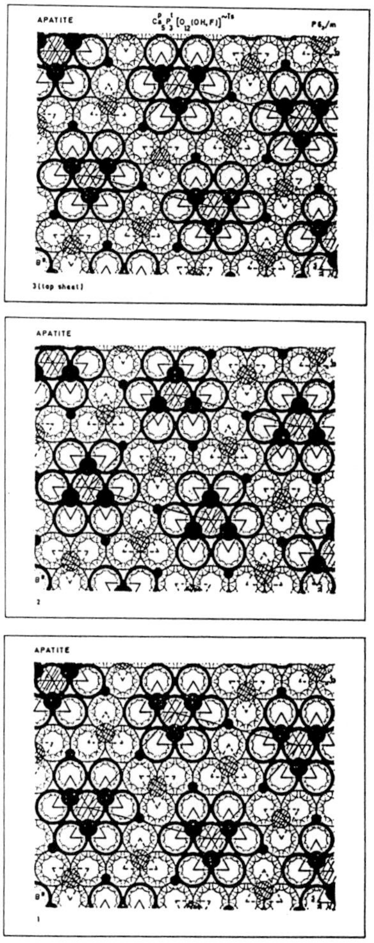

Fig. 4

8.4.2. Group structures

MARCASITE $Fe^{o}\{g\}[S_2]^{\ll h\gg c}$

Pnnm

a = 4.436Å
b = 5.414Å
c = 3.381Å
Z = 2

Fe (2a)
S (4g) u = 0.200
v = 0.378

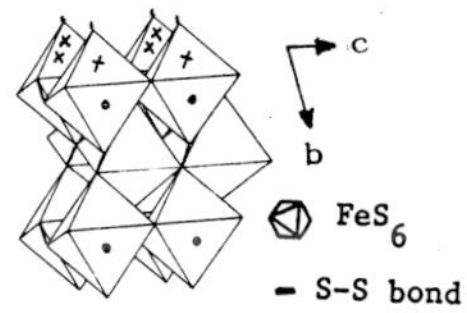

Fig. 5

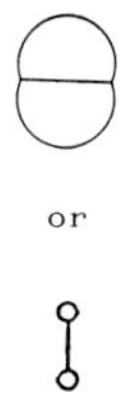

Fig. 1

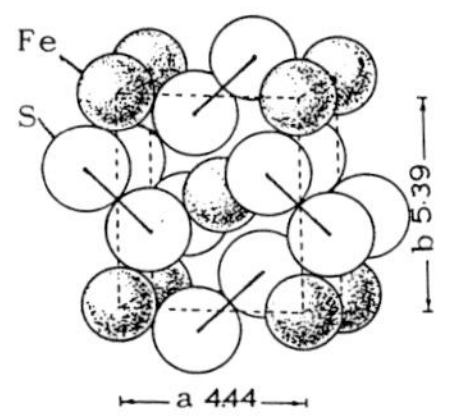

Fig. 2

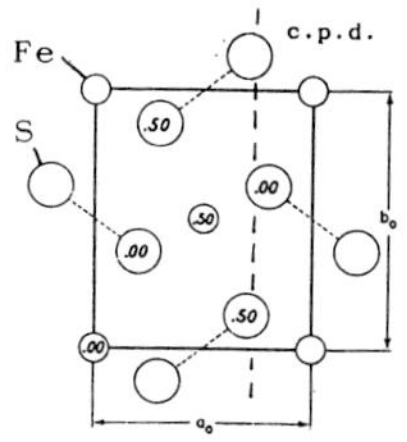

Fig. 3

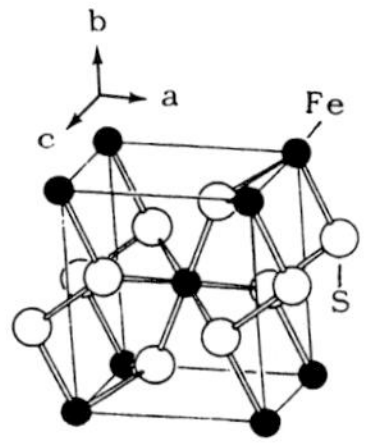

Fig. 4

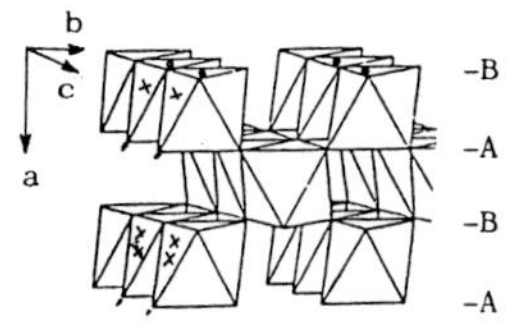

Fig. 6

Properties

Habit	Cleav.	Fract.	Twin.	Hardn.	Dens.	Colour	Transp.
tabular, pyramidal	distinct (101)	uneven	(101)	6 - 6.5	4.887	bronze- -yellow, green	opaque

Refr. index/Reflect.	Birefr.	Luster	Streak	Melt.p.	CPI
	45,5%	metallic	greyish, brownish black		

Population

Ferroselite	$Fe^{o}\{g\}[Se_2]^{\ll h\gg c}$
Hastite	$Co^{o}\{g\}[Se_2]^{\ll h\gg c}$
Kullerudite	$Ni^{o}\{g\}[Se_2]^{\ll h\gg c}$
Frohbergite	$Fe^{o}\{g\}[Te_2]^{\ll h\gg c}$
Rammelsbergite	$Ni^{o}\{g\}[As_2]^{\ll h\gg c}$
Löllingite	$Fe^{o}\{g\}[As_2]^{\ll h\gg c}$
Nisbite	$Ni^{o}\{g\}[Sb_2]^{\ll h\gg c}$

Distortion derivative

Safflorite $Co^{o}\{g\}[As_2]^{\ll h\gg c}$ monoc.

References

Kostov (1968) 130.
Povarennykh (1972) 206, 213, 243-244.
Wyckoff (1963) Vol. 1, 355, 356.
Zoltai (1974) 50.
R. Burns + V. Burns (1979) 13.
Strunz (1982) 136.
Wuensch (1974) 25.
Palache et al. (1944) Vol. 1, 312,313.

Figures

Fig. 1. The S_2 group of the marcasite structure.

Fig. 2. Packing drawing of the marcasite structure (adapted from Kostov, 1968).

Fig. 3. Projection along the c axis of the marcasite structure (after Wyckoff, 1963, Vol. 1). The dashed lines indicate the trace of the close packing direction (c.p.d.).

Fig. 4. Ball and spoke model of the marcasite structure (after Wuensch,1974). The close packing direction is parallel to the (100) plane.

Fig. 5. Polyhedral description of the marcasite structure (adapted from Zoltai, 1974).

Fig. 6. Polyhedral description of the packing analogue of the marcasite structure (see hydrophilite, $Ca^{o}[Cl_2]^{h}$) (adapted from R. Burns + V, Burns, 1969). The crosses and the black points relate faces of Fig. 5.

Fig. 7. Condensed model of the packing analogue of the marcasite structure. The large open circles represent S atoms, and the small black circles Fe atoms in 1/2 of the octahedral voids. The S-S bonds link S atoms of different layers (marked with double circles) and would be very difficult to indicate all on the condensed model.

Description

The marcasite structure is formed by the cubic (c) packing of S_2 groups with Fe atoms with octahedral coordination (o). The S atoms, if imagined separately, form an hexagonal closest packing (h). The packing analogue has the same structure as hydrophilite, $Ca^{o}[Cl_2]^{h}$, and is a coalescence derivative of it.

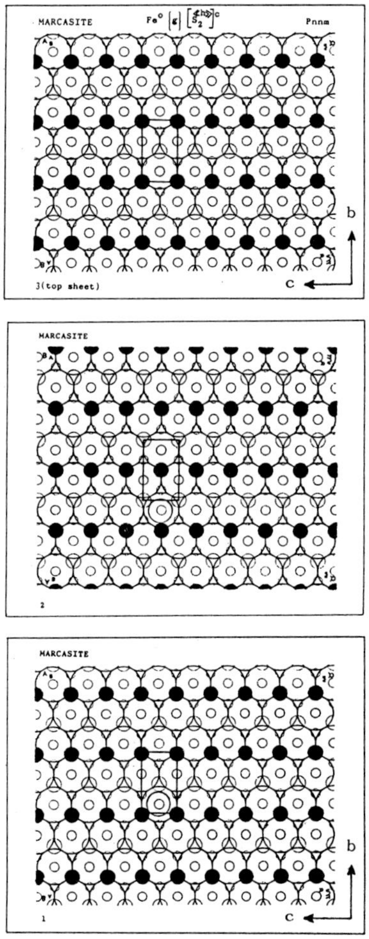

Fig. 7

ARSENOPYRITE

$B2_1/d$

Fe^{o} {g} [As S ≪h≫]

a = 9.15Å
b = 5.65Å
c = 6.42Å
β = 90°
Z = 8

Fe(8e) x = 0, y = 0, z = 0.275
As(8e) x = 0.147, y = 0.128, z = 0
S (8e) x = 0.167, y = 0.132, z = 0.500

Fig. 1

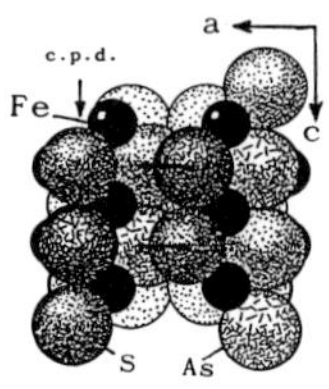

Fig. 2

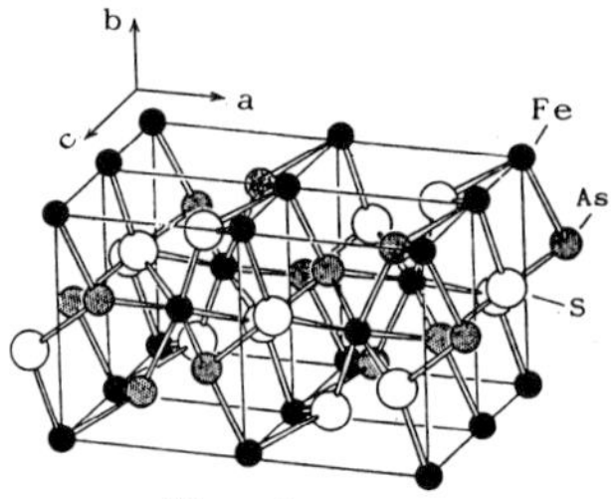

Fig. 3

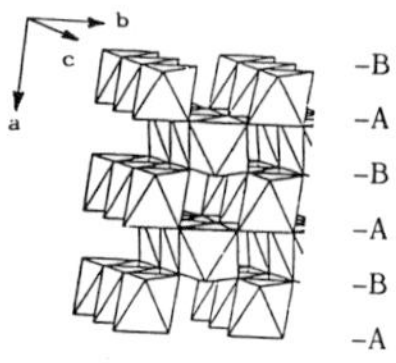

Fig. 4

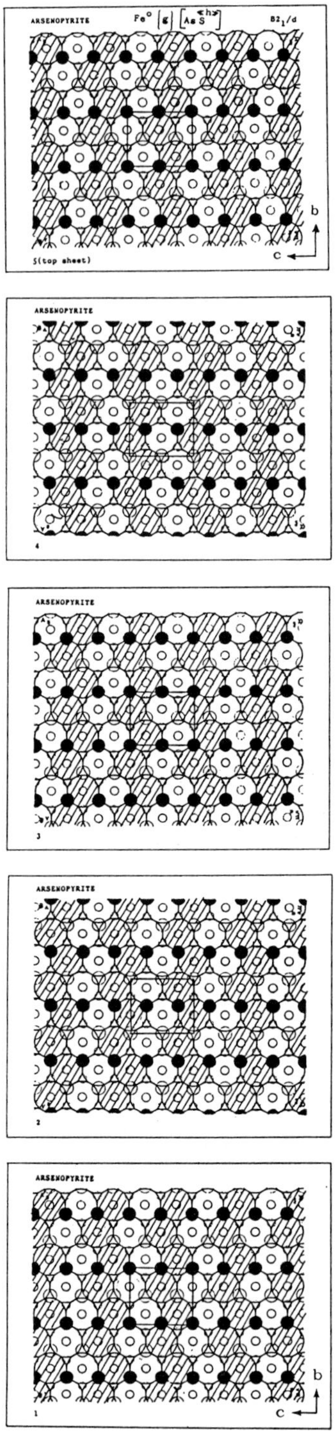

Fig. 5

Properties

Habit	Cleav.	Fract.	Twin.	Hardn.	Dens.	Colour	Transp.
prismatic columnar, compact	distinct (101)	uneven	(100),(001) (101),(012)	5.5 - 6	6.07	silver--white, steel-grey	opaque

Refr. index /Reflect.	Birefr.	Luster	Streak	Melt.p.	CPI
48,5%		metallic	greyish black		

Population

Gudmundite Fe^{o} {g} [Sb S ≪h≫]

Description

The arsenopyrite structure is a packing substitution derivative of the marcasite structure. It is formed by the packing of AsS groups, with Fe atoms in voids with octahedral coordination.

References

Kostov (1968) 132,133.
Povarennykh (1972) 245.
Wyckoff (1963) Vol. 1, 356-358.
Wuensch (1974) 25.
R. Burns + V. Burns (1979) 13.
Palache et al. (1944) Vol. 1, 317.

Figures

Fig. 1. The AsS group of the arsenopyrite structure.

Fig. 2. Packing drawing of the arsenopyrite structure viewed along the b axis (after Wyckoff, 1963, Vol. 1).

Fig. 3. Representation of the arsenopyrite structure (after Wuensch,1974).

Fig. 4. Polyhedral description of the packing analogue of the arsenopyrite structure (adapted from R. Burns + V. Burns, 1979).

Fig. 5. Condensed model of the packing analogue of the arsenopyrite structure. The large open circles represent oxygen atoms which with the lined large circles, corresponding to arsenium atoms, form an hexagonal closest packing. The small black circles represent iron atoms in octahedral voids.

PYRITE $Fe^{o}\{g\}[S_2]^{c}$

Pa3

a = 5.40667Å
Z = 4

Fe(4a)
S(8c) u = 0.386

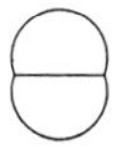

Fig. 1

Fig. 2

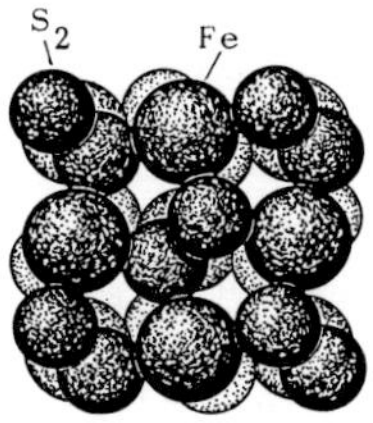

Fig. 3

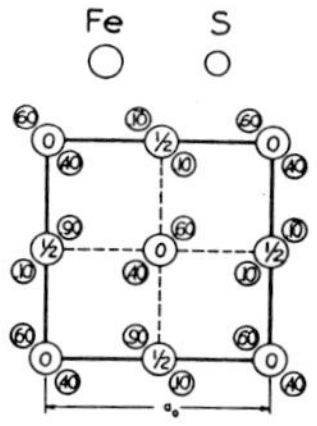

Fig. 4

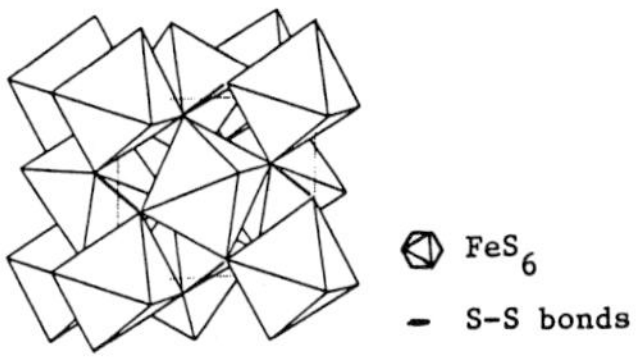

Fig. 5

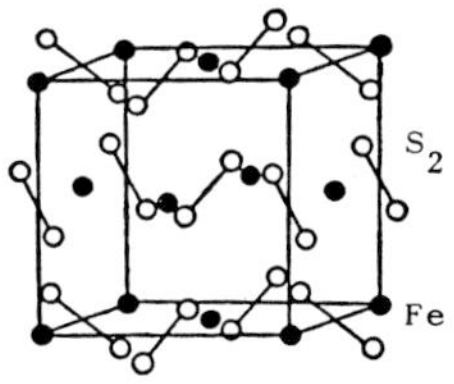

Fig. 6

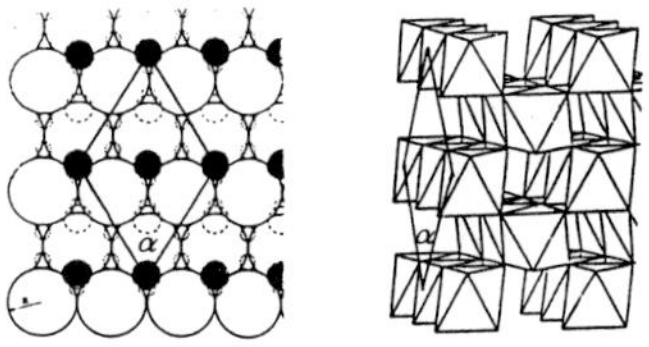

(a) (b)

Fig. 7

Properties

Habit	Cleav.	Fract.	Twin.	Hardn.	Dens.	Colour	Transp.
cubic, pyrito-hedral	indistinct (001)	conchoi-dal, uneven	(011)	6 - 6.5	5.018	pale brass-yellow	opaque

Refr. index/Reflect.	Birefr.	Luster	Streak	Melt.p.	CPI
53,5 %		metallic, splendent	greenish black	642° C	

Population

Vaesite	$Ni^{o}\{g\}[S_2]^{c}$
Cattierite	$Co^{o}\{g\}[S_2]^{c}$
Laurite	$Ru^{o}\{g\}[S_2]^{c}$
Hauerite	$Mn^{o}\{g\}[S_2]^{c}$
Villamaninite	$(Ni,Cu)^{o}\{g\}[S_2]^{c}$
Fukuchilite	$Cu_3^{o}Fe^{o}\{g\}[S_2]_4^{c}$
Penroseite	$Ni^{o}\{g\}[Se_2]^{c}$
Trogtalite	$Co^{o}\{g\}[Se_2]^{c}$
Cobaltite	$Co^{o}\{g\}[As\,S]^{c}$
Platarssulite	$Pt^{o}\{g\}[As\,S]^{c}$
Hollingworthite	$Rh^{o}\{g\}[As\,S]^{c}$
Irarsite	$Ir^{o}\{g\}[As\,S]^{c}$
Sperrylite	$Pt^{o}\{g\}[As_2]^{c}$
Geversite	$Pt^{o}\{g\}[Sb_2]^{c}$
Michenerite	$Pd^{o}\{g\}[Bi_2]^{c}$
Aurostibite	$Au^{o}\{g\}[Sb_2]^{c}$

Figures

Fig. 1. The S_2 group of the pyrite structure.

Fig. 2. Packing representation of the pyrite structure showing the packing of the S_2 groups (after Machatschki, 1953).

Fig. 3. Another packing drawing of the pyrite structure. The Fe ions are very largely re-represented (after Wyckoff, 1963, Vol. 1).

Fig. 4. Projection on a cubic face of the unit cell of pyrite (after Wyckoff, 1963, Vol. 1).

Fig. 5. Polyhedral description of the pyrite structure (after Zoltai, 1974).

Fig. 6. Unit cell atomic content of the pyrite structure emphasizing the formation of S_2 groups (after Machatschki, 1953).

Fig. 7. Two aspects of the marcasite structure showing the distortion process which could convert marcasite into pyrite with change of the angle $\alpha = 63^{o}$ to 90° a) packing drawing parallel to the (010) plane, b) polyhedral representation showing the (010) plane.

Description

The pyrite structure is based on an approximately cubic (c) close packing of S_2 groups, with Fe atoms with octahedral coordination. It can be imagined derived by a strong distortion of the marcasite structure. The distortion mechanism would involve the changing of the angle $\alpha = 63^{o}$ to 90° in the (010) plane (Fig. 7).

References

Kostov (1968) 125,126.

Povarennykh (1972) 206, 243,244.

Palache et al. (1944) Vol. 1, 282-284.

Wyckoff (1963) Vol. 1, 346-349.

Zoltai (1974) 52.

Machatschki (1953) 229.

CALCITE

$Ca^{ap}\left\{g\right\}\left[C^{tr}O_3^{\ll h\gg}\right]^{c}$

$R\bar{3}c$

$a_R = 6.361$
$\alpha = 46^\circ\ 6'$
$Z_R = 2$
$a_H = 4.99008$ Å
$c = 17.05951$ Å
$Z_H = 6$

(hexagonal description)
Ca(6b)
C (6a)
O(18e) u = 0.2593

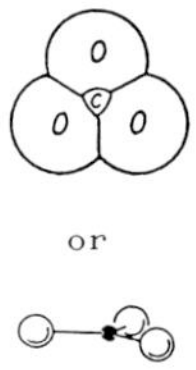

Fig. 1

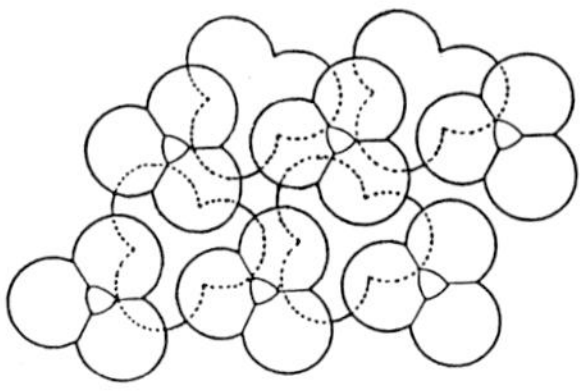

Fig. 2

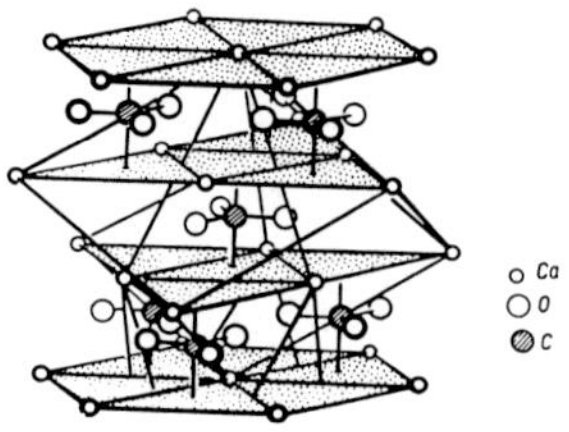

Fig. 3

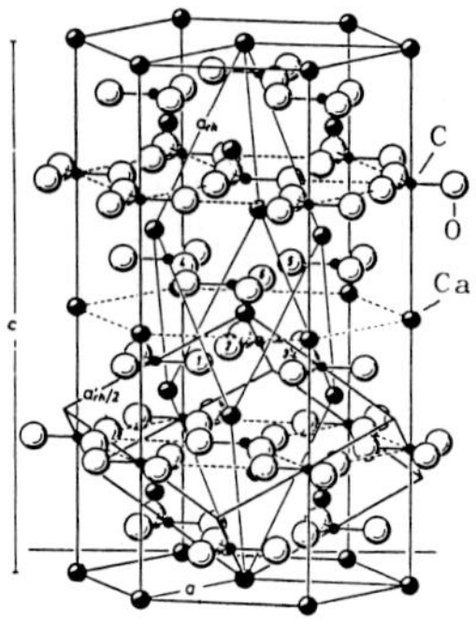

Fig. 4

Properties

Habit	Cleav.	Fract.	Twin.	Hardn.	Dens.	Colour	Transp.
prismatic rhombohedral, lenticular	perfect (10$\bar{1}$1)	conchoidal	(0001) (10$\bar{1}$0) (01$\bar{1}$2)	3	2.71	colourless, variable	transparent

Refr. index/Reflect.	Birefr.	Luster	Streak	Melt.p.	CPI
$n_\omega = 1.658$ $n_\epsilon = 1.486$	(–)	vitreous	white	1339°C	(SPI) 51

Figures

Fig. 1. The Co_3 group of the calcite structure.

Fig. 2. Projection of calcite normal to the triad axis showing the CO_3 groups and their packing.

Fig. 3. Structure of calcite (after Povarennykh, 1972).

Fig. 4. Ball and spoke model of the calcite structure (after Ramdohr + Strunz, 1980).

Fig. 5. (a) Structure of calcite and (b) position of the calcium atoms around the CO_3 groups (after Kostov, 1968).

Fig. 6. Condense model of the packing analogue of the calcite structure. The large open circles represent oxygen atoms in hexagonal closest packing and the very small black circles in triangular voids the carbon atoms. The double lines indicate the CO_3 groups. The other black circles represent the calcium atoms, and are located in octahedral voids. However the large size of the calcium makes the oxygen layers more apart and the coordination of calcium is really antiprismatic (ap) (Lima-de-Faria, 1990).

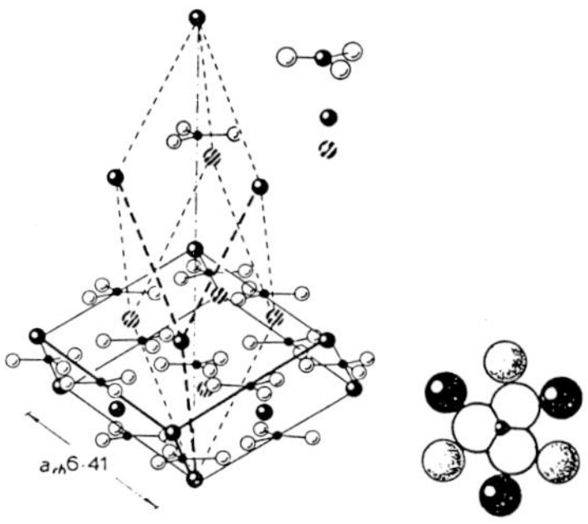

(a) (b)

Fig. 5

References

Kostov (1968) 529-533.

Povarennykh (1972) 182, 609, 610, 632.

Wyckoff (1964) Vol. 2, 359--364.

Ramdohr + Strunz (1980) 565.

Zoltai + Stout (1984) 426.

Tendeloo et al. (1985) 334.

Palache et al. (1951) Vol. 2, 143-151

Ingerson (1955) 351.

Lima-de-Faria (1990).

CALCITE (continued)

Population

Magnesite	$Mg^{ap}\{g\}[C^{tr} O_3^{\ll h \gg}]^c$
Siderite	$Fe^{ap}\{g\}[C^{tr} O_3^{\ll h \gg}]^c$
Rhodochrosite	$Mn^{ap}\{g\}[C^{tr} O_3^{\ll h \gg}]^c$
Gaspeite	$Ni^{ap}\{g\}[C^{tr} O_3^{\ll h \gg}]^c$
Smithsonite	$Zn^{ap}\{g\}[C^{tr} O_3^{\ll h \gg}]^c$
Otavite	$Cd^{ap}\{g\}[C^{tr} O_3^{\ll h \gg}]^c$
Nitronatrite	$Na^{ap}\{g\}[N^{tr} O_3^{\ll h \gg}]^c$

Description

The calcite structure is based on the packing of CO_3 groups, having Ca atoms in the voids of this packing with antiprismatic coordination (ap). The CO_3 groups form planar layers, and the structure can be imagined derived from a distorted hexagonal closest packing (h) of the oxygen atoms with carbon atoms occupying triangular voids, and the calcium atoms with octahedral coordination. The packing of the oxygens would be ideal if on Fig. 3a the atoms in positions zero superimposed the atoms with positions one third.

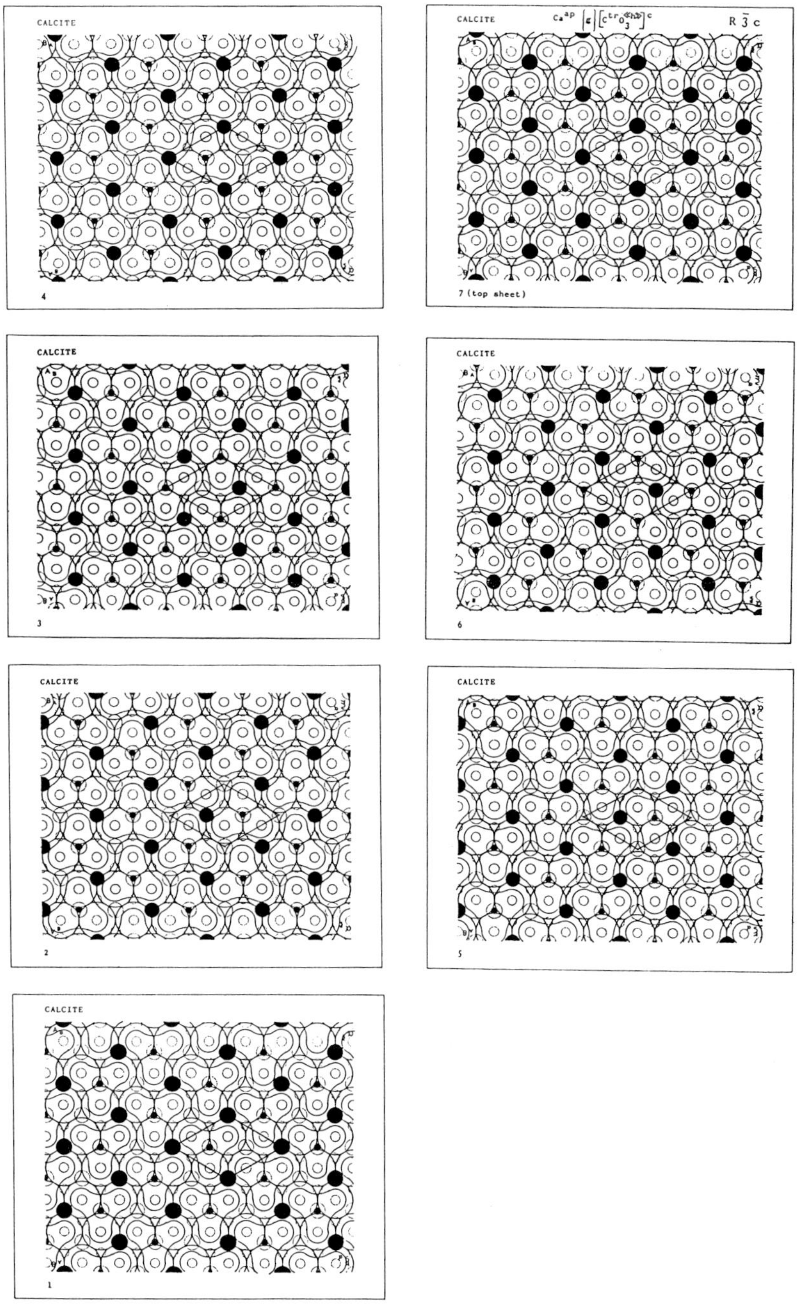

Fig. 6

DOLOMITE

R $\bar{3}$

$Ca^{ap}\ (Mg,Fe)^{ap}\ \{g\}\ [C^{tr}\ O_3^{\ll h \gg}]_2^c$

a = 6.0154Å
α = 47° 7'
Z_R = 1
a_H = 4.8079Å
c = 16.010
Z_R = 3

(hexagonal description)
Ca(3a)
Mg(3b)
C (6c) u ∼ 1/4
O(18f) x = 0.2374, y = -0.0347, z = 0.2440

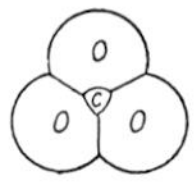

Fig. 1

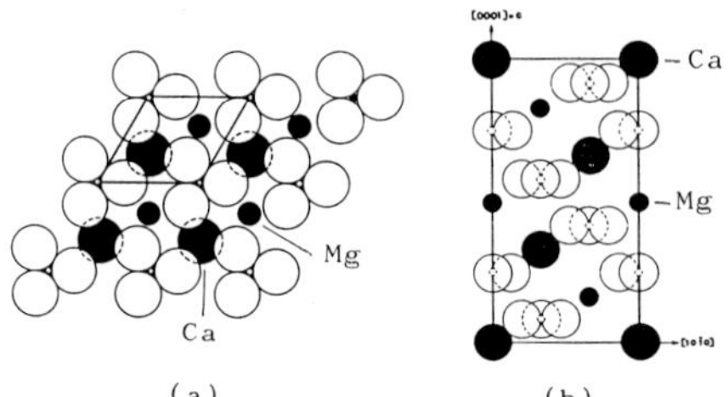

Fig. 2

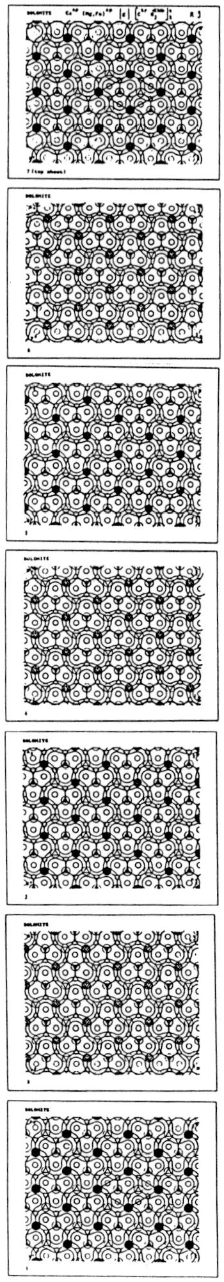

Fig. 3

Properties

Habit	Cleav.	Fract.	Twin.	Hardn.	Dens.	Colour	Transp.
rhombohedral, prismatic, tabular	perfect (10$\bar{1}$1)	subconchoidal	(0001) (10$\bar{1}$0) (11$\bar{2}$0)	3.5 - 4	2.85	colourless, greenish, variable	transparent to translucent

Refr. index /Reflect.	Birefr.	Luster	Streak	Melt.p.	CPI
n_ω = 1.679, n_ϵ = 1.500	(-)	vitreous	white		(SPI) 53

Population

Ankerite	$Ca^{ap}Fe^{ap}\ \{g\}\ [C^{tr}\ O_3^{\ll h \gg}]_2^c$
Kutnahorite	$Ca^{ap}Mn^{ap}\ \{g\}\ [C^{tr}\ O_3^{\ll h \gg}]_2^c$
Nordenskiöldine	$Ca^{ap}Sn^{ap}\ \{g\}\ [B^{tr}\ O_3^{\ll h \gg}]_2^c$

Figures

Fig. 1. The CO_3 group in the dolomite structure.

Fig. 2. (a) Projection on a plane normal to the triad axis, and (b) elevation along the c axis (after Tendeloo et al., 1985).

Fig 3. Condensed model of the packing analogue of the dolomite structure. The large open circles represent oxygen atoms in hexagonal closest packing, and the very small black circles the carbon atoms. The double lines indicate the CO_3 groups. The other black circles represent the calcium atoms which are located in octahedral voids. The cross lined circles represent the magnesium atoms. The large size of the calcium atoms makes the oxygen layers more apart and the coordination of the calcium and also of magnesium becomes trigonal antiprismatic (ap).

Description

Dolomite is an interstitial substitution derivative of calcite. The substitution of calcium by magnesium occurs in alternate layers

References

Kostov (1968) 540.
Povarennykh (1972) 465, 612.
Wyckoff (1964) Vol. 2, 361-364, 466-468.
Zoltai + Stout (1984) 429.
Tendeloo et al. (1985) 334.
Palache et al. (1951) Vol. 2, 208-211.

ARAGONITE $Ca^{[6/9]} \{g\} [C^{tr} O_3]$

Pbnm

a = 7.968Å
b = 5.741Å
c = 4.959Å
Z = 4

Ca(4c)	x = 0.417 y = 0.750 z = 1/4	
C (4c)	x = 0.75 y = -0.083 z = 1/4	
O_I(4c)	x = 0.917 y = -0.083 z = 1/4	
O_{II}(8d)	x = 0.67 y = -0.083 z = 0.48	

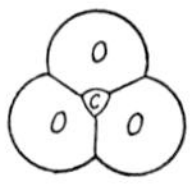

Fig. 1

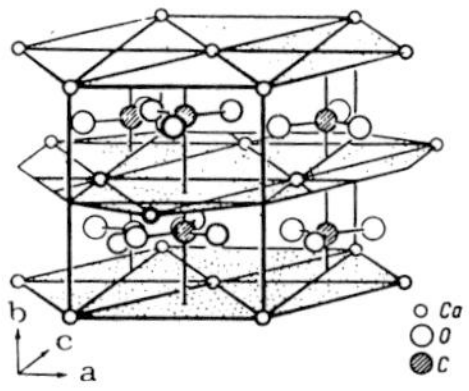

Fig. 2

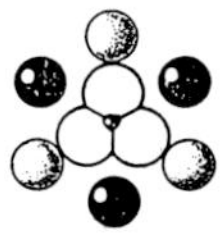

Fig. 3

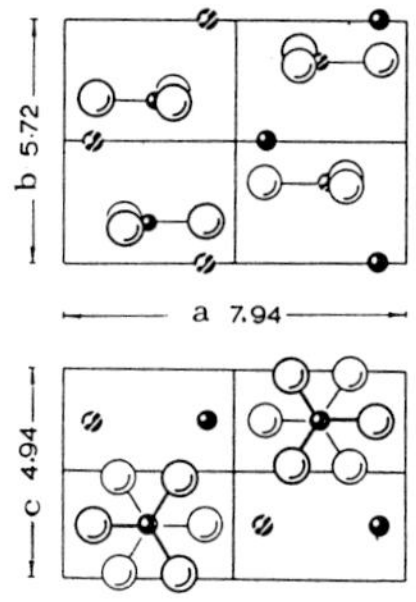

Fig. 4

Properties

Habit	Cleav.	Fract.	Twin.	Hardn.	Dens.	Colour	Transp.
acicular, pyramidal	good (100) (101)	subconchoidal	(101)	3.5 - 4	2.94	colourless, white	transparent to translucent

Refr. index/Reflect.	Birefr.	Luster	Streak	Melt.p.	CPI
n_α = 1.530 n_β = 1.681 n_γ = 1.685	(-) 2V= 18°	vitreous	white		(SPI) 57

Population

Strontianite $Sr^{[6/9]} \{g\} [C^{tr} O_3]$
Witherite $Ba^{[6/9]} \{g\} [C^{tr} O_3]$
Cerussite $Pb^{[6/9]} \{g\} [C^{tr} O_3]$
Nitre $K^{[6/9]} \{g\} [N^{tr} O_3]$

Description

The aragonite structure is based on the packing of CO_3 groups, with Ca in voids with a strong distorted cuboctahedral coordination from 6 to 9. Calcium is "geometrically" distributed in an hexagonal closest packing.

References

Kostov (1968) 529, 549.
Povarennykh (1972) 182, 610, 632.
Wyckoff (1964) Vol. 2, 364-368.
Zoltai + Stout (1984) 430.
Palache et al. (1951) Vol. 2, 184-187.

Figures

Fig. 1. The CO_3 group in the aragonite structure.

Fig. 2. Aragonite structure after Povarennykh, 1972).

Fig. 3. Position of the calcium atoms around the CO_3 groups in the aragonite structure (after Kostov, 1968).

Fig. 4. Two projections of the aragonite structure (after Kostov, 1968).

SKUTTERUDITE

$Co^{o}_{4}\{g\}\left[As_{4}\right]_{3}$

Im3

$a = 8.189$ Å

$Z = 2$ (for Co_4As_{12})

Co (8c)

As (24d) $u = 0.35$, $v = 0.15$

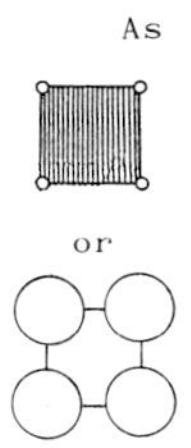

Fig. 1

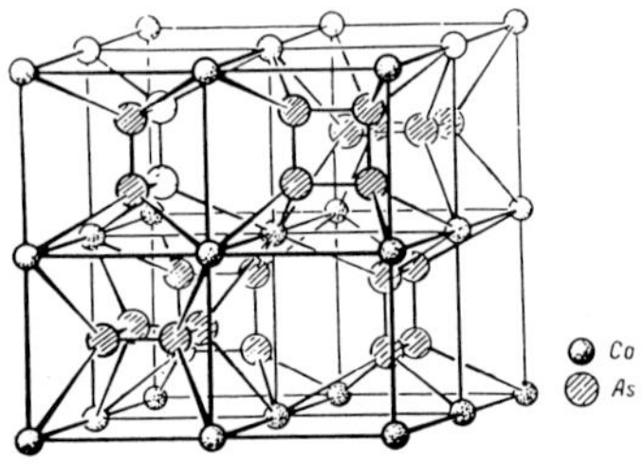

Fig. 2

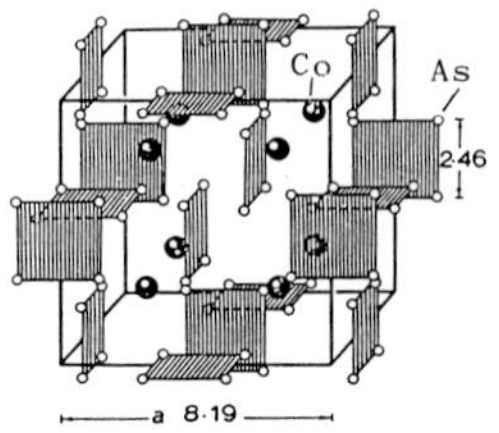

Fig. 3

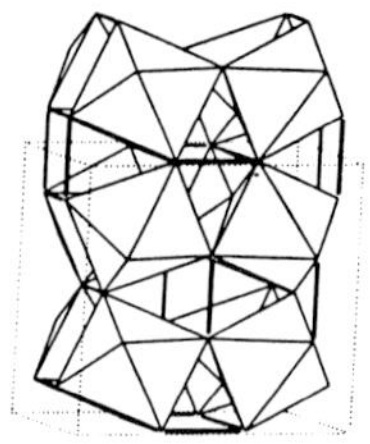

$CoAs_6$

– As-As bond

Fig. 4

Properties

Habit	Cleav.	Fract.	Twin.	Hardn.	Dens.	Colour	Transp.
cubic, cubo-octahedral	distinct (001) (111)	conchoidal to uneven	(112)	5.5 - 6	6.5	tin-white, silver-grey	opaque

Refr. index / Reflect.	Birefr.	Luster	Streak	Melt.p.	CPI
57,5%		metallic			

Population

Smaltite $(Co,Ni)^{o}_{4}\{g\}\left[As_{4}\right]_{3-x}$

Chloanthite $(Ni,Co)^{o}_{4}\{g\}\left[As_{4}\right]_{3-x}$

Description

The skutterudite structure is based on the packing of As_4 groups with a square shape. The cobalt atoms occupy octahedral voids in this packing. The size of the As_4 groups is 2.46Å, and in terms of As covalent radius is 2R = 2.38Å, an approximate value.

Figures

Fig. 1. The As_4 group of the skutterudite structure.

Fig. 2. Skutterudite structure (after Povarennykh, 1972).

Fig. 3. Skutterudite structure (after Kostov, 1968).

Fig. 4. Polyhedral representation of the skutterudite structure (after Zoltai, 1974).

References

Kostov (1968) 134,135.

Povarennykh (1972) 207, 758.

Wyckoff (1964) Vol. 2, 108-110.

Zoltai (1974) 48.

Palache et al. (1944) Vol. 1, 342,343.

SULPHUR

(α -form)

{g} [S_8]

Fddd

a = 10.467Å
b = 12.870Å
c = 24.493Å
Z = 16

S_I (32h)	x = 0.8554 y = 0.9526 z = 0.9516
S_{II} (32h)	x = 0.7844 y = 0.0301 z = 0.0763
S_{III} (32h)	x = 0.7069 y = 0.9795 z = 0.0040
S_{IV} (32h)	x = 0.7862 y = 0.9073 z = 0.1290

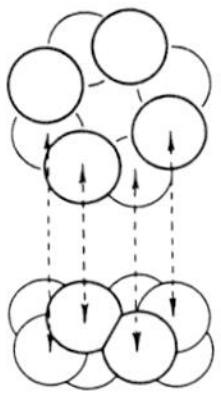

Fig. 1

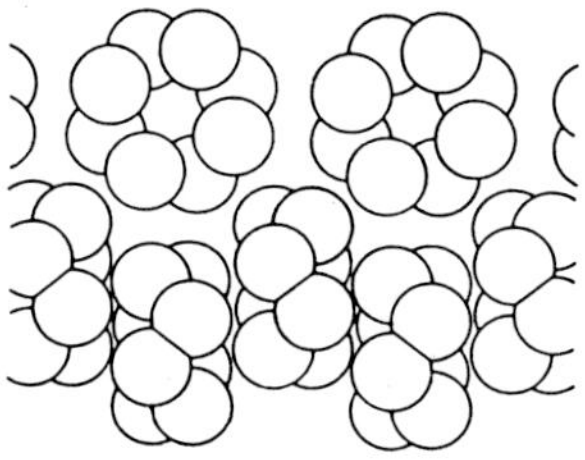

Fig. 2

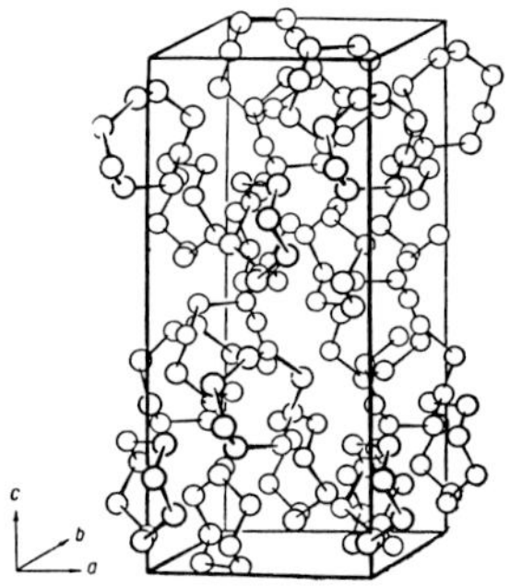

Fig. 3

Properties

Habit	Cleav.	Fract.	Twin.	Hardn.	Dens.	Colour	Transp.
pyramidal, tabular	poor (001) (110)	conchoi-dal		1.5 - 2.5	2.07	sulphur--yellow	transparent, translucent

Refr. index/Reflect.	Birefr.	Luster	Streak	Melt.p.	CPI
n_α = 1.958 n_β = 2.038 n_γ = 2.245	(+) 2V = 68° 58'	resinous, greasy	white	112.8°C	(SPI) 67

Figures

Fig. 1. The structural unit of the sulphur structure, the S_8 group, in two perspectives (after Kostov, 1968).

Fig. 2. Representation of the packing of the S_8 molecules (after Bunn, 1964).

Fig. 3. Representation of the sulphur structure (after Povarennykh, 1972).

Description

The structure of the α-form of sulphur is based on the packing of S_8 ring molecules. The bonds between the sulphur atoms within the molecules are covalent bonds (S-S = 2.14 Å), and between sulphur atoms of distinct molecules are residual (S - S = 3.3 Å). The β-form (stable above 95.6°C) of sulphur is probably based on the same S_8 molecules but with a different packing.

References

Kostov (1968) 99,100.

Povarennykh (1972) 195,196.

Wyckoff (1963) Vol. 1, 33-36.

Zoltai + Stout (1984) 377.

Bunn (1964) 58.

Palache et al. (1944) Vol. 1, 140-142.

PROUSTITE R3c

$\left[\left[Ag\right]_3 \{g\} \left[\overset{[3n]}{As} S_3\right]\right]$

a_R = 6.87Å	(hex. description)	
α = 103°31'	Ag(18b)	x = 0.246
Z_R = 2		y = 0.298
		z = 0.235
a_H = 10.80Å	As (6a)	u = 0
c = 8.69	Sb(18b)	x = 0.220
Z_H = 6		y = 0.095
		z = 0.385

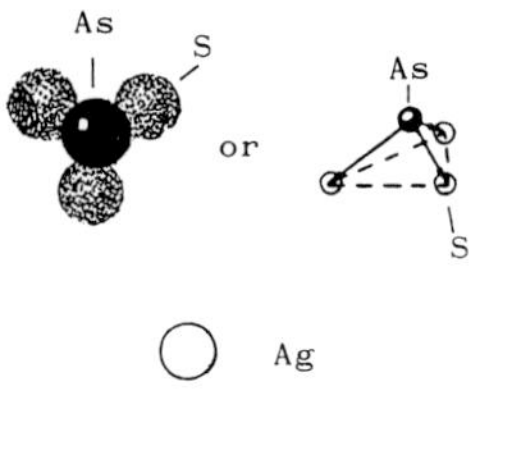

Fig. 1

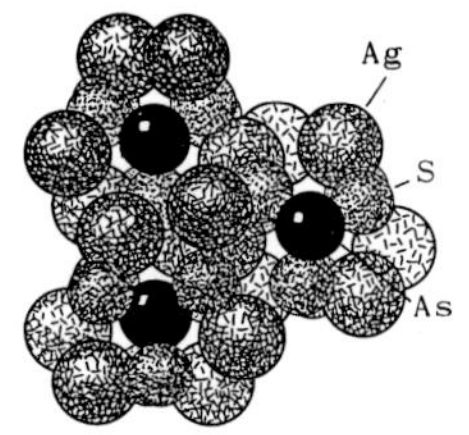

Fig. 2

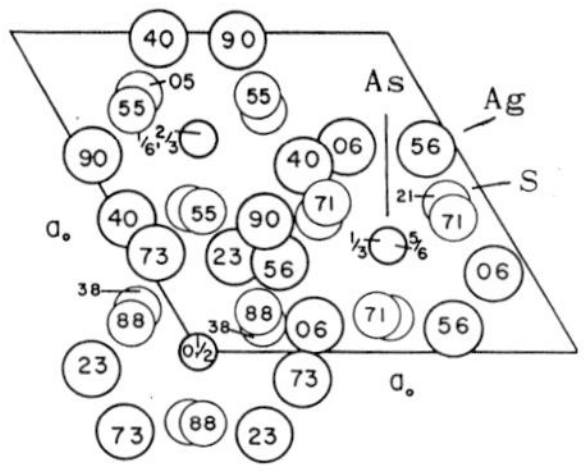

Fig. 3

Properties

Habit	Cleav.	Fract.	Twin.	Hardn.	Dens.	Colour	Transp.
prismatic, rhombohedral	distinct (10$\bar{1}$1)	conchoidal to uneven	(10$\bar{1}$4)	2 - 2.5	5.57	scarlet--vermilion	translucent

Refr. index	/Reflect.	Birefr.	Luster	Streak	Melt.p.	CPI
n_ω = 2.979 n_ϵ = 2.711	21,5%	(-)	adamantine	vermilion	490°C	

Population

Pyrargyrite $\left[\left[Ag\right]_3 \{g\} \left[\overset{[3n]}{Sb} S_3\right]\right]$

Figures

Fig. 1. Structural units of the proustite structure: AsS_3 pyramidal groups and Ag atoms.

Fig. 2. Packing representation of the AsS_3 groups and the Ag atoms in the proustite structure (after Wyckoff, 1964, Vol. 2).

Fig. 3. Projection of the proustite structure along the c axis corresponding to the representation of Fig. 2 (after Wyckoff,1964, Vol. 2).

Description

The proustite structure is based on a complex packing of AsS_3 pyramidal groups and Ag atoms. The Ag atoms are linked to S atoms forming spiral chains along the c axis.

References

Kostov (1968) 175.

Povarennykh (1972) 239.

Wyckoff (1964) Vol. 2, 508-511.

Palache et al. (1944) Vol. 1, 366,367.

REALGAR $P\,2_1/n$	$a = 9.27Å$	As_I (4e)	x = 0.118	As_{IV} (4e)	x = 0.038	S_{III} (4e)	x = 0.245
	$b = 13.50Å$		y = 0.024		y = -0.161		y = -0.225
	$c = 6.56Å$		z = -0.241		z = -0.290		z = -0.363
$\{g\}\ [As_4^{[3]}\,S_4]$	$\beta = 106°\ 37'$	As_{II} (4e)	x = 0.425	S_I (4e)	x = 0.346	S_{IV} (4e)	x = 0.115
	$Z = 4$		y = -0.140		y = 0.008		y = -0.215
			z = -0.142		z = -0.295		z = 0.048
		As_{III} (4e)	x = 0.318	S_{II} (4e)	x = 0.213		
			y = -0.127		y = 0.024		
			z = 0.181		z = 0.120		

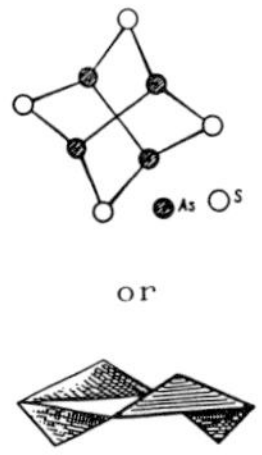

Fig. 1

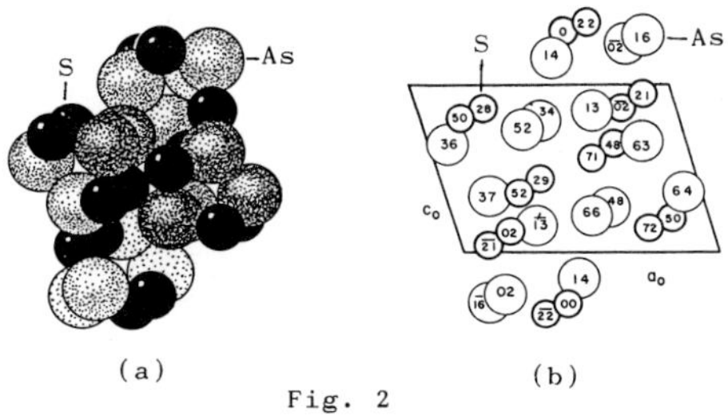

Fig. 2

Properties

Habit	Cleav.	Fract.	Twin.	Hardn.	Dens.	Colour	Transp.
short prismatic, granular	good (010)	conchoidal	(100)	1.5 - 2	3.56	red, orange	transparent to opaque

Refr. index/Reflect.	Birefr.	Luster	Streak	Melt.p.	CPI
$n_\alpha = 2.538$ $n_\beta = 2.864$ $n_\gamma = 2.704$	(-) $2V=40°34'$	resinous, greasy	orange red	307-314°C	

Figures

Fig. 1. Configuration of the As_4S_4 molecules and the link between S and As in the realgar structure (after Povarennykh, 1972).

Fig. 2. (a) Packing drawing of the realgar structure, and (b) unit cell content projected along b axis (after Wyckoff, 1963, Vol. 1).

Description

The realgar structure is based on the packing of As_4S_4 groups.

References

Kostov (1968) 167.

Povarennykh (1972) 242.

Wyckoff (1963) Vol. 1, 174-177.

Palache et al. (1944) Vol. 1, 256,257.

WARWICKITE   Pnam

$Mg_3^{o}\ Ti^{o}\left[\left[O\right]_2\ \{g\}\left[B^{tr}\ O_3\right]_2\right]^{«Ts/'h»}$

a = 9.20Å	$(Mg,Ti)_I$ (4c) u = 0.370, v = 0.201	O_I (4c) u = 0.000, v = 0.125	
b = 9.45Å			
c = 3.01Å	$(Mg,Ti)_{II}$ (4c) u = 0.390, v = 0.556	O_{II} (4c) u = 0.250, v = 0.000	
Z = 2	B(4c) u = 0.340, v = 0.875	O_{III} (4c) u = 0.264, v = 0.750 ...	

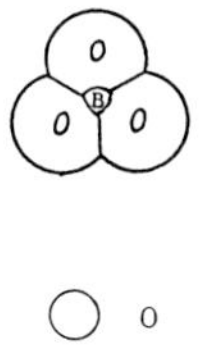

Fig. 1

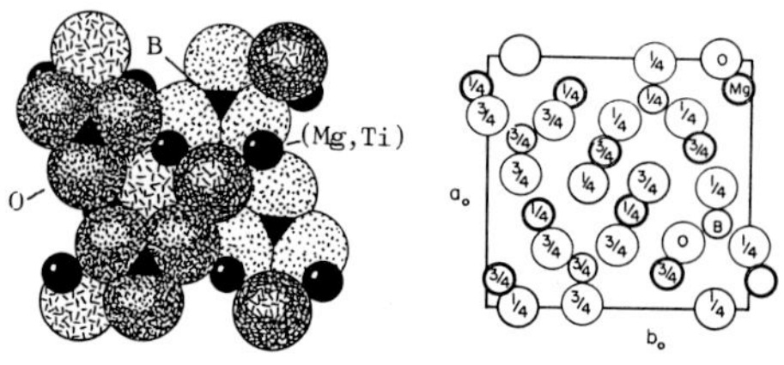

(a) Fig. 2 (b)

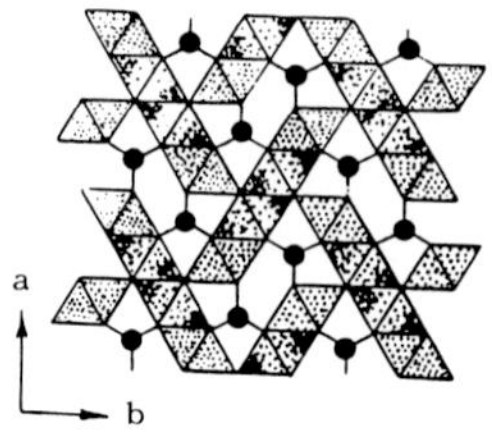

Fig. 3

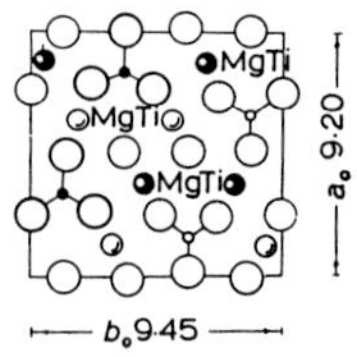

Fig. 4

Properties

Habit	Cleav.	Fract.	Twin.	Hardn.	Dens.	Colour	Transp.
slender prismatic	perfect (100)	uneven		3.5 - 4	3.35	dark brown, dull black	transparent in very small grains

Refr. index/Reflect.	Birefr.	Luster	Streak	Melt.p.	CPI
n_α = 1.806 n_β = 1.809 n_γ = 1.830	(+)	dull, pearly, subvitreous	bluish black		

Figures

Fig. 1. The structural units of the warwickite structure: BO_3 groups and oxygen atoms.

Fig. 2 (a) Packing drawing of the warwickite structure showing the way the BO_3 groups pack with the oxygen atoms and (b) correspon- ding projection along the c axis (after Wyckoff, 1964, Vol. 2).

Fig. 3. Polyhedral representation of the warwickite structure (after Povarennykh, 1972).

Fig. 4. Structure of warwickite (after Kostov, 1968).

Fig. 5. Condensed model of the packing analogue of the warwic- kite structure based on puckered simple hexagonal packing layers. The large open circles represent oxygen atoms which form the pack- ing layers, and the medium black circles correspont to Mg and Ti atoms in octahedral voids. The boron atoms (very small black circles) are placed in trian- gular voids and linked to the three neighbouring oxygen atoms,- forming BO_3 groups (embraced by a curved line). In each layer the packing atoms are at two diffe- rent levels (0 and 1/2) which are distinguished by the heavy and light lines on the corresponding circles. The Mg and Ti atoms are distribute at random (‿) in the corresponding equivalent positions.

Description

The warwickite structure is based on a complex packing of BO_3 groups with oxygen atoms, with Mg and Ti atoms in ictahedral voids. The pac- king can be imagined formed by pucke red simple hexagonal layers (∼Ts) where the packing atoms are at two different levels (0 and 1/2, thick and thin circles) the lower and higher parts fiting together in a closest packing way, giving rise to the octa- hedral voids.

References

Kostov (1968) 430,431.

Povarennykh (1972) 467,468.

Wyckoff (1964) Vol. 2, 518-521.

Palache et al (1951) Vol. 2, 326,327.

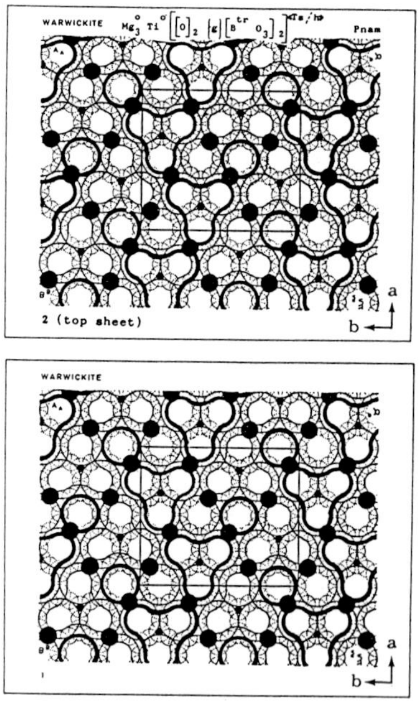

Fig. 5

Crystallographic data (continued)

O_{IV}(4c) u = 0.484, v = 0.875

PINAKIOLITE

P2/m

$\overset{o}{Mg}_3\ \overset{o}{Mn}\ \overset{o}{Mn}_2 \left[[O]_4 \{g\} [B^{tr}\ O_3]_2 \right] \ll Ts\,/\,h \gg$

a = 5.36Å		
b = 5.98Å		
c = 12.73Å		
β = 120° 34'		
Z = 2		

Mg_I	(2a)	x = 0, y = 0, z = 0	Mn_I^{3+}	(2b)	x = 1/2, y = 0, z = 0
Mg_{II}	(2e)	x = -0.053, y = 1/4, z = 0.197	Mn_{II}^{3+}	(2d)	x = 1/2, y = 0, z = 1/2
Mg_{III}	(2e)	x = 0, y = 1/4, z = 0.50	Mn_{III}^{2+}	(2e)	x = 0.053, y = 1/4, z = -0.197
			...		

Fig. 1

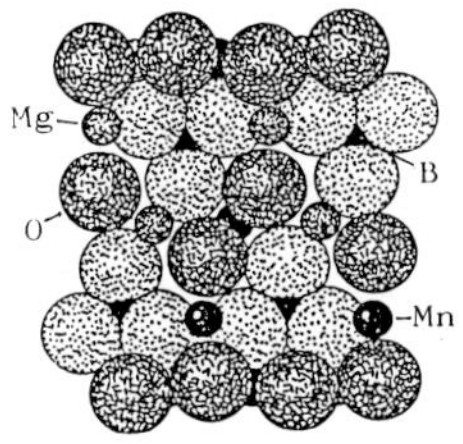

(a)

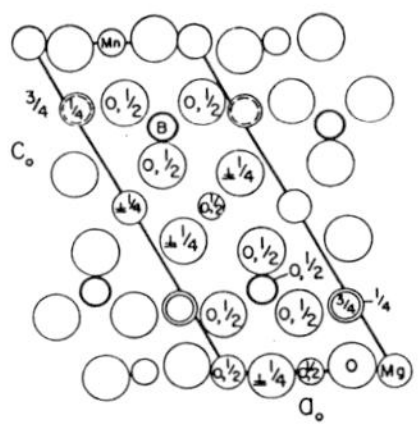

(b)

Fig. 2

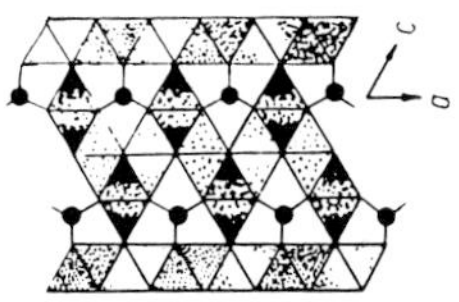

Fig. 3

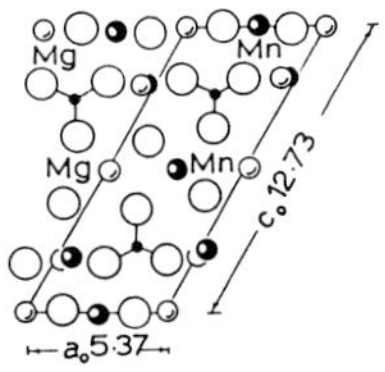

Fig. 4

Properties

Habit	Cleav.	Fract.	Twin.	Hardn.	Dens.	Colour	Transp.
thin tablets	good (010)		(011)	6	3.88	black	opaque

Refr. index / Reflect.	Birefr.	Luster	Streak	Melt.p.	CPI
n_α = 1.908 n_β = 2.05 n_γ = 2.065	(-) 2V = 32°	metallic	brownish grey		

Figures

Fig. 1. The structural units of the pinakiolite structure: BO_3 groups and O atoms.

Fig. 2. (a) Packing drawing of the pinakiolite structure, showing the way the BO_3 groups pack with the oxygen atoms, and (b) unit cell content projected along the b axis (after Wyckoff, 1964, Vol. 2).

Fig. 3. Polyhedral representation of the pinakiolite structure (after Povarennykh, 1972).

Fig. 4. Structure of pinakiolite (after Kostov, 1968).

Fig. 5. Condensed model of the packing analogue of the pinakiolite structure based on a puckered simple hexagonal packing. The large open circles represent oxygen atoms which form the packing layers, and the medium black and dashed circles correspond atoms in octahedral voids. The boron atoms (very small black circles) are placed in triangular voids and linked to the three neighbouring oxygen atoms forming BO_3 groups (embraced by a curved line). In each layer the packing atoms are at two different levels (0 and 1/2) which are distinguished by the heavy and light lines of the corresponding circles.

Description

The pinakiolite structure is based on the complex packing of BO_3 groups and O atoms. This packing is formed by two different parts of simple hexagonal packing, at four levels (0, 1/4, 1/2 and 3/4) which fit in the closest way, giving rise to octahedral voids. These octahedral voids are occupied by Mg and Mn atoms. The packing of the oxygens may also be imagined as forming a puckered simple hexagonal packing (~Ts), or an interpenetration of simple hexagonal with closest hexagonal packing (Ts/h).

Crystallographic data (continued)

B	(4f)	x = 0.500, y = 0.000, z = 0.250	O_{IV}	(4f)	x = 0.638, y = 0, z = 0.366
O_I	(2e)	x = 0.239, y = 1/4, z = -0.011	O_V	(4f)	x = 0.643, y = 0, z = 0.180
O_{II}	(2e)	x = -0.239, y = 1/4, z = 0.011	O_{VI}	(2e)	x = 0.202, y = 1/4, z = 0.396
O_{III}	(4f)	x = 0.175, y = 0, z = 0.175	O_{VII}	(2e)	x = -0.202, y = 1/4, z = -0.396

References

Kostov (1968) 430, 431.
Povarennykh (1972) 467,468.
Wyckoff (1964) Vol. 2, 523,524.
Palache et al. (1951) Vol. 2, 324,325.

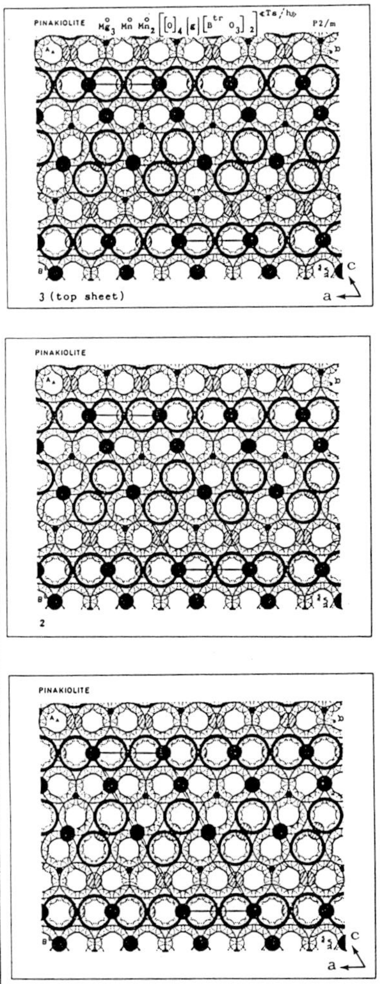

Fig. 5

FLUOBORITE

$C6_3/m$

$Mg^{o}_{3}\ [[(OH,F)]_{3}\ \{g\}\ [B^{tr}\ O_{3}]]$ «Ts/'h»

a = 9.06Å
c = 3.06Å
Z = 2

Atom	Position		
Mg	(6h)	u = 0.381	v = 0.038
(OH,F)	(6h)	u = 0.310	v = 0.218
B	(2c)	u = 1/3	v = 2/3
O	(6h)	u = 0.381	v = 0.537

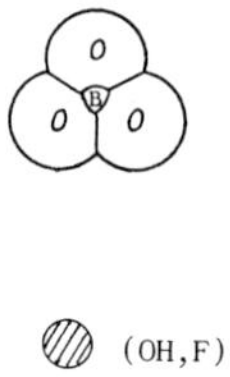

Fig. 1

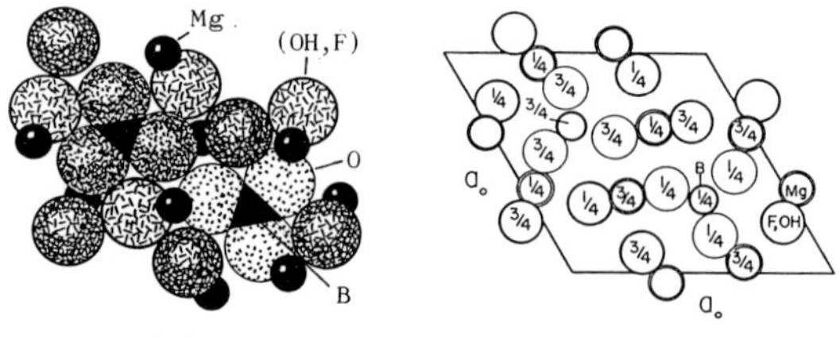

Fig. 2

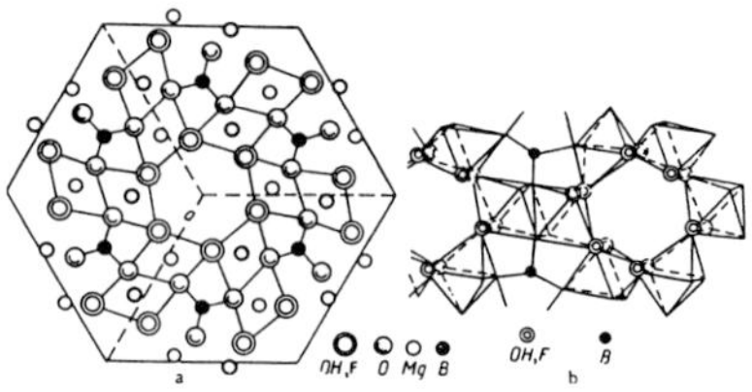

Fig. 3

Properties

Habit	Cleav.	Fract.	Twin.	Hardn.	Dens.	Colour	Transp.
acicular, felted aggregates	indistinct (0001)			3.5	2.85-2.98	colourless to white	
Refr. index/Reflect.		**Birefr.**		**Luster**	**Streak**	**Melt.p.**	**CPI**
n_ω = 1.577 n_ϵ = 1.522		(-)					

Figures

Fig. 1. The structural units of the fluoborite structure: BO_3 groups and OH ions.

Fig. 2 (a) Packing drawing of the fluoborite structure showing the way the BO_3 groups pack with OH ions and (b) corresponding projection along the c axis (after Wyckoff, 1964, Vol. 2).

Fig. 3. (a) Projection along the hexad axis and (b) polyhedral description of the fluoborite structure (after Povarennykh, 1972).

Fig. 4. The crystal structure of fluoborite (after Makovicky, 1985).

Fig. 5. Condensed model of the packing analogue of the fluoborite structure based on puckered simple hexagonal packing layers. The large open circles represent oxygen atoms which form the BO_3 groups, and the hydroxyls are part of the packing atoms and are represented by large lined circles. The medium black circles correspond to atoms in octahedral voids. The boron atoms (very small black circles) are placed in triangular voids and linked to the three neighbouring oxygen atoms, forming BO_3 groups (embraced by a curved line). In each layer the packing atoms are at two different levels (0 and 1/2) which are distinguished by the heavy and light lines of the corresponding circles. The crosses correspond to vacant packing atoms.

Description

The fluoborite structure is based on a complex packing of BO_3 and (OH) groups. This packing is formed by two parts of simple hexagonal packing at two different levels (0 and 1/2) which fit together in the closest way, giving rise to the formation of octahedral voids. The octahedral voids are occupied by Mg atoms. This packing may be imagined as a puckered simple hexagonal packing (~Ts) or an interpenetration of a simple hexagonal with an hexagonal closest packing (Ts/'h).

References

Kostov (1968) 428.
Povarennykh (1972) 469,470.
Wyckoff (1964) Vol. 2, 518,519.
Palache et al. (1951) Vol. 2, 369.
Makovicky (1985) 14.

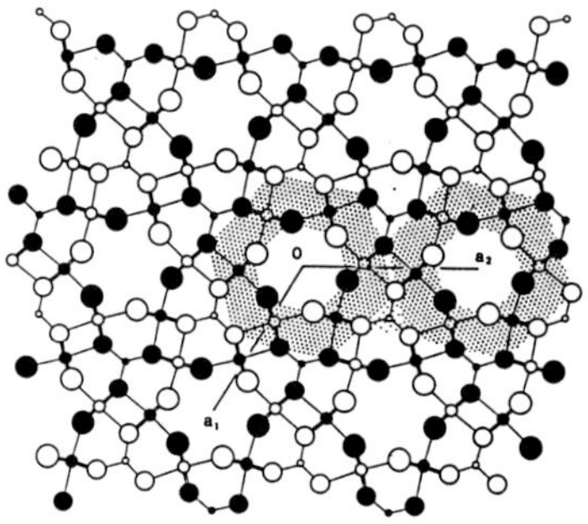

Fig. 4

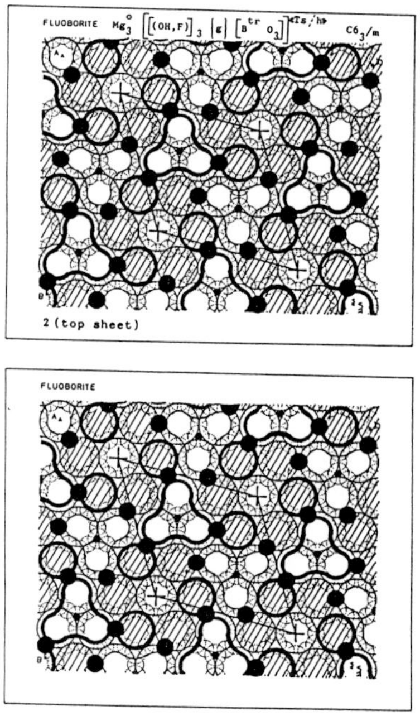

Fig. 5

LUDWIGITE

Pbam

$(Mg,Fe)_2^{o}\ Fe^{o}\left[[O]_2\ \{g\}\left[B^{tr}\ O_3\right]\right]$ ≪Ts/h≫

a = 9.14 Å
b = 12.45 Å
c = 3.05 Å
Z = 4

Atom	Position	x	y	z
Mg_I	(2b)			
Mg_{II}	(2c)			
Mg_{III}	(4h)	0	0.275	1/2
Fe	(4g)	0.250	0.114	0
B	(4g)	0.288	0.352	0
O_I	(4g)	0.375	0.444	0
O_{II}	(4g)	0.361	0.250	0

...

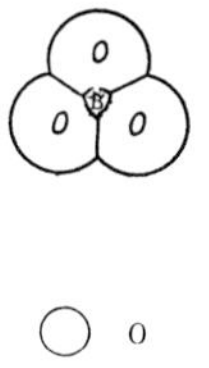

Fig. 1

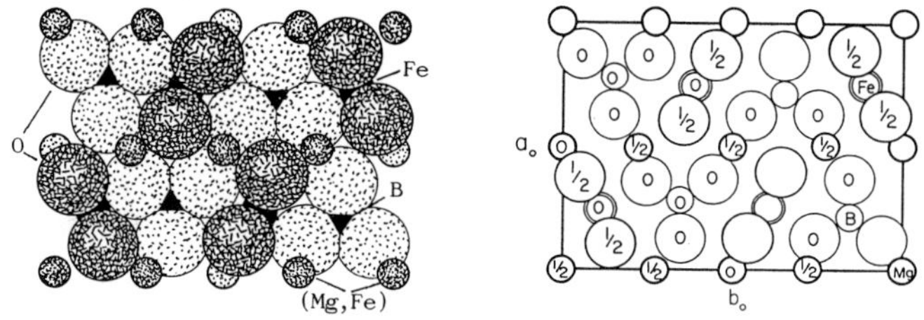

(a) Fig. 2 (b)

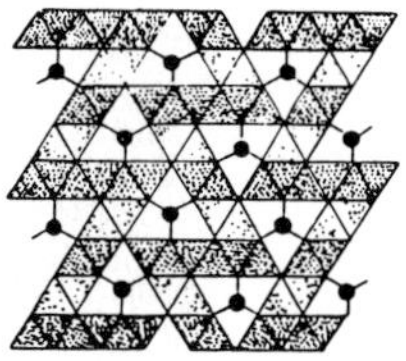

Fig. 3

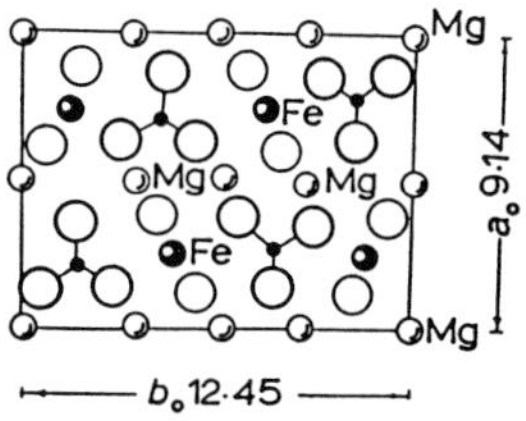

Fig. 4

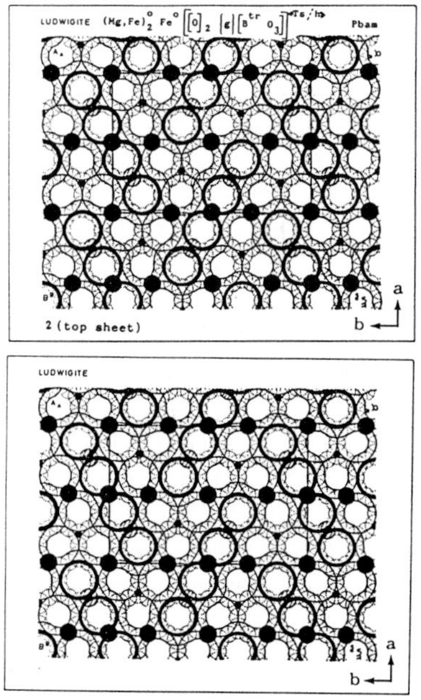

Fig. 5

Properties

Habit	Cleav.	Fract.	Twin.	Hardn.	Dens.	Colour	Transp.
acicular to fibrous	none			5	3.6 - 4.7	coal--black, greenish black	opaque

Refr. index/Reflect.	Birefr.	Luster	Streak	Melt.p.	CPI
n_α = 1.85 n_β = 1.85 n_γ = 2.02	(+)	silky	black, blackish green		

Figures

Fig. 1. The BO_3 group in the structure of ludwigite and the oxygen ions.

Fig. 2 (a) Packing drawing of the ludwigite structure showing the way the BO_3 groups pack with the oxygen atoms, and (b) corresponding projection along the c axis (after Wyckoff, 1964, Vol. 2).

Fig. 3. Polyhedral representation of the ludwigite structure (after Povarennykh, 1972).

Fig. 4. Structure of ludwigite (after Kostov, 1968).

Fig. 5. Condensed model of the packing analogue of the ludwigite structure based on puckered simple hexagonal packing layers. The large open circles represent oxygen atoms which form the packing layers, and the medium black and dashed circles correspond to atoms in octahedral voids. The boron atoms (very small black circles) are placed in triangular voids and linked to the three neighbouring oxygen atoms forming BO_3 groups (embraced by a curved line). In each layer the packing atoms are at two different levels (0 and 1/2) which are distinguished by the heavy and light lines of the corresponding circles.

Description

The ludwigite structure is based on the complex packing of BO_3 groups and O atoms. This packing is formed by two fragments of simple hexagonal packing at two different levels (0 and 1/2, thick and light circles), which fit in the closest way giving rise to octahedral voids. These octahedral voids are occupied by Mg and Fe atoms; the packing may be imagined as a puckered simple hexagonal packing (~Ts) or an interpenetration of simple hexagonal and closest packings (Ts/h).

References

Kostov (1968) 430,431.

Povarennykh (1972) 467,468.

Wyckoff (1964) Vol. 2, 521,522.

Palache et al. (1951) Vol. 2, 321, -322.

Crystallographic data (continued)

Atom	Position	x	y	z
O_{III}	(4g)	0.136	0.349	0
O_{IV}	(4h)	0.375	0.058	1/2
O_V	(4h)	0.112	0.145	1/2

MALACHITE

$P2_1/a$

$Cu_2^{o}\left[\left[(OH)\right]_2 \{g\} \left[C^{tr} O_3\right]\right]^{\ll Ts/h \gg}$

a =	9.48Å
b =	12.03Å
c =	3.21Å
β =	98°
Z =	4

Atom	Position	x	y	z
Cu_I	(4e)	0.00	0.21	0.89
Cu_{II}	(4e)	0.235	0.39	0.38
$(OH)_I$	(4e)	0.09	0.36	0.92
$(OH)_{II}$	(4e)	0.39	0.43	0.86
C	(4e)	0.27	0.13	0.47
O_I	(4e)	0.14	0.13	0.28
...				

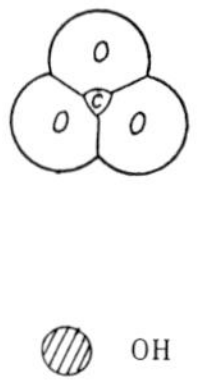

Fig. 1

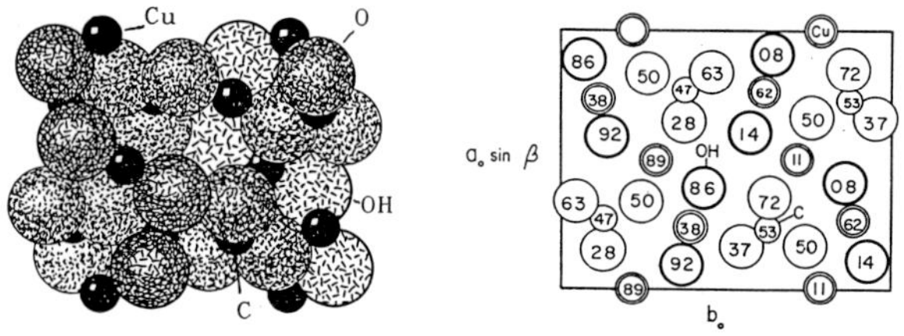

(a) Fig. 2 (b)

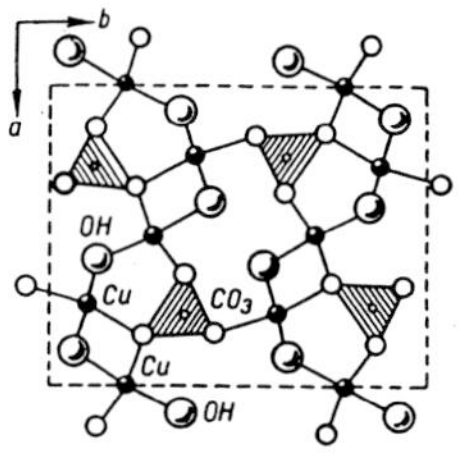

Fig. 3

Properties

Habit	Cleav.	Fract.	Twin.	Hardn.	Dens.	Colour	Transp.
prismatic, massive, fibrous	perfect ($\bar{2}01$)	subconchoidal	(100)	3.5 - 4	4.05	grass green	translucent

Refr. index/Reflect.	Birefr.	Luster	Streak	Melt.p.	CPI
n_α = 1.655	(-)	adamantine	pale green		(SPI)
n_β = 1.875	2V = 43°				48
n_γ = 1.909					

Figures

Fig. 1. Structural units of the malachite structure: CO_3 groups and OH ions.

Fig. 2. (a) Packing drawing of the malachite structure showing the way the CO_3 groups pack together with OH groups, and (b) corresponding projection along the c axis (after Wyckoff, 1964. Vol. 2)

Fig. 3. Malachite structure (after Kostov, 1968).

Fig. 4. Condensed model of the packing analogue of the malachite structure based on puckered simple hexagonal packing layers. The large open circles represent oxygen atoms, and the large lined circles hydroxyls which form the packing layers; the medium black circles correspond to atoms in octahedral voids. The boron atoms (very small black circles) are placed in triangular voids and linked to the three neighbouring oxygen atoms forming BO_3 groups (embraced by a curved line). In each layer the packing atoms are at two different levels (0 and 1/2) which are distinguished by the heavy and light lines of the corresponding circles.

Description

The malachite structure is based on a complex packing of CO_3 and OH ions, with copper atoms in octahedral voids. The packing can be imagined formed by strong distortion of puckered simple hexagonal layers (∼Ts) where the packing atoms are at two different levels (0 and 1/2, thick and thin circles) the lower and the higher parts fitting together in a closest way, giving rise to the octahedral voids.

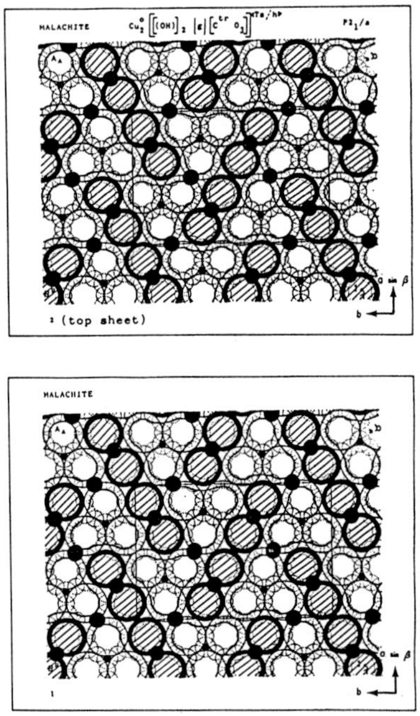

Fig. 4

References

Kostov (1968) 544.
Povarennykh (1972) 614,615.
Wyckoff (1964) Vol. 2, 476,477.
Zoltai + Stout (1984) 431.
Palache et al. (1951) Vol. 2, 253.

Crystallographic data (continued)

Atom	Position	x	y	z
O_{II}	(4e)	0.34	0.24	0.50
O_{III}	(4e)	0.33	0.05	0.63

AZURITE $P2_1/c$

$Cu^{sq}Cu_2^{[5y]}[[(OH)]_2\{g\}[C^{tr}O_3]_2]$

a = 5.00Å
b = 5.85Å
c = 10.35Å
β = 92° 20'
Z = 2

Cu_I (2a)	x = 0, y = 0, z = 0		C (4e)	x = 0.329, y = 0.298, z = 0.319
Cu_{II}(4e)	x = 0.252, y = 0.495, z = 0.085		O_I (4e)	x = 0.098, y = 0.390, z = 0.338
(OH)(4e)	x = 0.092, y = 0.812, z = 0.444		O_{II}(4e) ...	x = 0.447, y = 0.224, z = 0.421

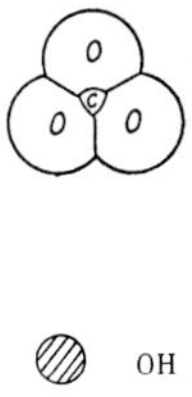

Fig. 1

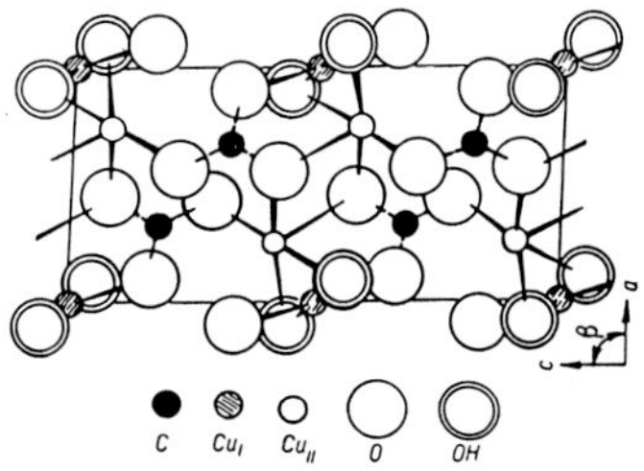

Fig. 2

Properties

Habit	Cleav.	Fract.	Twin.	Hardn.	Dens.	Colour	Transp.
tabular, lenticular, massive	perfect (011) good (100)	conchoidal		3.5-4	3.773	azure-blue, variable	transparent to translucent

Refr. index/Reflect.	Birefr.	Luster	Streak	Melt.p.	CPI
n_α = 1.730 n_β = 1.754 n_γ = 1.836	(+) 2V = 67°	vitreous, adamantine	blue		(SPI) 46

Figures

Fig. 1. The structural units of the azurite structure are CO_3 and (OH) ions.

Fig. 2. Structure of the azurite structure, showing the orientation of the CO_3 groups (after Povarennykh, 1972).

Description

The azurite structure is based on the packing of CO_3 and OH ions with two types of Cu atoms in the voids of this packing: one with nearly square coordination (2O+2OH) and another Cu with tetragonal pyramidal coordination [5y] (3O+2OH).

References

Kostov (1968) 545.

Povarennykh (1972) 615.

Wyckoff (1964) Vol. 2, 478, 479.

Zoltai + Stout (1984) 431.

Palache et al. (1951) Vol. 2, 265-267.

Crystallographic data (continued)

O_{III}(4e) x = 0.431, y = 0.303, z = 0.212

HYDROZINCITE C2/m

$Zn_3^o\ Zn_2^t \left[\left[(OH)\right]_6 \left\{g\right\} \left[C^{tr}\ O_3\right]_2\right]$

a = 13.62Å
b = 6.30Å
c = 5.42Å
β = 95° 50'
Z = 2

Atom	Position	x	y	z
Zn_I^o	(2a)	0	0	0
Zn_{II}^o	(4h)	0	0.263	0.5
Zn_{III}^o	(4i)	0.129	0.5	0.038
C	(4i)	0.322	0.5	0.309
O_I	(4i)	0.421	0.5	0.318
O_{II}	(4i)	0.266	0.5	0.069
...				

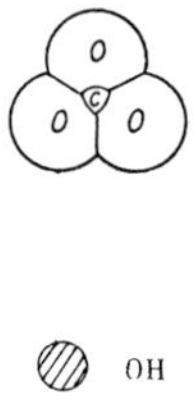

Fig. 1

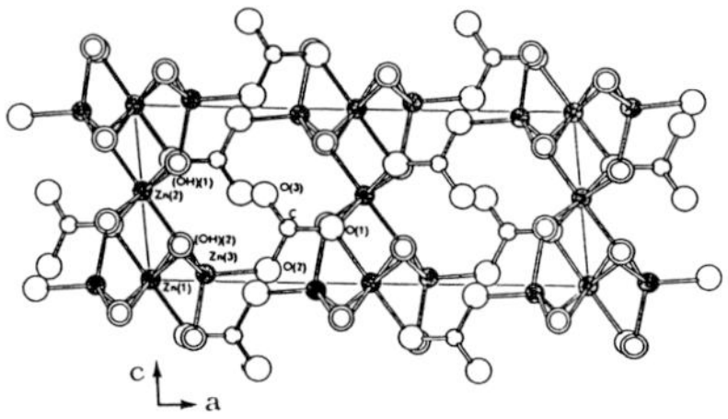

Fig. 2

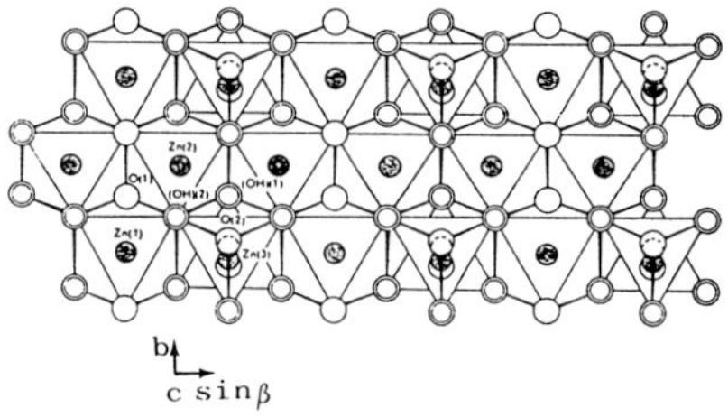

Fig. 3

Properties

Habit	Cleav.	Fract.	Twin.	Hardn.	Dens.	Colour	Transp.
flattened or massive aggregates	perfect (100)			2-2.5	3.5-3.8	pure white to grey, variable	transparent

Refr. index /Reflect.	Birefr.	Luster	Streak	Melt.p.	CPI
n_α = 1.640 n_β = 1.736 n_γ = 1.750	(-) 2V = 40°	dull earth	dull to shining		

Figures

Fig. 1. The structural units of the hydrozincite structure are CO_3, and OH ions.

Fig. 2. Projection along the b axis of the hydrozincite structure (after Ghose, 1964).

Fig. 3. A single oxy-hydroxy-zinc layer of the hydrozincite structure, viewed along the a axis (after Ghose, 1964).

Description

The structure of hydrozincite is formed by the packing of CO_3 and OH ions . The Zn atoms have octahedral and tetrahedral coordinations. There are layers of O and OH which contain the Zn atoms in octahedral coordination, with some holes. The Zn atoms with tetrahedral coordination occur above and below the holes in the octahedral layers. The complex layers are held together by CO_3 groups situated normal to the layers.

References

Kostov (1968) 542.
Povarennykh (1972) 615.
Ghose (1964) 1051-1057.
Palache et al. (1951) Vol. 2, 247.

Crystallographic data (continued)

Atom	Position	x	y	z
O_{III}	(4i)	0.275	0.5	0.500
$(OH)_I$	(4i)	0.410	0	0.324
$(OH)_{II}$	(8j)	0.078	0.245	0.198

HYDROMAGNESITE

$P2_1/c$

$Mg^{o}_{5}\left[\left[(OH)\right]_{2}\left[(H_{2}O)\right]_{4}\ \{g\}\ \left[C^{tr}O_{3}\right]_{4}\right]$

a = 10.105Å
b = 8.954Å
c = 8.378Å
β = 114.44°
Z = 2

Atom		Coordinates
Mg_I	(4e)	x = 0.34502, y = 0.06865, z = 0.35897
Mg_{II}	(4e)	x = 0.34474, y = 0.43518, z = 0.49177
Mg_{III}	(2a)	x = 0, y = 0, z = 0

...

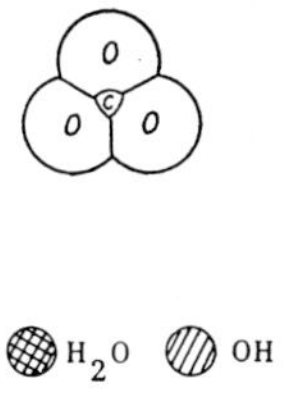

Fig. 1

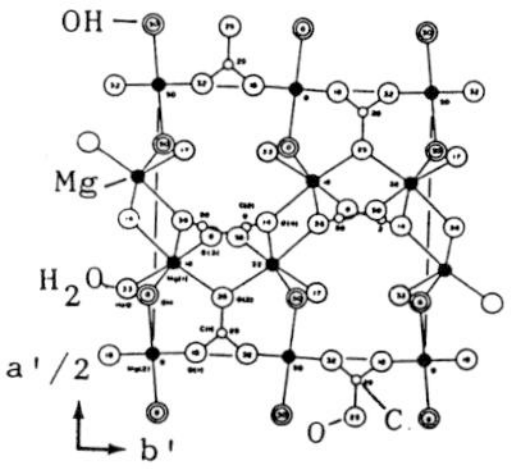

Fig. 2

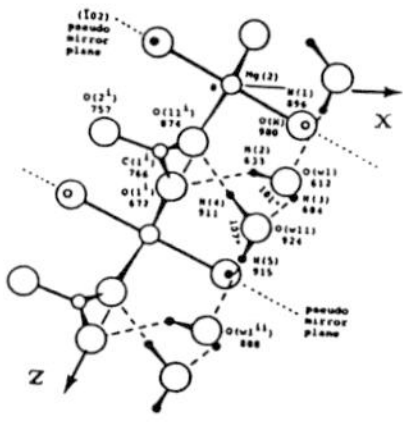

Fig. 3

Properties

Habit	Cleav.	Fract.	Twin.	Hardn.	Dens.	Colour	Transp.
acicular, massive	perfect (010)		(100)	3.75	2.236	colourless to white	transparent

Refr. index/Reflect.	Birefr.	Luster	Streak	Melt.p.	CPI
n_α = 1.523 n_β = 1.527 n_γ = 1.545	(+)	vitreous			

Figures

Fig. 1. The structural units of the hydromagnesite structure are CO_3, (H_2O) and OH ions.

Fig. 2. Hydromagnesite structure viewed along the c axis of the pseudo-orthorhombic cell (after Akao et al. 1974).

Fig. 3. Hydrogen-bonding scheme of hydromagnesite projected along the b axis. Hydrogen bonds are marked with dashed lines (after Akao + Iwai, 1977).

Description

The hydromagnesite is based on the packing of CO_3, H_2O and OH ions. Magnesium atoms occupy voids in this packing with octahedral coordination.

Crystallographic data (continued)

Atom		x	y	z
O(H)	(4e)	0.22455	0.97967	0.11721
O(W1)	(4e)	0.24791	0.61189	0.30117
O(W11)	(4e)	0.23813	0.92381	0.45921
O(1)	(4e)	0.00878	0.17176	0.16220
O(11)	(4e)	0.01677	0.37440	0.30587
O(2)	(4e)	0.22288	0.25692	0.35464
O(3)	(4e)	0.43075	0.19566	0.21587
O(4)	(4e)	0.49221	0.40165	0.37848
O(44)	(4e)	0.49778	0.39688	0.11400
C_I	(4e)	0.08223	0.26599	0.27463
C_{II}	(4e)	0.47277	0.33128	0.23558
H(1)	(4e)	0.230	0.896	0.116
H(2)	(4e)	0.163	0.633	0.278
H(3)	(4e)	0.289	0.684	0.356
H(4)	(4e)	0.145	0.911	0.385
H(5)	(4e)	0.243	0.915	0.566

References

Kostov (1968) 526.
Povarennykh (1972) 620.
Akao et al. (1974) 2670-2672.
Akao + Iwai (1977) 1273-1275.
Palache et al. (1951) Vol. 2, 271, 272.

NATRON

Cc

$$\left[\{g\}\left[\overset{o}{Na_2}(H_2O)_{10}\right]\{g\}\left[\overset{tr}{C}\,O_3\right]\right]^{c}$$

(synthetic material)

a = 12.83Å
b = 9.03Å
c = 13.44Å
β = 123.0°
Z = 4

Atom	Position	x	y	z
Na_I	(4a)	0.240	0.225	0.128
Na_{II}	(4a)	0.252	0.270	-0.128
O_I	(4a)	0.125	-0.012	0.091
O_{II}	(4a)	0.376	0.440	0.208
O_{III}	(4a)	0.093	0.312	0.175
O_{IV}	(4a)	0.374	0.121	0.322
O_V	(4a)	0.139	0.374	-0.058
O_{VI}	(4a)	0.355	0.125	0.047
O_{VII}	(4a)	0.128	0.049	-0.215
O_{VIII}	(4a)	0.372	0.502	-0.080
O_{IX}	(4a)	0.120	0.376	-0.332
O_X	(4a)	0.399	0.187	-0.186
C	(4a)	0.250	0.754	0
O(12')	(4a)	0.344	0.793	-0.003
O(11")	(4a)	0.253	0.834	-0.073
O(13")	(4a)	0.366	0.809	0.093
O(11')	(4a)	0.266	0.718	0.097
O(13')	(4a)	0.139	0.794	-0.096
O(12")	(4a)	0.151	0.675	-0.001

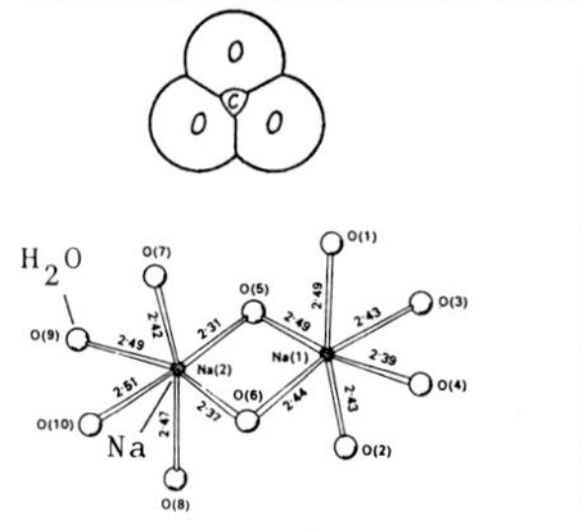

Fig. 1

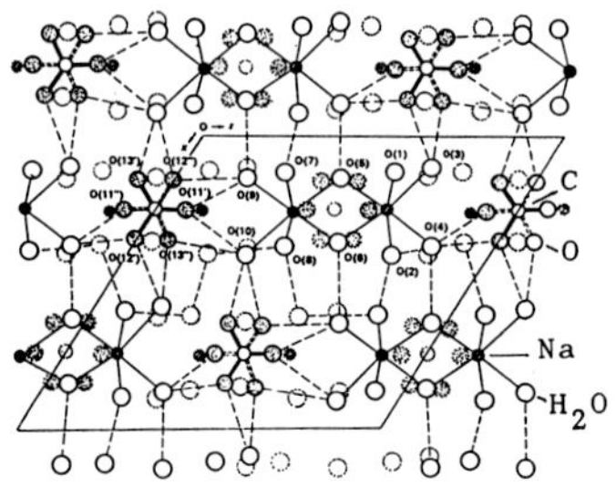

Fig. 2

Properties

Habit	Cleav.	Fract.	Twin.	Hardn.	Dens.	Colour	Transp.
tabular, granular	distinct (001)	conchoidal	(001)	1-1.5	1.478	colourless to white, grey, yellow	

Refr. index/Reflect.	Birefr.	Luster	Streak	Melt.p.	CPI
n_α = 1.405 n_β = 1.425 n_γ = 1.440	(-)	vitreous		34.5° C	

Figures

Fig. 1. The structural units of the natron structure: CO_3 and $Na_2(H_2O)_{10}$ groups (after Taga, 1969).

Fig. 2. Projection of the natron structure along the b axis (after Taga, 1969).

Description

The structure of natron may be imagined as a distorted $Na^o[Cl]^c$ type structure, composed of $[Na_2(H_2O)_{10}]^{2+}$ and CO_3^{2-} groups. The $[Na_2(H_2O)_{10}]^{2+}$ consist of two sodium-water octahedra which share an edge (Fig. 1). These pairs of octahedra are linked by hydrogen bonds to form a three-dimensional framework in the crystal structure (Fig. 2).

References

Kostov (1968) 529.

Povarennykh (1972) 618.

Taga (1969) 2656-2658.

Palache et al (1951) Vol. 2, 230,231.

MELILITE

$P\bar{4}2_1m$

$\overset{[8]}{Ca}\overset{[8]}{Na}(Al,Mg)^{o}\{g\}[(Si,Al)^{t}_{2}O_{7}]$

a = 7.74Å
c = 5.02Å
Z = 2

Atom	Coordinates	Atom	Coordinates
(Ca,Na)(4e')	x = 0.3355, y = 0.1645, z = 0.5073	O_I (2c)	x = 1/2, y = 0, z = 0.1804
(Al,Mg)(2a)	x = 0, y = 0, z = 0	O_{II} (4e')	x = 0.1450, y = 0.3550, z = 0.2583
(Si,Al)(4e')	x = 0.1396, y = 0.3604, z = 0.9412	O_{III} (8f)	x = 0.0820, y = 0.1820, z = 0.7909

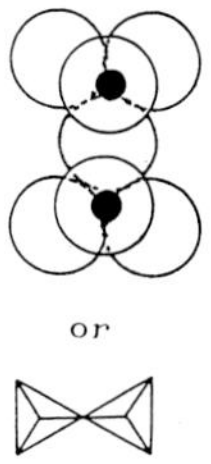

or

Fig. 1

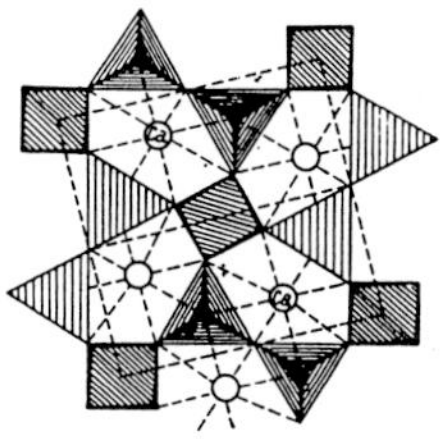

Fig. 2

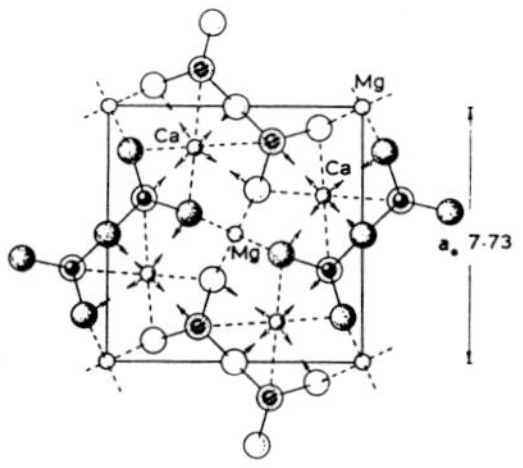

Fig. 3

Properties

Habit	Cleav.	Fract.	Twin.	Hardn.	Dens.	Colour	Transp.
tabular, short prismatic	moderate (001)	conchoidal	(100) (001)	5-6	2.95-3.05	honey-yellow, brown	

Refr. index/Reflect.	Birefr.	Luster	Streak	Melt.p.	CPI
n_{ω} = 1.624-1.666 n_{ϵ} = 1.616-1.661	(-)(+)	vitreous, resinous			

Population

Gehlenite $\overset{[8]}{Ca_2}Al^{o}\{g\}[\overset{t}{Si}\overset{t}{Al}O_7]$

Hardystonite $\overset{[8]}{Ca_2}Zn^{o}\{g\}[\overset{t}{Si_2}O_7]$

Description

Melilite is a group structure formed by the packing of $(Si,Al)^{t}_{2}O_7$ groups, with Ca and Na in eightfold coordination, and Al and Mg in octahedral voids.

Figures

Fig. 1. Structural unit of the melilite structure $(Si,Al)^{t}_{2}O_7$ group

Fig. 2. Polyhedral representation of the melilite structure (after Povarennykh, 1972).

Fig. 3 Melilite structure projected along the c axis (after Kostov, 1968).

References

Kostov (1968) 309.
Povarennykh (1972) 440.
Wyckoff (1968) Vol. 4, 225-227.
Deer et al. (1962) Vol. 1, 236.
Ford (1932) 606,607.

EPIDOTE

$P2_1/m$

$\overset{[9]}{Ca}\overset{[10]}{(Ca,Ce)}\overset{o}{Al_2}\overset{o}{Fe}\overset{t}{Si}\left[[O]_5[OH]\{g\}[\overset{t}{Si_2}O_7]\right]$

a = 8.96Å
b = 5.63Å
c = 10.30Å
β = 115° 24'
Z = 2

Atom	Position	x	y	z
Ca_I	(2e)	0.755	3/4	0.150
Ca_{II}	(2e)	0.603	3/4	0.421
Al_I	(2a)	0	0	0
Al_{II}	(2c)	0	0	1/2
Fe	(2e)	0.297	1/4	0.227
Si_I	(2e)	0.342	3/4	0.055

...

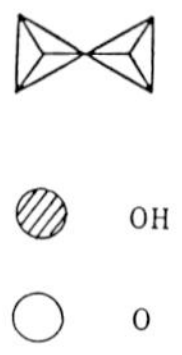

Fig. 1

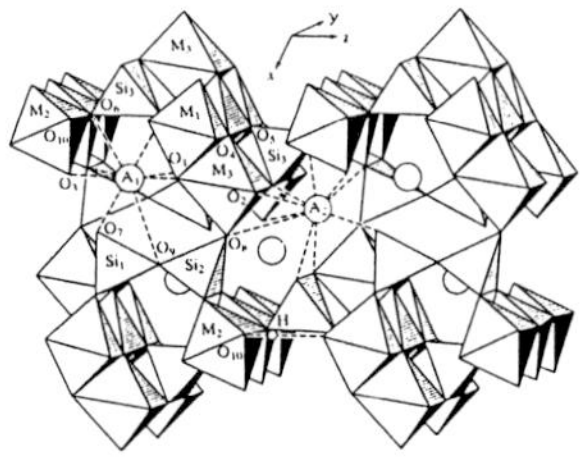

Fig. 2

Properties

Habit	Cleav.	Fract.	Twin.	Hardn.	Dens.	Colour	Transp.
prismatic	perfect (001) poor (100)	uneven	(100)	6.5	3.4-3.5	yellow,, green	transparent to translucent

Refr. index /Reflect.	Birefr.	Luster	Streak	Melt.p.	CPI
n_α = 1.71-1.75	(-)	vitreous	white		(SPI)
n_β = 1.72-1.78	2V = 90°-115°				56
n_γ = 1.73-1.80					

Population

Allanite

$\overset{[9/10]}{(Ca,Ce)_2}\overset{o}{(Al,Fe)_3}\overset{t}{Si}\left[[O]_5[OH]\{g\}[\overset{t}{Si_2}O_7]\right]$

Orthite = variety of allanite

Figures

Fig. 1. Structural units of the epidote structure: Si_2O_7 groups, OH and O atoms.

Fig. 2. Polyhedral representation of the epidote structure (after Dollase, 1971).

Description

The epidote structure is based on the packing of Si_2O_7 groups, OH and O atoms. The interstitial atoms are Ca, Al, Fe and Si. Ca has two kinds of coordination [10] and [9], Al and Fe have octahedral coordination, and Si is tetrahedraly coordinated.

References

Kostov (1968) 310, 312.
Povarennykh (1972) 405.
Wyckoff (1968) Vol. 4, 249-251.
Zoltai + Stout (1984) 373.
Dollase (1971) 458.

Crystallographic data (continued)

Atom	Position	x	y	z
Si_{II}	(2e)	0.682	1/4	0.270
Si_{III}	(2e)	0.183	3/4	0.314
(OH)	(2e)	0.086	1/4	0.413
O_I	(4f)	0.235	0.992	0.053
O_{II}	(4f)	0.300	0.987	0.353
O_{III}	(4f)	0.798	0.008	0.338
O_{IV}	(2e)	0.047	1/4	0.142
O_V	(2e)	0.047	3/4	0.142
O_{VI}	(2e)	0.080	3/4	0.413
O_{VII}	(2e)	0.522	3/4	0.183
O_{VIII}	(2e)	0.530	1/4	0.315
O_{IX}	(2e)	0.625	1/4	0.098

ZOISITE

Pnma

$Ca_2^{[7]}Al_3^{o}Si^{t}\left[[O]_5 [OH] \{g\} [Si_2^{t} O_7]\right]$

a = 16.23Å
b = 5.51Å
c = 10.16Å
Z = 4

Atom		Coordinates
Ca_I	(4c)	x = 0.372, y = 1/4, z = 0.436
Ca_{II}	(4c)	x = 0.454, y = 1/4, z = 0.112
Al_I	(8d)	x = 0.250, y = 0, z = 0.1905
Al_{II}	(4c)	x = -0.107, y = 1/4, z = -0.300
Si_I	(4c)	x = 0.0845, y = 1/4, z = 0.102
Si_{II}	(4c)	x = -0.412, y = 1/4, z = -0.281

...

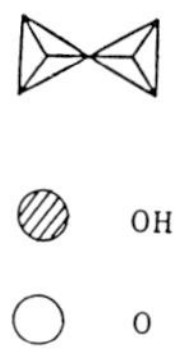

Fig. 1

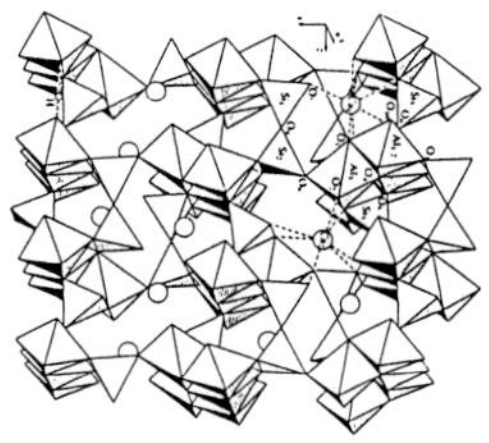

Fig. 2

Properties

Habit	Cleav.	Fract.	Twin.	Hardn.	Dens.	Colour	Transp.
prismatic, massive	perfect (100)	uneven, subconchoidal		6-7	3.15--3.365	grey, green--brown	transparent to translucent

Refr. index/Reflect.	Birefr.	Luster	Streak	Melt.p.	CPI
n_α = 1.685-1.705 n_β = 1.688-1.710 n_γ = 1.697-1.725	(+) 2V=20°30°	vitreous	uncoloured		

Figures

Fig. 1. Structural units of the zoisite structure: Si_2O_7 groups, OH and O atoms.

Fig. 2. Polyhedral representation of the zoisite structure (after Dollase, 1968).

Description

Zoisite is based on Si_2O_7 groups which pack together with OH and O atoms. The interstitial Al and Ca atoms have octahedral and 7 coordination respectively. Some Si atoms are also interstitial, having tetrahedral coordination.

References

Kostov (1968) 312.
Wyckoff (1968) Vol. 4, 251-253.
Dollase (1968) 1889.
Roberts et al. (1974) 691.

Crystallographic data (continued)

Atom		Coordinates
Si_{III}	(4c)	x = 0.162, y = 1/4, z = 0.435
O_I	(8d)	x = 0.118, y = 0.000, z = 0.150
O_{II}	(8d)	x = 0.101, y = 0.000, z = 0.431
O_{III}	(8d)	x = 0.375, y = 0.000, z = 0.245
O_{IV}	(4c)	x = 0.233, y = 1/4, z = 0.304
O_V	(4c)	x = -0.223, y = 1/4, z = -0.304
O_{VI}	(4c)	x = -0.2685, y = 1/4, z = -0.057
O_{VII}	(4c)	x = 0.997, y = 1/4, z = 0.163
O_{VIII}	(4c)	x = -0.997, y = 1/4, z = -0.296
O_{IX}	(4c)	x = -0.435, y = 1/4, z = -0.438
(OH)	(4c)	x = 0.2685, y = 1/4, z = 0.075

ILVAITE

Pbnm

$Ca^{[7]}\overset{II}{Fe}{}^{o}_{2}(\overset{III}{Fe},Mn)^{o}\left[[O][OH]\{g\}[Si^{t}_{2}O_{7}]\right]$

a = 8.80Å
b = 13.07Å
c = 5.86Å
Z = 4

Atom	Position	x	y	z
Ca	(4c)	-0.370	-0.185	1/4
Fe^{2+}	(8d)	0.052	0.110	0.010
Fe^{3+}	(4c)	0.744	0.058	1/4
Si_I	(4c)	0.366	0.038	1/4
Si_{II}	(4c)	0.227	0.319	1/4
O_I	(4c)	-0.033	0.000	1/4
...				

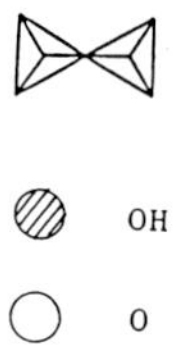

Fig. 1

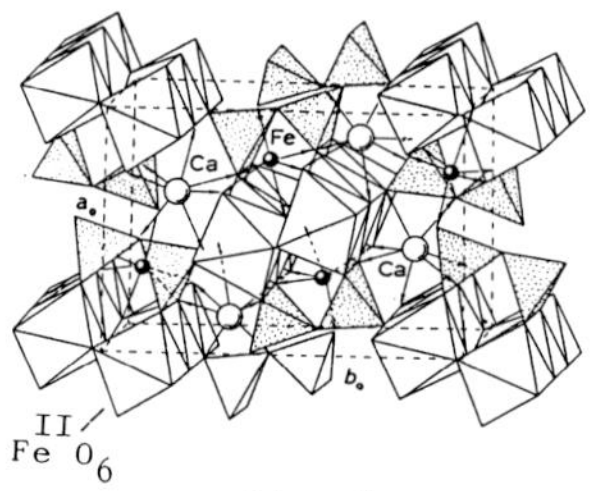

Fig. 2

Properties

Habit	Cleav.	Fract.	Twin.	Hardn.	Dens.	Colour	Transp.
prismatic, columnar, massive	distinct (010) (001)	uneven		5.5-6	3.8-4.1	black, greyish--black	opaque

Refr. index/Reflect.	Birefr.	Luster	Streak	Melt.p.	CPI
n_α = 1.727 n_β = 1.870 n_γ = 1.883	(-) 2V=20°-30°	submetallic	black		

Figures

Fig. 1. Structural units of the ilvaite structure: Si_2O_7 group, OH and O atoms.

Fig. 2. Polyhedral representation of the ilvaite structure (after Kostov, 1968).

Description

Ilvaite is a group structure formed by the packing of Si_2O_7 groups, OH and O atoms, with Ca, Fe and Si in the interstices. Fe has octahedral coordination, and Ca is [7] coordinated. Bivalent iron forms columns of octahedra parallel to the c axis.

Crystallographic data (continued)

Atom	Position	x	y	z
O_{II}	(8d)	0.273	0.055	0.020
O_{III}	(4c)	0.104	0.220	1/4
O_{IV}	(8d)	0.335	0.336	0.020
O_V	(4c)	0.087	0.410	1/4
O_{VI}	(4c)	-0.026	-0.391	1/4
O_{VII}	(4c)	-0.120	-0.213	1/4

References

Kostov (1968) 314.
Wyckoff (1968) Vol. 4, 235.
Roberts et al. (1974) 299.

IDOCRASE

(Vesuvianite)

P4/nnc

$Ca_{10}^{[6/8]}(Mg,Fe)_2^{o}Al_4^{o}Si_5^{t}\left[[O]_{20}[OH]_4\{g\}[Si_2^{t}O_7]_2\right]$

a = 15.55Å
c = 11.81Å
Z = 4

Ca_I (4c) x = 1/2, y = 0, z = 0
Ca_{II} (4e) x = 0, y = 0, z = 0.125
Ca_{III}(16k) x = 0.06, y = 0.30, z = 0.11
...

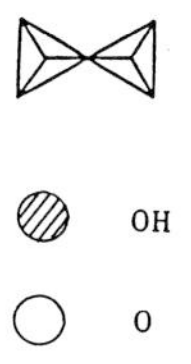

Fig. 1

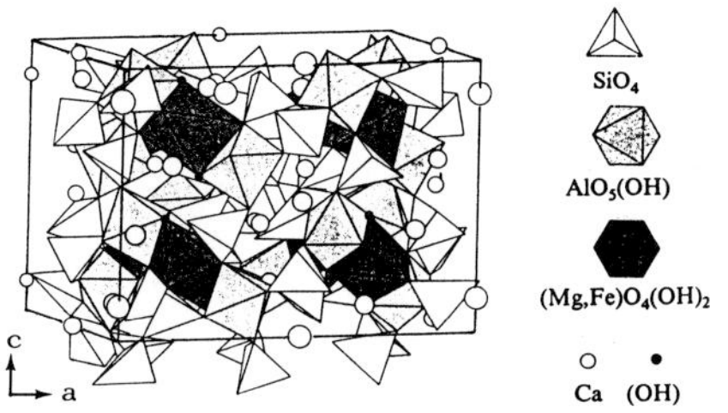

Fig. 2

Properties

Habit	Cleav.	Fract.	Twin.	Hardn.	Dens.	Colour	Transp.
prismatic	poor (100)	subcon-choidal		6.5	3.4	green, brown, yellow	transparent

Refr. index/Reflect.	Birefr.	Luster	Streak	Melt.p.	CPI
n_ω = 1.706 n_ϵ = 1.701	(-)	vitreous	white		(SPI) 50

Figures

Fig. 1. Structural units of the idocrase structure: $Si_2^tO_7$ groups, OH and O atoms.

Fig. 2. Polyhedral representation of the idocrase structure (after Zoltai + Stout, 1984).

Description

Idocrase is a group structure formed by the packing of $Si_2^tO_7$ groups, OH and oxygen atoms, with Ca, Mg, Al and Si atoms in its interstices. 3/4 of the Ca atoms have 8 coordination while the rest of the Ca, and the Mg and Al atoms have 6 coordination.

References

Kostov (1968) 318.
Wyckoff (1968) Vol. 4, 256-258.
Zoltai+Stout (1984) 370,371.

Crystallographic data (continued)

Atom	Position	x	y	z
Ca_{IV}	(16k)	0.42	0.034	0.13
Mg	(8f)	1/4	1/4	1/4
Al	(16k)	0.36	0.14	0.125
O_I	(16k)	0.08	0.47	0.33
O_{II}	(16k)	0.12	0.41	0.03
O_{III}	(16k)	0.47	0.19	0.17
O_{IV}	(16k)	0.18	0.38	0.23
O_V	(16k)	0.26	0.08	0.07
Si_I	(4d)	1/2	0	1/4
Si_{II}	(16k)	0.20	0.44	0.125
Si_{III}	(16k)	0.16	0.08	0.125
O_{VI}	(16k)	0.03	0.12	0.18
O_{VII}	(16k)	0.20	0.07	0.43
O_{VIII}	(16k)	0.17	0.15	0.18
O_{IX}	(8h)	0.09	0.09	0
(OH)	(16k)	0.31	0.24	0.08

HEMIMORPHITE

(Calamine)

Imm2

$Zn^t_4\left[\left[H_2O\right]\left[OH\right]_2\{g\}\left[Si^t_2O_7\right]\right]$

a = 8.37Å
b = 10.719Å
c = 5.120Å
Z = 2

Zn (8e) x = 0.204, y = 0.160, z = 0.000

(H_2O)(2b) u = 0.558

(OH) (4c) u = 0.308, v = 0.952

Si (4d) u = 0.148, v = 0.503

O_I (4d) u = 0.155, v = 0.190

O_{II}(8e) x = 0.160, y = 0.205, z = 0.637

...

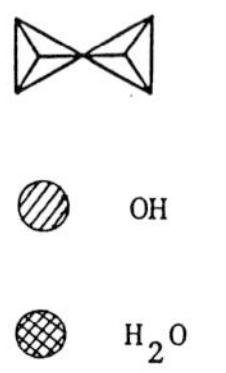

Fig. 1

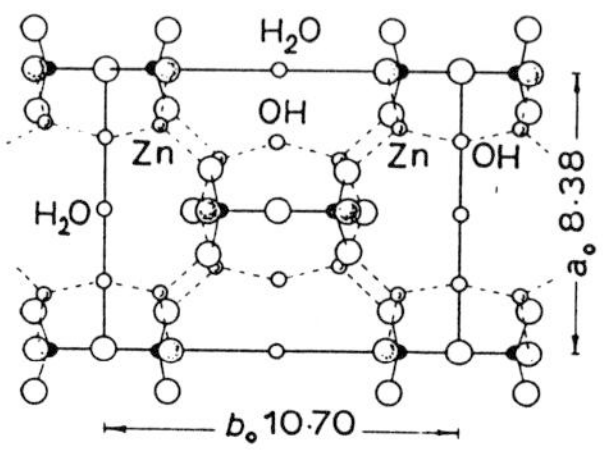

Fig. 2

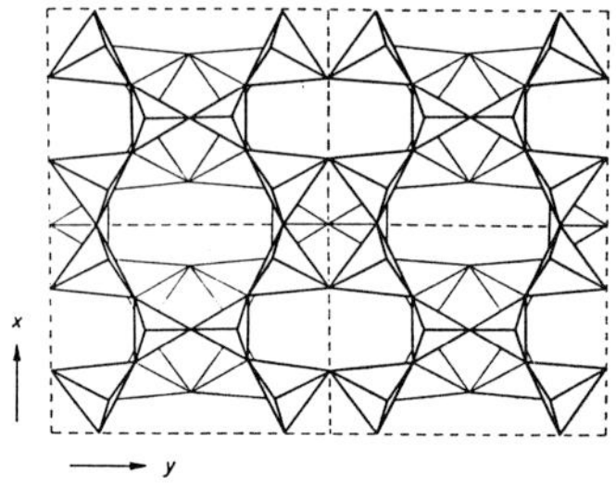

Fig. 3

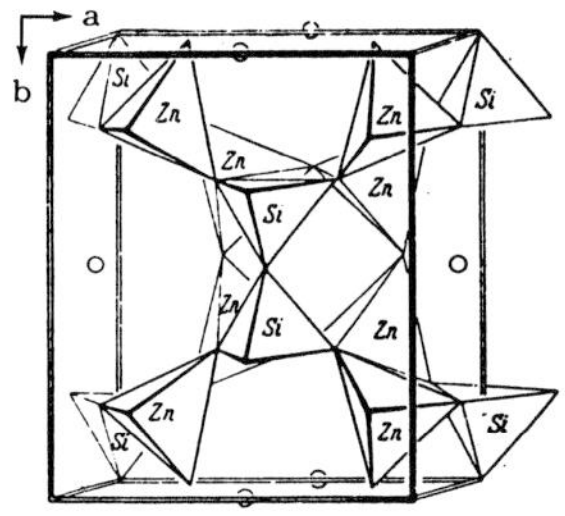

Fig. 4

Properties

Habit	Cleav.	Fract.	Twin.	Hardn.	Dens.	Colour	Transp.
tabular, striated, massive	perfect (110)	uneven, subconchoidal	(001)	4.5-5.0	3.4-3.5	white, colourless, pale blue	

Refr. index /Reflect.	Birefr.	Luster	Streak	Melt.p.	CPI
n_α = 1.614 n_β = 1.617 n_γ = 1.636	(+) 2V = 46°	vitreous, dull	uncoloured		

Figures

Fig. 1. Structural units of the hemimorphite structure: Si_2O_7 groups, H_2O and OH.

Fig. 2. Hemimorphite structure (after Kostov, 1968).

Fig. 3. Polyhedral description of the hemimorphite structure projected along the c axis (after McDonald + Cruischank, 1967).

Fig. 4. Polyhedral representation of the unit cell content of the hemimorphite structure (after Hermann et al., 1936, Vol. 2).

Description

Hemimorphite is a group structure formed by the packing of Si_2O_7 groups, H_2O and OH, with Zn atoms in tetrahedral voids.

It has also been considered as a framework (McDonald + Cruischank, 1967) and a close-packed structure (Moore, 1993).

References

Kostov (1968) 324, 325.
Povarennykh (1972) 403, 659.
Wyckoff (1968) Vol. 4, 215-217.
Hermann et al. (1937) 126.
McDonald + Cruischank (1967) 180-191.
Moore (1993)
Roberts et al. (1974) 267, 268.

Crystallographic data (continued)

O_{III}(2a) u = 0.637

TOURMALINE R3m

$\overset{[6+3]}{Na}\,\overset{o}{Mg_3}\,\overset{o}{Al_6}\,(OH)_4 \left[\left\{g\right\}\left[\overset{tr}{B}O_3\right]_3\left\{g\right\}\left[\overset{t}{Si_6}O_{18}\right]\right]$

a_R = 9.50 Å	α = 113°56'	Na (3a)	x = 0, y = 0, z = 0.8576	B (9b)	x = 0.2206, y = 0.1103, z = 0.6362
Z_R = 1	a_H = 15.951 Å	Mg (9b)	x = 0.0635, y = 0.1270, z = 0.4618	Si(18c)	x = 0.1898, y = 0.1922, z = 0.0900
c = 7.24 Å	Z_H = 3	Al(18c)	x = 0.2615, y = 0.2976, z = 0.4762	O_I (3a) ...	x = 0, y = 0, z = 0.3117

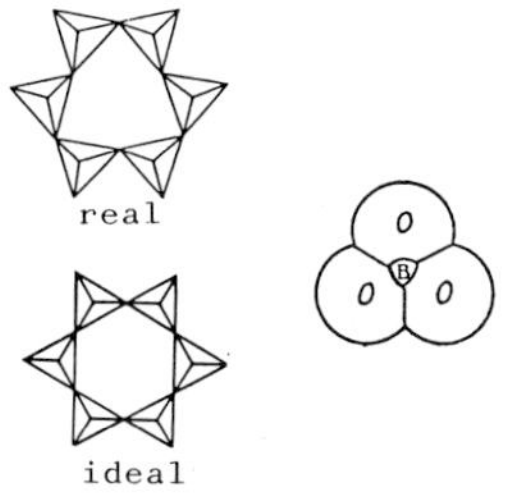

Fig. 1

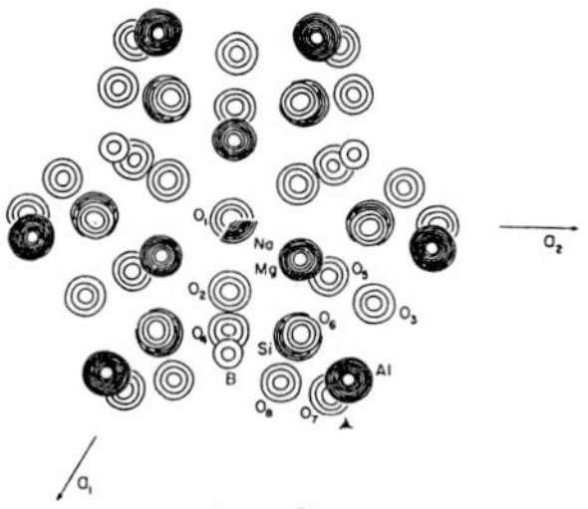

Fig. 2

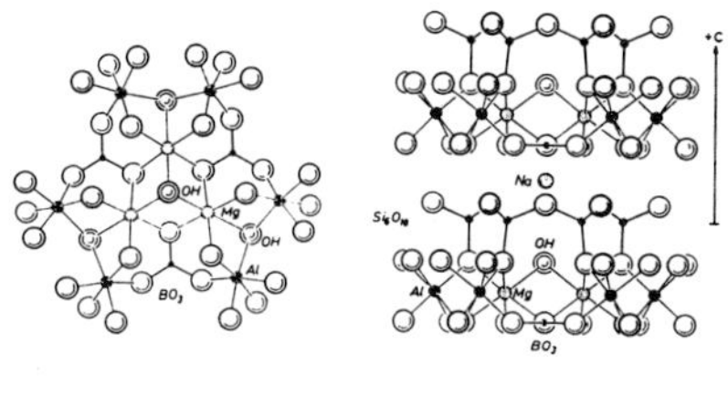

Fig. 3

Properties

Habit	Cleav.	Fract.	Twin.	Hardn.	Dens.	Colour	Transp.
prismatic	poor (101)	subconchoidal	(101)	7-7.5	3.02	brown, yellow	translucent

Refr. index/Reflect.	Birefr.	Luster	Streak	Melt.p.	CPI
n_ω = 1.650 n_ϵ = 1.628	(-)	resinous	white		(SPI) 56

Figures

Fig. 1. Structural units of the tourmaline structure: $Si_6^tO_{18}$ and $B^{tr}O_3$ groups (adapted from Kostov, 1968) .

Fig. 2. Electron density map of the tourmaline structure projected on (0001) (after Buerger et al., 1962).

Fig. 3. Two aspects of the tourmaline structure (after Randohr+ +Strunz, 1980).

Fig. 4. Condensed model of the symmetrical analogue of the tourmaline structure. The large open circles represent oxygen atoms, the dashed large circles OH, the medium circles represent Al (black) and Mg (dashed), the smaller circles correspond to Si (black) and the still smaller to B atoms (black).

Description

The tourmaline structure is a group structure formed by the packing of $Si_6^tO_{18}$ and $B^{tr}O_3$ groups, with Mg,Al and Na atoms and OH in the interstices.

References

Kostov (1968) 319-322.
Povarennykh (1972) 377.
Wyckoff (1968) Vol. 4, 281-283.
Donnay et al. (1963) 645.
Randohr + Strunz (1980) 709.
Zoltai + Stout (1984) 354.
Buerger et al. (1962) 587.

Crystallographic data (continued)

Atom	Coordinates	Atom	Coordinates
O_{II} (9b)	x = 0.1218, y = 0.0609, z = 0.6113	O_{VI} (18c)	x = 0.1866, y = 0.1952, z = 0.3111
O_{III} (9b)	x = 0.1340, y = 0.2680, z = 0.5755	O_{VII} (18c)	x = 0.2851, y = 0.2844, z = 0.0090
O_{IV} (9b)	x = 0.1870, y = 0.0935, z = 0.0176	O_{VIII} (18c)	x = 0.2698, y = 0.2085, z = 0.6455
O_V (9b)	x = 0.0906, y = 0.1812, z = 0.0012		

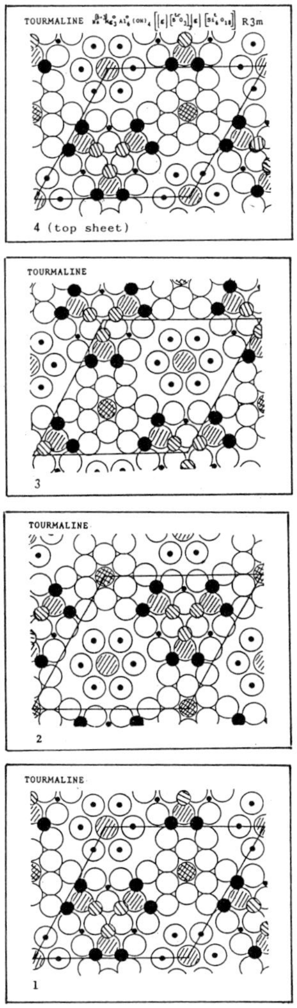

Fig. 4

DIOPTASE $R\bar{3}$

$Cu^{o}_{6}\left[\left[H_2O\right]_6\{g\}\left[Si^{t}_{6}\,O_{18}\right]\right]$

a_R = 8.8 Å	Cu (18f)	x = 0.406 y = 0.403 z = 0.068	O_I (18f)	x = 0.066 y = 0.183 z = -0.074
α = 111°42'				
Z_R = 1	(H_2O)(18f)	x = 0.133 y = 0.174 z = 0.586	O_{II} (18f)	x = 0.284 y = 0.292 z = -0.055
a_H = 14.61Å				
c = 7.80 Å	Si (18f)	x = 0.175 y = 0.217 z = 0.042	O_{III}(18f)	x = 0.159 y = 0.263 z = 0.223
Z_H = 3				

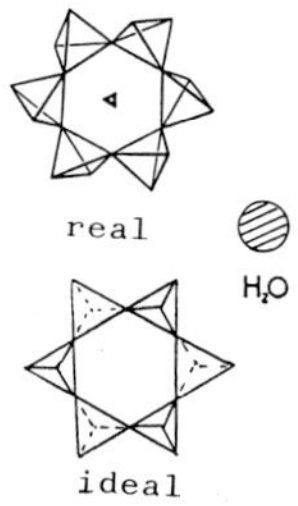

Fig. 1

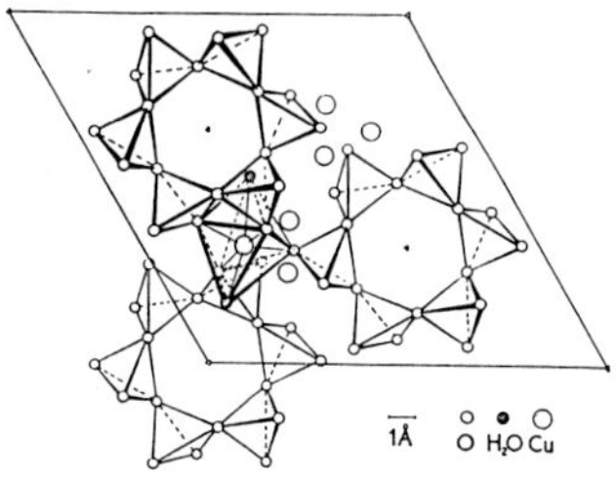

Fig. 2

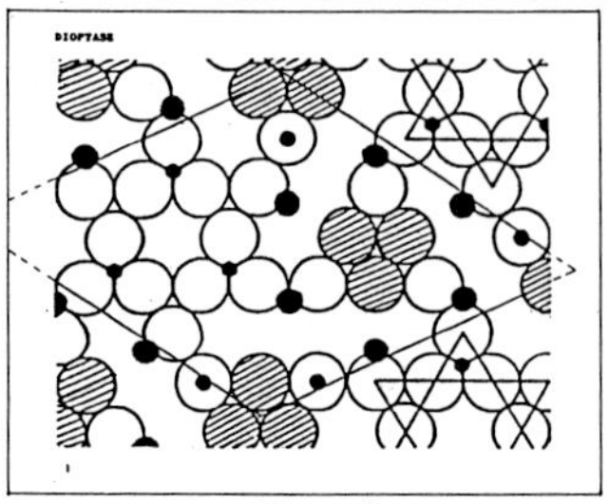

Fig. 3

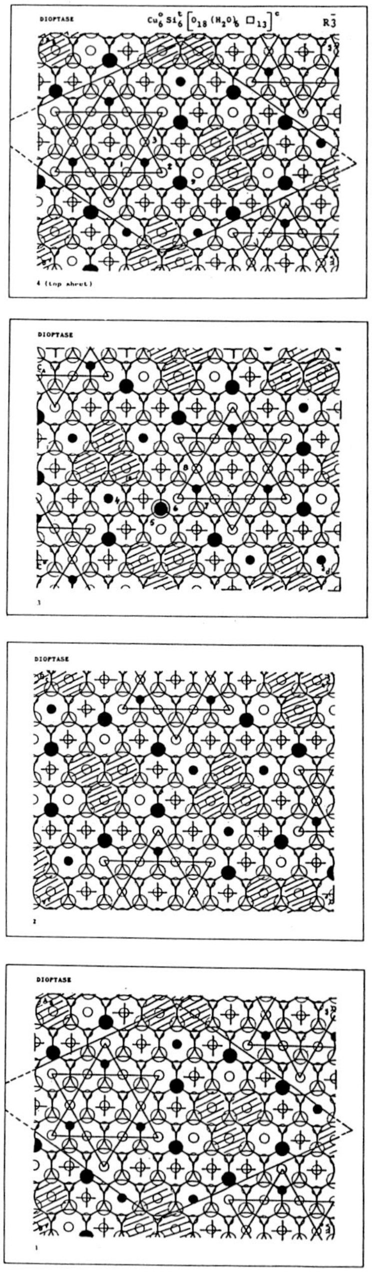

Fig. 4

Properties

Habit	Cleav.	Fract.	Twin.	Hardn.	Dens.	Colour	Transp.
prismatic, massive	perfect (10$\bar{1}$1)	uneven, conchoidal		5	3.28-3.35	emerald green, bluish green	transparent to translucent

Refr. index/Reflect.	Birefr.	Luster	Streak	Melt.p.	CPI
n_ω = 1.644-1.658 n_ϵ = 1.607-1.7094	(+)	vitreous			

Figures

Fig. 1. Structural unit of the dioptase structure: $Si^t_6O_{18}$ group (after Strunz, 1982) and H_2O molecule.

Fig. 2. Polyhedral representation of the dioptase structure showing the coordination polyhedra of Cu atoms (after Bragg + Claringbull, 1985).

Fig. 3. First sheet of the condensed model of the symmetrical analogue of the dioptase structure (Lima--de-Faria, 1988).

Fig. 4. Condensed model of the defect packing analogue of the dioptase structure. In this case the defect packing analogue is equal to the symmetrical analogue.

Description

The dioptase is a group structure formed by the packing of $Si^t_6O_{18}$ groups and H_2O molecules, with Cu atoms in octahedral interstices.

References

Kostov (1968) 325,326.
Wyckoff (1968) Vol. 4, 284,285.
Bragg + Claringbull (1965) 220.
Donnay et al. (1963) 648.
Strunz (1982) 81.
Lima-de-Faria (1988) 39.

AXINITE

$P\bar{1}$

$Ca_4^{ap}(Fe,Mn)_2^{o}\,Al_4^{o}(OH)_2\{g\}\left[B_2^{t}Si_8^{t}O_{30}\right]$

a = 7.1566 Å
b = 9.1995 Å
c = 8.959 Å
α = 91.8°
β = 98.14°
γ = 77.30°
Z = 1

		x	y	z
Ca_I	(2i)	0.7465	0.3480	0.3956
Ca_{II}	(2i)	0.1831	0.1006	0.0837
Fe	(2i)	0.7687	0.5904	0.1120
Al_I	(2i)	0.0529	0.8009	0.2543
Al_{II}	(2i)	0.3520	0.9362	0.4212
H	(2i)	0.0023	0.9697	0.6259

...

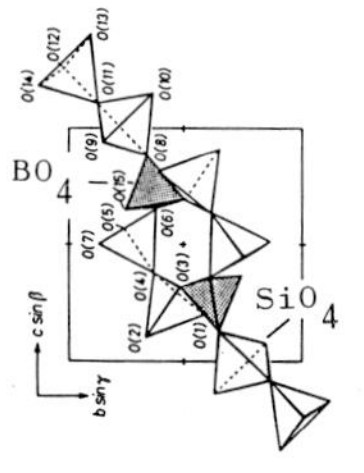

Fig. 1

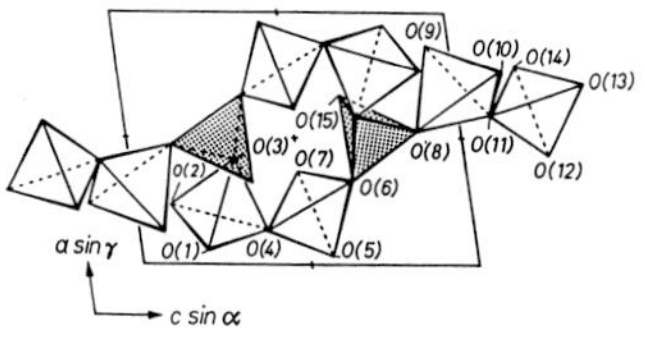

Fig. 2

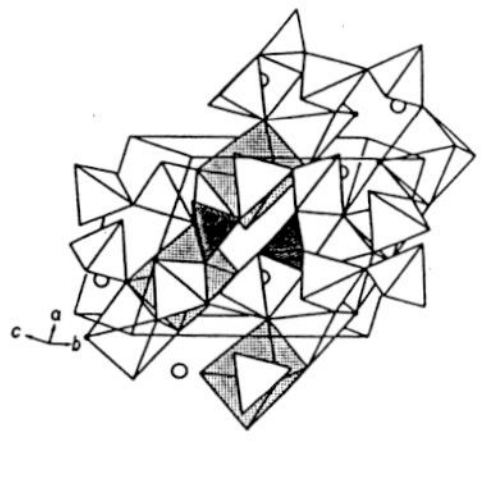

Fig. 3

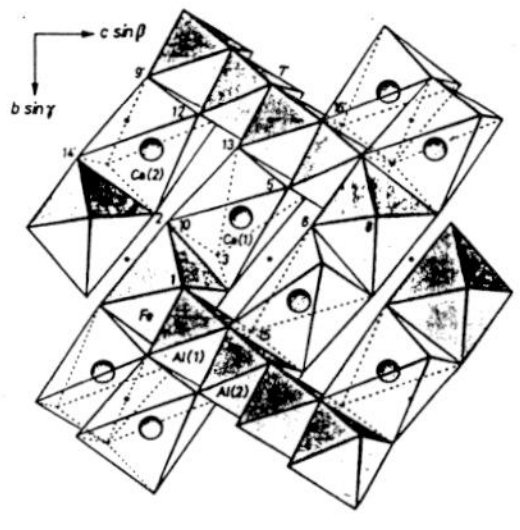

Fig. 4

Properties

Habit	Cleav.	Fract.	Twin.	Hardn.	Dens.	Colour	Transp.
tabular wedge--shaped	good (100)	uneven conchoi-dal		6.5-7	3.26--3.36	violet--brown, colourless	transparent to translu-cent

Refr. index /Reflect.	Birefr.	Luster	Streak	Melt.p.	CPI
n_α = 1.674-1.693 n_β = 1.681-1.701 n_γ = 1.684-1.704	(-) 2V=63°-80°	vitreous	uncoloured		

Figures

Fig. 1. Structural unit of the axinite structure: $B_2^tSi_8^tO_{30}$ group (after Takeuchi et al., 1974).

Fig. 2. Projection of the axinite structure viewed along the b axis (after Takeuchi et al., 1974).

Fig. 3. Crystal structure of axinite (after Takeuchi et al., 1974).

Fig. 4. Polyhedra around Al, Fe and Ca in the axinite structure (after Takeuchi et al., 1974).

Description

The axinite structure is a group structure formed by the packing of $B_2^tSi_8^tO_{30}$ groups, with Al, Fe and Ca in the interstices. The Ca atoms have trigonal anti-prismatic coordination, and Al and Fe are located in octahedral voids.

References

Kostov (1968) 322.
Takéuchi et al. (1974) 289-312.
Roberts et al. (1974) 44.
Deer et al. (1962) Vol. 1, 320.

Crystallographic data (continued)

		x	y	z
B	(2i)	0.4619	0.6346	0.2860
Si_I	(2i)	0.2120	0.4502	0.2356
Si_{II}	(2i)	0.2189	0.2748	0.5242
Si_{III}	(2i)	0.6995	0.2553	0.0112
Si_{IV}	(2i)	0.6413	0.0189	0.2304
O_I	(2i)	0.0564	0.6033	0.1897
O_{II}	(2i)	0.2333	0.3386	0.0982
O_{III}	(2i)	0.4202	0.4864	0.3135
O_{IV}	(2i)	0.1357	0.3739	0.3713
O_V	(2i)	0.0218	0.2419	0.5638
O_{VI}	(2i)	0.3261	0.3791	0.6455
O_{VII}	(2i)	0.3802	0.1274	0.4956
O_{VIII}	(2i)	0.5371	0.3433	0.8773
O_{IX}	(2i)	0.8759	0.1543	0.9334
O_X	(2i)	0.7693	0.3655	0.1394
O_{XI}	(2i)	0.6037	0.1348	0.0863
O_{XII}	(2i)	0.4359	0.9817	0.2442
O_{XIII}	(2i)	0.7204	0.0998	0.3842
O_{XIV}	(2i)	0.7943	0.8735	0.1783
O_{XV}	(2i)	0.3256	0.7464	0.3545
O_{XVI}	(2i)	0.0968	0.9954	0.3232

8.4.3. Chain structures

TELLURIUM

$P3_221$

$^1_\infty\left[Te^{[2]}\right]^{I^{11}}_{by}$

a = 4.44693Å
c = 5.91492Å
Z = 3

Te(3a) u = 0.269

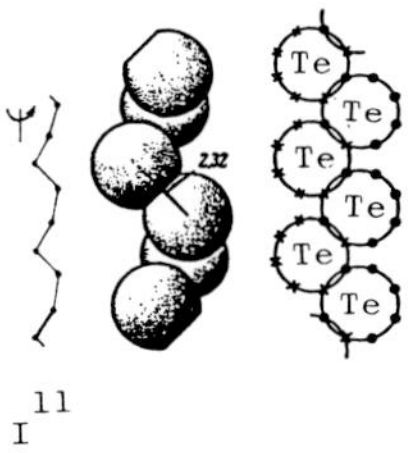

Fig. 1

(a)

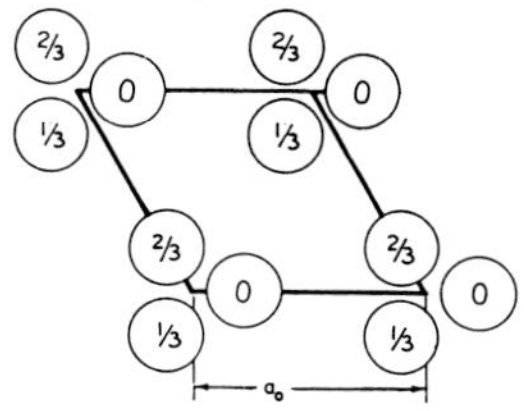

(b)

Fig. 2

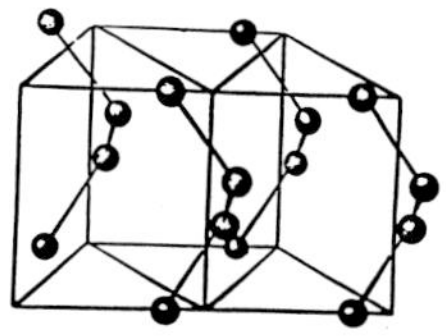

Fig. 3

Properties

Habit	Cleav.	Fract.	Twin.	Hardn.	Dens.	Colour	Transp.
prismatic, acicular, massive	perfect (10$\bar{1}$0)			2-2.25	6.1--6.3	tin--white	opaque

Refr. index/Reflect.	Birefr.	Luster	Streak	Melt.p.	CPI
63%		metallic	grey	450°C	

Figures

Fig. 1. Representation of the chain of the selenium structure, which is isotypic with tellurium structure, and corresponding covalent bonding scheme (adapted from Grigoriev, 1964).

Fig. 2. (a) Packing drawing of the tellurium structure, and (b) basal projection of the unit cell content (after Wyckoff, 1963, Vol. 1).

Fig. 3. Tellurium structure (after Povarennykh, 1972).

Description

The tellurium structure is formed by spiral chains of tellurium atoms.

References

Kostov (1968) 98.
Povarennykh (1972) 196.
Wyckoff (1963) Vol. 1, 36-38.
Palache et al. (1944) Vol. 1, 138.
Ingerson (1955) 350.

CINNABAR

$P3_121$

$\overset{1}{\infty}\left[Hg^{o}\ \overset{\ll Tv \gg}{S}\right]^{I^{13}}_{by}$

a = 4.149Å
c = 9.495Å
Z = 3

Hg(3a) u = 0.720

S (3b) u = 0.485

Fig. 1

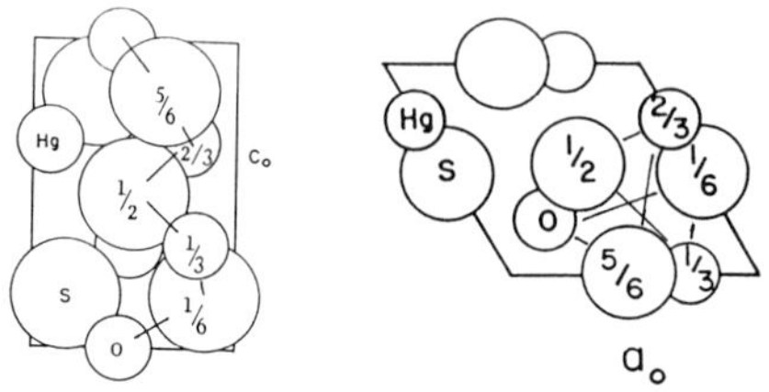

(a) Fig. 2 (b)

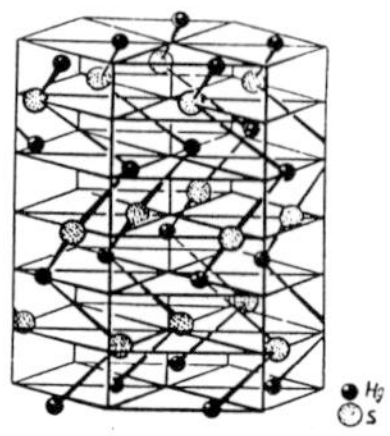

Fig. 3

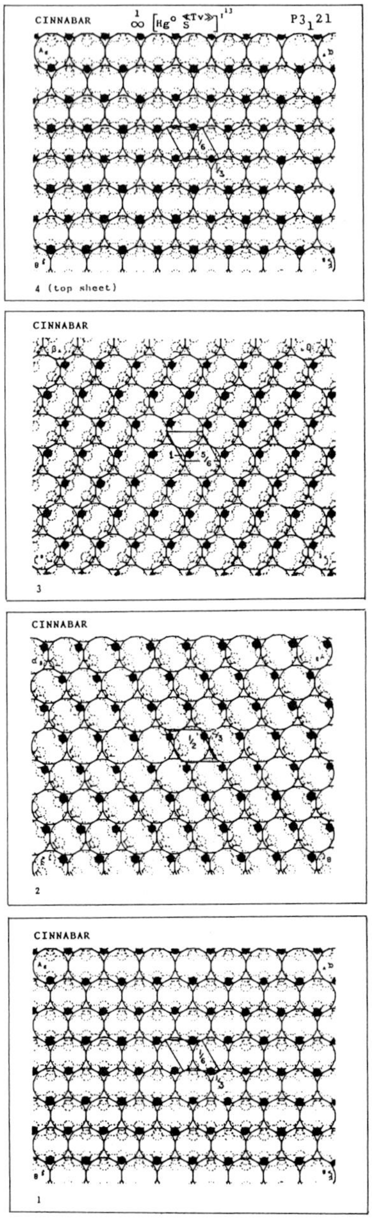

Fig. 4

Properties

Habit	Cleav.	Fract.	Twin.	Hardn.	Dens.	Colour	Transp.
rhombo-hedral, tabular	perfect $(10\bar{1}0)$	subcon-choidal, uneven	(0001)	2-2.5	8.090	cochineal--red, brownish	transparent to opaque

Refr. index	Reflect.	Birefr.	Luster	Streak	Melt.p.	CPI
n_ω = 2.90 n_ϵ = 3.25	25%	(+)	adaman-tine	scarlet	580°C (sublimes)	

Figures

Fig. 1. Spiral chain of the cinnabar structure.

Fig. 2. (a) Packing drawing of the cinnabar structure (the indicated positions are along the c axis) and (b) unit cell content projected along the c axis (adapted from Wyckoff, 1963, Vol. 1).

Fig. 3. Cinnabar structure showing the S-Hg-S-Hg- chains (after Povarennykh, 1972).

Fig. 4. Condensed model of the packing analogue of the cinnabar structure, which is based on a Tv packing of S (large open circles) with Hg (small black circles) in distorted octahedral voids. Compare with Fig. 2 (b).

Description

The cinnabar structure is build of Hg S chains, which can be derived from a Tv packing of S with Hg in octahedral voids.

References

Kostov (1968) 161, 162.
Povarennykh (1972) 248.
Wyckoff (1963) Vol. 1, 98, 99.
Zoltai + Stout (1984) 390.
Ingerson (1955) 350.
Palache et al.(1944) Vol.1, 251-253.

CALOMEL

I4/mmm

$\frac{1}{\infty}\left[\mathrm{Hg}^{[5]}\ \mathrm{Cl}^{\ll(2Qs)f\gg}\right]$

a = 4.478Å
c = 10.91 Å
Z = 4

Hg(4e) u = 0.116
Cl(4e) u = 0.347

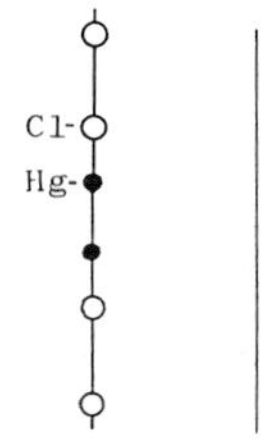

Fig. 1

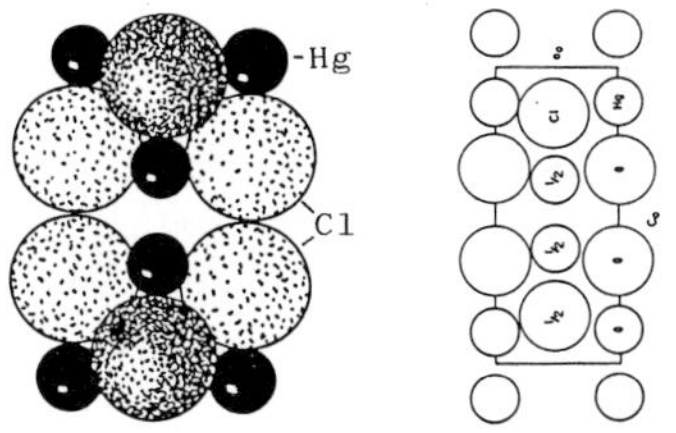

(a) (b)

Fig. 2

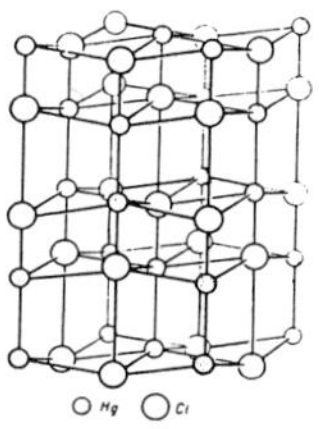

Fig. 3

Properties

Habit	Cleav.	Fract.	Twin.	Hardn.	Dens.	Colour	Transp.
tabular, prismatic, pyramidal	good (110)	conchoidal	(112)	1.5	7.15	colourless, white, yellowish grey	transparent

Refr. index /Reflect.	Birefr.	Luster	Streak	Melt.p.	CPI
n_ω = 1.973 n_ϵ = 2.656	(+)	adamantine	yellowish	302°C	(SPI) 62

Figures

Fig. 1. Structural unit of the calomel structure: the -Cl-Hg-Hg-Cl- chain.

Fig. 2. (a) Packing drawing of the calomel structure, and (b) unit cell content projected along the a axis (adapted from Wyckoff, 1963, Vol. 1).

Fig. 3. Representation of the calomel structure showing the HgCl chains (after Povarennykh, 1972).

Fig. 4. Condensed model of the packing analogue of the calomel structure. The packing is (2Qs)f, with Hg in square voids, but fifth coordination.

Description

The calomel structure is based on the packing of HgCl chains, which can be derived from a (2Qs)f packing of square layers Q of Cl atoms with Hg atoms in 5th coordination voids.

References

Kostov (1968) 204
Povarennykh (1972) 642, 643.
Wyckoff (1963) Vol. 1, 151, 152.
Palache et al. (1951) Vol. 2, 25-27.
Zoltai + Stout (1984) 402.

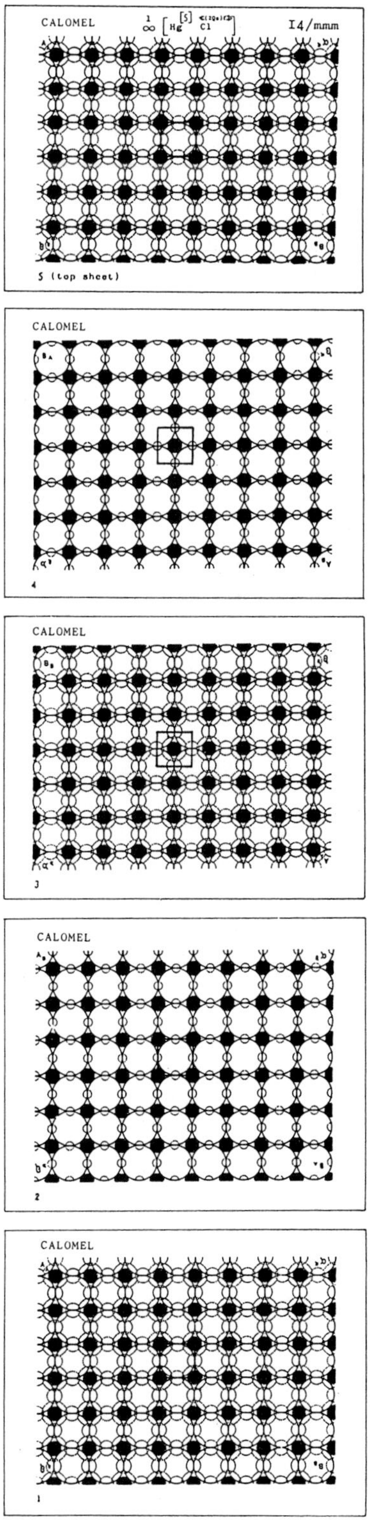

Fig. 4

MINIUM

P4_2/mbc

$^{II[3n]}Pb_2 \; \overset{1}{\infty} \left[{}^{IV}Pb^{O} \; O_4 \right]$

a = 8.788Å
c = 6.551Å
Z = 4

Pb^{+4}(4d)

Pb^{+2}(8h) u = 0.153
v = 0.153

O_I (8h) u = 0.114
v = 0.614

O_{II}(8g) u = 0.672

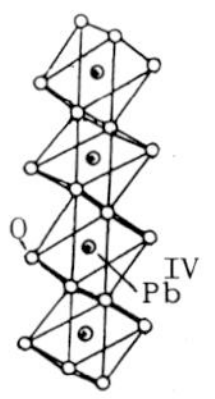

Fig. 1

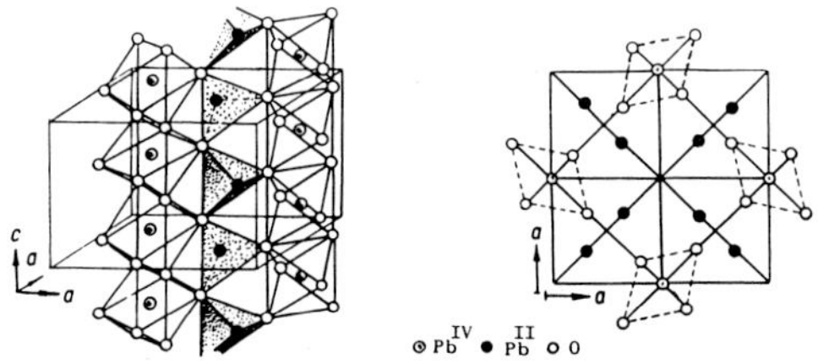

(a) Fig. 2 (b)

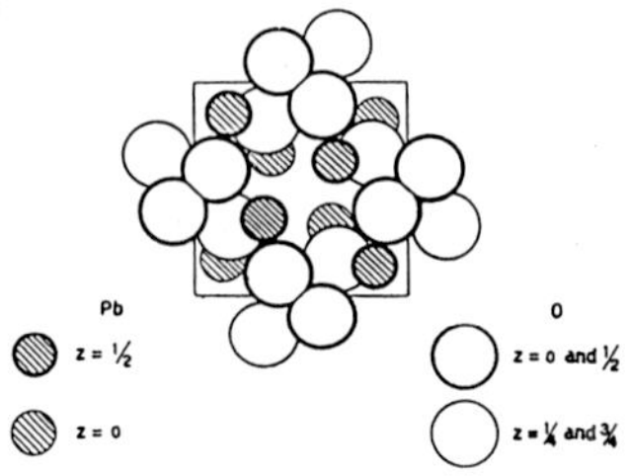

Fig. 3

Properties

Habit	Cleav.	Fract.	Twin.	Hardn.	Dens.	Colour	Transp.
massive, pulveru- lent	perfect (110)			2.5	8.9- -9.2	scarlet- -red, brownish red	opaque

Refr. index/Reflect.	Birefr.	Luster	Streak	Melt.p.	CPI
n = 2.4	(-)	greasy, dull	orange- -yellow		

Population

Schafarzikite $^{[3n]}Sb_2 \; \overset{1}{\infty} \left[Fe^{o} \; O_4 \right]$

Trippkeite $^{[3n]}As_2 \; \overset{1}{\infty} \left[Cu^{o} \; O_4 \right]$

Description

The minium structure is based on the packing of $Pb^{IVo}O_4$ chains, with Pb^{II} ions in the interstices, having [3n] coordination.

This description is in agreement with the perfect cleavage along the (110) planes, and also with the strong valence of the Pb atoms in these chains.

Figures

Fig. 1. Representation of the structural unit of minium: $Pb^{o}_2O_4$ chains (after Povarennykh, 1972).

Fig. 2. (a) Representation of the minium structure showing the way the $Pb^{o}_2O_4$ chains pack together, and (b) projection along the c axis (after Povarennykh, 1972).

Fig. 3. Representation of the minium structure projected along the c axis (after Structure Reports 1951 , Vol. 11).

References

Kostov (1968) 259.
Povarennykh (1972) 306, 307.
Wyckoff (1964) Vol. 2, 149, 150.
Structure Reports (1951) Vol. 11, 241.
Palache et al. (1944) Vol. 1, 517.

DIOPSIDE

C2/c

$Ca^{[8]}Mg^{o}\ {}^{1}_{\infty}\left[Si^{t}_{2}\ O_{6}\right]^{II^{11}\overline{cx}}$

a = 9.750Å
b = 8.930Å
c = 5.249Å
β = 74° 10'
Z = 4

Atom	x	y	z
Ca(4e)	0	-0.306	1/4
Mg(4e)	0	0.084	1/4
Si(8f)	0.211	0.407	0.236
O_{I} (8f)	0.375	0.419	0.139
O_{II} (8f)	0.142	0.253	0.318
O_{III} (8f)	0.145	0.481	0.000

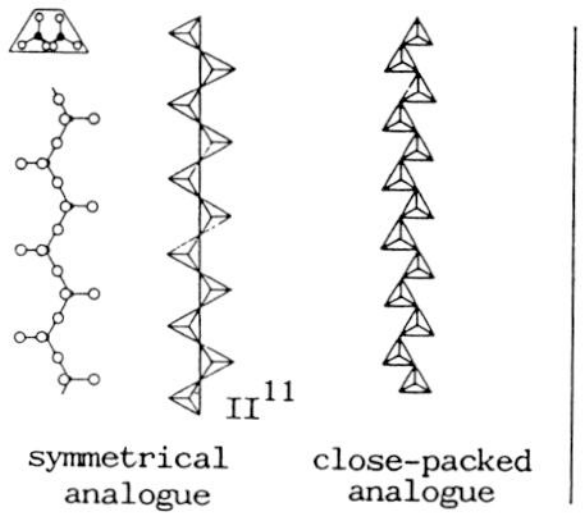

Fig. 1

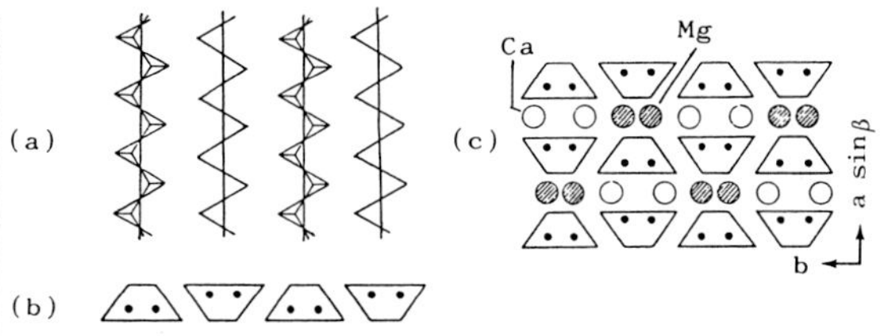

Fig. 2

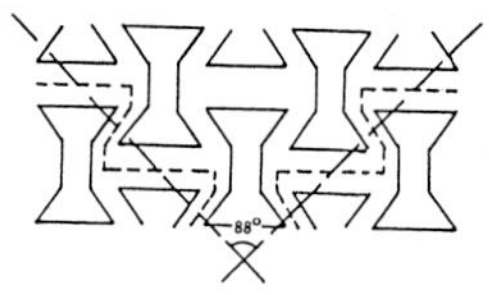

Fig. 3

Properties

Habit	Cleav.	Fract.	Twin.	Hardn.	Dens.	Colour	Transp.
prismatic	perfect (110)	uneven	(100), (001)	5.5-6.5	3.2-3.5	green, variable	transparent to translu-cent

Refr. index/Reflect.	Birefr.	Luster	Streak	Melt.p.	CPI
n_{α} = 1.665 n_{β} = 1.672 n_{γ} = 1.695	(+) 2V= 56°-62°	vitreous	white, grey	1391°C	(SPI) 73

Distortion derivatives

Clinoenstatite $Mg^{[6]}Mg^{o}\ {}^{1}_{\infty}\left[Si_{2}\ O_{6}\right]^{II^{11}\overline{cx}}$

Pigeonite $(Ca,Fe)^{[7]}(Mg,Fe)^{o}\ {}^{1}_{\infty}\left[Si^{t}_{2}O_{6}\right]^{II^{11}\overline{cx}}$

$P2_{1}/c$

Figures

Fig. 1. Representations of the $Si^{t}_{2}\ O_{6}$ infinite chain: two aspects of the ideal chain and one corresponding to closest packing (after Kostov, 1968). The real chain is intermediate between these two extreme cases.

Fig. 2. Representation of the packing of the $Si^{t}_{2}O_{6}$ infinite chains: (a) parallel to the chains, and (b) in cross section (c) more complete representation of the cross section.

Fig. 3. Structural interpretation of the cleavage of diopside.

Fig. 4. Condensed model of the symmetrical analogue o the diopside structure. The large open circles represent oxygens, the small black circles Si atoms in tetrahedral voids, and the medium circles Ca (hatched) and Mg (black) atoms with [8] and octahedral coordination, respectively.

Fig. 5. The two first sheets of the condensed model of the packing analogue of the diopside structure. Compare it with the symmetrical packing analogue (Fig. 4).

Fig. 6. Polyhedral description of the diopside structure (after Zoltai + Stout, 1984).

References

Kostov (1968) 330, 334, 335.
Povarennykh (1972) 409.
Wyckoff (1968) Vol. 4, 295, 296.
Zoltai + Stout (1984) 336-339.
Ingerson (1955) 352.

DIOPSIDE (continued)

Description

Is built of $Si^{t}_{2}O_{6}$ infinite chains, with Mg and Ca in octahedral and 8 coordination voids, respectively. The symbol II^{11} means infinite chain derived by translation from II^{1} group, that is the group of two silica tetrahedra, $Si^{t}_{2}O_{7}$. $\bar{c}x$ means that the stacking vector cx is 1/3 along the chain, and the dash over the c means that the chains inverte alternately.

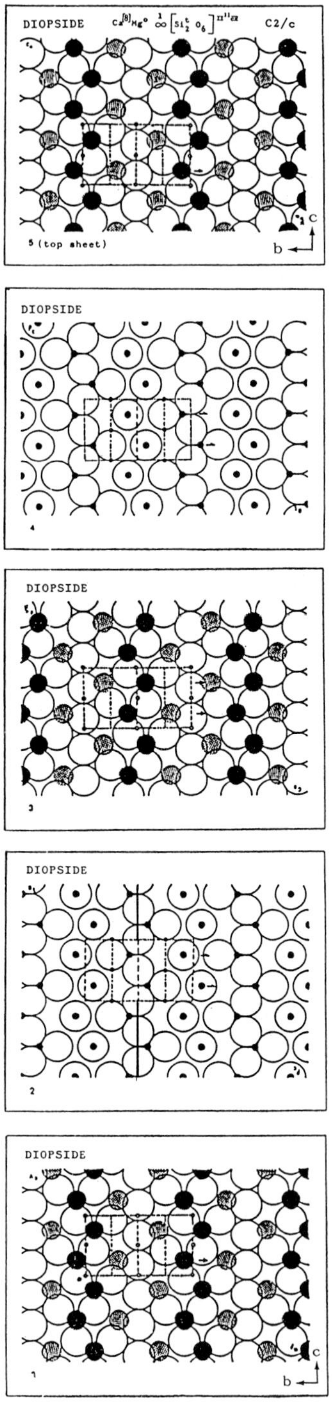

Fig. 4

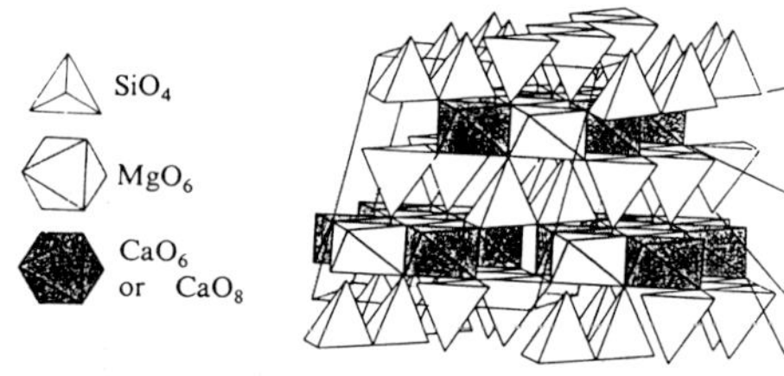

Fig. 6

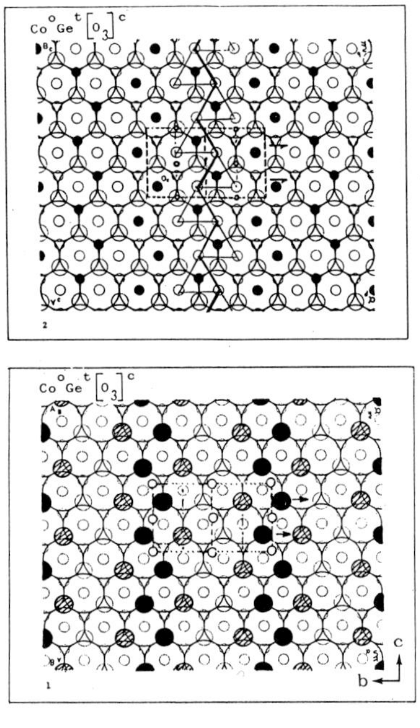

Fig. 5

ENSTATITE

Pbca

$$Mg^{[6]}Mg^{o}\,{}^{1}_{\infty}\left[Si^{t}_{2}O_{6}\right]^{11}_{II\,\overline{chx}}$$

a = 18.310 Å
b = 8.927 Å
c = 5.226 Å
Z = 8

Atom	Position	x	y	z
Mg^{o}	(8c)	0.12484	0.3453	0.3740
$Mg^{[6]}$	(8c)	0.12155	-0.0167	0.3666
Si_{I}	(8c)	0.02611	-0.3365	0.2973
Si_{II}	(8c)	0.22879	-0.1593	0.0512
O_{I}	(8c)	0.06210	0.1643	0.2069
O_{II}	(8c)	0.06555	0.5163	0.2034

...

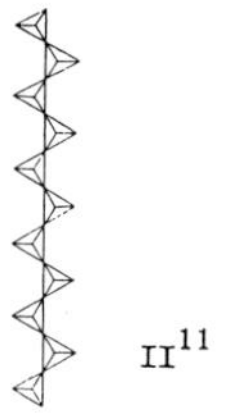

symmetrical analogue

Fig. 1

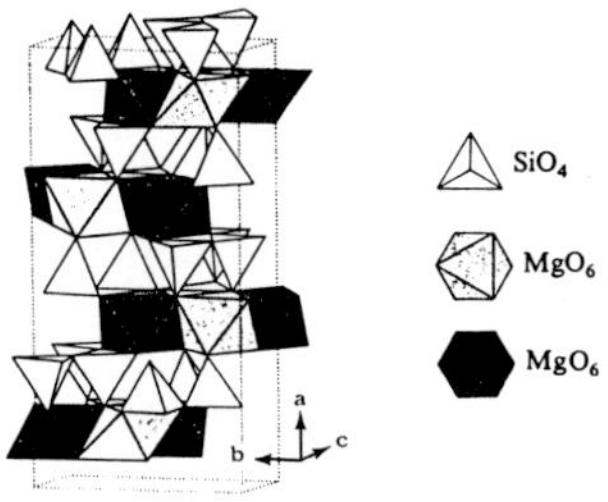

Fig. 2

Properties

Habit	Cleav.	Fract.	Twin.	Hardn.	Dens.	Colour	Transp.
prismatic acicular	perfect (210)	uneven		5-6	3.2-3.5	grey, green	translucent

Refr. index/Reflect.	Birefr.	Luster	Streak	Melt.p.	CPI
n_α = 1.657 n_β = 1.659 n_γ = 1.665	(+) 2V= 54°	vitreous, pearly	white, grey		(SPI) 66

Figures

Fig. 1. Polyhedral representation of the symmetrical analogue of the $Si^{t}O_3$ pyroxene infinite chain.

Fig. 2. Polyhedral description of the enstatite structure (after Zoltai + Stout, 1984).

Fig. 3. Condensed model of the symmetrical analogue of the enstatite structure. The large open circles represent oxygen atoms, the small black circles Si atoms in tetrahedral voids and the medium circles Mg atoms, Mg^{o} (black) and $Mg^{[6]}$ (grey).

Description

Enstatite is a chain structure, a pyroxene, based on the packing of $Si^{t}O_3$ infinite chains, with interstitial Mg atoms in [6] and octahedral coordination. The layers of chains are of the same type as diopside, but are packed in a different way: $\overline{chx}$. Protoenstatite is an artificial material, which corresponds to a polytype of enstatite, with formula $Mg^{[8]}Mg^{o}\,{}^{1}_{\infty}\left[Si^{t}_{2}O_{6}\right]^{11}_{II\,\overline{hx}}$.

Crystallogr. data (continued)

Atom	Position	x	y	z
O_{III}	(8c)	0.05347	-0.2023	0.0928
O_{IV}	(8c)	0.18402	0.3375	0.0444
O_{V}	(8c)	0.18948	0.0009	0.0524
O_{VI}	(8c)	0.19764	-0.2344	0.3176

References

Kostov (1968) 331, 332.

Povarennykh (1972) 408.

Wyckoff (1968) Vol. 4. 301-303.

Zoltai + Stout (1984) 335, 336.

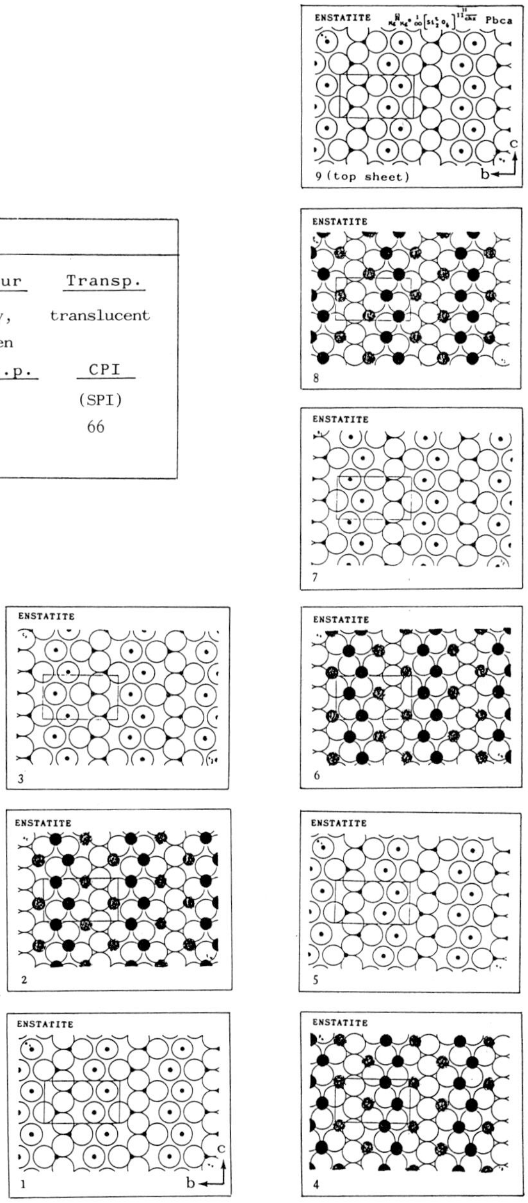

Fig. 3

TREMOLITE

C2/m

$Ca_2^{[8]}Mg_5^{o}\,(OH,F)_2\;{}^{1}_{\infty}\left[Si_8^{t}\,O_{22}\right]^{11}_{2-II\overline{cx}}$

a = 9.80Å
b = 17.74Å
c = 5.27Å
β = 106° 2'
Z = 2

Atom		x	y	z
Ca	(4h)	0	0.9	1/2
Mg_I	(4h)	0	0.17	0
Mg_{II}	(2a)	0	0.00	0
Mg_{III}	(4h)	0	0.28	1/2
Si_I	(8j)	0.29	0.08	0.01
Si_{II}	(8j)	0.29	0.18	0.51

...

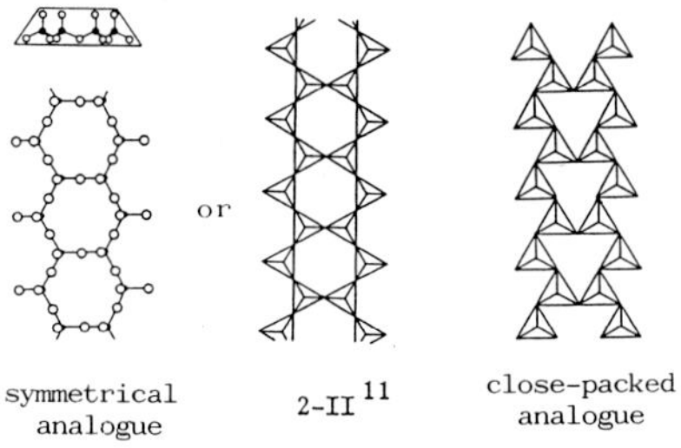

Fig. 1

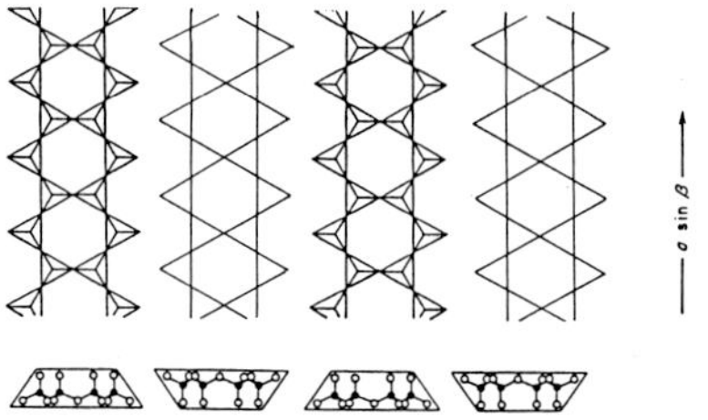

Fig. 2

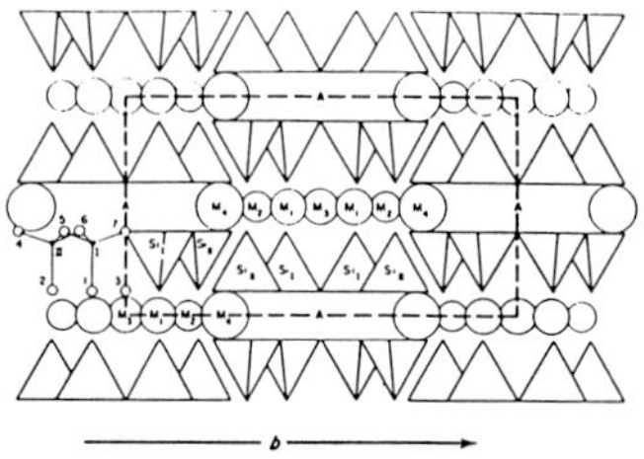

Fig. 3

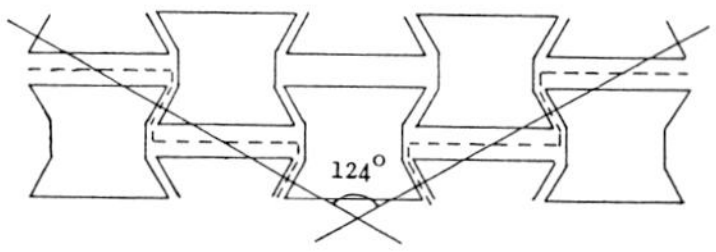

Fig. 4

Properties

Habit	Cleav.	Fract.	Twin.	Hardn.	Dens.	Colour	Transp.
prismatic, fibrous	perfect (110)	uneven	(100), (001)	5-6	3.0-3.3	white, green	transparent to translucent

Refr. index/Reflect.	Birefr.	Luster	Streak	Melt.p.	CPI
n_α = 1.608 n_β = 1.618 n_γ = 1.630	(-) 2V = 85°	vitreous	white		(SPI) 64

Population

Actinolite
$(Ca,Mg)_2^{[8]}\,(Mg,Fe)_5^{o}\,(OH)_2\;{}^{1}_{\infty}\left[Si_8^{t}\,O_{22}\right]^{11}_{2-II\overline{cx}}$

Tschermakite
$Ca_2^{[8]}\underbrace{(Mg,Fe)_3^{o}(Al,Fe)_2^{o}}(OH)_2\;{}^{1}_{\infty}\left[\underbrace{Si_6^{t}Al_2^{t}}O_{22}\right]^{11}_{2-II\overline{cx}}$

Glaucophane
$Na_2^{[8]}\underbrace{(Mg,Fe)_3^{o}(Al,Fe)_2^{o}}(OH)_2\;{}^{1}_{\infty}\left[Si_8^{t}O_{22}\right]^{11}_{2-II\overline{cx}}$

Description

Tremolite is a chain structure, an amphibole, based on the packing of $Si_4^{t}O_{11}$ infinite chains, with interstitial Ca and Mg atoms in [8] and octahedral coordination respectively. The amphibole chains form layers which are packed with a $\overline{cx}$ sequence.

Figures

Fig. 1. Representations of the $Si_4^{t}O_{11}$ amphibole chain: two drawings of the symmetrical analogue and one corresponding to the close-packed analogue (after Kostov, 1968). The real chain is intermediate between these two extreme cases.

Fig. 2. Representation of the packing of the $Si_4^{t}O_{11}$ infinite chains: (a) parallel to the chains, and (b) in cross section.

Fig. 3. Packing drawing of the tremolite structure (after Ernst, 1968).

Fig. 4. Structural interpretation of the cleavage of tremolite.

References

Kostov (1968) 330, 342, 343, 346.
Povarennykh (1972) 415, 416, 425.
Wyckoff (1968) Vol. 4, 304-306.
Zoltai + Stout (1984) 342-345.
Ernst (1968) 5.

Atom		x	y	z
O_{IV}	(8j)	0.36	0.14	0.75
O_V	(8j)	0.36	0.11	0.25
O_{VI}	(8j)	0.36	0.00	-0.10
(OH)	(4i)	0.14	0.00	0.60
O_I	(8j)	0.14	0.08	0.10
O_{II}	(8j)	0.14	0.18	0.60
O_{III}	(8j)	0.36	0.25	0.40

TREMOLITE (continued)

Fig. 5. Condensed model of the symmetrical analogue of the tremolite structure. The large open circles represent oxygen atoms; and the hatched large circles OH, the small black circles Si atoms, in tetrahedral voids, and the medium circles Ca (grey) and Mg (black) atoms with [8] and octahedral coordination, respectively .

Fig. 6. Polyhedral description of the tremolite structure (adapted from Zoltai + Stout, 1984).

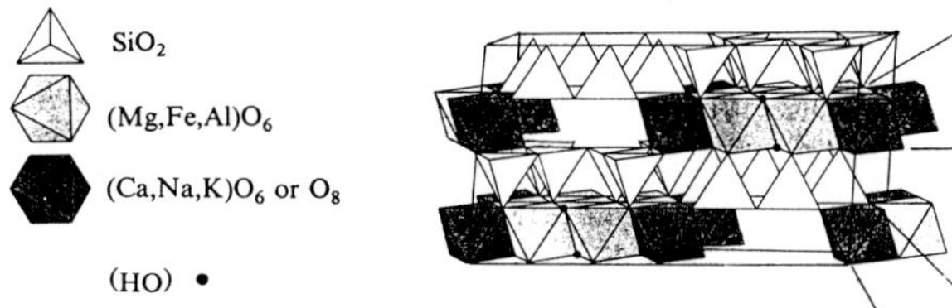

Fig. 6

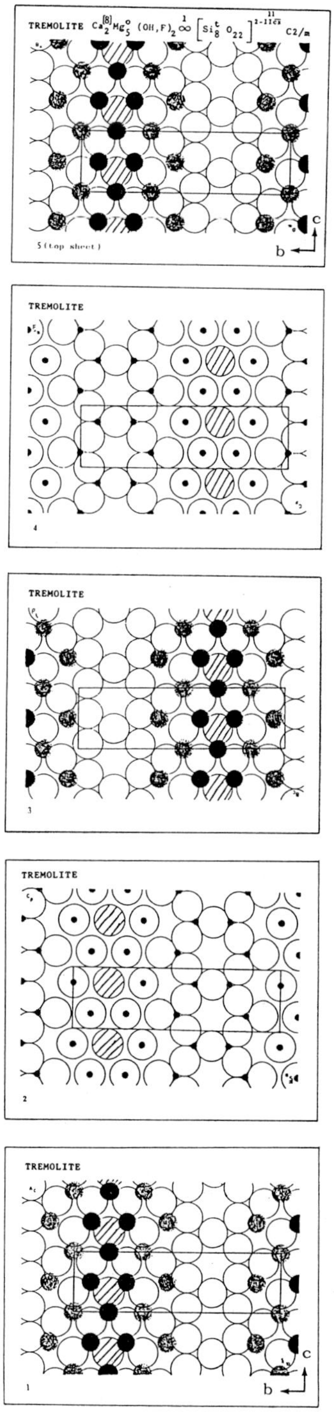

Fig. 5

HORNBLENDE

C2/m

$Na^{[10]}Ca_2^{[8]}\,Mg_5^{o}(OH,F)_2\,{}^{1}_{\infty}\left[Si_7^{t}\,Al^{t}\,O_{22}\right]^{2\text{-}II^{11}_{cx}}$

a = 9.87Å
b = 18.01Å
c = 5.333Å
β = 105°58'
Z = 2

Atom	Site	x	y	z
Na	(2b)	0	1/2	0
Ca	(4h)	0	0.281	1/2
Mg_I	(4h)	0	0.090	1/2
Mg_{II}	(4g)	0	0.177	0
Mg_{III}	(2a)	0	0	0
(OH,F)	(4i)	0.114	0	0.60
...				

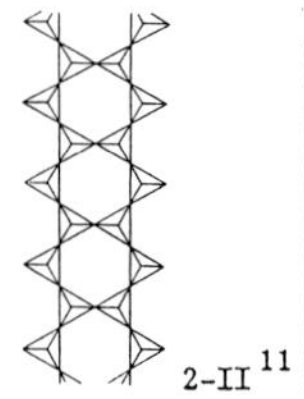

symmetrical analogue

Fig. 1

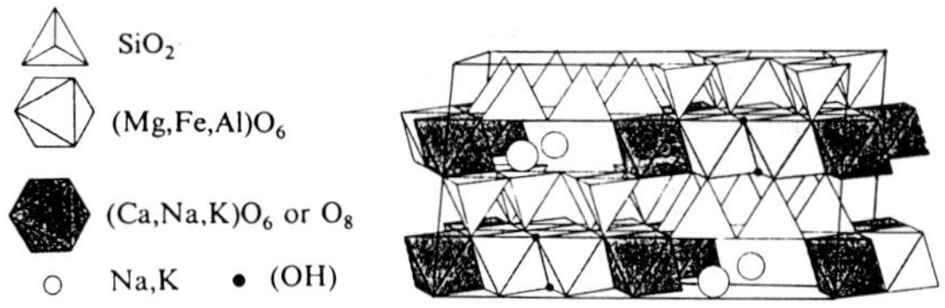

Fig. 2

Properties

Habit	Cleav.	Fract.	Twin.	Hardn.	Dens.	Colour	Transp.
prismatic	perfect (110)	uneven	(100)	5-6	3.0-3.5	green, black	translucent

Refr. index/Reflect.	Birefr.	Luster	Streak	Melt.p.	CPI
n_α = 1.65	(-)	vitreous	white		(SPI)
n_β = 1.66					65
n_γ = 1.67	2V = 50°-80°				

Population

Arfvedsonite
$Na^{[10]}Na_2^{[8]}Fe_5^{o}(OH)_2\,{}^{1}_{\infty}\left[Si_8^{t}O_{22}\right]^{2\text{-}II^{11}_{cx}}$

Kaersutite
$Na^{[10]}Ca_2^{[8]}(Mg,Fe)_4^{o}Ti^{o}(OH)_2\,{}^{1}_{\infty}\left[Si_8^{t}O_{22}\right]^{2\text{-}II^{11}_{cx}}$

Eckermannite
$Na^{[10]}Na_2^{[8]}Na^{o}Mg_2^{o}Al^{o}Fe^{o}(OH)_2\,{}^{1}_{\infty}\left[Si_8^{t}O_{22}\right]^{2\text{-}II^{11}_{cx}}$

Pargasite
$Na^{[10]}Ca_2^{[8]}(Mg,Fe)_5^{o}(OH)_2\,{}^{1}_{\infty}\left[Si_7^{t}Al^{t}O_{22}\right]^{2\text{-}II^{11}_{cx}}$

Közulite
$Na^{[10]}Na_2^{[8]}Mn_4^{o}(Al,Fe)^{o}(OH)_2\,{}^{1}_{\infty}\left[Si_8^{t}O_{22}\right]^{2\text{-}II^{11}_{cx}}$

Figures

Fig. 1. Polyhedral drawing of the $Si_4^tO_{11}$ amphibole chain (symmetrical analogue).

Fig. 2. Polyhedral representation of the hornblende structure (after Zoltai + Stout, 1984).

Fig. 3. Condensed model of the symmetrical analogue of hornblende. The large open circles represent oxygen atoms, the large hatched circles OH, the small black circles Si atoms in tetrahedral voids, the medium circles Ca (grey) and Mg (black) atoms in [8] and [6] coordination respectively. The Na atoms (line crossed) have [10] coordination.

Description

Hornblende is an amphibole, which is a chain structure based on the packing of $Si_4^tO_{11}$ infinite chains. Hornblende can be considered as a interstitial derivative of tremolite, with Na occupying voids of [10] coordination.

Crystallographic data (continued)

Atom	Site	x	y	z
Si_I	(8j)	0.279	0.087	0.06
Si_{II}	(8j)	0.294	0.172	0.56
O_I	(8j)	0.098	0.093	0.11
O_{II}	(8j)	0.115	0.177	0.60
O_{III}	(8j)	0.366	0.251	0.43
O_{IV}	(8j)	0.353	0.139	0.79
O_V	(8j)	0.337	0.115	0.31
O_{VI}	(4i)	0.338	0	0.00

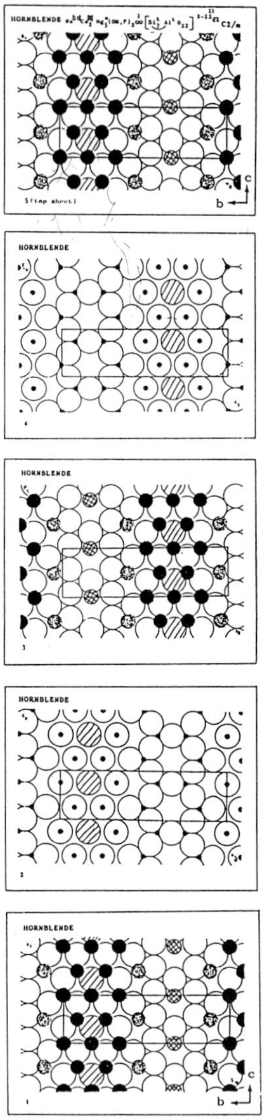

Fig. 3

References

Kostov (1968) 345,346.
Povarennykh (1972) 416, 425, 426.
Wyckoff (1968) Vol. 4, 306-310.
Zoltai + Stout (1984) 343, 345.

ANTHOPHYLLITE Pnma

$Mg_2^{[8]} Mg_5^{o} (OH,F)_2 \; {}^{1}_{\infty}\left[Si_8^{t} O_{22}\right]^{2\text{-}II^{11}\overline{chx}}$

a = 18.56Å
b = 18.08Å
c = 5.28Å
Z = 4

Atom	Site	Coordinates	Atom	Site	Coordinates
$Mg^{[8]}$	(8d)	x = 0.13, y = 0.17, z = 0.37	Mg^{o}_{III}	(4c)	x = 0.13, y = 1/4, z = -0.13
Mg^{o}_{I}	(8d)	x = 0.13, y = 0.08, z = -0.13	$(OH)_I$	(4c)	x = 0.06, y = 1/4, z = 0.20
Mg^{o}_{II}	(8d)	x = 0.13, y = -0.02, z = 0.37	$(OH)_{II}$	(4c)	x = 0.19, y = 1/4, z = -0.45

...

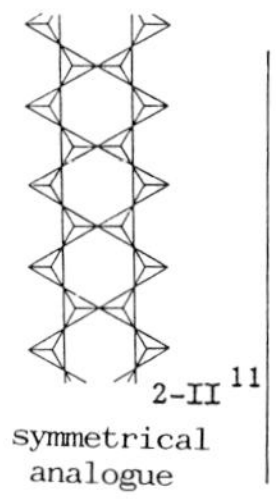

Fig. 1

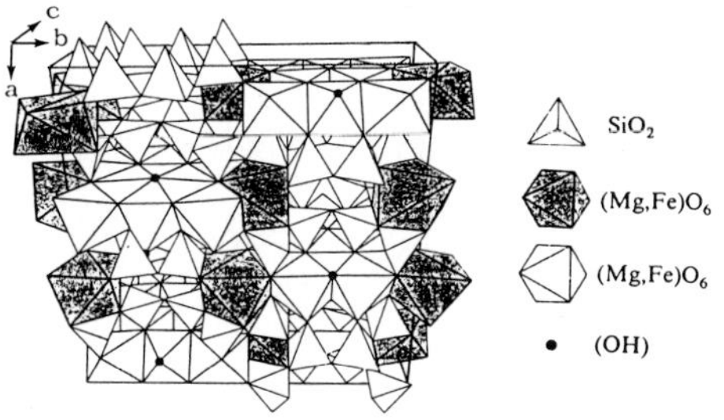

Fig. 2

Properties

Habit	Cleav.	Fract.	Twin.	Hardn.	Dens.	Colour	Transp.
prismatic, fibrous	perfect (210)	uneven		5.5-6	2.9-3.2	grey, brown	transparent to translucent

Refr. index/Reflect.	Birefr.	Luster	Streak	Melt.p.	CPI
n_α = 1.60 n_β = 1.62 n_γ = 1.63	(-) 2V = 65°-90°	vitreous	white		(SPI) 60

Figures

Fig. 1. Polyhedral drawing of the $Si_4^tO_{11}$ amphibole chain (symmetrical analogue).

Fig. 2. Polyhedral representation of the anthophyllite structure (after Zoltai + Stout, 1984).

Fig. 3. Condensed model of the symmetrical analogue of anthophyllite. The large open circles represent oxygen atoms, the large hatched circles OH, the small black circles Si atoms in tetrahedral voids, and the medium circles Mg atoms with [8] (grey) and octahedral (black) coordinations.

Description

Anthophyllite is an amphibole which is a chain structure based on the packing of $Si_4^tO_{11}$ infinite chains. Mg atoms occupy the interstices, with [8] and octahedral coordination. The amphibole chains form layers which are packed in a $\overline{chx}$ sequence.

(Crystallographic data (continued)

Atom	Site	Coordinates	Atom	Site	Coordinates
Si_I	(8d)	x = 0.03, y = -0.18, z = 0.29	O_V	(8d)	x = 0.05, y = -0.13, z = 0.05
Si_{II}	(8d)	x = 0.03, y = -0.08, z = -0.21	O_{VI}	(8d)	x = 0.05, y = -0.13, z = -0.46
Si_{III}	(8d)	x = 0.22, y = -0.08, z = 0.04	O_{VII}	(8d)	x = 0.19, y = 0.18, z = 0.05
Si_{IV}	(8d)	x = 0.22, y = -0.18, z = -0.46	O_{VIII}	(8d)	x = 0.19, y = 0.07, z = -0.44
O_I	(8d)	x = 0.06, y = 0.07, z = 0.20	O_{IX}	(8d)	x = 0.19, y = 0.00, z = 0.05
O_{II}	(8d)	x = 0.06, y = 0.18, z = -0.30	O_X	(4c)	x = -0.19, y = 1/4, z = -0.45
O_{III}	(4c)	x = -0.06, y = 1/4, z = -0.20	O_{XI}	(8d)	x = 0.20, y = -0.13, z = 0.30
O_{IV}	(8d)	x = 0.60, y = 0.00, z = -0.30	O_{XII}	(8d)	x = 0.20, y = -0.13, z = -0.20

References

Kostov (1968) 340-342, 330.
Povarennykh (1972) 414.
Wyckoff (1968) Vol. 4, 310-312.
Zoltai + Stout (1984) 341, 342.

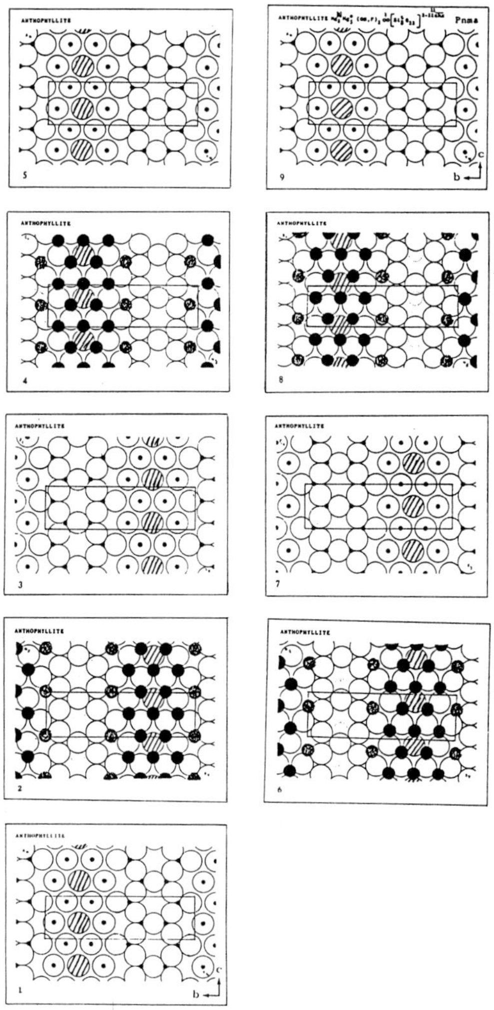

Fig. 3

WOLLASTONITE
(α - form)

$Ca^{o}\ {}^{1}_{\infty}\left[Si^{t}\ O_3\right]^{61}_{\text{III}\ my}$

$P2_1/a$

a = 15.36Å
b = 7.29Å
c = 7.08Å
β = 95° 24'
Z = 12

Atom	Site	x	y	z
Ca_I	(4e)	0.236	0.125	0.000
Ca_{II}	(4e)	0.053	0.125	0.775
Ca_{III}	(4e)	0.444	0.125	0.189
Si_I	(4e)	0.344	0.125	0.478
Si_{II}	(4e)	0.183	0.125	0.714
Si_{III}	(4e)	0.183	0.125	0.303
...				

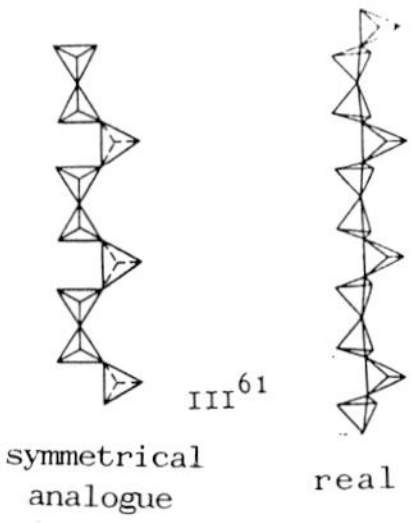

Fig. 1

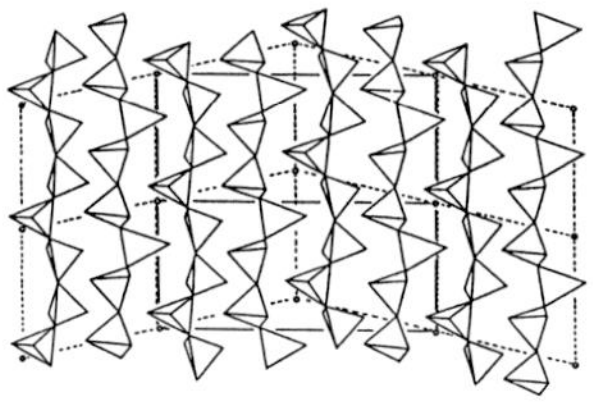
Fig. 2

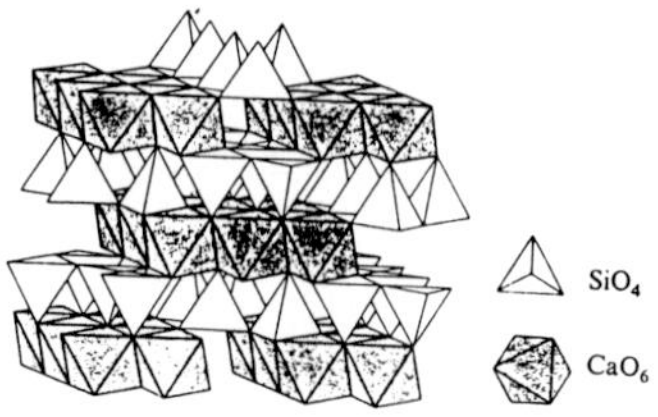

Fig. 3

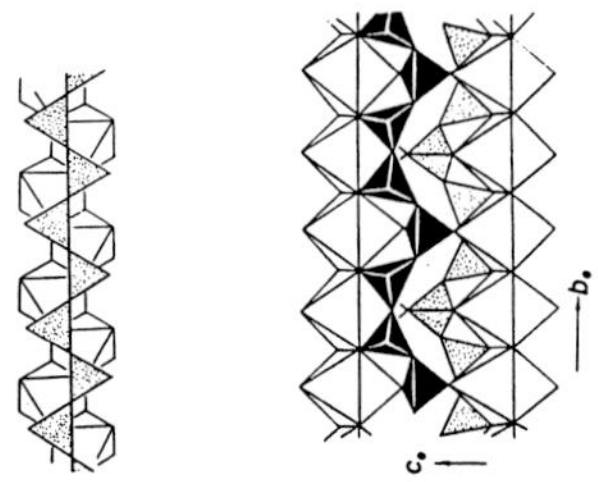
(a) (b)

Fig. 4

Properties

Habit	Cleav.	Fract.	Twin.	Hardn.	Dens.	Colour	Transp.
prismatic needle-like	perfect (100) (001)	uneven	(100)	4.5-5	3.1	white	transparent to translucent

Refr. index/Reflect.	Birefr.	Luster	Streak	Melt.p.	CPI
n_α = 1.620 n_β = 1.532 n_γ = 1.634	(-) 2V= 39°	silky	white		(SPI) 65

Figures

Fig. 1. Polyhedral drawings of the Si^tO_3 wollastonite infinite chain: symmetrical analogue and real chain (after Bragg + Claringbull, 1965).

Fig. 2. Projection of the structure of parewollastonite showing the way the wollastonite chains pack together (after Bragg + Claringbull, 1965).

Fig. 3. Polyhedral representation of the wollastonite structure (after Zoltai + Stout, 1984).

Fig. 4. (a) Pyroxene chain adjustement to a zig-zag chain of magnesium octahedra and (b) wollastonite chain adjustment to a rectilinear chain of calcium octahedra (after Kostov, 1968).

Description

Wollastonite is a pyroxenoid based on an infinite chain Si^tO_3 with a similar chemical composition of the pyroxene chain but different geometry. The reason for the different shape of the chains is due to their adjustment to the calcium octahedra which is much bigger than the magnesium octahedra of the pyroxenes.

References

Kostov (1968) 329, 348.
Wyckoff (1968) Vol. 4, 265, 266.
Zoltai + Stout (1984) 347, 348.
Bragg + Claringbull (1965) 245, 248.

Crystallographic data (continued)

Atom	Site	x	y	z
O_I	(4e)	0.286	0.125	0.672
O_{II}	(4e)	0.286	0.125	0.308
O_{III}	(4e)	0.408	0.306	0.467
O_{IV}	(4e)	0.150	0.306	0.833
O_V	(4e)	0.150	0.306	0.203
O_{VI}	(4e)	0.136	0.125	0.522
O_{VII}	(4e)	0.408	-0.056	0.467
O_{VIII}	(4e)	0.150	-0.056	0.833
O_{IX}	(4e)	0.150	-0.056	0.203

PECTOLITE

$P\bar{1}$

$Na^{[8]}Ca^{o}_{2}\ {}^{1}_{\infty}\left[Si^{t}_{3}\ O_{8}(OH)\right]\ III^{61}_{by}$

a = 7.99Å
b = 7.04Å
c = 7.02Å
α = 90° 31'
β = 95° 11'
γ = 102° 28'
Z = 2

Atom	Site	x	y	z
Na	(2i)	0.552	0.265	0.344
Ca_I	(2i)	0.857	0.596	0.146
Ca_{II}	(2i)	0.843	0.074	0.139
Si_I	(2i)	0.221	0.402	0.337
Si_{II}	(2i)	0.210	0.954	0.344
Si_{III}	(2i)	0.451	0.735	0.148

...

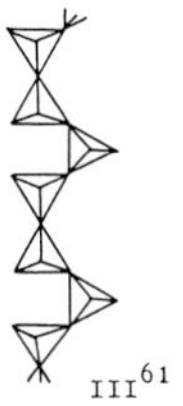

Fig. 1

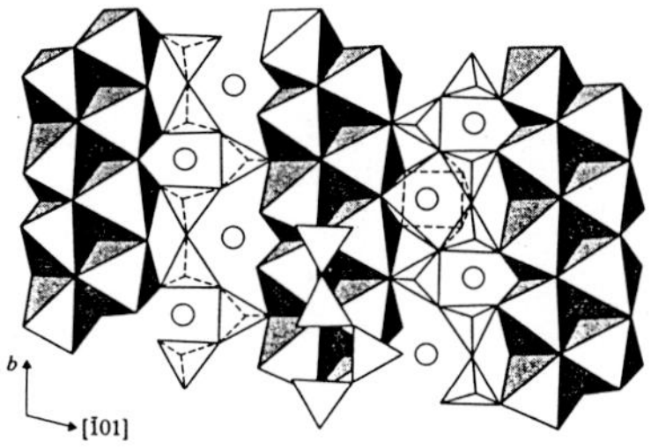

Fig. 2

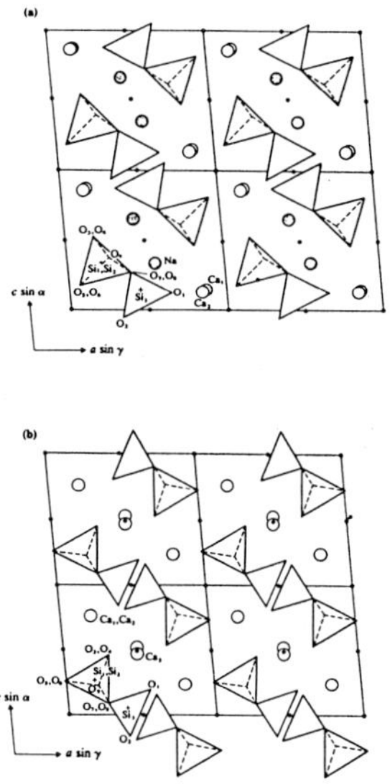

Fig. 3

Properties

Habit	Cleav.	Fract.	Twin.	Hardn.	Dens.	Colour	Transp.
radiating, fibrous	perfect (100), (001)	uneven	(100)	4.5-5	2.9	white	translucent

Refr. index/Reflect.	Birefr.	Luster	Streak	Melt.p.	CPI
n_α = 1.59 n_β = 1.61 n_γ = 1.63	(+) 2V = 35º63°	silky	white		(SPI) 63

Figures

Fig. 1. Polyhedral drawings of the Si^tO_3 pyroxenoide infinite chain: symmetrical analogue and real chain.

Fig. 2. Projection of a portion of an octahedral layer of pectolite structure onto (101). The circles represent Na ions and the octahedra contain Ca (after Prewitt, 1967).

Fig. 3. Projections along the y axis of the structures of pectolite (a), and of wollastonite (b) (after Prewitt + Buerger, 1963).

Description

Pectolite is a pyroxenoid based on infinite chains Si^tO_3 with a similar chemical composition of the pyroxene chains but different geometry. In the pectolite structure the interstices formed by the packing of the infinite tetrahedral chains are occupied by Na and Ca with coordination [8] and octahedral, respectively.

Crystallographic data (continued)

Atom	Site	x	y	z
O_I	(2i)	0.652	0.788	0.125
O_{II}	(2i)	0.322	0.702	-0.057
O_{III}	(2i)	0.185	0.496	0.538
O_{IV}	(2i)	0.171	0.839	0.541
O_V	(2i)	0.070	0.393	0.171
O_{VI}	(2i)	0.053	0.896	0.179
O_{VII}	(2i)	0.396	0.533	0.275
O_{VIII}	(2i)	0.402	0.906	0.275
O_{IX}	(2i)	0.260	0.182	0.381

References

Kostov (1968) 305.
Wyckoff (1968) Vol. 4, 316,317.
Zoltai + Stout (1984) 348.
Prewitt (1967) 307.
Prewitt + Buerger (1963) 293-302.

XONOTLITE

P2/a

$Ca^{p}_{4}\, Ca^{o}_{2}\, (OH)_2\, {}^{1}_{\infty}\left[Si^{t}_{6}\, O_{17}\right]$ 61 2-IIIs

a = 16.53Å
b = 7.33Å
c = 7.04Å
β = 90°
Z = 2

Ca^{o}_{I} (2e) u = 0.375

Ca^{o}_{II} (2e) u = 0.875

Ca^{p}_{I} (4g) x = 0.069, y = 0.125, z = 0.160

Ca^{p}_{II} (4g) x = 0.069, y = 0.625, z = 0.160

(OH)(4g) x = 0.146, y = 0.375, z = 0.223

Si_{I} (4g) x = 0.392, y = 0.170, z = 0.271

...

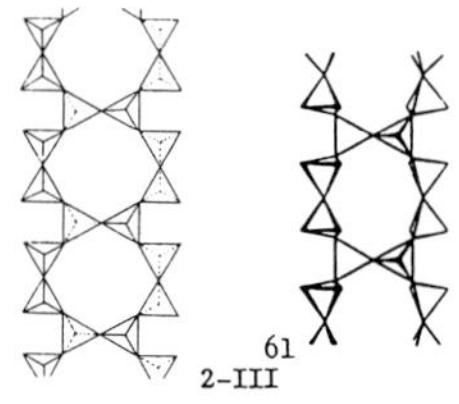

Fig. 1

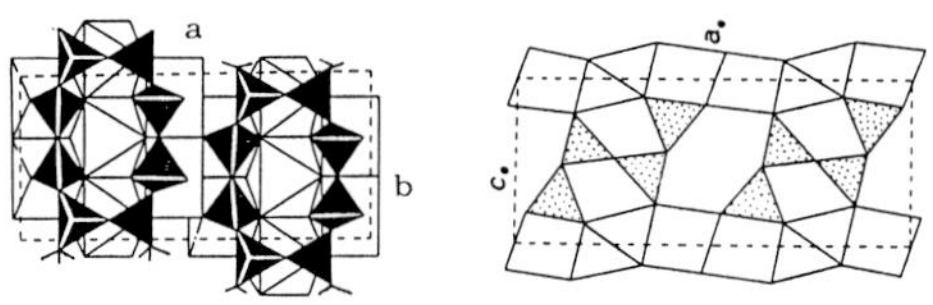

Fig. 2

Properties

Habit	Cleav.	Fract.	Twin.	Hardn.	Dens.	Colour	Transp.
acicular, massive	good			6	2.71	white, colourless	transparent

Refr. index/Reflect.	Birefr.	Luster	Streak	Melt.p.	CPI
n_α = 1.583 n_β = 1.583 n_γ = 1.593	(+)	greasy, pearly			

Figures

Fig. 1. Polyhedral drawing of the $Si^{t}_{6}O_{17}$ xonotlite infinite chain: symmetrical analogue and real chain (after Kostov, 1968, and Eitel, 1964).

Fig. 2. Two projections of the structure of xonotlite (after Belov + Mamedov, quoted by Kostov, 1968).

Description

The xonotlite structure is based on the packing of $Si^{t}_{6}O_{17}$ double infinite chains, which can be imagined derived from the condensation of two wollastonite chains. Ca atoms are located in the interstices, some with octahedral and others with trigonal prismatic coordination.

References

Kostov (1968) 330, 349.
Povarennykh (1972) 417.
Wyckoff (1968) Vol. 4, 337-339.
Eitel (1964) 97.

Crystallographic data (continued)

Si_{II} (4g) x = 0.392, y = 0.580, z = 0.271

Si_{III} (4g) x = 0.158, y = 0.875, z = 0.443

O_{I} (2f) u = 0.875

O_{II} (4g) x = 0.392, y = 0.375, z = 0.312

O_{III} (4g) x = 0.325, y = 0.625, z = 0.116

O_{IV} (4g) x = 0.325, y = 0.125, z = 0.116

O_{V} (4g) x = 0.385, y = 0.052, z = 0.460

O_{VI} (4g) x = 0.385, y = 0.698, z = 0.460

O_{VII} (4g) x = 0.478, y = 0.625, z = 0.171

O_{VIII} (4g) x = 0.478, y = 0.125, z = 0.171

O_{IX} (4g) x = 0.146, y = 0.875, z = 0.223

RHODONITE

$P\bar{1}$

$$(Mn,Ca)^{[7]}(Mn,Ca)^{o}_{4}\ {}^{1}_{\infty}\left[Si^{t}_{5}O_{15}\right]^{11}_{V\,my}$$

a = 7.6816Å
b = 11.8180Å
c = 6.7073Å
α = 92.355°
β = 93.948°
γ = 105.665°
Z = 2

Atom	Site	x	y	z
$(Mn,Ca)^{o}_{I}$	(2i)	0.8819	0.8517	0.9697
$(Mn,Ca)^{o}_{II}$	(2i)	0.6827	0.5548	0.8748
$(Mn,Ca)^{o}_{III}$	(2i)	0.4916	0.2700	0.8109
...				

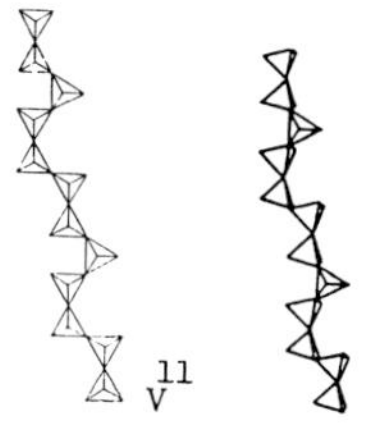

symmetrical analogue real

Fig. 1

Fig. 2

Properties

Habit	Cleav.	Fract.	Twin.	Hardn.	Dens.	Colour	Transp.
tabular, massive	perfect (110)	conchoidal	rare (010)	5.5-6	3.5-3.7	pink, red	transparent to translucent

Refr. index/Reflect.	Birefr.	Luster	Streak	Melt.p.	CPI
n_α = 1.717 n_β = 1.720 n_γ = 1.730	(+) 2V= 63°-76°	vitreous	white	2197°C	(SPI) 63

Figures

Fig. 1. Polyhedral representation of the $Si^{t}O_3$ rhodonite infinite chain: symmetrical analogue and real chain (after Kostov, 1968, and Povarennykh, 1972).

Fig. 2. Rhodonite chain adjustment to the calcium octahedra (after Belov, 1963).

Description

The rhodonite is based on the packing of $Si^{t}O_3$ infinite chains. Some of the Ca atoms have seven coordination, and others octahedral coordination.

References

Kostov (1968) 329, 350.
Povarennykh (1972) 152.
Wyckoff (1968) Vol. 4, 319, 320.
Zoltai + Stout (1984) 348.
Ingerson (1955) 352.
Belov (1963) 16.

Crystallographic data (continued)

Atom	Site	x	y	z
$(Mn,Ca)^{o}_{IV}$	(2i)	0.3018	0.9767	0.7967
$(Mn,Ca)^{[7]}$	(2i)	0.0457	0.6938	0.6389
Si_{I}	(2i)	0.2191	0.1246	0.4956
Si_{II}	(2i)	0.2687	0.4701	0.6375
Si_{III}	(2i)	0.4610	0.7393	0.7092
Si_{IV}	(2i)	0.7446	0.0891	0.7538
Si_{V}	(2i)	0.9263	0.3466	0.8450
O_{I}	(2i)	0.9544	0.6772	0.9628
O_{II}	(2i)	0.6011	0.7312	0.8955
O_{III}	(2i)	0.7485	0.3895	0.8858
O_{IV}	(2i)	0.3981	0.4375	0.8074
O_{V}	(2i)	0.5485	0.0975	0.8053
O_{VI}	(2i)	0.1970	0.1318	0.7374
O_{VII}	(2i)	0.3218	0.8149	0.7438
O_{VIII}	(2i)	0.9337	0.8524	0.6591
O_{IX}	(2i)	0.2560	0.9962	0.4459
O_{X}	(2i)	0.7457	0.5871	0.5846
O_{XI}	(2i)	0.8430	0.414	0.9433
O_{XII}	(2i)	0.5819	0.0414	0.5154
O_{XIII}	(2i)	0.3191	0.6142	0.6300
O_{XIV}	(2i)	0.0546	0.4360	0.7006
O_{XV}	(2i)	0.8607	0.2221	0.7024

SILLIMANITE

P b n m

$Al^{o}\ \overset{1}{\infty}\left[Si^{t}\ Al^{t}\ O_{5}\right]^{2-I^{11}}by$

a=7.4856 Å
b=7.6738 Å
c=5.7698 Å
Z=4

Al^{o}_{I}	(4a)	
Al^{t}_{II}	(4c)	x=0.1419, y=0.3447, z=1/4
Si	(4c)	x=-0.1535, y=-0.3404, z=1/4
O_{I}	(4c)	x=-0.3599, y=-0.4078, z=1/4
O_{II}	(4c)	x=0.3576, y=0.4352, z=1/4
O_{III}	(4c)	x=-0.4753, y=0.0008, z=1/4
...		

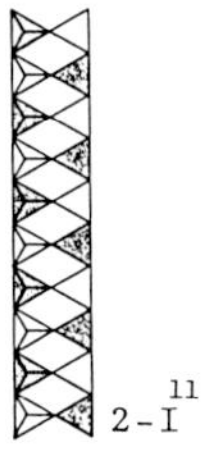

Fig. 1

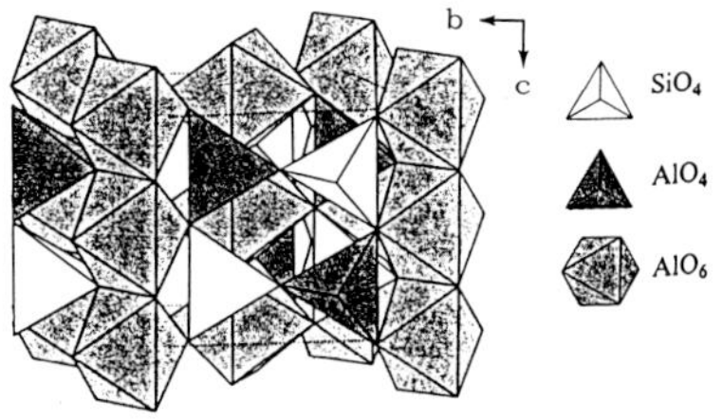

Fig. 2

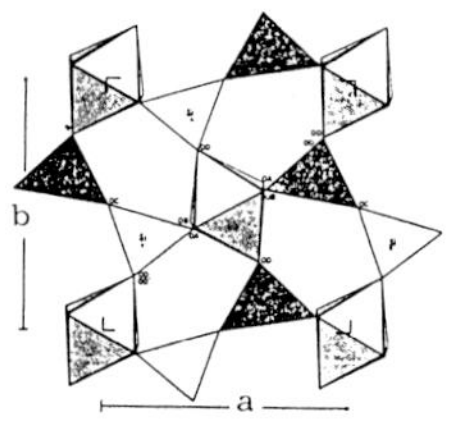

Fig. 3

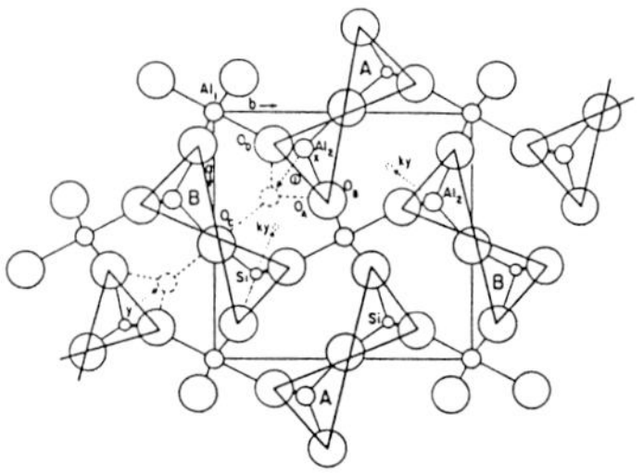

Fig. 4

Properties

Habit	Cleav.	Fract.	Twin.	Hardn.	Dens.	Colour	Transp.
prismatic	perfect (010)	uneven		6-7	3.23	white, brown	transparent to translucent

Refr. index/Reflect.	Birefr.	Luster	Streak	Melt.p.	CPI
n_{α} = 1.658 n_{β} = 1.662 n_{γ} = 1.680	(+) 2V = 20°-30°	vitreous	white		(SPI) 63

Disordered derivative

Mullite $Al^{o}\ \overset{1}{\infty}\left[Si^{t}_{2-x}Al^{t}_{x}O_{5.5-0.5x}\right]^{2-I^{11}}by$

P b a m

Figures

Fig. 1. The silicate chain of the sillimanite structure.

Fig. 2. Polyhedral representation of the sillimanite structure (after Zoltai + Stout, 1984).

Fig. 3. Sillimanite structure projected along the c axis (after Papike, 1987, quoted by Kerrick, 1990).

Fig. 4. Projection of the sillimanite structure down the c axis showing the possible shifts of Al_2 and Si atoms in the transformation of sillimanite to andalusite (denoted x and y) and to kyanite (denoted ky) (after Winter + Ghose, 1979).

Fig. 5. Structural schemes of sillimanite (a), andalusite (b) and kyanite (c) (after Kostov, 1968, 283).

Description

The sillimanite structure is formed by the packing of $Si\,Al\,O_5$ tetrahedral chains, which generate infinite chains of $Al\,O_6$ octahedra parallel to them. Sillimanite is related to andalusite, mullite and kyanite.

Crystallographic data (continued)

O_{IV} (8d) x=0.1248, y=0.2237, z=0.5164

References

Kostov (1968) 282-286.
Povarennykh (1972) 423, 424.
Wyckoff (1968) Vol. 4, 186-188.
Zoltai + Stout (1984) 363, 364.
Winter + Ghose (1979) 582.
Papike (1987) 1483-1526.
Kerrick (1990) 27.

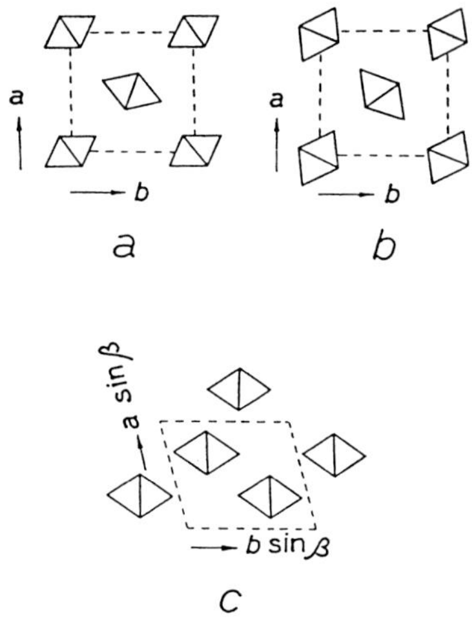

Fig. 5

COLEMANITE

$P2_1/a$

$Ca^{[7]}(H_2O)\ {}^{1}_{\infty}\left[B^{tr}\ B^{t}_{2}\ O_4\ (OH)_3\right]^{31}_{VI\ my}$

a = 8.743Å
b = 11.264Å
c = 6.102Å
β = 110° 7'
Z = 4

Atom	Position	x	y	z
Ca	(4c)	0.6345	0.2112	0.7366
(H_2O)	(4e)	0.616	0.018	0.900
B	(4e)	0.336	0.212	0.315
B	(4e)	0.040	0.174	0.160
B	(4e)	0.219	0.052	0.493
O_I	(4e)	-0.018	0.247	0.314
...				

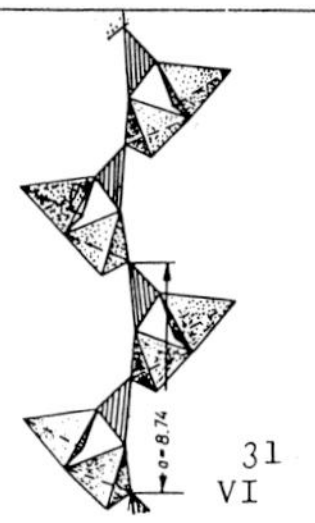

Fig. 1

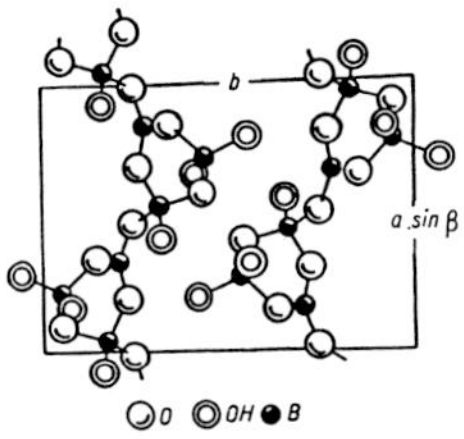

Fig. 2

Properties

Habit	Cleav.	Fract.	Twin.	Hardn.	Dens.	Colour	Transp.
short prismatic	perfect (010)	subconchoidal		4-4.5	2.42	colourless, white	transparent to translucent

Refr. index/Reflect.	Birefr.	Luster	Streak	Melt.p.	CPI
n_α = 1.586 n_β = 1.592 n_γ = 1.614	(+) 2V = 56°	vitreous	white		(SPI) 41

Figures

Fig. 1. Polyhedral drawing of the $B^{tr}B^{t}_{2}O_4(OH)_3$ infinite chain of the colemanite structure (after Povarennykh, 1972).

Fig. 2. Partial representation of the colemanite structure projected along the c axis showing the way the chains pack together (adapted from Povarennykh, 1972).

Description

Colemanite is a chain structure build of $B^{tr}B^{t}_{2}O_4(OH)_3$ infinite chains. Within the packing of these chains are located H_2O molecules and Ca atoms with coordination 7 . The chains run along the a axis, what explains the (010) cleavage.

References

Kostov (1968) 434, 435.
Povarennykh (1972) 159, 482.
Wyckoff (1965) Vol. 3, 583-585.
Zoltai + Stout (1984) 437.
Christ et al. (1958) 761-770.

Crystallographic data (continued)

Atom	Position	x	y	z
O_{II}	(4e)	0.188	0.233	0.139
O_{III}	(4e)	0.344	0.149	0.510
O_{IV}	(4e)	0.088	0.055	0.270
$(OH)_I$	(4e)	-0.080	0.164	-0.071
$(OH)_{II}$	(4e)	0.298	-0.066	0.506
$(OH)_{III}$	(4e)	0.165	0.078	0.699

KERNITE

$P2_1/c$

$Na_2^{[5]}(H_2O)_3 \; {}^{1}_{\infty}\left[B_2^{t}\; B_2^{tr}\; O_6\; (OH)_2\right]$

a = 7.0172 Å
b = 9.1582 Å
c = 15.6724 Å
β = 108.861°
Z = 4

Atom	x	y	z
Na(1)(4e)	0.31928	0.46486	0.31089
Na(2)(4e)	0.18557	0.36637	0.07245
O (1)(4e)	0.51273	0.02504	0.39728
O(2)(4e)	0.44023	0.09882	0.24094
O(3)(4e)	0.76727	0.20590	0.26849
O(4)(4e)	0.46493	0.35087	0.20889
...			

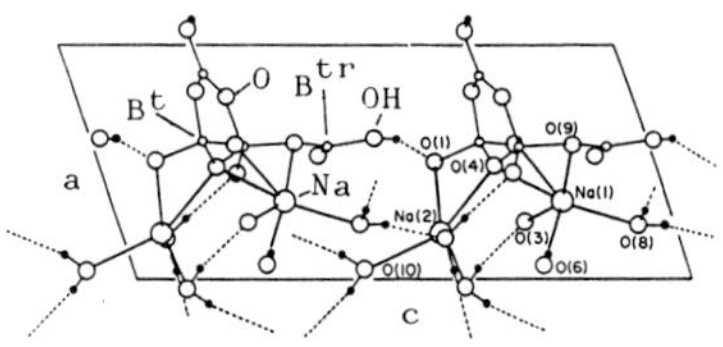

Fig. 1

Properties

Habit	Cleav.	Fract.	Twin.	Hardn.	Dens.	Colour	Transp.
coarse agregates, massive	perfect (001), (100)	uneven		3	1.90	colourless, white	transparent

Refr. index/Reflect.	Birefr.	Luster	Streak	Melt.p.	CPI
n_α = 1.454 n_β = 1.472 n_γ = 1.488	(-) 2V = 80°	vitreous, pearly	white		(SPI) 36

Figures

Fig. 1. Projection of the kernite structure on the ac plane showing the hydrogen bonding and sodium coordination (after Cooper et al., 1973).

Description

The kernite structure consists of infinite chains $B_2^tB_2^{tr}O_6(OH)_2$, which run parallel to the b axis, linked to Na and H_2O.

Kernite has two perfect cleavages (001) and (100), and both involve breaking of hydrogen bonds and sodium water molecules bonds.

References

Kostov (1968) 437.
Povarennykh (1972) 482, 483.
Wyckoff (1965) Vol. 3, 751-753.
Cooper et al. (1973) 21-31.
Zoltai + Stout (1984) 436.

Crystallogr. data (continued)

Atom	x	y	z
O(5) (4e)	0.79170	0.44864	0.21540
O(6) (4e)	0.06216	0.29094	0.24052
O(7) (4e)	0.58479	0.29347	0.00999
O(8) (4e)	0.77580	0.03282	0.06828
O(9) (4e)	0.56568	0.27262	0.36309
O(10)(4e)	0.03920	0.06215	0.41534
O(11)(4e)	0.16711	0.11744	0.07617
B(1) (4e)	0.55313	0.16779	0.42154
B(2) (4e)	0.55340	0.23267	0.26884
B(3) (4e)	0.57414	0.48048	0.19969
B(4)(4e)	0.86829	0.31396	0.24180
H(1)(4e)	0.2506	0.1008	0.1206
H(2)(4e)	0.0934	0.2134	0.2498
H(3)(4e)	0.9720	0.1114	0.3818
H(4)(4e)	0.5666	0.3580	0.0360
H(5)(4e)	0.1086	0.0233	0.3802
H(6)(4e)	0.7212	0.1005	0.0512
H(7)(4e)	0.2233	0.4689	0.4882
H(8)(4e)	0.0314	0.0928	0.0812

BORAX

C2/c

$$\left[\{g\}\left[B_2^{tr}\,B_2^{t}\,O_5\,(OH)_4\right]^{IV\,4}\;{}^{1}_{\infty}\left[Na_2^{o}\,(H_2O)_8\right]^{III\,11}\right]$$

a = 11.858Å
b = 10.674Å
c = 12.197Å
β = 106° 41'
Z = 4

Atom	Position	Coordinates
B	(8f)	x = 0.085, y = 0.345, z = 0.217
B	(8f)	x = 0.095, y = 0.454, z = 0.390
O_I	(4e)	u = 0.265
O_{II}	(8f)	x = 0.154, y = 0.419, z = 0.314
O_{III}	(8f)	x = 0.018, y = 0.435, z = 0.123
$(OH)_I$	(8f)	x = 0.161, y = 0.269, z = 0.167

...

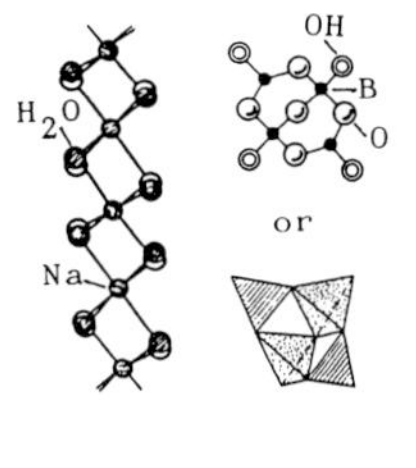

Fig. 1

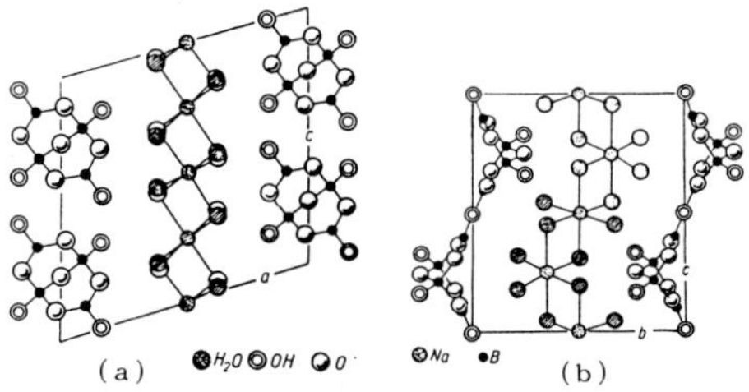

Fig. 2

Properties

Habit	Cleav.	Fract.	Twin.	Hardn.	Dens.	Colour	Transp.
prismatic	perfect (100) (110)	conchoi-dal		2-2.5	1.7-1.9	white, grey	translucent

Refr. index /Reflect.	Birefr.	Luster	Streak	Melt.p.	CPI
n_α = 1.447	(-)	vitreous,	white		(SPI)
n_β = 1.469		resinous			24
n_γ = 1.472	2V = 40°				

Figures

Fig. 1. Representation of the $Na_2^o(H_2O)_8$ infinite chain and the $B_2^{tr}B_2^tO_5(OH)_4$ group of the borax structure (after Povarennykh, 1972).

Fig. 2. Structure of borax in two projections with the c axis vertical: (a) along the b axis, and (b) along the a axis (after Povarennykh, 1972).

Description

The borax structure is based on the packing of $Na_2^o(H_2O)_8$ infinite chains and $B_2^{tr}B_2^tO_5(OH)_4$ groups. The orientation of the infinite chains parallel to the c axis explains the perfect cleavage parallel to (100).

References

Kostov (1968) 437, 438.

Povarennykh (1972) 158, 478.

Wyckoff (1965) Vol. 3, 864--867.

Zoltai + Stout (1984) 436.

Crystallographic data (continued)

Atom	Position	Coordinates
$(OH)_{II}$	(8f)	x = 0.163, y = 0.511, z = 0.491
Na_I	(4a)	
Na_{II}	(4e)	u = 0.845
$(H_2O)_I$	(8f)	x = 0.123, y = 0.846, z = 0.450
$(H_2O)_{II}$	(8f)	x = 0.125, y = 0.002, z = 0.196
$(H_2O)_{III}$	(8f)	x = 0.119, y = 0.165, z = 0.459
$(H_2O)_{IV}$	(8f)	x = 0.116, y = 0.707, z = 0.171

8.4.4. Sheet structures

GRAPHITE

$P6_3mc$

$^{2}_{\infty}\left[C^{tr}\right]^{Hh}$

a = 2.456 Å
c = 6.696 Å
Z = 4

C_I (2a) u = 0
C_{II}(2b) v ≃ 0

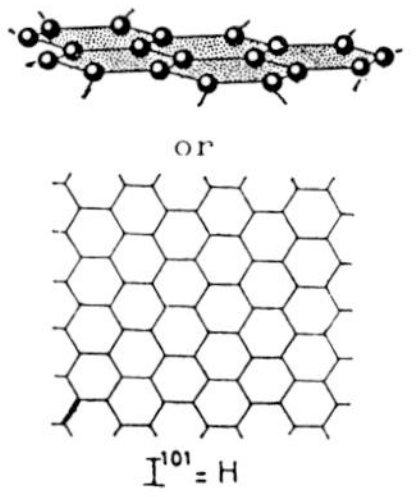

Fig. 1

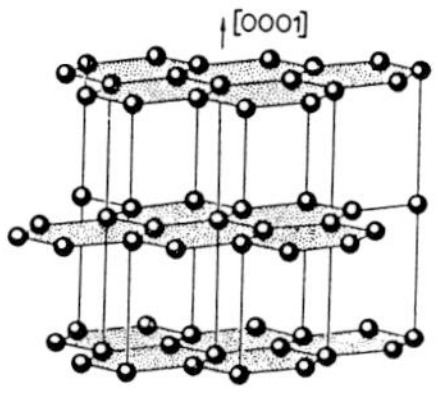

Fig. 2

Properties

Habit	Cleav.	Fract.	Twin.	Hardn.	Dens.	Colour	Transp.
tabular, foliated masses	perfect (0001)			1-2	2.09-2.23	iron black, steel grey	opaque

Refr. index/Reflect.	Birefr.	Luster	Streak	Melt.p.	CPI
n_ω = 1.93-2.07	(-)	metallic, dull	black		(SPI) ~ 19

Figures

Fig. 1. Schematic representation of the H layer of graphite.

Fig. 2. Packing representation of the graphite structure (after Kostov, 1968).

Fig. 3. Defect packing analogue of the graphite structure. It is based on a defect simple hexagonal packing, and the corresponding structural formula is $\left[C_2 \square\right]^{Ts}$. However, the difference between the real structure and its analogue is quite important: c/a ideal 4R/2R = 2, and c/a real 2.7.

Description

Graphite is a sheet structure, formed by the packing of carbon sheets with an honeycomb hexagonal pattern (H), packed in a h sequence. This structure corresponds to a 2H layer sequence to which is also attributed the space group $P6_3mmc$ (Strunz, 1982). There is another graphite with rhombohedral symmetry, correspondent to a 3R layer sequence, with space group R3m, a = 2.456Å and c = 10.044Å.

References

Kostov (1968) 102.
Wyckoff (1963) Vol. 1, 26-28.
Zoltai + Stout (1984) 377.
Roberts et al. (1974) 246.
Strunz (1982) 100.

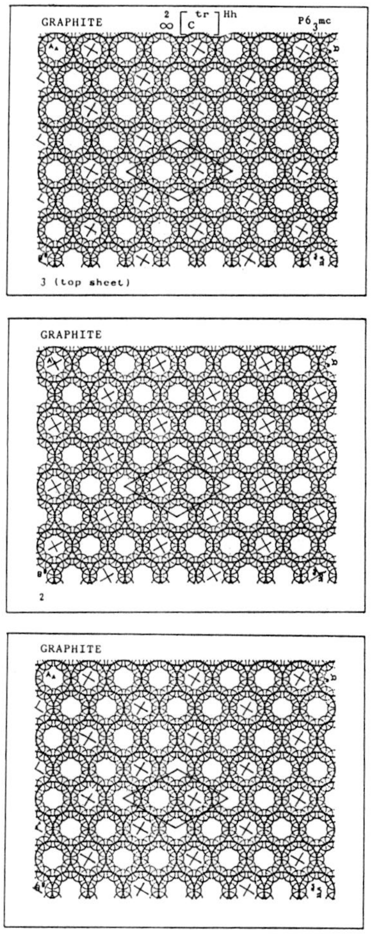

Fig. 3

ARSENIC

$R\bar{3}m$

$^{2}_{\infty}\left[{}^{[3n]}_{As}\right]\tilde{H}c$

a_R = 4.131 Å
α = 54° 10'
Z_R = 2
a_H = 3.760 Å
c = 10.548 Å
Z_H = 6

(Hexagonal description)

As(6c) z = 0.226

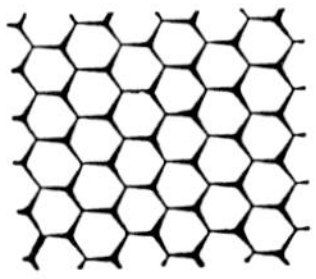

$I^{102} = \tilde{H}$

Fig. 1

Fig. 2

Properties

Habit	Cleav.	Fract.	Twin.	Hardn.	Dens.	Colour	Transp.
rhombohedral acicular	perfect (0001)	uneven		3.5	5.7	tin white, dark grey	opaque

Refr. index / Reflect.	Birefr.	Luster	Streak	Melt.p.	CPI
62%		metallic			(SPI) 43

Population

Bismuth $^{2}_{\infty}\left[{}^{[3n]}_{Bi}\right]\tilde{H}c$

Antimony $^{2}_{\infty}\left[{}^{[3n]}_{Sb}\right]\tilde{H}c$

Allemontite $^{2}_{\infty}\left[{}^{[3n]}_{As}\ {}^{[3n]}_{Sb}\right]\tilde{H}c$

Description

The arsenic structure is based on corrugated honeycomb hexagonal sheets packed in a c sequence. The arsenium atoms have trigonal pyramidal coordination [3n] within the sheets. $\tilde{H}$ means a corrugated layer with a honeycomb pattern.

Figures

Fig. 1. Schematic representation of the corrugated layers $\tilde{H}$ of the arsenic structure.

Fig. 2. Expanded view of the As structure, showing the corrugated layers parallel to (0001) (after Bokii, 1954)

References

Kostov (1968) 98.

Povarennykh (1972) 198.

Wyckoff (1963) Vol. 1, 32, 33.

Zoltai + Stout (1984) 376.

Roberts et al. (1974) 35.

Bokii (1954) 248.

ORPIMENT

$P2_1/n$

$^{2}_{\infty}\left[{}^{[3n]}As_2\,S_3\right]\,II_b^{201}$

a = 11.46 Å	As_I (4e) x = 0.267, y = 0.190, z = 0.143	S_{II} (4e) x = 0.355, y = 0.397, z = 0.013
b = 9.57 Å	As_{II} (4e) x = 0.484, y = 0.323, z = 0.643	S_{III} (4e) x = 0.125, y = 0.293, z = 0.410
c = 4.22 Å	S_I (4e) x = 0.395, y = 0.120, z = 0.500	
β = 90° 30'		
Z = 4		

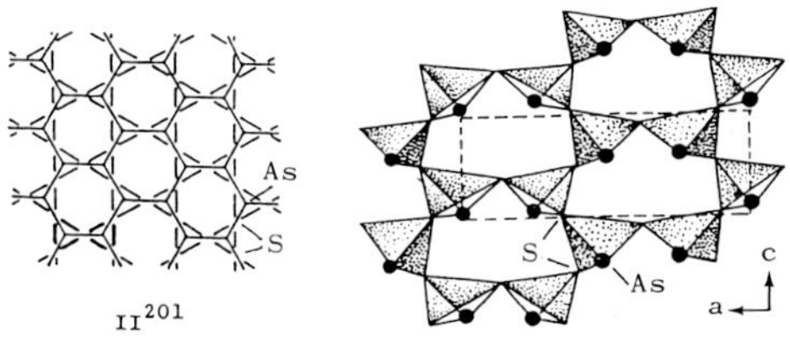

ideal real

Fig. 1

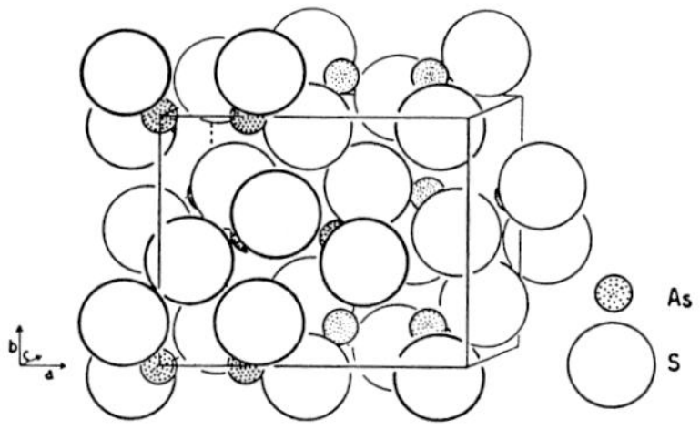

Fig. 2

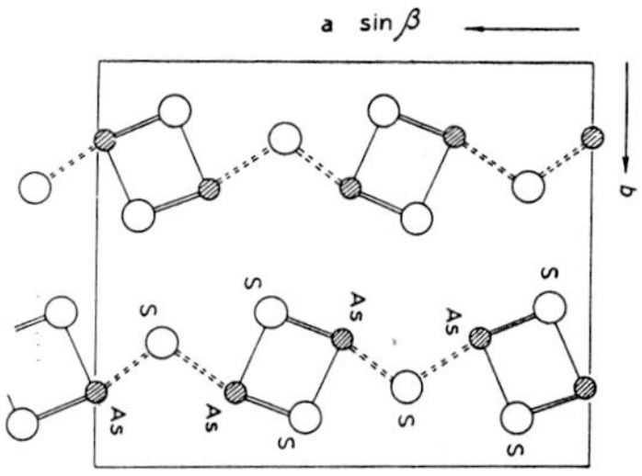

Fig. 3

Properties

Habit	Cleav.	Fract.	Twin.	Hardn.	Dens.	Colour	Transp.
prismatic, foliated	perfect (010)		(100)	1.5-2	3.49	yellow orange	transparent to opaque

Refr. index/Reflect.	Birefr.	Luster	Streak	Melt.p.	CPI
n_α = 2.40, n_β = 2.81, n_γ = 3.02	(-) 2V = 76°	resinous	yellow	310°C	

Figures

Fig. 1. (a) Schematic ideal representation of the corrugated sheet of the orpiment structure, and (b) real sheet (after Povarennykh, 1972).

Fig. 2. Structure of orpiment (after Structure Reports, 1952, Vol. 12).

Fig. 3. Structure of orpiment projected along the c axis (after Kostov, 1968).

Description

The orpiment structure is based on As_2S_3 sheets where As is trigonal pyramidal coordinated by S atoms. The As_2S_3 sheets are packed along the b axis, and repeat at the third sheet. The perfect cleavage parallel to (010) is in agreement with this structural description.

References

Kostov (1968) 160, 166.
Povarennykh (1972) 259,260.
Wyckoff (1964) Vol. 2, 26,27.
Structure Reports (1952) Vol. 12, 176.
Roberts et al. (1974) 450.
Ingerson (1955) 350.

PYROPHYLLITE (2M) C2/c

$Al^{o}_{2}(OH)_{2}\ ^{2}_{\infty}\left[Si^{t}_{4}O_{10}\right]^{(2.II^{101}_{h})_{c}}$

a = 5.16 Å
b = 8.90 Å
c = 18.60 Å
β = 100° 45'
Z = 4

Atom		Coordinates	Atom		Coordinates
Al	(8f)	x = 0.000, y = 0.333, z = 0.000	O_I	(8f)	x = 0.203, y = 0.500, z = 0.058
Si_I	(8f)	x = -0.239, y = 0.000, z = 0.143	O_{II}	(8f)	x = 0.203, y = 0.167, z = 0.058
Si_{II}	(8f)	x = 0.261, y = 0.167, z = 0.143	O_{III}	(8f)	x = 0.025, y = 0.084, z = 0.176

...

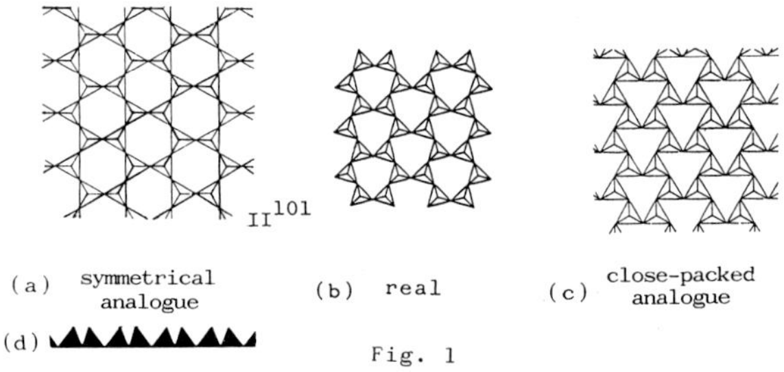

(a) symmetrical analogue (b) real (c) close-packed analogue
(d)

Fig. 1

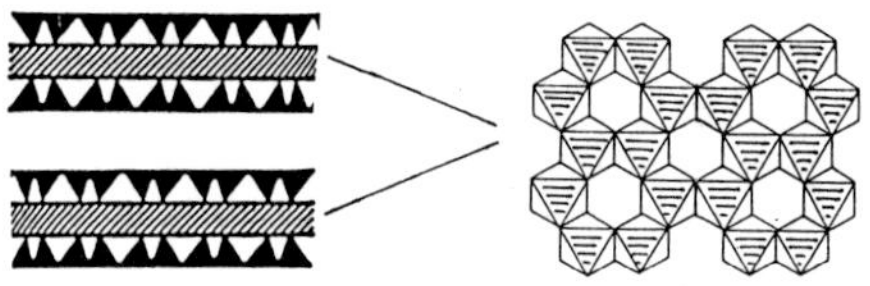

(a) (b)

Dioctahedral

Fig. 2

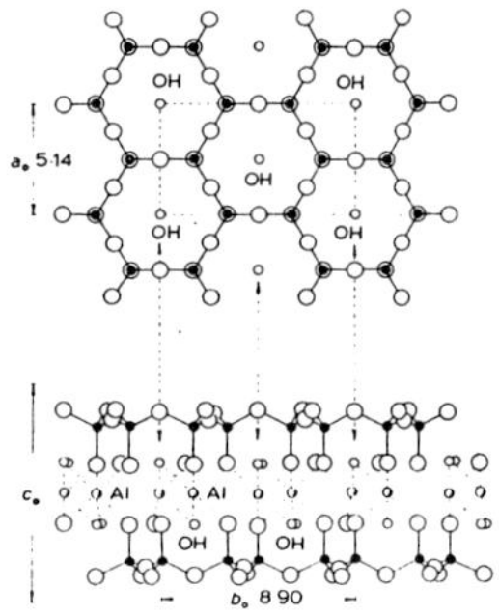

Fig. 3

Properties

Habit	Cleav.	Fract.	Twin.	Hardn.	Dens.	Colour	Transp.
tabular, foliated	perfect (001)	flexible		1.5	2.65- -2.90	white, greyish white	transparent to translucent

Refr. index/Reflect.	Birefr.	Luster	Streak	Melt.p.	CPI
n_α = 1.553-1.556 n_β = 1.586-1.589 n_γ = 1.596-1.601	(-) 2V = 52°-62°	pearly, dull	white		(SPI) 51

Figures

Fig. 1. Polyhedral representations of part of the structural unit of pyrophyllite, the $Si^{t}_{2}O_{5}$ infinite sheet: a) symmetrical analogue, b) real sheet (Liebau, 1985) and c) close-packed analogue (adapted from Belov, 1951), d) cross section of the sheet symmetrical analogue (adapted from Liebau, 1985).

Fig. 2. Polyhedral description of (a) the way the $Si^{t}_{2}O_{5}$ infinite sheets pack together (after Liebau, 1985), and (b) the dioctahedral layer located in between the $Si^{t}_{2}O_{5}$ sheets. These two silicate sheets and the dioctahedral layer in between form a structural module.

Fig. 3. Ball and spoke representation of the pyrophyllite structure (after Kostov, 1968).

Description

Pyrophyllite is a sheet structure based on the packing of $Si^{t}_{2}O_{5}$ infinite sheets. Pyrophyllite is a 2M layer silicate, where M means monoclinic and 2 the number of structural modules per unit cell.

References

Kostov (1968) 352-356.
Povarennykh (1972) 432.
Wyckoff (1968) Vol. 4, 365,366.
Zoltai + Stout (1984) 321.
Zoltai (1975) IV-8.
Liebau (1985) 116, 215, 218.
Belov (1951) 19.

Crystallographic data (continued)

Atom		Coordinates
O_{IV}	(8f)	x = -0.475, y = 0.084, z = 0.176
O_V	(8f)	x = 0.275, y = 0.333, z = 0.176
(OH)	(8f)	x = 0.203, y = -0.167, z = 0.058

TALC (2M) C2/c

$$Mg_3^o\,(OH)_2\,{}^2_\infty\left[Si_4^t\,O_{10}\right]^{(2II_h^{101})_c}$$

a = 5.28 Å
b = 9.15 Å
c = 18.92 Å
β = 100° 15'
Z = 4

Atom	x	y	z
Mg_I (8f)	0.000	0.333	0.000
Mg_{II} (4a)			
Si_I (8f)	-0.239	0.000	0.143
Si_{II} (8f)	0.261	0.167	0.143
O_I (8f)	0.203	0.500	0.058
O_{II} (8f)	0.203	0.167	0.058
O_{III} (8f)	0.025	0.084	0.176
O_{IV} (8f)	-0.475	0.084	0.176
O_V (8f)	0.275	0.333	0.176
(OH) (8f)	0.203	-0.167	0.058

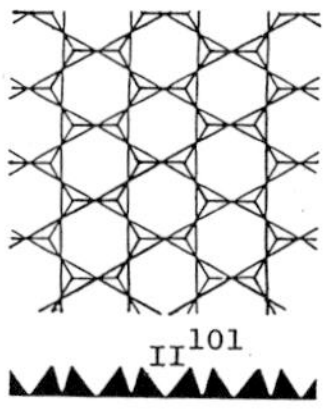

Fig. 1

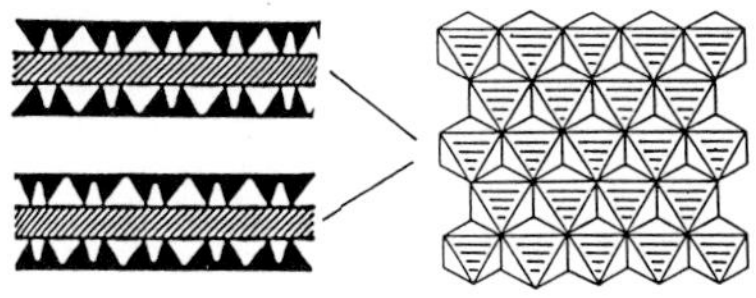

Fig. 2

Properties

Habit	Cleav.	Fract.	Twin.	Hardn.	Dens.	Colour	Transp.
platy, massive	perfect (001)	flexible		1	2.8	white, grey	translucent

Refr. index/Reflect.	Birefr.	Luster	Streak	Melt.p.	CPI
n_α = 1.54 n_β = 1.58 n_γ = 1.58	(-) 2V = 6°-30°	resinous, silky	white		(SPI) 55

Population

Willemseite

$$Ni_3^o\,(OH)_2\,{}^2_\infty\left[Si_4^t\,O_{10}\right]^{(2II_h^{101})_c}$$

Description

Talc is a sheet structure based on the packing of $Si_2^tO_5$ infinite sheets. The 2M notation means monoclinic (M), and (2) two structural modules per unit cell.

Figures

Fig. 1. Polyhedral representation of the structural unit of talc (symmetrical analogue) and its cross section (adapted from Liebau, 1985).

Fig. 2. Polyhedral description of: (a) the way the $Si_2^tO_5$ infinite sheets pack together (after Liebau, 1985) and (b) the trioctahedral layer located in between the $Si_2^tO_5$ sheets.

Fig. 3. Condensed model of the symmetrical analogue of the talc structure. Large open circles represent oxygen and large hatched circles hydroxyls. Black medium circles correspond to Mg atoms, and very small black circles to Si atoms (after Figueiredo, 1979).

Fig. 4. Polyhedral representation of the talc structure (after Zoltai, 1975).

References

Kostov (1968) 355.
Povarennykh (1972) 432.
Wyckoff (1968) Vol. 4, 365, 366.
Zoltai + Stout (1984) 322.
Figueiredo (1979 b).
Zoltai (1975) IV-9.
Liebau (1985) 116, 218.

TALC (2M) (continued)

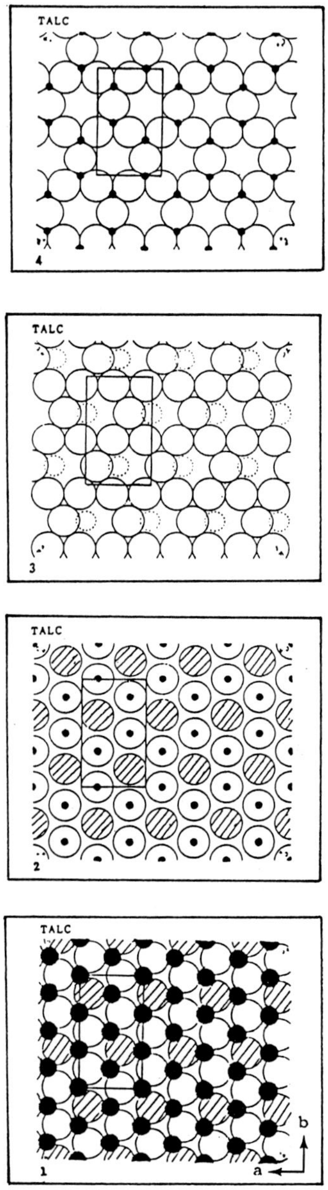

Fig. 3

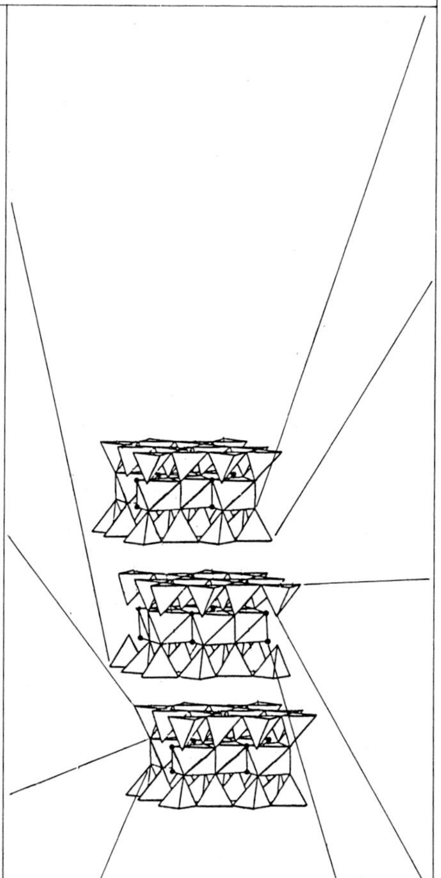

Fig. 4

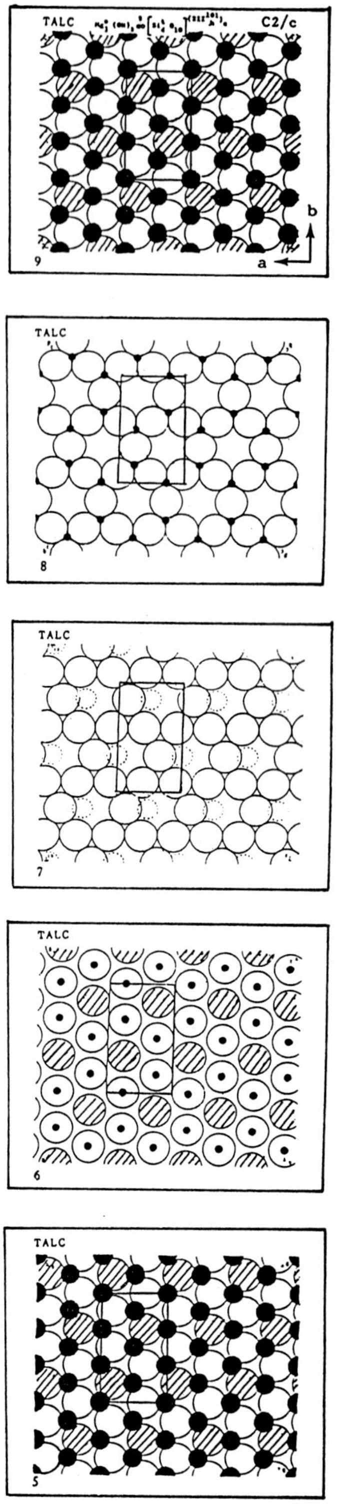

MUSCOVITE ($2M_1$) C2/c

$K^{[12]} Al_2^o (OH)_2 \ {}^2_\infty\left[Si_3^t Al^t O_{10}\right] \ (2.II^{101}_s)_c$

a = 5.189 Å
b = 8.995 Å
c = 20.097 Å
β = 95° 11'
Z = 4

Atom	Position	Coordinates
K	(4e)	x = 0, y = 0.1016, z = 1/4
Al	(8f)	x = 0.2484, y = 0.0871, z = 0.0016
$(Si,Al)_I$	(8f)	x = 0.4625, y = 0.9242, z = 0.1372
...		

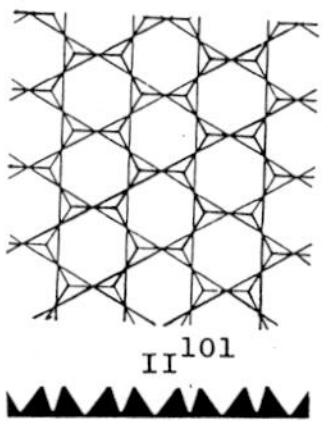

Fig. 1

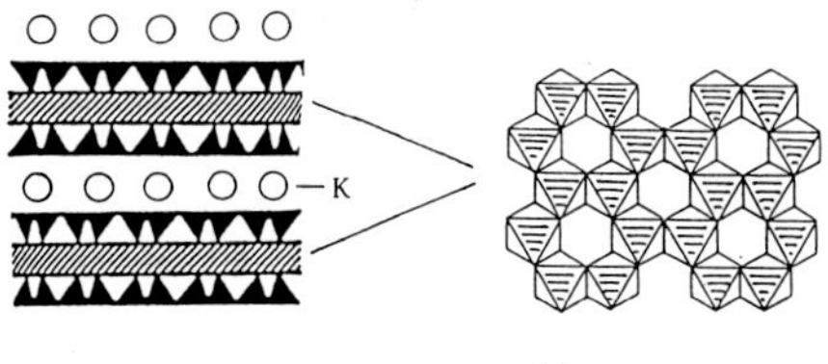

Dioctahedral

(a) Fig. 2 (b)

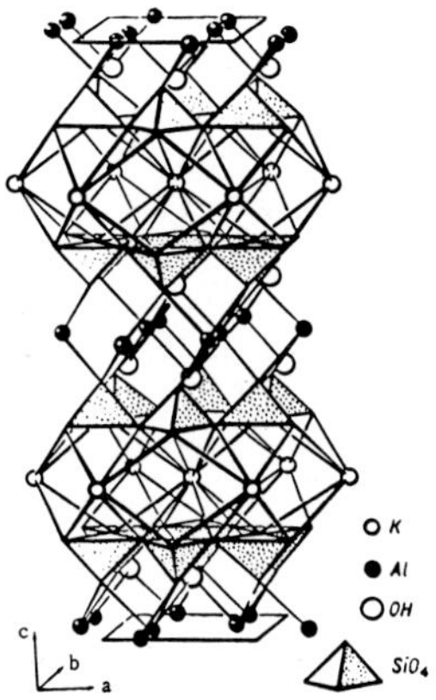

Fig. 3

Properties

Habit	Cleav.	Fract.	Twin.	Hardn.	Dens.	Colour	Transp.
lamellar	perfect	elastic	(001)	2.5	2.8	white, grey brown	transparent

Refr. index/Reflect.	Birefr.	Luster	Streak	Melt.p.	CPI
$n\alpha$ = 1.565 $n\beta$ = 1.596 $n\gamma$ = 1.600	(-) 2V = 30°-40°	vitreous	white		(SPI) 58

Population

Paragonite

$Na^{[12]} Al_2^o (OH)_2 \ {}^2_\infty\left[Si_3^t Al^t O_{10}\right] (2.II^{101}_s)_c$

Glauconite

$K^{[12]} (Al,Fe)_2^o (OH)_2 \ {}^2_\infty\left[(Si,Al)_4^t O_{10}\right] (2.II^{101}_s)_c$

Figures

Fig. 1. Polyhedral representation of the structural unit of muscovite (symmetrical analogue) and its cross section (adapted from Liebau, 1985). 218).

Fig. 2. Polyhedral description of: (a) the way the $Si_2^tO_5$ infinite sheets pack together (adapted from Liebau, 1985), and (b) the dioctahedral layer between the $Si_2^tO_5$ sheets.

Fig. 3. Ball and spoke representation of the muscovite structure (after Povarennykh, 1972).

Fig. 4. More complete polyhedral description of the muscovite structure (after Zoltai, 1975).

Fig. 5. Condensed model of the symmetrical analogue of the muscovite structure. Large open circles represent oxygen atoms, and large hatched circles hydroxyls. Black medium circles represent Mg atoms, large crossed lined circles correspond to K atoms, and very small black circles to Si atoms (after Figueiredo (1979 b).

Description

Muscovite is a sheet structure based on the packing of $Si_2^tO_5$ infinite sheets, with K atoms in between the structural modules.

Crystallographic data (continued)

$(Si,Al)_{II}$	(8f)	x = 0.4593, y = 0.2550, z = 0.1365
O_I	(8f)	x = 0.2629, y = 0.3713, z = 0.1674
O_{II}	(8f)	x = 0.2450, y = 0.8020, z = 0.1620
O_{III}	(8f)	x = 0.4080, y = 0.0960, z = 0.1680
O_{IV}	(8f)	x = 0.4650, y = 0.9450, z = 0.0527
O_V	(8f)	x = 0.4250, y = 0.2600, z = 0.0542
(OH)	(8f)	x = 0.4530, y = 0.5580, z = 0.0520

MUSCOVITE ($2M_1$) (continued)

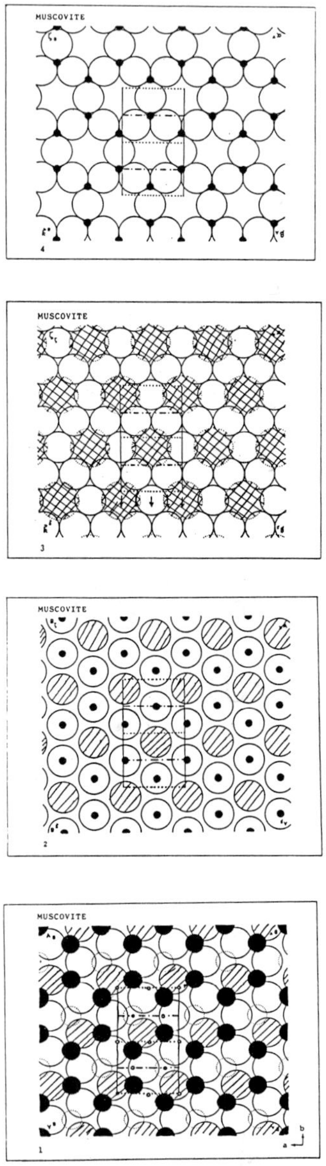

Fig. 5

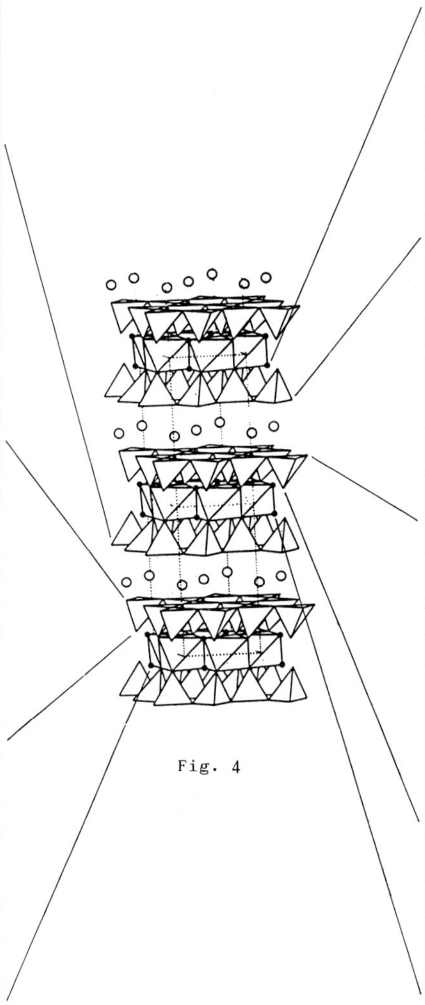

Fig. 4

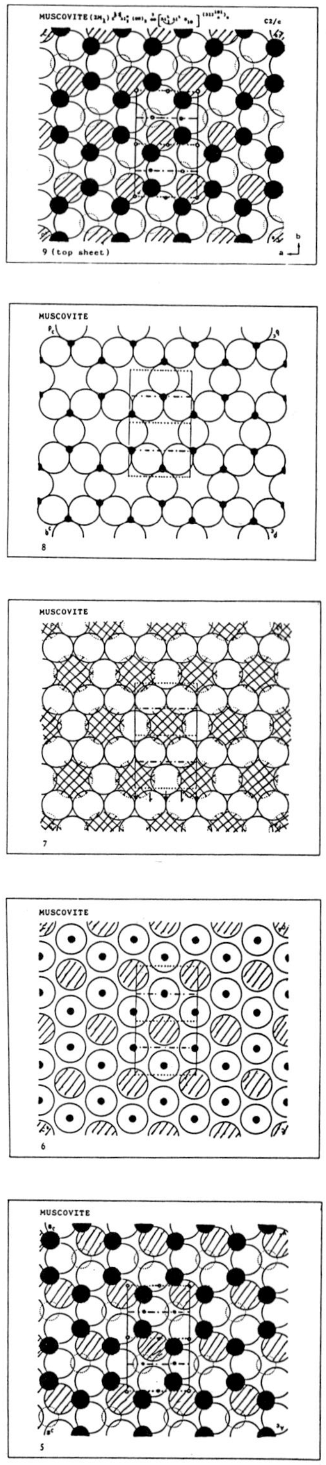

References

Kostov (1968) 356-360.
Povarennykh (1972) 442.
Wyckoff (1968) Vol. 4, 345-348.
Zoltai + Stout (1984) 321, 326.
Liebau (1985) 116, 218.
Figueiredo (1979b).
Zoltai (1975) IV-10.

PHLOGOPITE (1M) C2/m

(Magnesiobiotite)

$$\overset{[12]}{K}\ Mg_3^{o}\ (OH)_2\ \overset{2}{\infty}\left[Si_3^{t}\ Al^{t}\ O_{10}\right]^{(2II_s^{101})_c}$$

a = 5.31 Å
b = 9.21 Å
c = 10.13 Å
β = 100° 1'
Z = 2

K (2b)
Mg_I (2c)
Mg_{II} (4h) u = 0.3308
(Si,Al) (8j) x = 0.0753, y = 0.1665, z = 0.2250
...

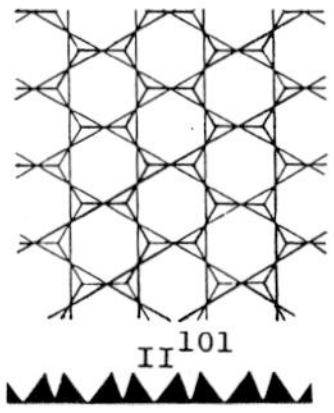

Fig. 1

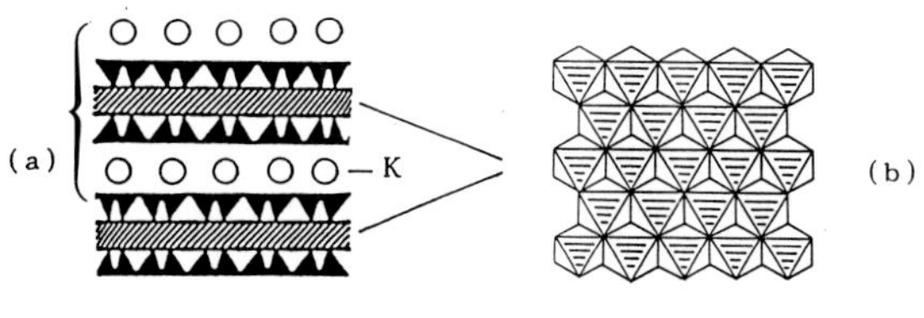

Trioctahedral

Fig. 2

Properties

Habit	Cleav.	Fract.	Twin.	Hardn.	Dens.	Colour	Transp.
platy, foliated	perfect (001)		(001)	2.5	2.8	yellow brown	transparent

Refr. index/Reflect.	Birefr.	Luster	Streak	Melt.p.	CPI
n_α = 1.56 n_β = 1.60 n_γ = 1.60	(-) 2V = 0°-20°	pearly	white		(SPI) 62

Population

Biotite

$$\overset{[12]}{K}\ (Mg,Fe)_3^{o}\ (OH)_2\ \overset{2}{\infty}\left[(Si,Al)_4^{t}O_{10}\right]^{(2II_s^{101})_c}$$

Lepidolite

$$\overset{[12]}{K}\ (Li,Al)_3^{o}\ (OH)_2\ \overset{2}{\infty}\left[(Si,Al)_4^{t}O_{10}\right]^{(2II_s^{101})_c}$$

Description

Phlogopite is a sheet structure based on the packing of $Si_2^tO_5$ infinite sheets, with some K atoms in between the structural modules.

Figures

Fig. 1. Polyhedral representation of the structural unit of phlogopite (symmetrical analogue) and corresponding cross section (adapted from Liebau, 1985).

Fig. 2. Polyhedral description of: (a) the way the $Si_2^tO_5$ infinite sheets pack together (adapted from Liebau, 1985) and (b) trioctahedral layer between the $Si_2^tO_5$ sheets.

Fig. 3. Condensed model of the symmetrical analogue of the phlogopite structure. Large open circles represent oxygen atoms, and large hatched circles hydroxyls. Black medium circles represent Mg atoms, large crossed lined circles correspond to K atoms, and very small black circles to Si atoms (after Figueiredo, 1979 b). The condensed model corresponds to the part of the structure marked with curly brackets (Fig. 2).

Cryst. data (continued)

O_I	(4i)	u = 0.0235 v = 0.1664
O_{II}	(8j)	x = 0.3218 y = 0.2346 z = 0.1670
O_{III}	(8j)	x = 0.1305 y = 0.1665 z = 0.3910
(OH,F)	(4i)	x = -0.3671 v = 0.4020

References

Kostov (1968) 358-359.
Povarennykh (1972) 443.
Wyckoff (1968) Vol. 4, 344-346.
Zoltai + Stout (1984) 332.
Liebau (1985) 116, 218.
Figueiredo (1979 b).

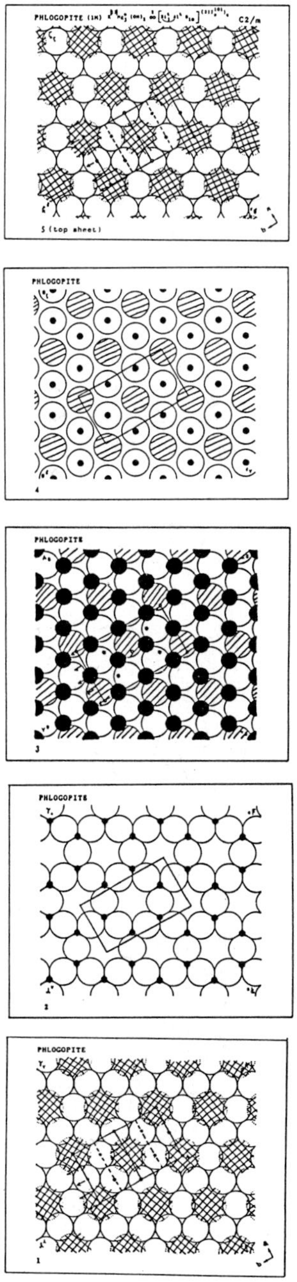

Fig. 3

MONTMORILLONITE

C2/m

$n(H_2O)\ Al_2^o\ (OH)_2\ {}^2_\infty\left[Si_4^t\ O_{10}\right](2.II_s^{101})_c$

a = 5.18 Å
b = 8.96 Å
c = 9.97 Å
β = 99° 54'
Z = 2

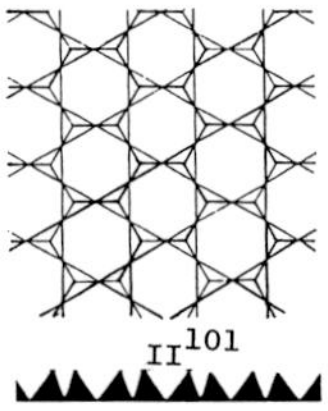

Fig. 1

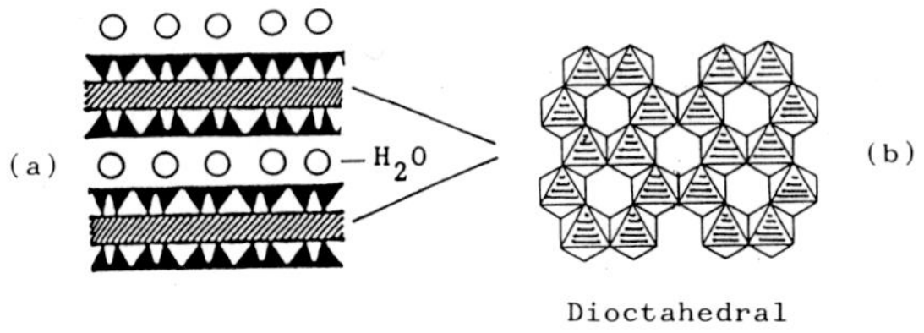

Fig. 2

Properties

Habit	Cleav.	Fract.	Twin.	Hardn.	Dens.	Colour	Transp.
massive, fine grained	perfect (001)			1-2	2-3	white, grey, yellowish	

Refr. index/Reflect.	Birefr.	Luster	Streak	Melt.p.	CPI
n_α = 1.48-1.57 n_β = 1.50-1.60 n_γ = 1.50-1.60	(-)	dull			

Population

Nontronite

$Fe_2^o\ n(H_2O)\ (OH)_2\ {}^2_\infty\left[Si_4^t\ O_{10}\right](2.II_s^{101})_c$

Figures

Fig. 1. Polyhedral representation of the structural unit of montmorillonite (symmetrical analogue) and corresponding cross section (adapted from Liebau, 1985).

Fig. 2. Polyhedral description of: (a) the way the $Si_2^tO_5$ infinite sheets pack together (adapted from Liebau, 1985), and (b) dioctahedral layer between the $Si_2^tO_5$ sheets.

Description

Montmorillonite is a sheet structure based on the packing of $Si_2^tO_5$ infinite sheets, with water molecules in between the structural modules.

References

Kostov (1968) 371.
Povarennykh (1972) 445.
Wyckoff (1968) Vol. 4, 372, 373.
Liebau (1985) 116, 218.
Roberts et al. (1974) 417.

VERMICULITE

C2/c

$(H_2O)_4\ Mg^{o}_3(OH)_2\ \overset{2}{\infty}\left[Si^{t}_3\ Al^{t}\ O_{10}\right]^{(2,II^{101}_{s})_c}$

a = 5.349 Å	
b = 9.255 Å	
c = 28.89 Å	
β = 97° 7'	
Z = 4	

Atom	Position	Coordinates
Mg_I	(4e)	x = 0, y = 0,1638, z = 1/4
Mg_{II}	(4e)	x = 0, y = 0.4997, z = 1/4
Mg_{III}	(4e)	x = 0, y = 0.8332, z = 1/4
...		

Monoclinic Unit Cell
Space group C2/c

$\alpha = \gamma = 90°$
$\beta = 97°$

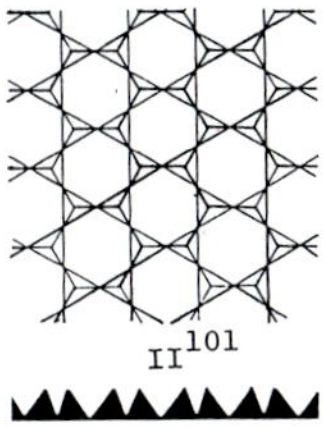

Fig. 1

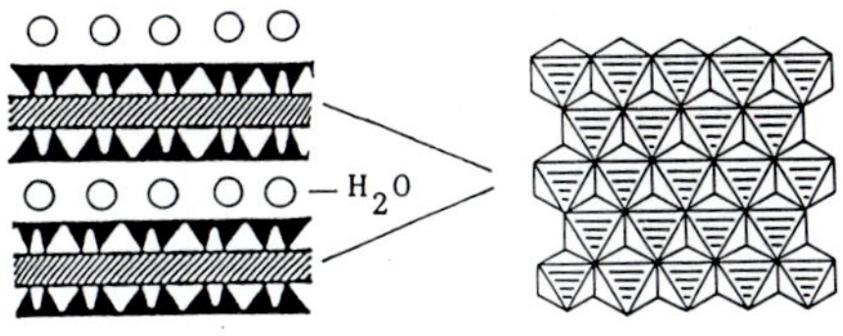

Trioctahedral

(a) Fig. 2 (b)

Properties

Habit	Cleav.	Fract.	Twin.	Hardn.	Dens.	Colour	Transp.
massive	perfect (001)			≃1.5	≃2.3	colourless, yellow, green	

Refr. index/Reflect.	Birefr.	Luster	Streak	Melt.p.	CPI
n_α = 1.525-1.564 n_β = 1.545-1.583 n_γ = 1.545-1583	(-) 2V = 0°-8°				

Crystallographic data (continued)

Atom	Position	Coordinates
$(Si,Al)_I$	(8f)	x = 0.1042, y = -0.0003, z = 0.1545
$(Si,Al)_{II}$	(8f)	x = 0.1026, y = 0.6647, z = 0.1547
O_I	(8f)	x = 0.1424, y = 0.0039, z = 0.2132
O_{II}	(8f)	x = 0.1410, y = 0.6683, z = 0.2113
O_{III}	(8f)	x = 0.3579, y = 0.0697, z = 0.1338
O_{IV}	(8f)	x = 0.3529, y = 0.5964, z = 0.1346
O_V	(8f)	x = 0.5593, y = 0.3316, z = 0.1339
(OH)	(8f)	x = 0.1420, y = 0.3380, z = 0.2129
H_2O_I	(8f)	x = 0.3363, y = -0.0264, z = 0.0397
H_2O_{II}	(8f)	x = 0.3515, y = 0.3274, z = 0.0412
H_2O_{III}	(8f)	x = 0.3832, y = 0.6520, z = 0.0414

(for H_2O only 0.62 are filled)

Population

Saponite

$(H_2O)_4\ Mg^{o}_3(OH)_2\ \overset{2}{\infty}\left[Si^{t}_4\ O_{10}\right]^{(2,II^{101}_{s})_c}$

Description

Vermiculite is a sheet structure based on the packing of $Si^{t}_2O_5$ infinite sheets, with water molecules in between the structural modules.

Figures

Fig. 1. Polyhedral representation of the structural unit of vermiculite (symmetrical analogue) and corresponding cross section (adapted from Liebau, 1985).

Fig. 2. Polyhedral description of: (a) the way the $Si^{t}_2O_5$ infinite sheets pack together (adapted from Liebau, 1985), and (b) trioctahedral layer located between the silicate sheets.

References

Kostov (1968) 371.
Povarennykh (1972) 446.
Wyckoff (1968) Vol. 4, 359.
Deer et al. (1962) Vol. 3, 246.

CHLORITE (1M) C 2/c

$(Mg,Al)_2^o\ (OH)_2 \left[{}^2_\infty \left[(Si,Al)_4^t\ O_{10} \right]^{2,II^{101}_s}\ {}^2_\infty \left[(Mg,Al)_3^o\ (OH)_6 \right] \right]$

a = 5.314
b = 9.205
c = 28.363
β = 97° 10'
Z = 4

Atom	Site	x	y	z
$(Mg,Al)_I$	(4a)	0	0	0
$(Mg,Al)_{II}$	(4e)	0	0.500	1/4
$(Mg,Al)_{III}$	(8f)	0	0.333	0
...				

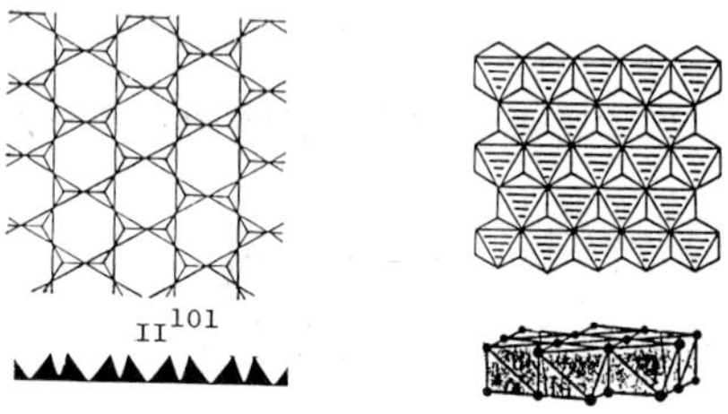

Fig. 1

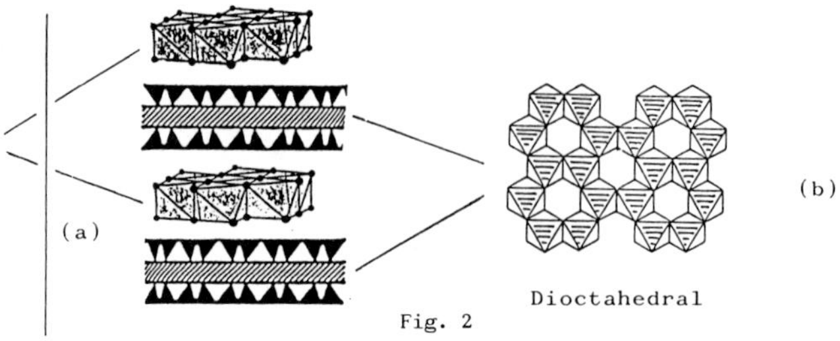

Fig. 2

Properties

Habit	Cleav.	Fract.	Twin.	Hardn.	Dens.	Colour	Transp.
foliated scaly	perfect (001)	flexible	[310] comp. plane (001)	2-3	3.0	green, variable	transparent to translu-cent

Refr. index/Reflet.	Birrefr.	Luster	Streak	Melt.p.	CPI
n_α= 1.56-1.60 n_β= 1.57-1.61 n_γ= 1.58-1.61	(-) δ = 0.006-0.020 2V= 0-40°	vitreous	white, green		(SPI) 52

Crystallographic data (continued)

Atom	Site	x	y	z
$(Mg,Al)_{IV}$	(4e)	0	0.167	1/4
$(Si,Al)_I$	(8f)	-0.269	0.000	0.094
$(Si,Al)_{II}$	(8f)	-0.269	-0.333	0.094
$(OH)_I$	(8f)	-0.308	0.333	0.039
$(OH)_{II}$	(8f)	0.142	0.000	0.211
$(OH)_{III}$	(8f)	0.142	0.333	0.211
$(OH)_{IV}$	(8f)	0.142	-0.333	0.211
O_I	(8f)	-0.308	-0.333	0.039
O_{II}	(8f)	-0.308	0.000	0.039
O_{III}	(8f)	-0.006	0.083	0.114
O_{IV}	(8f)	-0.006	-0.417	0.114
O_V	(8f)	-0.256	0.167	0.114

Population

Clinochlore

$(Mg,Al)_3^o\ (OH)_2 \left[{}^2_\infty \left[Al^t Si_3^t\ O_{10} \right]^{2,II^{101}_s}\ {}^2_\infty \left[Mg_3^o\ (OH)_6 \right] \right]$

Figures

Fig. 1. Polyhedral representation of the structural units of the chlorite structure, and corresponding cross sections (adapted from Zoltai, 1975, and Liebau, 1985).

Fig. 2. Polyhedral description of: (a) the way the mica structural modules and the brucite type layers pack together (adapted from Liebau, 1985) and (b) the dioctahedral layer between the silicate sheets.

Fig. 3. Polyhedral description of the chlorite structure (adapted from Zoltai, 1975).

Fig. 4. Condensed model of the symmetrical analogue of the chlorite structure. The large open circles represent oxygen atoms, and the hatched circles hydroxyls. The black medium circles correspond to (Mg, Al) atoms, and the small black circles to Si atoms. The black points indicate the position of the hydrogen ions (after Figueiredo, 1979 b).

Description

Chlorite is a sheet structure based on the packing of $Si_2^tO_5$ infinite sheets and Mg $(OH)_2$ sheets (brucite type layers).

References

Kostov (1968) 366.
Povarennykh (1972) 444.
Wyckoff (1968) Vol. 4, 358-361.
Zoltai (1975) IV-10.
Zoltai + Stout (1985) 323.
Liebau (1985) 218.
Figueiredo (1979 b).

CHLORITE (1M) (continued)

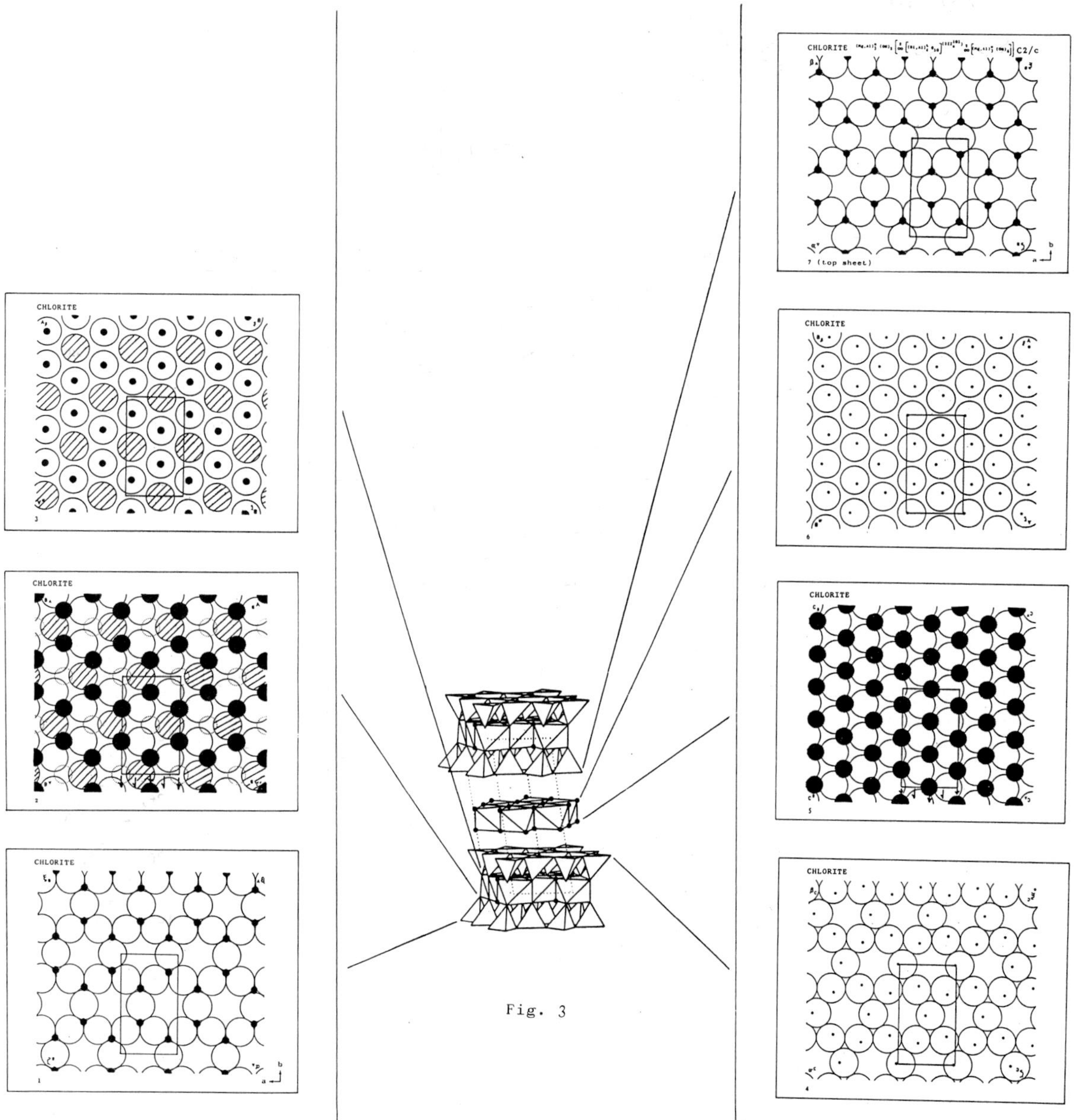

Fig. 3

Fig. 4

KAOLINITE (1Tc) C1

$$\overset{2}{\infty}\left[Al_2^{o}\,(OH)_4\;\overset{2}{\infty}\left[Si_2^{t}\,O_5\right]^{II^{101}_{c}}\right]$$

a = 5.14 Å
b = 8.93 Å
c = 7.37 Å
α = 91° 48'
β = 104° 30'
γ = 90° 0'
Z = 2

Atom		Coordinates
Al_I	(2a)	x = 0.502, y = 0.172, z = 0.003
Al_{II}	(2a)	x = 0.002, y = 0.330, z = 0.002
Si_2	(2a)	x = 0.800, y = 0.322, z = 0.382
Si_{II}	(2a)	x = 0.800, y = 0.000, z = 0.385
O_I	(2a)	x = 0.754, y = 0.315, z = 0.155
O_{II}	(2a)	x = 0.690, y = 0.004, z = 0.157
...		

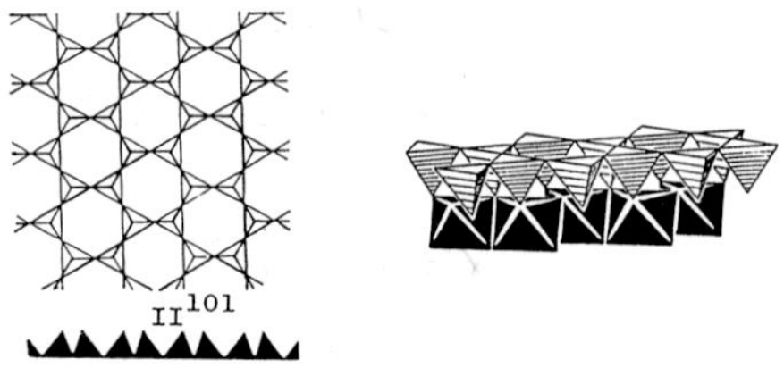

Fig. 1

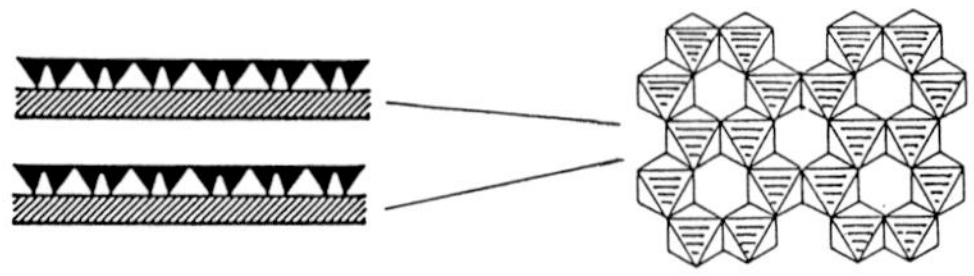

Dioctahedral

(a) (b)

Fig. 2

Properties

Habit	Cleav.	Fract.	Twin.	Hardn.	Dens.	Colour	Transp.
platy, massive	perfect (001)	flexible		2-2.5	2.6	white	translucent

Refr. index/Reflect.	Birefr.	Luster	Streak	Melt.p.	CPI
n_α = 1.556 n_β = 1.563 n_γ = 1.565	(-) 2V = 40°	dull	white		(SPI) 45

Distortion derivative

Dickite Cc

$$\overset{2}{\infty}\left[Al_2^{o}\,(OH)_4\;\overset{2}{\infty}\left[Si_2^{t}\,O_5\right]^{II^{101}_{c}}\right]$$

Figures

Fig. 1. Polyhedral representation of the structural units of the chlorite structure: $Si_2^tO_5$ infinite sheet and corresponding cross section, and $Al(OH)_2$ dioctahedral sheet. Both sheets form the structural unit (Liebau, 1985).

Fig. 2. Polyhedral description of: (a) the way the kaolinite structural units pack together (adapted from Liebau, 1985), and (b) the dioctahedral layer linked to the silicate sheets.

Fig. 3. Polyhedral description of the kaolinite structure (after Zoltai, 1975).

Fig. 4. Condensed model of the symmetrical analogue of the kaolinite structure. The large open circles represent oxygen atoms, and the hatched circles hidroxyls. The black medium circles correspond to Al atoms, and the small black circles to Si atoms. The black points indicate the position of the hydrogen ions (after Figueiredo (1979b).

Description

Kaolinite is a clay mineral . It is a sheet structure based on the packing of infinite silicate sheets, $Si_2^tO_5$, and $Al_2(OH)_3$ sheets. Both sheets form the kaolinite structural unit. 1Tc means triclinic (Tc) with one module per unit cell.

References

Kostov (1968) 377-378.
Povarennykh (1972) 431.
Wyckoff (1968) Vol. 4, 371-373.
Zoltai + Stout (1984) 321-326.
Liebau (1985) 214, 216, 220.
Figueiredo (1979b).
Zoltai (1975) IV-8.

Crystallographic data (continued)

Atom		Coordinates
O_{III}	(2a)	x = 0.791, y = 0.165, z = 0.482
O_{IV}	(2a)	x = 0.612, y = -0.120, z = 0.455
O_V	(2a)	x = 0.108, y = -0.058, z = 0.455
$(OH)_I$	(2a)	x = 0.778, y = 0.180, z = -0.140
$(OH)_{II}$	(2a)	x = 0.278, y = 0.320, z = -0.138
$(OH)_{III}$	(2a)	x = 0.316, y = -0.008, z = -0.136
$(OH)_{IV}$	(2a)	x = 0.248, y = 0.184, z = 0.155

KAOLINITE (1Tc) (continued)

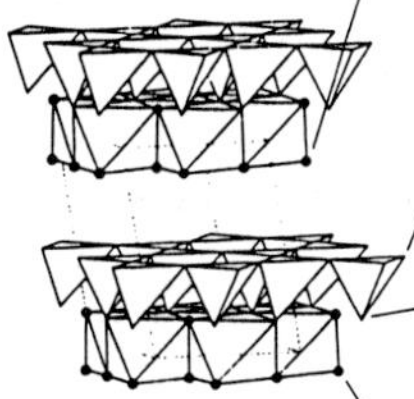

Fig. 3

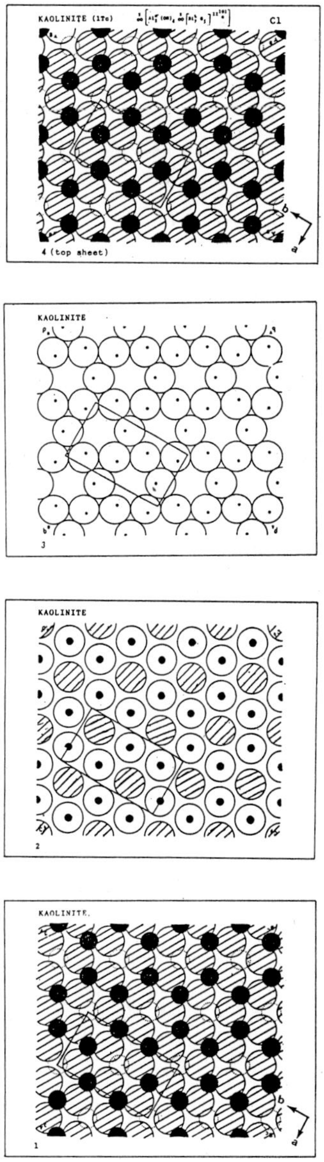

Fig. 4

CHRYSOTILE Cm

$$\overset{2}{\infty}\left[Mg_3^{o}(OH)_4 \overset{2}{\infty}\left[Si_2^{t}O_5\right]^{II^{101}}_{c}\right]$$

a = 14.65 Å
b = 9.25 Å
c = 5.34 Å
β = 93° 16'
Z = 2

Atom	Site	x	y	z
Mg_I	(2a)	0.227	0	0.873
Mg_{II}	(4b)	0.227	0.167	0.373
$(OH)_I$	(2a)	0.150	0	0.523
$(OH)_{II}$	(2a)	0.292	0	0.217
$(OH)_{III}$	(4b)	0.292	0.167	0.717
Si	(4b)	0.033	0.167	0.006
...				

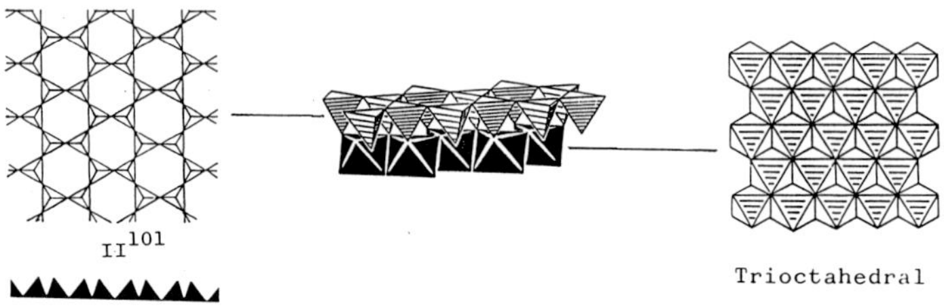

Fig. 1

Fig. 2

Properties

Habit	Cleav.	Fract.	Twin.	Hardn.	Dens.	Colour	Transp.
massive, fibrous	fibrous // [001]			2.5	≃ 2.55	yellow, white, grey	translucent

Refr. index/Reflect.	Birefr.	Luster	Streak	Melt.p.	CPI
	(-)	silky, greasy			

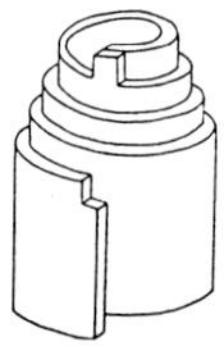

Fig. 3

Figures

Fig. 1. Silicate infinite sheet, $Si_2^tO_5$, and $Mg_3^o(OH)_4$ sheet, which form the chrysotile structural unit (after Liebau, 1985).

Fig. 2. Schematic representation of the way the chrysotile structural units pack together (adapted from Liebau, 1985).

Fig. 3. The manner of growth of chrysotile crystals (after Kostov, 1968).

Fig. 4. Spiral structure in cross-section of a chrysotile tube crystal (after Liebau, 1985).

Description

Chrysotile is a kind of distorted kaolinite structure. It is based on curved structural units, similar to kaolinite but with a trioctahedral sheet linked to the silicate sheet, instead of a dioctahedral one.

Fig. 4

Crystallographic data (continued)

Atom	Site	x	y	z
O_I	(2a)	-0.013	0	0
O_{II}	(4b)	0	0.250	0.250
O_{III}	(4b)	0.150	0.167	0.023

References

Kostov (1968) 374, 375.

Liebau (1985) 116, 214, 220, 224, 226.

Whittaker (1956) 855-862.

Deer et al. (1962) Vol. 3, 171.

Roberts et al. (1974) 131.

HALLOYSITE Cm

$$(H_2O)_2 \; {}^{2}_{\infty}\left[Al^{o}_{2}\,(OH)_4 \; {}^{2}_{\infty}\left[Si^{t}_{2}\,O_5\right]^{II^{101}_{c}}\right]$$

a = 5.20 Å	Al (4b)	x = 0.250 y = 0.167 z = 0	$(OH)_{II}$ (2a)	x =-0.046 y = 0 z = 0.118
b = 8.92 Å				
c =10.25 Å	Si (4b)	x = 0.0083 y = 0.167 z =-0.445	$(OH)_{III}$(4b)	x =-0.046 y = 0.333 z = 0.394
β = 100°				
Z = 2	$(OH)_{I}$(4b)	x =-0.046 y =-0.167 z = 0.118	$(OH)_{IV}$ (4b)	x =-0.046 y =-0.167 z =-0.118
			...	

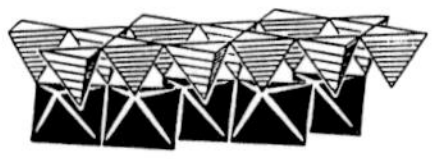

Fig. 1

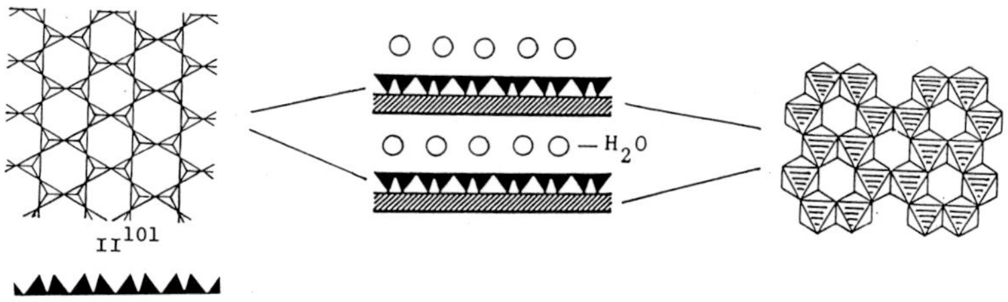

Fig. 2

Properties

Habit	Cleav.	Fract.	Twin.	Hardn.	Dens.	Colour	Transp.
microscopic tubular, mealy masses				2-2.5	2.0-2.2	colourless, white, yellowish	transparent to translucent

Refr. index/Reflect.	Birefr.	Luster	Streak	Melt.p.	CPI
		pearly, dull			

Fig. 1. The structural unit of halloysite (adapted from Liebau, 1985).

Fig. 2. Schematic representation of the way the structural modules pack together, and their constitution (adapted from Liebau, 1985).

Description

Halloysite is a sheet structure based on the packing of kaolinite structural modules, with water molecules occupying voids in between them.

Crystallographic data (continued)

$(OH)_V$	(2a)	x =-0.046 y = 0 z =-0.118
O_I	(2a)	x = 0.248 y = 0 z =-0.375
O_{II}	(4b)	x = 0.033 y = 0.250 z =-0.375

References

Kostov (1968) 379

Strukturbericht (1937) Vol. 3, 544, 545.

Roberts et al. (1974) 256.

Liebau (1985) 116, 214, 218.

ANTIGORITE Cm

$$Mg^{o}_{6}\,(OH)_{8}\;{}^{2}_{\infty}\left[Si^{t}_{4}\,O_{10}\right]$$

a = 5.32 Å	Mg_{I} (2a)	x = 0.270, y = 0, z = 0.161	Mg_{IV} (4b)	x = 0.025, y = 0.167, z = 0.661
b = 9.50 Å				
c = 14.90 Å	Mg_{II} (2a)	x = −0.475, y = 0, z = 0.661	Si_{I} (4b)	x = 0.008, y = 0.333, z = 0.364
β = 101° 54'	Mg_{III} (4b)	x = −0.230, y = 0.167, z = 0.161	Si_{II} (4b) ...	x = 0.008, y = 0.333, z = 0.864
Z = 2				

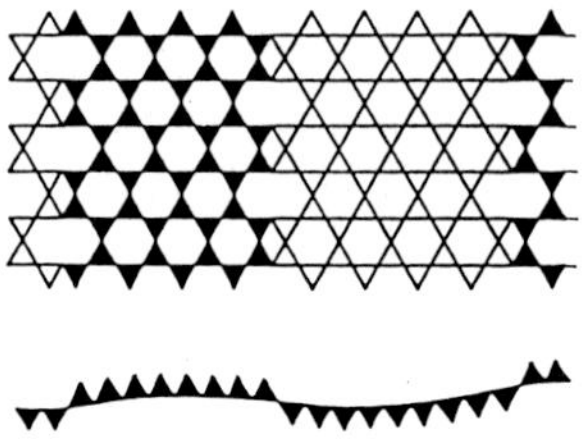

Fig. 1

Fig. 2

Properties

Habit	Cleav.	Fract.	Twin.	Hardn.	Dens.	Colour	Transp.
platy, massive	perfect (001)	flexible		3-4	2.6	green yellow	translucent

Refr. index/Reflect.	Birefr.	Luster	Streak	Melt.p.	CPI
n_{α} = 1.56, n_{β} = 1.57, n_{γ} = 1.57	(-) 2V = 20°-60°	resinous, silky	white		(SPI) 48

Figures

Fig. 1. Structural unit of antigorite: the $Si^{t}_{2}O_{5}$ infinite puckered silicate sheet and its cross-section (after Liebau, 1985).

Fig. 2. Schematic drawing of the way the silicate sheets pack together (adapted from Liebau, 1985).

Description

Palygorskite is a sheet structure formed by the packing of puckered infinite silicate sheets of composition $Si^{t}_{2}O_{5}$, with brucite type layers located in between the silicate sheets. It could also be considered as a complex framework of structural formula:

$${}^{3}_{\infty}\left[Mg^{o}_{6}(OH)_{8}\;{}^{2}_{\infty}\left[Si^{t}_{4}O_{10}\right]\right]$$

References

Kostov (1968) 374,375.

Povarennykh (1972) 431.

Liebau (1985) 220-224.

Zoltai + Stout (1984) 322.

Structure Reports (1954), 13, 380.

Crystallographic data (continued)

Atom	Site	x	y	z
$(OH)_{I}$	(2a)	−0.046	0	0.262
$(OH)_{II}$	(2a)	−0.046	0	0.762
$(OH)_{III}$	(4b)	0.046	0.167	0.100
$(OH)_{IV}$	(4b)	0.250	0.333	0.600
$(OH)_{V}$	(2a)	−0.455	0	0.100
$(OH)_{VI}$	(2a)	0.250	0	0.600
O_{I}	(4b)	−0.046	0.333	0.252
O_{II}	(4b)	−0.046	0.333	0.752
O_{III}	(2a)	0.533	0	0.405
O_{IV}	(2a)	0.533	0	0.905
O_{V}	(4b)	0.283	0.250	0.405
O_{VI}	(4b)	0.283	0.250	0.905

PALYGORSKITE

A 2/m

$Mg_5^o (H_2O)_8 (OH)_2 \overset{2}{\infty} \left[Si_8^t O_{20}\right]$

a = 5.2Å
b = 18.0Å
c = 13.4Å
β = 90°-93°
Z = 2

Fig. 1

Fig. 2

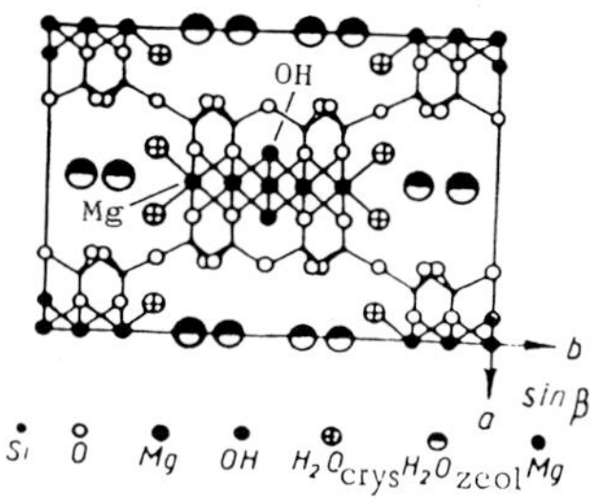

Fig. 3

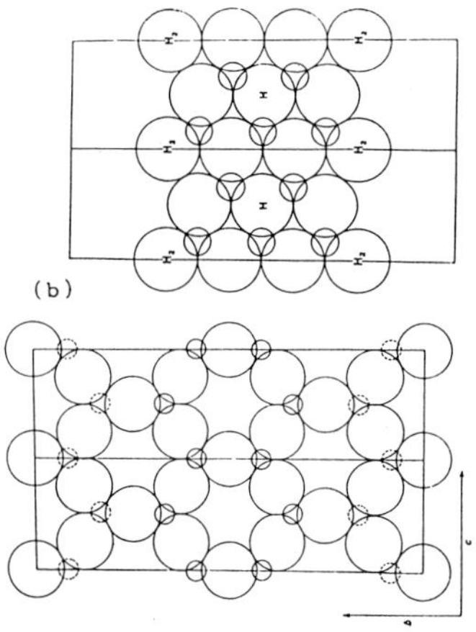

Fig. 4

Properties

Habit	Cleav.	Fract.	Twin.	Hardn.	Dens.	Colour	Transp.
fibrous, platy	good (110)	subconchoidal		2-2.5	2.2	white, grey	transparent to translucent

Refr. index /Reflect.	Birefr.	Luster	Streak	Melt.p.	CPI
n_α = 1.52 n_β = 1.53 n_γ = 1.55	(-) 2V = 0°-60°	vitreous, dull	white		(SPI) 16

Figures

Fig. 1. Structural unit of palygorskite, the $Si_2^tO_5$ infinite puckered silicate sheet (after Liebau, 1985).

Fig. 2. Schematic drawing of the way the silicate sheets pack together (adapted from Liebau, 1985) and polyhedral representation of the octahedral infinite chain which exist in between the silicate sheets.

Fig. 3. Structure of palygorskite projected along the c axis (after Povarennykh, 1972).

Fig. 4. Structure of palygorskite viewed along a normal to (001): (a) arrangement of silicons with respect to continuous oxygen sheet; (b) octahedral chain (adapted from Caillère + Hénin, 1961).

Description

Palygorskite is a sheet structure formed by the packing of puckered infinite silicate sheets of composition $Si_2^tO_5$, with Mg atoms, OH and H_2O within the interstices. The infinite silicate sheets are formed by amphibole infinite chains linked together in a reverse and alternate way. Certain water molecules are bounded to the extremities of the octahedral chains, and another possible structural description of the palygorskite structure would be to consider it as a framework with structural formula:

$$(OH_2)_4 \overset{3}{\infty} \left[Mg_5^o(OH_2)_4(OH)_2 \overset{2}{\infty} \left[Si_8^tO_{20}\right]\right]$$

References

Kostov (1968) 352.
Povarennykh (1972) 420.
Liebau (1985), 217, 219.
Caillère+Hénin (1961) 347.

APOPHYLLITE

P4/mnc

$Ca_4^{[7]} K^{[8]} F(H_2O)_8 \; {}^{2}_{\infty}\left[Si^{t}_{8} O_{20} \right] IV^{601}_{s}$

a = 9.00 Å
c = 15.8 Å
Z = 2

Atom	x	y	z
Ca(8h)	0.120	0.243	0
K (2b)	0	0	1/2
F (2a)	0	0	0
(H_2O)(16i)	0.237	0.445	0.094
O_I (8g)	0.138	0.638	1/4
O_{II} (16i)	0.089	0.184	0.217
...			

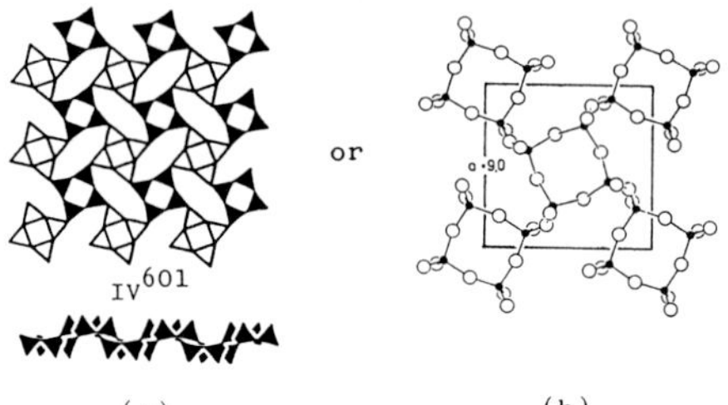

(a) Fig. 1 (b)

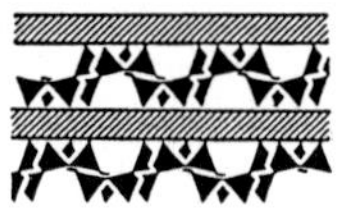

Fig. 2

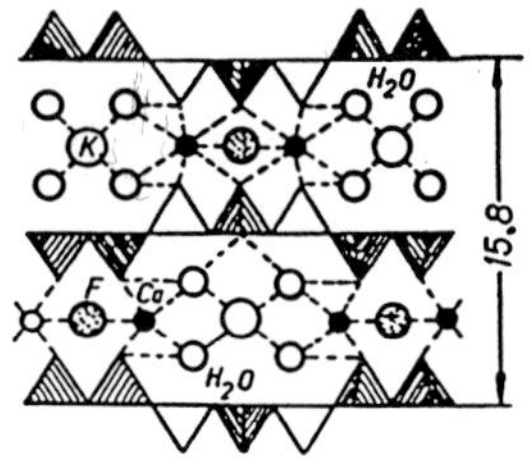

Fig. 3

Properties

Habit	Cleav.	Fract.	Twin.	Hardn.	Dens.	Colour	Transp.
short prismatic, tabular	perfect (001)	uneven		4.5-5	2.3	white, grey	transparent to translucent

Refr. index/Reflect.	Birefr.	Luster	Streak	Melt.p.	CPI
n_ω = 1.535 n_ϵ = 1.537	(+)	pearly	white		(SPI) 31

Figures

Fig. 1. Structural unit of apophyllite: the $Si^t_2O_5$ infinite puckered silicate sheet and its cross-section: (a) polyhedral drawing (adapted from Liebau, 1985), and (b) ball and spoke description (Ramdhor + Strunz, 1980).

Fig. 2. Schematic representation of the way the silicate sheets pack together (adapted from Liebau, 1985).

Fig. 3. Structure of apophyllite projected on (110) (after Povarennykh, 1972).

Description

Apophyllite is a sheet structure formed by the packing of puckered infinite silicate sheets of composition $Si^t_2O_5$, with Ca, K, and water molecules located in the interstices. It could also be considered as a complex framework of structural formula:

$${}^{3}_{\infty}\left[Ca_4^{[7]} K^{[8]} F(H_2O)_8 \; {}^{2}_{\infty}\left[Si^{t}_{8} O_{20} \right] \right]$$

References

Kostov (1968) 382.
Povarennykh (1972) 433.
Wyckoff (1968) Vol. 4, 393,394.
Zoltai + Stout (1984) 325
Liebau (1985) 221.
Ramdhor + Strunz (1980) 739.

Crystallographic data (continued)

Atom	x	y	z
O_{III}(16i)	0.287	0.117	0.094
Si(16i)	0.237	0.091	0.188

PREHNITE

Pncm

$Ca_2^{[7]} Al^{o} (OH)_2 \; {}^{2}_{\infty}\left[Si^{t}_{3} Al^{t} O_{10}\right]$

a = 4.646 Å	Ca(4e) x = 0, y = 1/2, z = 0.0992	(Si,Al)(4g) x = 0.1895, y = 1/4, z = 1/4
b = 5.483 Å	Al(2a) x = 0, y = 0, z = 0	O_I (8i) x = 0.7511, y = 0.1323, z = 0.0739
c = 18.486 Å		
Z = 2	Si(4f) x = 1/2, y = 0, z = 0.1195	O_{II} (8i) x = 0.3686, y = 0.2130, z = 0.1716
	...	

Fig. 1

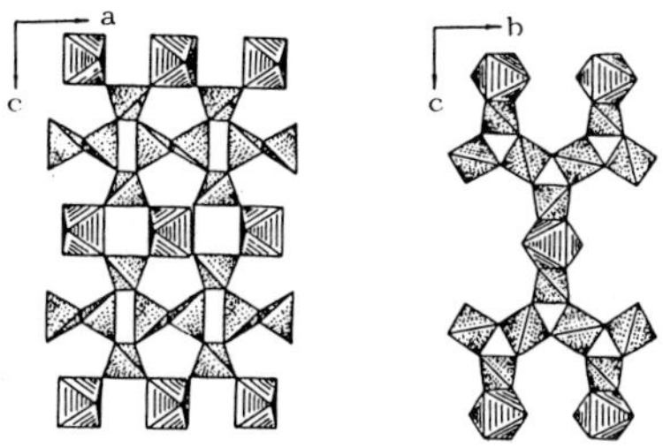

Fig. 2

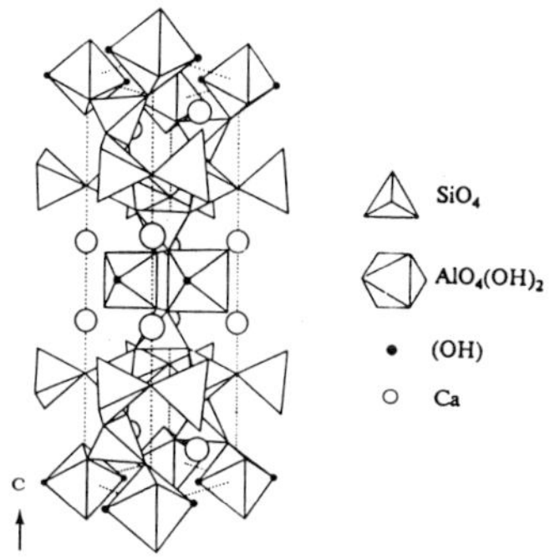

Fig. 3

Properties

Habit	Cleav.	Fract.	Twin.	Hardn.	Dens.	Colour	Transp.
tabular, prismatic, globular, reniform	good (001)	uneven		6-6.5	2.9	pale green	transparent to translucent

Refr. index /Reflect.	Birefr.	Luster	Streak	Melt.p.	CPI
n_α = 1.625, n_β = 1.635, n_γ = 1.655	(+) 2V = 65°-70°	vitreous	white		(SPI) 47

Figures

Fig. 1. Structural unit of prehnite: the $Si^t_3Al^tO_{10}$ infinite silicate sheet and its cross-section (after Liebau, 1985).

Fig. 2. Polyhedral drawing of the prehnite structure showing the way the silicate sheets pack together, in two projections. The calcium atoms are omited (adapted from Povarennykh, 1972).

Fig. 3. Polyhedral representation of the prehnite structure (after Papike + Zoltai, 1967).

Description

Prehnite is a sheet structure formed by the packing of infinite silicate sheets of composition $\left[Si^t_2(Si,Al)^t_2O_{10}\right]$, with Al, Ca, and OH located in the interstices.

Crystallographic data (continued)

O_{III}(4e)	x = 0, y = 0, z = 0.2687
OH (4h)	x = 0.2054, y = 0.3018, z = 0

References

Kostov (1968) 313.

Povarennykh (1972) 439, 440.

Zoltai + Stout (1984) 325, 331.

Liebau (1985) 120.

Papike + Zoltai (1967) 974-984.

URANOPHANE

P 2_1

$$^{[8]}Ca(H_2O)_3(H_2O)_3 \; ^{2}_{\infty}\left[^{[2+5]}(UO_2)_2(Si^{t}O_4)_2 \right]$$

a = 15.97 Å
b = 7.07 Å
c = 6.68 Å
β = 97° 18'
Z = 2

Atom	Position	x	y	z
Ca	(2a)	0.019	0.750	0.271
U_I	(2a)	0.256	0.787	0.138
U_{II}	(2a)	0.756	0.713	0.138
Si_I	(2a)	0.282	0.287	0.335
Si_{II}	(2a)	0.782	0.213	0.335
$(H_2O)_I$	(2a)	0.179	-	0.646

...

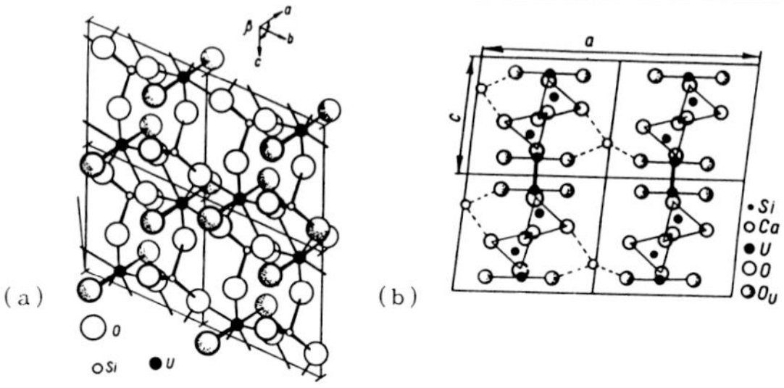

Fig. 1

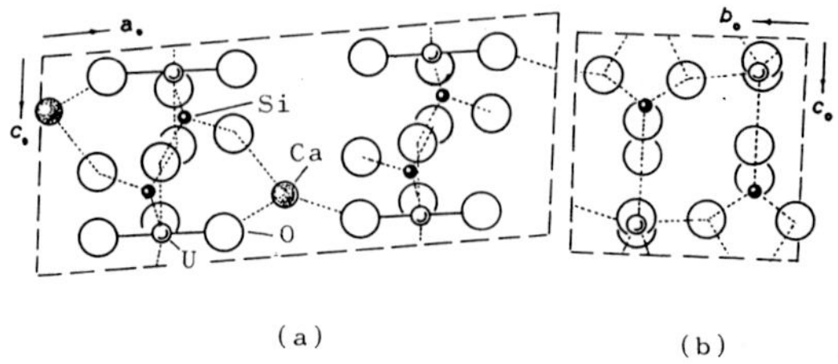

Fig. 2

Properties

Habit	Cleav.	Fract.	Twin.	Hardn.	Dens.	Colour	Transp.
prismatic, radiated, aggregations	perfect (100)			≃ 2.5	3.83	yellow, yellowish green	transparent to translucent

Refr. index /Reflect.	Birefr.	Luster	Streak	Melt.p.	CPI
n_α = 1.642-1.648 n_β = 1.661-1.667 n_γ = 1.667-1.675	(-) 2V = 32°	vitreous, greasy, pearly			

Figures

Fig. 1. (a) General view of the structure of uranophane, and (b) projection on (010) (after Povarennykh, 1972).

Fig. 2. Structure of uranophane (a) projected on (010), and (b) on (100) (after Smith et al., 1957, quoted by Kostov, 1968).

Description

Uranophane is a sheet structure. The sheets are formed by $[(UO_2)_2(SiO_4)_2]^{4-}$, are parallel to (100) and linked by Ca atoms and water molecules. This structure explains the perfect (100) cleavage.

References

Kostov (1968) 327.
Povarennykh (1972) 455, 456.
Wyckoff (1968) Vol. 4, 208-210.
Smith et al. (1957) 594-618.
Roberts et al. (1974) 641.

Crystallographic data (continued)

Atom	Position	x	y	z
$(H_2O)_{II}$	(2a)	0.317	-	0.542
$(H_2O)_{III}$	(2a)	0.264	-	0.771
$(H_2O)_{IV}$	(2a)	0.243	-	0.030
$(H_2O)_V$	(2a)	0.280	-	0.133
O_I	(2a)	0.374	0.787	0.138
O_{II}	(2a)	0.874	0.713	0.138
O_{III}	(2a)	0.136	0.787	0.129
O_{IV}	(2a)	0.636	0.713	0.129
O_V	(2a)	0.257	0.477	0.197
O_{VI}	(2a)	0.757	0.403	0.197
O_{VII}	(2a)	0.257	0.097	0.197
O_{VIII}	(2a)	0.757	0.023	0.197
O_{IX}	(2a)	0.231	0.287	0.533
O_X	(2a)	0.731	0.213	0.533
O_{XI}	(2a)	0.381	0.287	0.433
O_{XII}	(2a)	0.881	0.213	0.433

AUTUNITE   I4/mmm

$(H_2O)_{10} \left[{}^{[6]}Ca \; {}^{2}_{\infty} \left[{}^{[2+4]}UO_2 {}^{t}PO_4 \right]_2 \right]$

a = 6.989 Å
c = 20.63 Å
Z = 2

Ca(2a)
P (4d)
U (4e) u = 0.208
(Positions not given for oxygen atoms or water molecules)

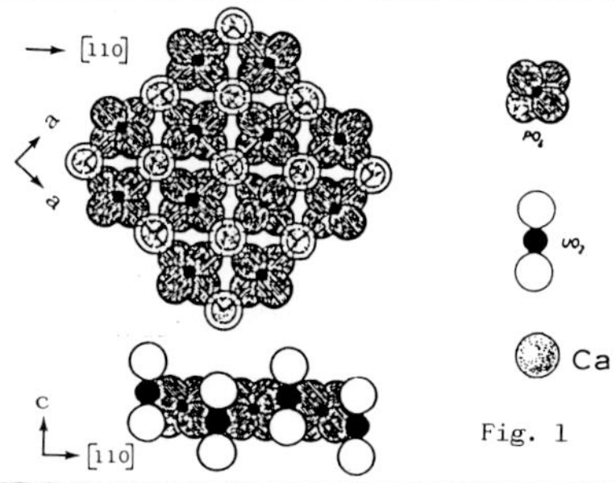

Fig. 1

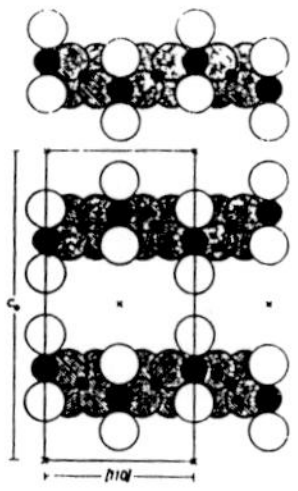

Fig. 2

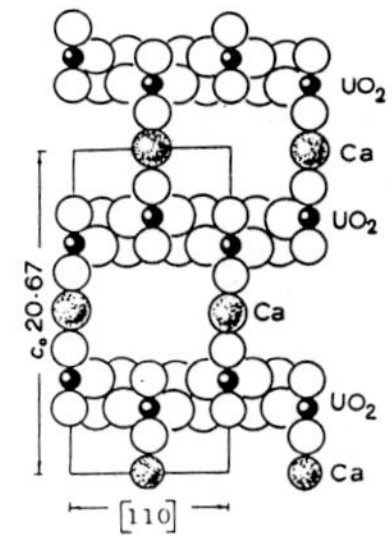

Fig. 3

Properties

Habit	Cleav.	Fract.	Twin.	Hardn.	Dens.	Colour	Transp.
tabular, foliated	perfect (001)	uneven		2-2.5	3.15	yellow	transparent to translucent

Refr. index/Reflect.	Birefr.	Luster	Streak	Melt.p.	CPI
n_ω = 1.577 n_ϵ = 1.553	(-)	adamantine	yellow		(SPI) 32

Population

Torbernite

$(H_2O)_{10} \left[{}^{[6]}Cu \; {}^{2}_{\infty} \left[{}^{[2+4]}UO_2PO_4 \right]_2 \right]$

Description

Autunite is a sheet structure based on the packing of UO_2PO_4 sheets and Ca atoms, with some water molecules within the interstices. The perfect cleavage parallel to (001) is in agreement with this structural description.

Figures

Fig. 1. Structural units of the autunite structure: the UO_2PO_4 sheet, and Ca atoms which are also structural units because they are part of the packing. The corresponding cross-section of the UO_2PO_4 sheet is also represented (adapted from Ramdohr + Strunz, 1980).

Fig. 2. Packing drawing of the autunite structure only showing the way the UO_2PO_4 sheets pack together (after Ramdohr + Strunz, 1980).

Fig. 3. Packing drawing of the autunite structure. The water molecules are omitted (after Kostov, 1968).

References

Kostov (1968) 477-479.
Povarennykh (1972) 555,556.
Wyckoff (1965) Vol. 3, 869,870.
Zoltai + Stout (1984) 453.
Ramdohr + Strunz (1980) 653-655.

META-AUTUNITE

P4/nmm

$(H_2O)_6 \left[\overset{[6]}{Ca} \; \overset{2}{\infty} \left[\overset{[2+4]}{UO_2} \; \overset{t}{PO_4} \right] \right]_2$

a = 6.980 Å
c = 8.420 Å
Z = 1

Only one Ca atom statistically distributed over (2c) u = 0.612.
Six water molecules statistically distributed among (8j) u = 0.486 v = 0.392

U (2c) u = 0.106
O_I (2c) u = 0.343
O_{II} (2c) u = 0.893
O_{III}(8i) u = 0.584 v = 0.106
P(2a)

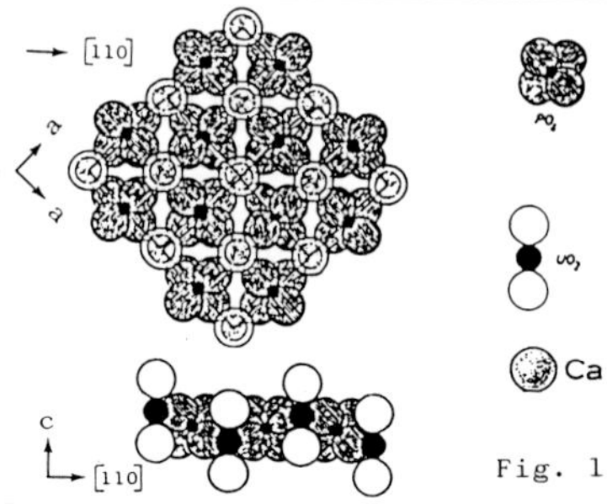

Fig. 1

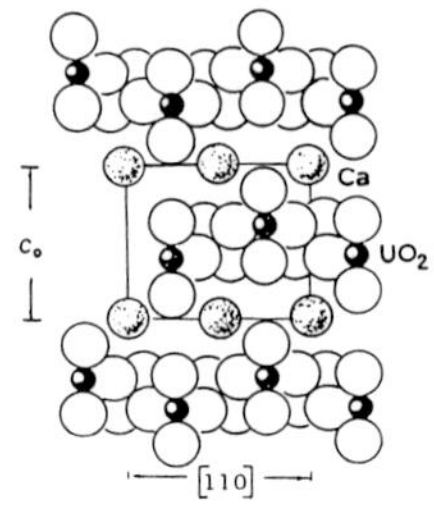

Fig. 2

Properties

Habit	Cleav.	Fract.	Twin.	Hardn.	Dens.	Colour	Transp.
tabular, platy	perfect (001)			2-2.5	3.45--3.55	lemon yellow, greenish yellow	translucent to opaque

Refr. index/Reflect.	Birefr.	Luster	Streak	Melt.p.	CPI
n_α = 1.585-1.600 n_β = 1.595-1.610 n_γ = 1.595-1.613	(-)	pearly, dull			

Figures

Fig. 1. Structural units of the meta-autunite structure: the UO_2PO_4 sheets, and Ca atoms which are also structural units because they are part of the packing. The corresponding cross-section of the UO_2PO_4 sheet is also represented (adapted from Ramdohr + Strunz. 1980).

Fig. 2. Packing drawing of the meta--autunite structure, showing the way the sheets pack together. The water molecules are omited (after Kostov, 1968).

Description

Meta-autunite is a sheet structure based on the packing of UO_2PO_4 sheets and Ca atoms, with some water molecules within the interstices. Meta-autunite is the result of dehydration of autunite. The perfect cleavage parallel to (001) is in agreement with this structural description.

References

Kostov (1968) 477.
Povarennykh (1972) 556, 557
Wyckoff (1965) Vol. 3, 869-871.
Ramdohr + Strunz (1980) 653-655.
Roberts et al. (1974) 395, 396.

CARNOTITE

P $2_1/a$

$K_2(H_2O)_3 \overset{2}{\infty}\left[\left[\overset{[2+5]}{UO_2}\right]_2\left[\overset{[5]}{V_2O_8}\right]\right]$

a = 10.47 Å
b = 8.41 Å
c = 6.59 Å
β = 103° 50'
Z = 2

Atomic positions for the Cs analogue of anhydrous carnotite.

Cs(4e) x = 0.6358, y = 0.7681, z = 0.4576

U(4e) x = 0.6813, y = 0.4795, z = 0.9873

V(4e) x = 0.9440, y = 0.6495, z = 0.8959

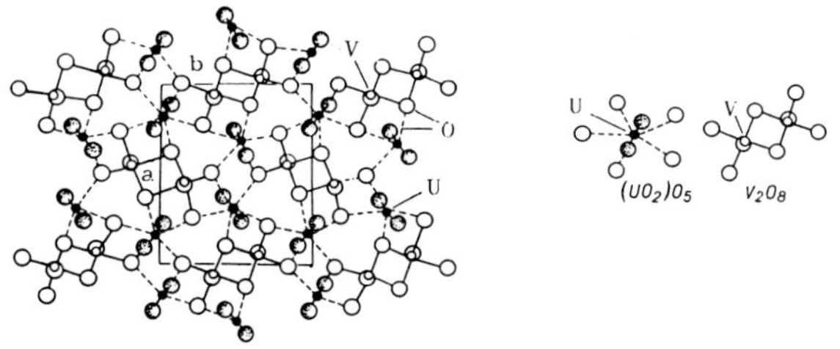

Fig. 1

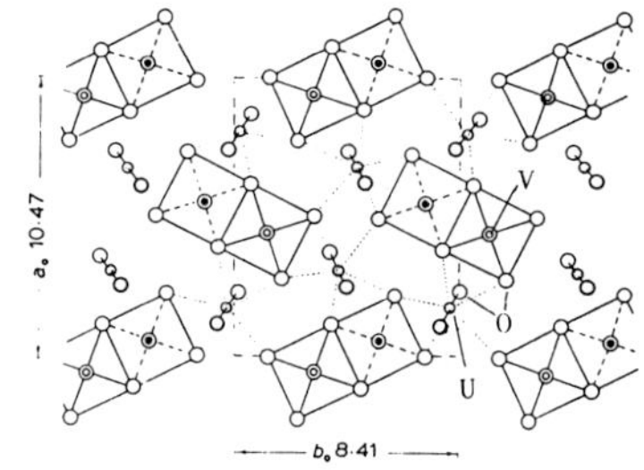

Fig. 2

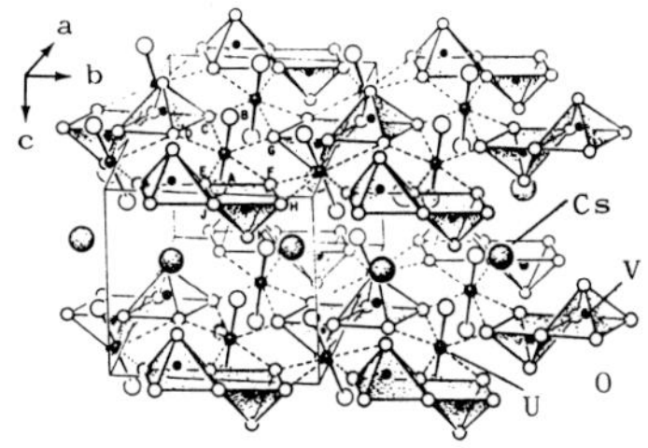

Fig. 3

Properties

Habit	Cleav.	Fract.	Twin.	Hardn.	Dens.	Colour	Transp.
fine powder	perfect (001)			2	4.5	yellow, green	translucent

Refr. index/Reflect.	Birefr.	Luster	Streak	Melt.p.	CPI
n_α = 1.75 n_β = 1.92 n_γ = 1.95	(-) 2V = 38°-44°	dull, earthy	yellow		(SPI) 30

Figures

Fig. 1. $\left[\left[UO_2\right]_2\left[V_2O_8\right]\right]$ sheet of the anhydrous carnotite structure, and coordination polyhedra of U atoms and V atoms (after Povarennykh, 1972).

Fig. 2. $\left[\left[UO_2\right]_2\left[V_2O_8\right]\right]$ anhydrous carnotite sheet parallel to (001) (after Kostov, 1968).

Fig. 3. Representation of the cesium analogue of the anhydrous carnotite structure (after Appleman + Evans, 1965).

Description

Carnotite is formed by $\left[\left[UO_2\right]_2\left[V_2O_8\right]\right]$ sheets laying parallel to (001), with K atoms and water molecules located in between these sheets.

Crystallographic data (continued)

Atom	x	y	z
O_I (4e)	0.6238	0.4382	0.7415
O_{II} (4e)	0.7406	0.5226	1.2330
O_{III} (4e)	0.5098	0.6525	0.9667
O_{IV} (4e)	0.9347	0.6342	0.6603
O_V (4e)	0.8953	0.4424	0.9820
O_{VI} (4e)	0.7786	0.7077	0.9418

References

Kostov (1968) 480.
Povarennykh (1972) 167, 503.
Zoltai + Stout (1984) 453.
Appleman + Evans (1965) 825-842.

GYPSUM A 2/n

$^{2}_{\infty}\left[\overset{[6+2]}{Ca}(H_2O)_2\, S^{t}O_4\right]$

a = 5.68 Å
b = 15.18 Å
c = 6.52 Å
β = 118°23'
Z = 4

(described in relation to I2/c)

Atom	Position	Coordinates
Ca	(4e)	u = 0.4200
O_I	(8f)	x = 0.9665, y = 0.1352, z = 0.5506
O_{II}	(8f)	x = 0.7602, y = 0.0220, z = 0.6735
O_{III}	(8f)	x = 0.3797, y = 0.1831, z = 0.4500
S	(4e)	u = -0.0782
H_I	(8f)	x = 0.236, y = 0.165, z = 0.4890

...

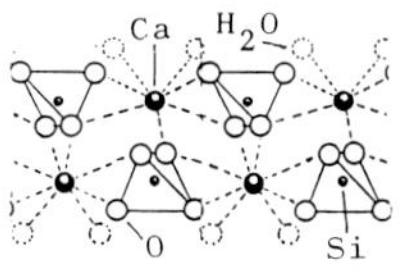

Fig. 1

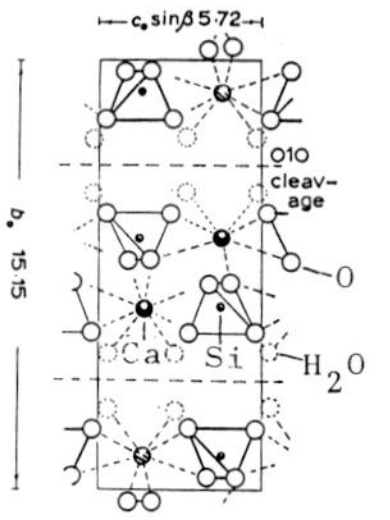

Fig. 2

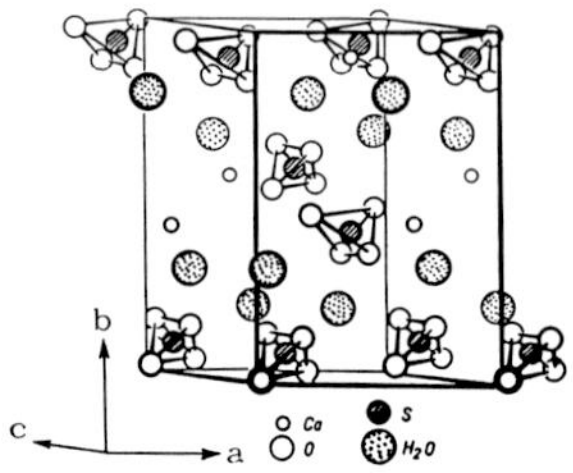

Fig. 3

Properties

Habit	Cleav.	Fract.	Twin.	Hardn.	Dens.	Colour	Transp.
tabular, variable	perfect (010) (100)	conchoidal	(100)	2	2.32	colourless, variable	transparent to translucent

Refr. index / Reflect.	Birefr.	Luster	Streak	Melt.p.	CPI
n_α = 1.520	(+)	vitreous,	white		(SPI)
n_β = 1.523	2V = 58°	pearly			44
n_γ = 1.529					

Population

Churchite

$^{2}_{\infty}\left[\overset{[6+2]}{Y}(H_2O)_2\, P^{t}O_4\right]$

Figures

Fig. 1. The double sheet of the gypsum structure, which is formed by sulphate groups, S^tO_4, calcium atoms and water molecules (adapted from Kostov, 1968).

Fig. 2. Polyhedral description of the gypsum structure ptojected along the a axis (after Kostov, 1968).

Fig. 3. Perspective of the gypsum structure (after Povarennykh, 1972).

Description

The gypsum structure is built of double sheets formed by sulphate groups, S^tO_4, linked to calcium atoms and water molecules. Ca atoms are irregularly coordinated by six oxygen atoms and two water molecules. The perfect cleavage parallel to (010) is in agreement with this structural description.

Crystallographic data (continued)

H_{II} (8f) x = 0.410, y = 0.246, z = 0.488

References

Kostov (1968) 506.
Povarennykh (1972) 178, 557.
Wyckoff (1965) Vol. 3, 642-644.
Zoltai + Stout (1984) 442.

TRONA C2/c

$$\overset{2}{\infty}\left[Na_3^{o/p}\,H\,(H_2O)_2\,\{g\}\left[\overset{tr}{C}\,O_3\right]_2\right]$$

a = 20.346 Å
b = 3.49 Å
c = 10.296 Å
β = 106° 26'
Z = 4

Atom	Position	x	y	z
Na_I	(4e)	0	0.748	1/4
Na_{II}	(8f)	0.159	0.165	0.4260
C	(8f)	0.0932	0.262	0.1040
O_I	(8f)	0.1514	0.373	0.1018
O_{II}	(8f)	0.0547	0.139	0.9892
O_{III}	(8f)	0.0720	0.257	0.2070

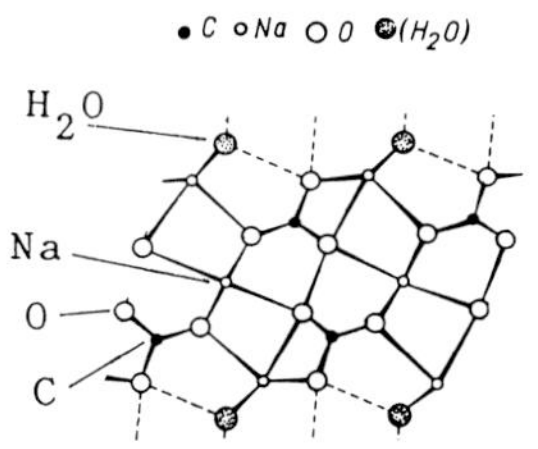

Fig. 1

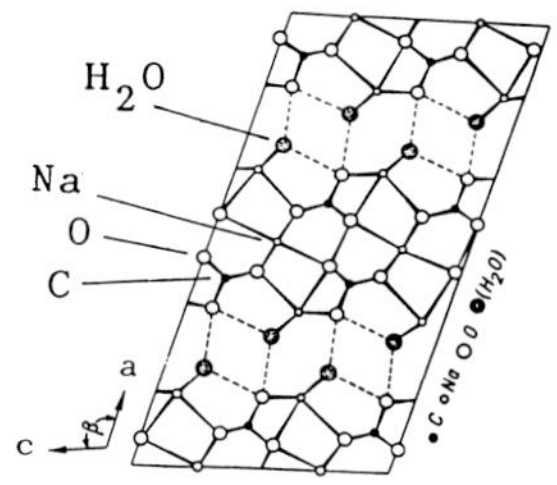

Fig. 2

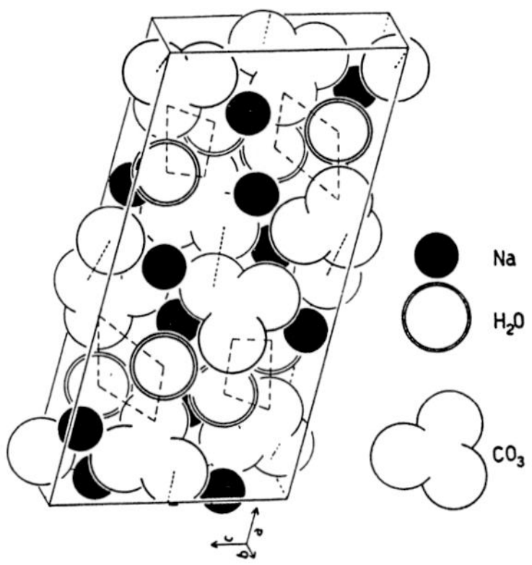

Fig. 3

Properties

Habit	Cleav.	Fract.	Twin.	Hardn.	Dens.	Colour	Transp.
tabular, platy, massive	perfect (100)			2.5-3	2.11	colourless, greyish	transparent to translucent

Refr. index/Reflect.	Birefr.	Luster	Streak	Melt.p.	CPI
n_α = 1.412 n_β = 1.492 n_γ = 1.540	(-) 2V ≃ 76°				

Figures

Fig. 1. Cross-section of the sheet of the trona structure, which consists of CO_3 triangular groups linked to Na atoms and water molecules (adapted from Povarennykh, 1972).

Fig. 2. Ball and spoke representation of the trona structure projected along the b axis (after Povarennykh, 1972).

Fig. 3. Packing drawing of the trona structure (after Brown et al., 1949, quoted in Structure Reports, 1952, Vol. 12).

Description

Trona is formed by sheets parallel to (100) consisting of CO_3 triangular groups, and Na atoms in octahedral and prismatic coordination, and water molecules. Between the sheets there are hydroxyl-hydrogen bonds. The perfect cleavage parallel to (100) is in agreement with this structural description.

Crystallographic data (continued)

Atom	Position	x	y	z
O_{IV}	(8f)	0.2120	0.669	0.3542
H_I	(4a)	0	0	0
H_{II}	(8f)	0.196	—	0.272
H_{III}	(8f)	0.257	—	0.385

References

Kostov, 529.
Povarennykh, 626, 627.
Structure Reports (1952) 12, 238-240.
Structure Reports (1963) 20, 389-392.
Brown et al. (1949) 167.
Bacon + Curry (1956) 82-85.
Roberts et al. (1974) 630.

8.4.5. Framework structures

DIAMOND

Fd3m

$^{3}_{\infty}\left[C^{t}\right]^{\tilde{H}_c}$

a = 3.56679 Å (at 20°C) C (8a)

Z = 8

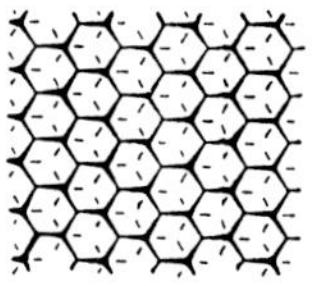

$I^{102}:\tilde{H}$

Fig. 1

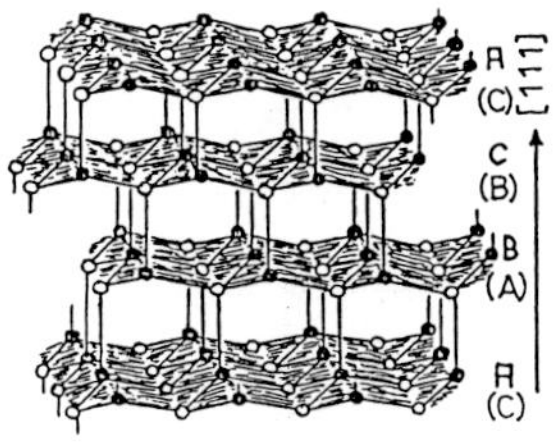

Fig. 2

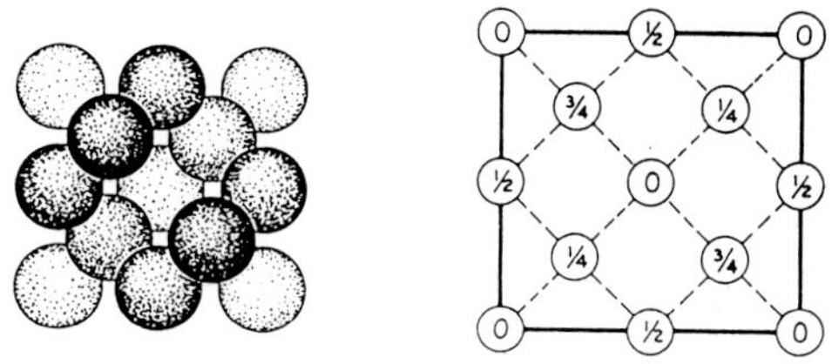

(a) (b)

Fig. 3

Properties

Habit	Cleav.	Fract.	Twin.	Hardn.	Dens.	Colour	Transp.
octahedral	perfect (111)	conchoidal	(111)	10	3.5	colourless	transparent

Refr. index/Reflect.	Birefr.	Luster	Streak	Melt.p.	CPI
n = 2.419		adamantine	white		(SPI) 30

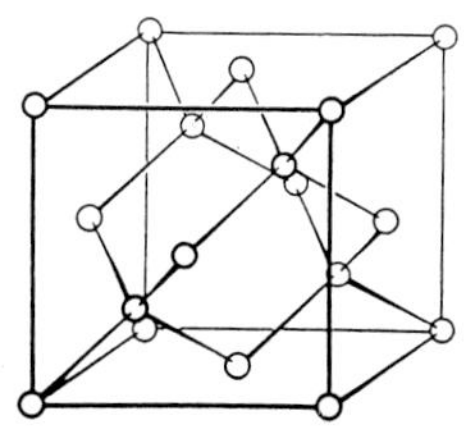

Fig. 4

Polytypes

Lonsdaleite $^{3}_{\infty}\left[C^{t}\right]^{\tilde{H}h}$

Figures

Fig. 1. Corrugated honeycombed layer of carbon atoms, perpendicular to a cube diagonal [111], which constitutes the connected unit of the diamond structure.

Fig. 2. Representation of the diamond structure showing the way the connected units are linked together in a ABC, or c, sequence (after Krebs, 1968).

Fig. 3. (a) Packing drawing of the diamond structure, and (b) corresponding unit cell content projected along an a axis (after Wyckoff, 1963, Vol. 1).

Fig. 4. Ball and spoke model of the diamond structure (after Povarennykh, 1972).

Fig. 5. Representation of the diamond structure sowing the location of the cleavage plane (111) parallel to the corrugated honeycombed carbon layers (after Zemann, 1969).

Description

The diamond structure is a typycal framework structure, with channels along [110].
The carbon atoms are all interlinked, and have a tetrahedral coordination.

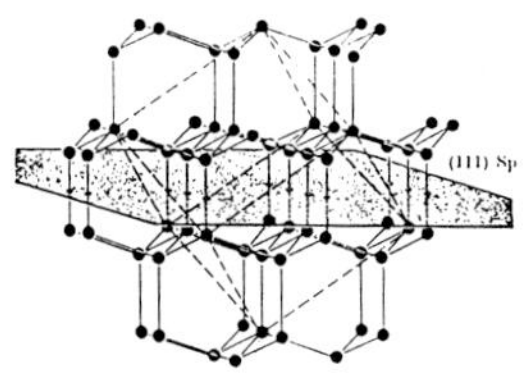

Fig. 5

References

Kostov (1968) 101, 102.
Povarennykh (1972) 118.
Wyckoff (1963) Vol. 1, 25,26.
Zoltai + Stout (1984) 378.
Krebs (1968) 153.
Zemann (1969) 109.

TIN (white)

I4/amd

$^{3}_{\infty}\left[Sn^{t}\right]\tilde{H}_{c}$

a = 5,8197 Å (at 25°C)
c = 3.17488 Å
Z = 4

Sn (4a)

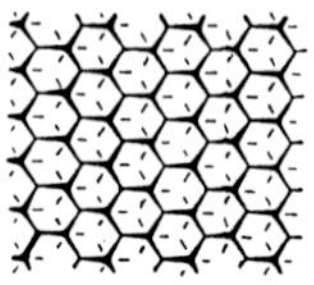

$I^{102} = \tilde{H}$

Fig. 1

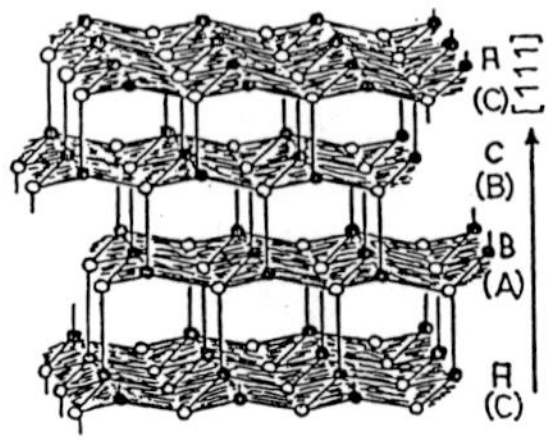

Fig. 2

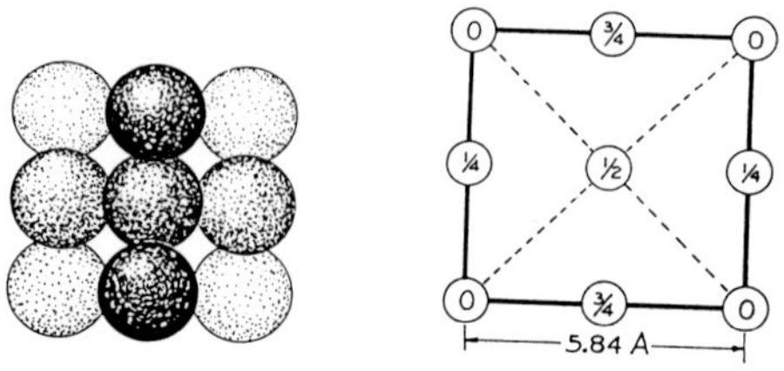

(a) (b)

Fig. 3

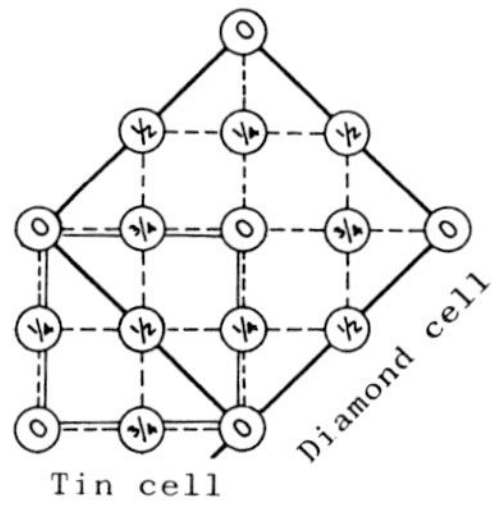

Fig. 4

Properties

Habit	Cleav.	Fract.	Twin.	Hardn.	Dens.	Colour	Transp.
embedded grains, rounded grains	not distinct	hackly		2	7.3	tin- -white	opaque

Refr. index /Reflect.	Birefr.	Luster	Streak	Melt.p.	CPI
		metallic		232°C	

Figures

Fig. 1. Corrugated honeycombed layer of tin atoms, which forms the connected unit of the tin structure.

Fig. 2. Schematic representation of the tin structure showing the way the connected únits are linked together in a sequence ABC (or c) sequence, like in diamond (after Krebs, 1968).

Fig. 3. (a) Packing drawing of the tin structure, and (b) corresponding unit cell content projected along the c axis (after Wyckoff, 1963, Vol. 1).

Fig. 4. Relation between the tin and the diamond units cells (adapted from Wyckoff, 1963, Vol. 1).

Description

The tin structure is a framework structure, where the tin atoms are all interlinked, with a tetrahedral coordination, and may be considered as a distorted diamond structure.

References

Kostov (1968) 91.
Povarennykh (1972) 194.
Wyckoff (1963) Vol. 1, 28, 29.
Roberts et al. (1974) 618, 619.
Ingerson (1955) 350.

β - CRISTOBALITE

(high-temperature form)

Fd3m

$\overset{3}{\infty}\left[Si^{t\,\ll c \gg}O_2^{[2]}\right]\ II_c^{201}$

a = 7.16 Å (at 290°C)
Z = 8

Si(8a)
O(16c)

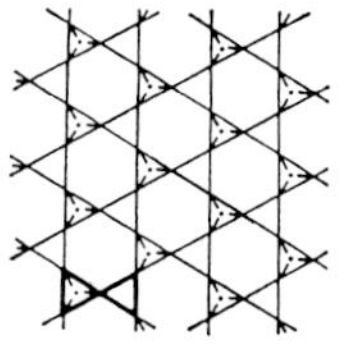

II^{201}

Fig. 1

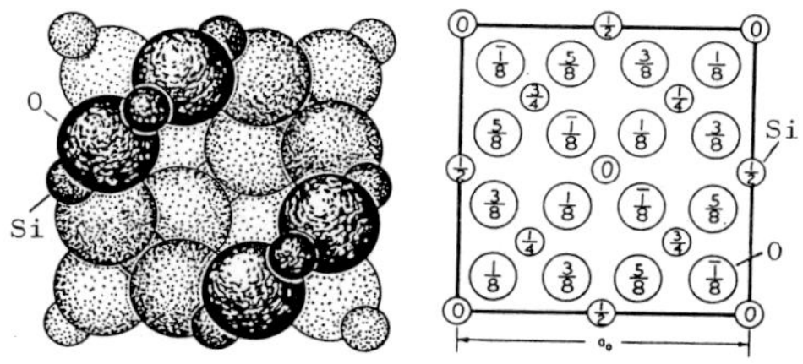

(a) Fig. 2 (b)

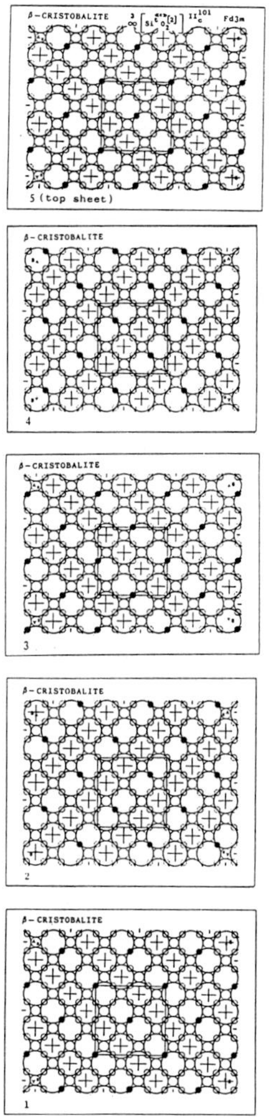

Fig. 7

Properties

Habit	Cleav.	Fract.	Twin.	Hardn.	Dens.	Colour	Transp.

Refr. index/Reflet.	Birrefr.	Luster	Streak	Melt.p.	CPI

Figures

Fig. 1. Corrugated tetrahedral silicate layer which constitutes the connected unit of the beta-cristobalite structure.

Fig. 2. (a) Packing drawing of the beta-cristobalite structure, and (b) corresponding unit cell content projected along an a axis (after Wyckoff, 1963, Vol.1)

Fig. 3. Unit cell content of beta-cristobalite projected on an a axis according to a more recent determination (space group $P2_13$, $Si_I(4a)$ u = 0.255, $Si_{II}(4a)$ u = -0.008, O(4a) u = 0.125 and the other twelve oxygen atoms in (12b) with x = y = 0.66 and z = 0.06) (after Wyckoff, 1963, Vol. 1).

Fig. 4. Ball and spoke model of the beta-cristobalite structure (after Povarennykh, 1972).

Fig. 5. Polyhedral description of the beta-cristobalite structure (after Povarennykh, 1972).

Fig. 6. Another polyhedral description of the beta-cristobalite structure projected along an a axis (after Kostov, 1968).

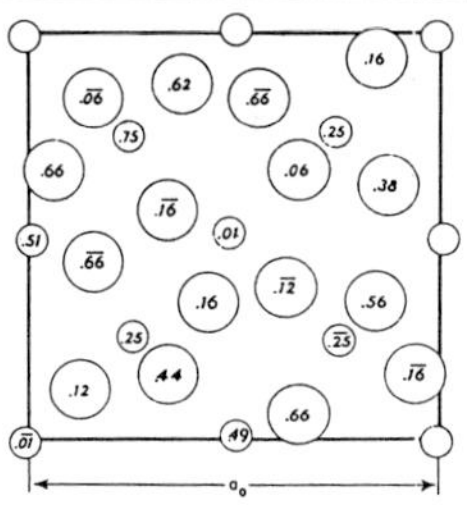

Fig. 3

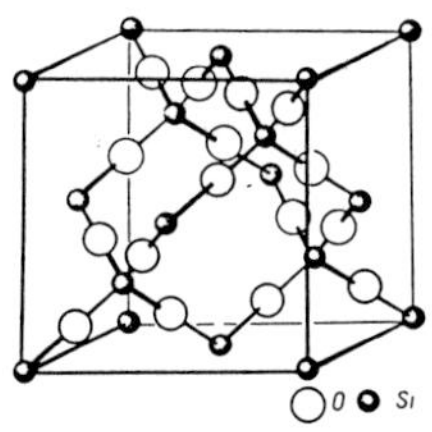

Fig. 4

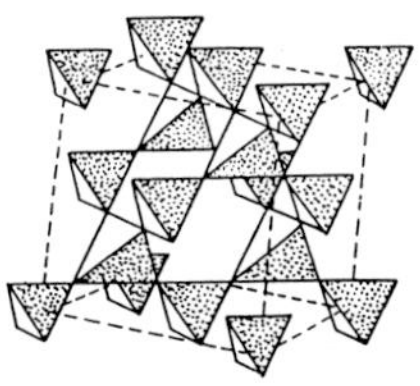

Fig. 5

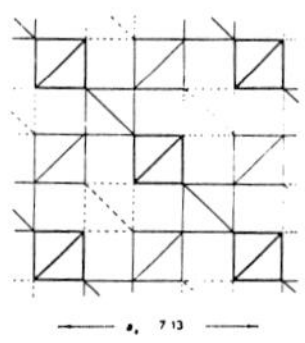

Fig. 6

Fig. 7. Condensed model of the defect packing analogue of beta-cristobalite, to which corresponds the formula $Si^{t}\left[O_2\square_2\right]^{c}$.

Description

Beta-cristobalite is a framework structure based on the linkage of corrugated silicate layers in a c sequence. It is the high-temperature form of cristobalite.

References

Kostov (1968) 390.

Povarennykh (1972) 292.

Wyckoff (1963) Vol. 1, 318,319.

Wells (1962) 1006.

α -CRISTOBALITE

(low-temperature form, below 200° C, ordinary)

$P\ 4_1 2_1$

$^3_\infty\left[Si^t\ O_2^{[2]}\right]\ II_c^{201}$

a = 4.9733 Å
c = 6.9262 Å
Z = 4

Si (4a) u = 0.30

O (8b) x = 0.245, y = 0.10, z = 0.175

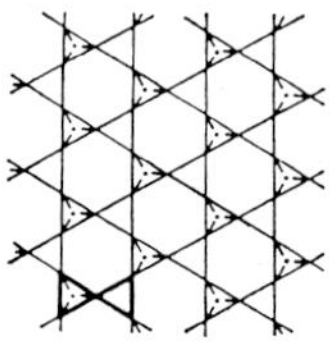

II^{201}

Fig. 1

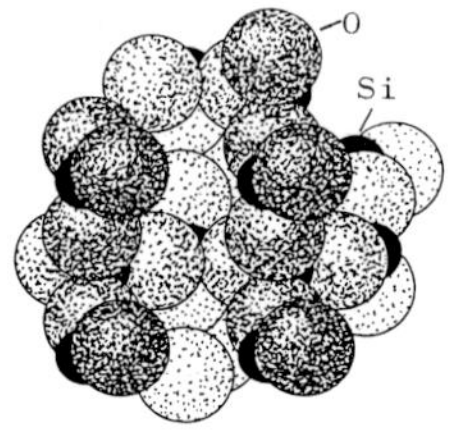

(a)

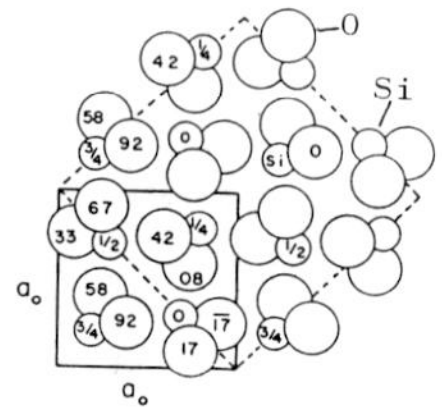

(b)

Fig. 2

Properties

Habit	Cleav.	Fract.	Twin.	Hardn.	Dens.	Colour	Transp.
cubic, coarse aggregates		conchoi-dal	spinel (111)	6-7	2.33	colourless	transparent

Refr. index/Reflect.	Birefr.	Luster	Streak	Melt.p.	CPI
n_ω = 1.489 n_ϵ = 1.482	(-)	vitreous	white	1710° C	(SPI) 43

Figures

Fig. 1. Corrugated tetrahedral silicate layer which constitutes the connected unit of the alfa-cristobalite structure.

Fig. 2. (a) Packing drawing of the alfa-cristobalite structure, and (b) corresponding unit cell content projected along the c axis (after Wyckoff, 1963, Vol. 1).

Description

Alfa-cristobalite is a framework structure based on the linkage of corrugate silicate layers in a c sequence. It may be considered as a distortion derivative of the beta-cristobalite structure.

References

Kostov (1968) 390.
Wyckoff (1963) Vol. 1, 316, 317.
Zoltai + Stout (1984) 299.
Ingerson (1955) 351.

ICE

(high-pressure form, antartic) $^{3}_{\infty}\left[O^{t}\ H_{2}^{[2]}\right]^{II}\ {}^{201}_{c}$

Fd3m

a = 6.350 (at -130°C)
Z = 8

O (8a)
H(16c)

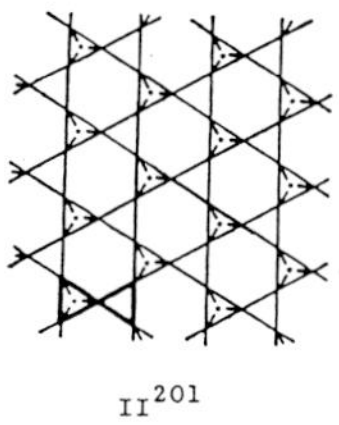

Fig. 1

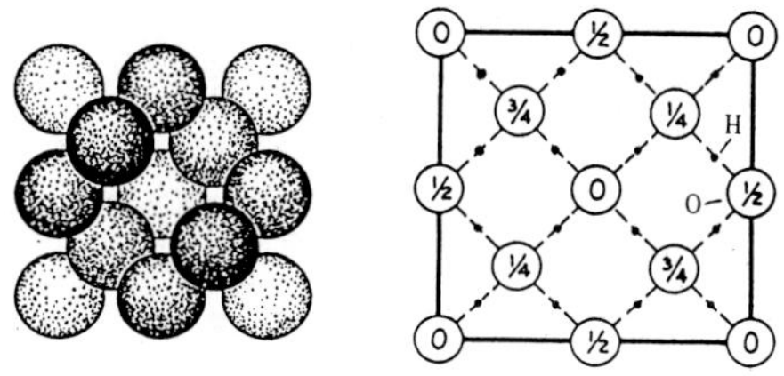

(a) Fig. 2 (b)

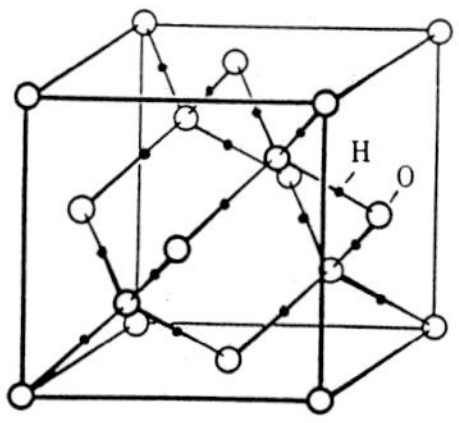

Fig. 3

Properties

Habit	Cleav.	Fract.	Twin.	Hardn.	Dens.	Colour	Transp.

Refr. index/Reflet.	Birrefr.	Luster	Streak	Melt.p.	CPI

Figures

Fig. 1. Corrugated layer formed by $O^{t}H_{4}$ tetrahedra, which constitutes the connected unit of the ice structure (high pressure form). It is similar to the cristobalite connected unit but the size of the atoms forming the tetrahedra are reversed.

Fig. 2. (a) Packing drawing of the ice structure (high pressure form), and (b) corresponding unit cell content projected along an a axis (adapted from Wyckoff, Vol. 1). These drawings are similar to those of diamond (Fig. 3 and 4) because the lattice complex D , of the carbon atoms in diamond is the same of the oxygen atoms in ice.

Fig. 3. Ball and spoke representation of the ice (high pressure form) (adapted from Povarennykh).

Description

The structure of ice (high-pressure form) is similar to that of cristobalite, but the size of the atoms forming the tetrahedra are reversed, that is instead of $Si^{t}O_{4}$ it is based on $O^{t}H_{4}$ tetrahedra. The sequence of the corrugated tetrahedral layers is also ABC (or c).

References

Kostov (1968) 210.
Povarennykh (1972) 118, 249-295.
Wyckoff (1963) Vol. 1, 25, 322-325.
Bokii (1954) 470.

β - TRIDYMITE

(high-temperature form)

$P6_3/mmc$

$^{3}_{\infty}\left[Si^{t}\,O_2^{\ll ch \gg [2]}\right] II^{201}_{h}$

a = 5.03 Å
c = 8.22 Å
Z = 4

Si(4f) u = 0.44
O_I (2c)
O_{II}(6g)

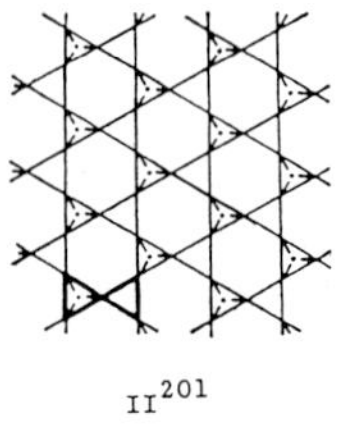

Fig. 1

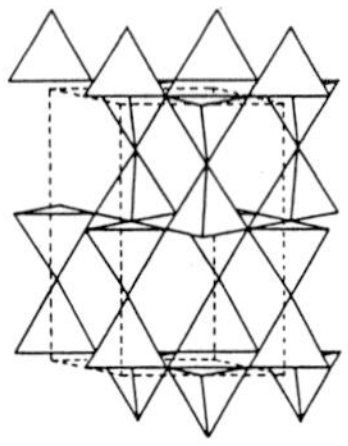

Fig. 2

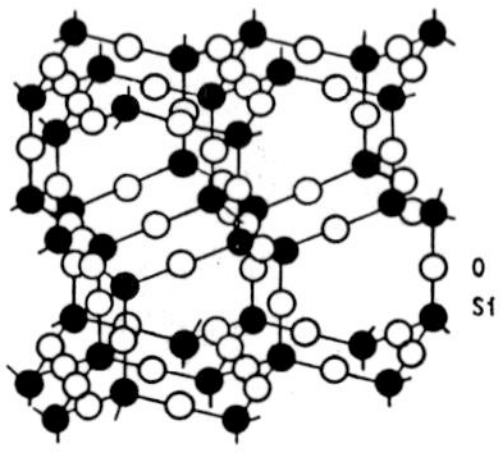

Fig. 3

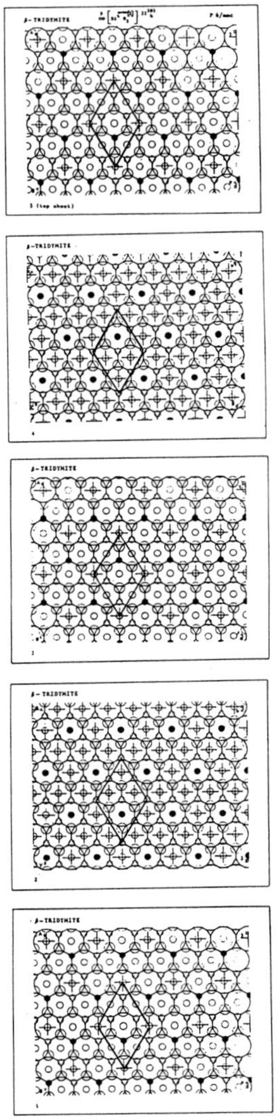

Fig. 4

Properties

Habit	Cleav.	Fract.	Twin.	Hardn.	Dens.	Colour	Transp.
tabular, flattened, rosettes			$(10\bar{1}6)$		2.197		

Refr. index/Reflect.	Birefr.	Luster	Streak	Melt.p.	CPI
	(+)				

Figures

Fig. 1. Corrugated tetrahedral silicate layer which constitutes the connected unit of beta-tridymite.

Fig. 2. Polyhedral description of the framework silicate tetrahedra of the beta-tridymite structure (after Zoltai, 1977).

Fig. 3. Ball and spoke representation of the beta-tridymite structure (after Mumpton, 1981).

Fig. 4. Condensed model of the defect packing analogue of beta-tridymite. The large open circles represent oxygen atoms, and the small black circles Si atoms. The crosses correspond to vacancies. It is based on a defect ch closest packing, and the corresponding structural formula is $Si^{t}[O_2\square]^{ch}$.

Description

The crystal structure of beta-tridymite is a framework, which can be decomposed on corrugated tetrahedral silicate layers linked in a AB (or h) sequence. The corresponding packing analogue is based on a defect ch closest packing. Beta-tridymite is the high-temperature form (> 870°C) of tridymite.

However there is also a orthorhombic form of high-tridymite (over 107°C and below 870°C) to which correspond the space group $C222_1$, a = 8.74 Å, b = 5.04 Å, c = 8.24 Å, and Z = 8 (Strunz, 1982).

References

Kostov (1968) 389.
Wyckoff (1963) Vol. 1, 313-316.
Zoltai (1977) 6-45.
Mumpton (1981) 3.
Roberts et al. (1974) 625.
Strunz (1982) 194.

α - TRIDYMITE

(low-temperature form, ordinary)

Cc

$^{3}_{\infty}\left[Si^{t}\ O_2^{\ll ch \gg [2]} \right]$ II^{201}_{h}

a = 18.524 Å	Si_I	(4a)	x = 0.5507 y = 0.541 z = 0.5642	O_V	(4a) x = 0.4155 y = 0.352 z = 0.6741
b = 5.003 Å c = 23.810 Å	Si_{II}	(4a)	x = 0.7036 y = 0.947 z = 0.7374	O_{VI}	(4a) x = 0.4127 y = 0.852 z = 0.6434
β = 105.82° Z = 48 (below 107°C)	Si_{III}	(4a)	x = 0.4198 y = 0.549 z = 0.6225	O_{VII} ...	(4a) x = 0.8773 y = 0.164 z = 0.9213

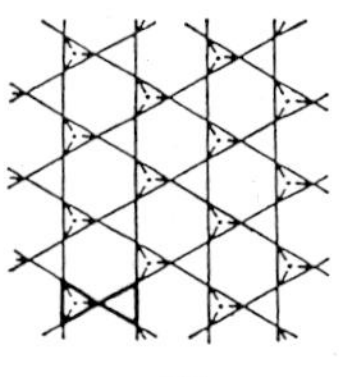

II^{201}

Fig. 1

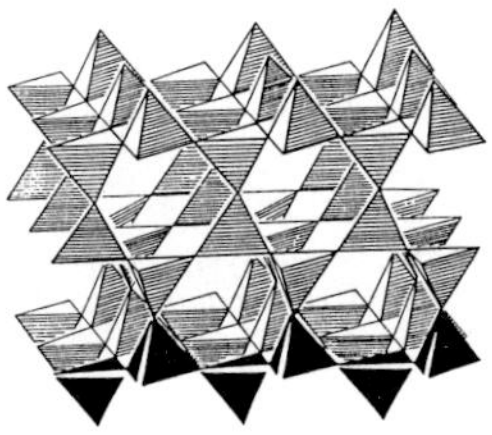

Fig. 2

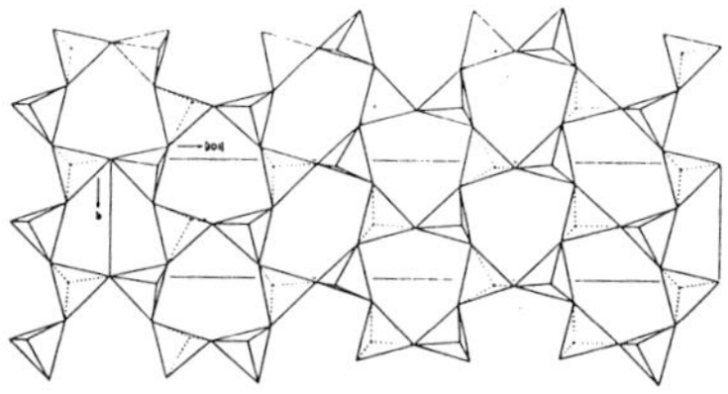

Fig. 3

Properties

Habit	Cleav.	Fract.	Twin.	Hardn.	Dens.	Colour	Transp.
wedge shaped		conchoidal	(110)	6 - 7	2.28	colourless	transparent to translucent

Refr. index /Reflect.	Birefr.	Luster	Streak	Melt.p.	CPI
n_α = 1.478 n_β = 1.479 n_γ = 1.481	(+) 2V = 70°	vitreous	white		(SPI) 42

Figures

Fig. 1. Schematic drawing of the ideal corrugated tetrahedral silicate connected unit.

Fig. 2. Polyhedral description of the alfa-tridymite structure (after Liebau, 1985).

Fig. 3. Projection onto (10$\bar{1}$) of a slab of the silicate layer of monoclinic alfa-tridymite, showing two types of six-member ring configurations - ditrigonal and oval (Dollase + Baur, 1976).

Description

The alpha-tridymite structure is a distortion derivative of beta--tridymite crystal structure. It is based on corrugated tetrahedral silicate layers linked in a <u>h</u> sequence.

References

Kostov (1968) 389.

Zoltai + Stout (1984) 297.

Dollase (1967) 617-623.

Dollase + Baur (1976) 971-978.

Liebau (1985) 127.

Crystallographic data (continued)

Si_{IV}	(4a)	x = 0.5751 y = 0.043 z = 0.7955	O_{VIII}	(4a)	x = 0.9160 y = 0.668 z = 0.9361
Si_V	(4a)	x = 0.9247 y = 0.548 z = 0.6971	O_{IX}	(4a)	x = 0.7741 y = 0.145 z = 0.7538
Si_{VI}	(4a)	x = 0.7643 y = 0.944 z = 0.5386	O_X	(4a)	x = 0.7336 y = 0.643 z = 0.7402
Si_{VII}	(4a)	x = 0.8023 y = 0.449 z = 0.7661	O_{XI}	(4a)	x = 0.6195 y = 0.341 z = 0.5847
Si_{VIII}	(4a)	x = 0.6453 y = 0.041 z = 0.6027	O_{XII}	(4a)	x = 0.5811 y = 0.843 z = 0.5661
Si_{IX}	(4a)	x = 0.8531 y = 0.463 z = 0.9016	O_{XIII}	(4a)	x = 0.4966 y = 0.511 z = 0.6062
Si_X	(4a)	x = 0.9493 y = 0.966 z = 0.9358	O_{XIV}	(4a)	x = 0.3533 y = 0.489 z = 0.5665
Si_{XI}	(4a)	x = 0.7244 y = 0.559 z = 0.9605	O_{XV}	(4a)	x = 0.6498 y = 0.005 z = 0.6711
Si_{XII}	(4a)	x = 0.5697 y = 0.549 z = 0.8702	O_{XVI}	(4a)	x = 0.6524 y = 0.996 z = 0.7805
O_I	(4a)	x = 0.5749 y = 0.351 z = 0.8182	O_{XVII}	(4a)	x = 0.8636 y = 0.499 z = 0.7328
O_{II}	(4a)	x = 0.5707 y = 0.840 z = 0.8476	O_{XVIII}	(4a)	x = 0.8361 y = 0.503 z = 0.8333
O_{III}	(4a)	x = 0.7315 y = 0.138 z = 0.4851	O_{XIX}	(4a)	x = 0.5052 y = 0.993 z = 0.7398
O_{IV}	(4a)	x = 0.7509 y = 0.644 z = 0.5152	O_{XX}	(4a)	x = 0.6369 y = 0.511 z = 0.9268
O_{XXI}	(4a)	x = 0.9927 y = 0.992 z = 0.8868	O_{XXIII}	(4a)	x = 0.7252 y = 0.991 z = 0.5898
O_{XXII}	(4a)	x = 0.5* y = 0.468 z = 0.5*	O_{XXIV}	(4a)	x = 0.7744 y = 0.521 z = 0.9155

KALSILITE

$P6_3$

$K^{[6]}\ {}^{3}_{\infty}\left[Si^{t}\ Al^{t}\ O_4^{\ll ch \gg [2]}\right]\ II^{201}_{h}$

$a = 5.161$ Å
$c = 8.693$ Å
$Z = 2$

K (2a) u = 0.2411

Al(2b) u = 0.0461

Si(2b) u = 0.4281

$O_I(6c)$ x = 0.6169, y = 0.0125, z = 0.9858

$1/3\ O_{II}(6c)$ x = 0.3339, y = 0.7155, z = 0.2500

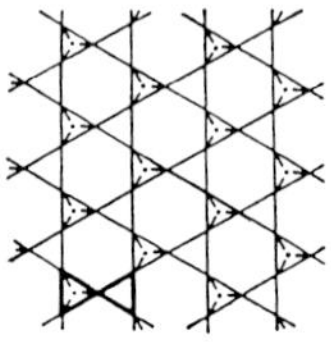

II^{201}

Fig. 1

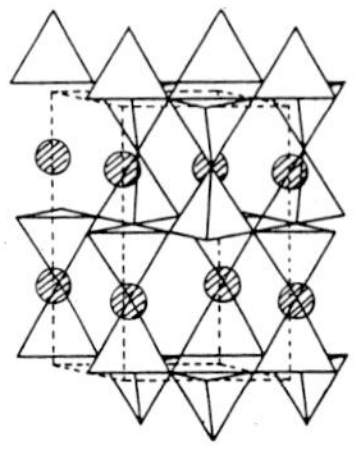

Fig. 2

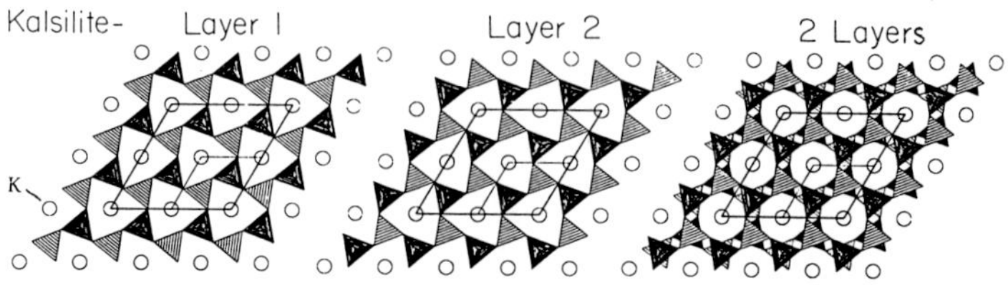

Fig. 3

Properties

Habit	Cleav.	Fract.	Twin.	Hardn.	Dens.	Colour	Transp.
massive, compact	poor (10$\bar{1}$0)	subconchoidal		6	2.59–2.625	colourless, white, grey	transparent to translucent

Refr. index/Reflect.	Birefr.	Luster	Streak	Melt.p.	CPI
$n_\omega = 1.538-1.543$ $n_\epsilon = 1.532-1.537$	(-)	vitreous, greasy			

Figures

Fig. 1. Schematic drawing of the ideal corrugated tetrahedral silicate layer which constitutes the connected unit of the kalsilite structure.

Fig. 2. Polyhedral representation of the way the silicate layers are linked together, following a AB (or h) sequence. The small dashed circles represent K atoms (adapted from Zoltai, 1977).

Fig. 3. Representation of two consecutive kalsilite layers, and their superimposition (after Steele + Pluth, 1990).

Description

The kalsilite structure is an insertion derivative of the tridymite structure. It is also distorted in relation to the ideal tridymite structure.

References

Kostov (1968) 401.
Povarennykh (1972) 345, 346.
Wyckoff (1968) Vol. 4, 169,170.
Steele + Pluth (1990) 1190.
Zoltai (1977) 6-45.
Roberts et al. (1974) 317.

NEPHELINE

P 6_3

$K^{[9]} Na_3^{[8]} \,^3_\infty \left[Si_4^t \, Al_4^t \, O_{16}^{\ll ch \gg [2]} \right] II^{201}_h$

a = 10.05 Å
c = 8.38 Å
Z = 2

K(2a)		x = 0, y = 0, z = 0.998
Na	(6c)	x = 0.000, y = 0.444, z = 0.989
Al_I	(2b)	x = 1/3, y = 2/3, z = 0.188
Al_{II}	(6c)	x = 0.090, y = 0.328, z = 0.686
Si_I	(2b)	x = 1/3, y = 2/3, z = 0.812
Si_{II}	(6c)	x = 0.090, y = 0.328, z = 0.313
1/3 O_I	(6c)	x = 0.290, y = 0.589, z = 0.990

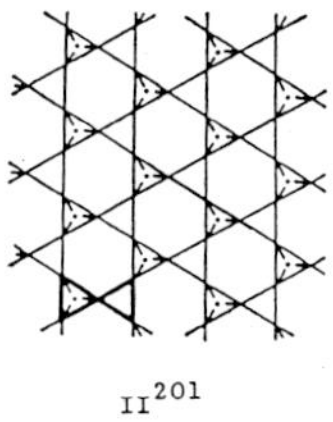

Fig. 1

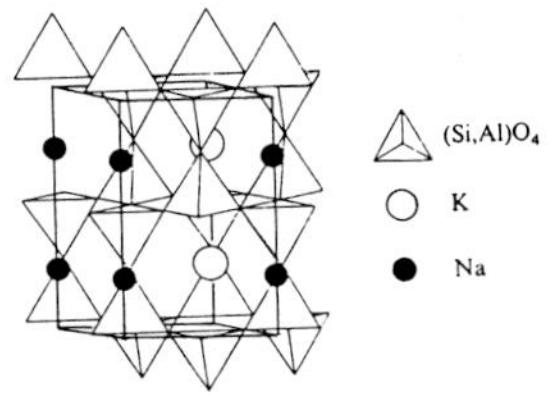

Fig. 2

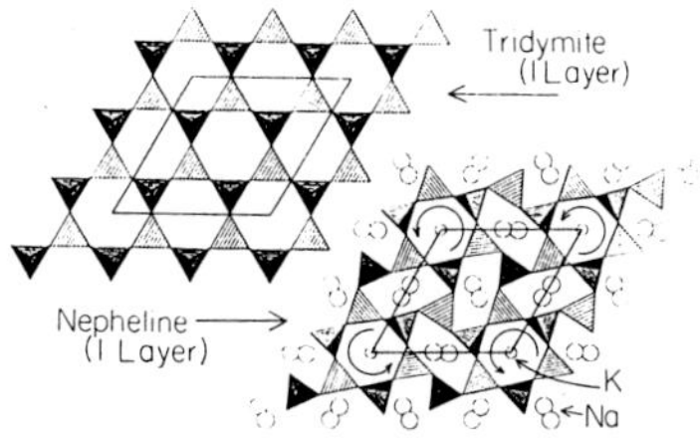

Fig. 3

Properties

Habit	Cleav.	Fract.	Twin.	Hardn.	Dens.	Colour	Transp.
prismatic	perfect (100) (001)	subconchoidal	(100) (112) (335)	5.5-6	2.60	colourless, turbid	transparent

Refr. index /Reflect.	Birefr.	Luster	Streak	Melt.p.	CPI
n_ω = 1.540 n_ϵ = 1.536	(-)	vitreous	white		(SPI) 43

Figures

Fig. 1. Schematical drawing of the ideal corrugated tetrahedral silicate layer which constitutes the connected unit of the nepheline structure.

Fig. 2. Polyhedral representation of the nepheline structure (after Zoltai + Stout, 1984).

Fig. 3. Comparison between the ideal tridymite and the nepheline tetrahedral layers (after Steele + Pluth, 1990).

Description

The structure of nepheline is a stuffed derivative of the tridymite structure. It is also distorted in relation to the ideal tridymite structure.

References

Kostov (1968) 401.
Povarennykh (1972) 345, 346.
Wyckoff (1968) Vol. 4, 453, 454.
Zoltai + Stout (1984) 310, 311.
Steele + Pluth (1990) 1189.

Crystallographic data (continued)

O_{II}	(6c)	x = 0.020, y = 0.312, z = 0.499
O_{III}	(6c)	x = 0.178, y = 0.516, z = 0.738
O_{IV}	(6c)	x = 0.156, y = 0.512, z = 0.260
O_V	(6c)	x = 0.215, y = 0.264, z = 0.314
O_{VI}	(6c)	x = 0.215, y = 0.264, z = 0.685

ICE (ordinary)

$P6_3/mmc$

$^{3}_{\infty}\left[O^{t\ \ll ch \gg [2]}\ H_2\right] II^{201}_{h}$

a = 4.5227 Å (at 0°C)
c = 7.3671 Å
Z = 4

O(4f) u = 1/16

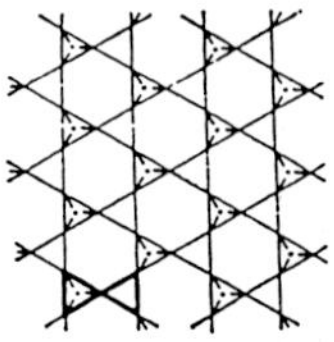

Fig. 1

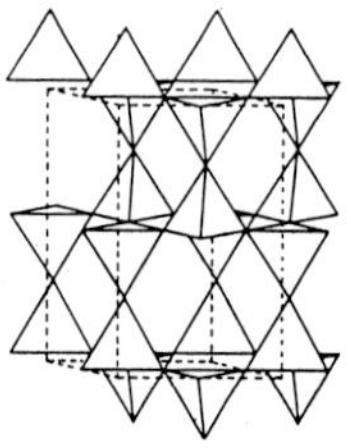

Fig. 2

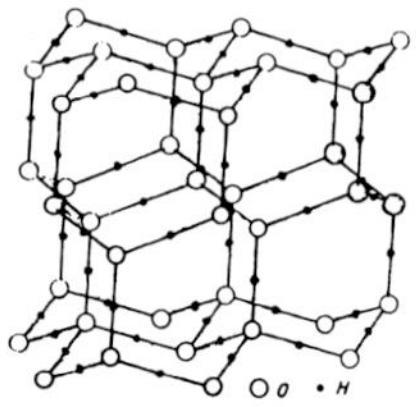

Fig. 3

Properties

Habit	Cleav.	Fract.	Twin.	Hardn.	Dens.	Colour	Transp.
flatened stellate forms	none	conchoi-dal	(0001) (000$\bar{1}$)	1.5	0.9167	colourless, pale blue	transparent

Refr. index /Reflect.	Birefr.	Luster	Streak	Melt.p.	CPI
n_ω = 1.30907 n_ϵ = 1.31052	(+)	vitreous	colourless	0° C	

Figures

Fig. 1. Schematic drawing of the ideal corrugated layers of O^tH_4 tetrahedra, which constitutes the connected unit of the ordinary ice structure.

Fig. 2. Polyhedral description of the structure of ordinary ice (after Zoltai, 1977). The tetrahedral layers follow a AB (or h) sequence.

Fig. 3. Ball and spoke representation of the structure of ordinary ice (adapted from Povarennykh, 1972),

Description

The structure of ordinary ice is similar to the structure of tridymite but the relative size of the atoms forming the tetrahedra is inverted, that is, instead of Si^tO_4 it is based on O^tH_4 tetrahedra. The sequence of the corrugated tetrahedral layers is also AB (or h).

References

Kostov (1968) 210.
Povarennykh (1972) 294.
Wyckoff (1963) Vol. 1, 322-324.
Zoltai (1977) 6-45.

β - QUARTZ

(high-temperature form)

$P6_22$

$\overset{3}{\infty}\left[Si^{t}\overset{\ll Tv \gg}{O_2}\right] I^{161}_{by}$

a = 5.01 Å (ca. 600°C)
c = 5.47 Å
Z = 3

Si(3c)
O (6j) u = 0.197

I^{161}

Fig. 1

Fig. 2

Properties

Habit	Cleav.	Fract.	Twin.	Hardn.	Dens.	Colour	Transp.
bipyra-midal	good (10$\bar{1}$1)	conchoi-dal	(10$\bar{1}$1) (11$\bar{2}$2)	7	2.53	colourless, white, grey	transparent to opaque

Refr. index/Reflect.	Birefr.	Luster	Streak	Melt.p.	CPI
n_ω = 1.5329 n_ϵ = 1.5405	(+)	vitreous	white		(SPI) 47

Figures

Fig. 1. Schematic drawing of the ideal corrugated tetrahedral silicate layer which constitutes the connected unit of the beta-quartz structure.

Fig. 2. Polyhedral drawing of the beta-quartz structure projected along the c axis (after Povarennykh, 1972).

Fig. 3. Condensed model of the defect packing analogue of the beta-quartz structure. The large open circles represent oxygen atoms, and the small black circles correspond to silicon atoms in distorted tetrahedral voids. The crosses indicate the vacancies of the packing atoms. The analogue is based on a Tv defect packing to which corresponds the structural formula $Si^{t}\left[O_2\square\right]^{Tv}$ (after Figueiredo, 1977).

Description

Beta-quartz is a framework of $Si^{t}O_4$ tetrahedra, and it can be considered as a defect Tv packing of oxygen atoms with silicon atoms in distorted tetrahedral voids.

References

Kostov (1968) 386.
Povarennykh (1972) 291.
Wyckoff (1963), Vol. 1, 312-314.
Zoltai + Stout (1984) 298.
Figueiredo (1977) 28.

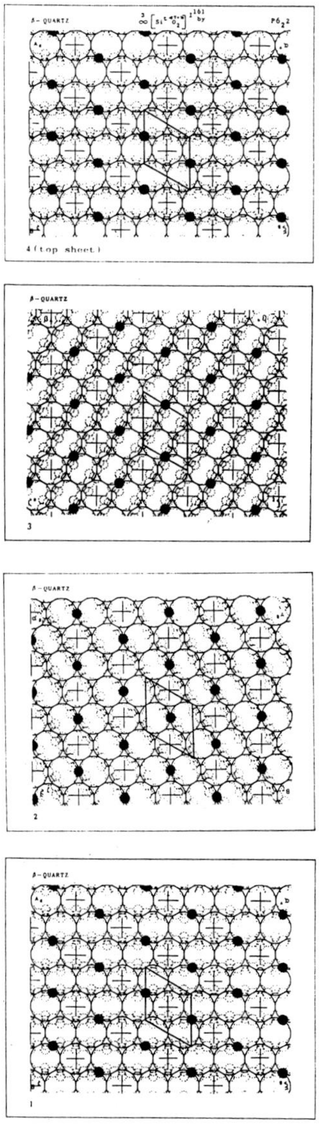

Fig. 4

α - QUARTZ $P3_12$

(low-temperature form, ordinary) $^{3}_{\infty}\left[Si^{t}\,O_2^{\ll Tv \gg}\right] I^{161}$ by

a = 4.91304 Å
c = 5.40463 Å
Z = 3

Si(3a) u = 0.465
O (6c) x = 0.415
y = 0.272
z = 0.120

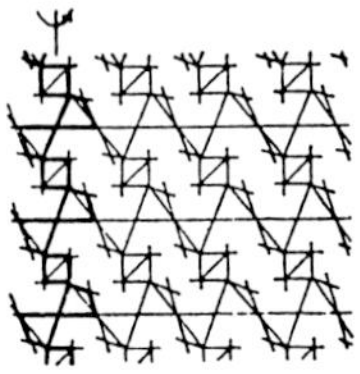

I^{161}

Fig. 1

Fig. 2

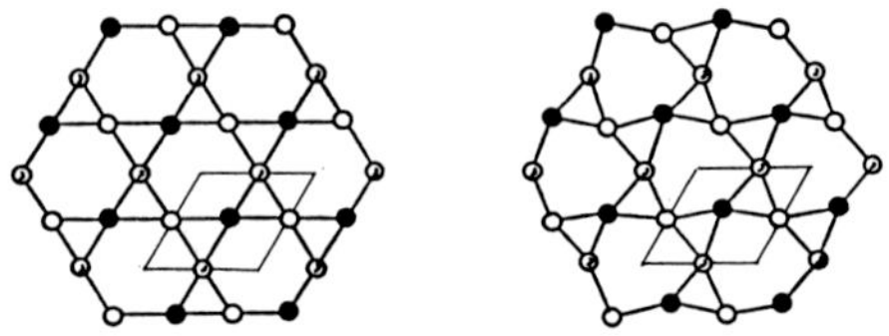

High Quartz (β - form) Low Quartz (α - form)

Fig. 3

Properties

Habit	Cleav.	Fract.	Twin.	Hardn.	Dens.	Colour	Transp.
hexagonal, prismatic, massive		conchoidal	Dauphiné [001]	7	2.65	colourless, white, variable	transparent to translucent

Refr. index / Reflect.	Birefr.	Luster	Streak	Melt.p.	CPI
$n_\omega = 1.544$ $n_\epsilon = 1.533$	(+)	vitreous, greasy	white		(SPI) 49

Figures

Fig. 1. Schematic drawing of the ideal corrugated tetrahedral silicate layer which constitutes the connected unit of the alfa-quartz structure. It is similar to that of beta-quartz.

Fig. 2. Polyhedral drawing of the alfa-quartz structure projected along the c axis (after Povarennykh, 1972).

Fig. 3. Ball and spoke representation of the beta and alfa-quartz structures in order to visualize their relationship (adapted from Povarennykh, 1972).

Description

Alfa-quartz (low-temperature form) is the ordinary quartz. It is a framework of Si^tO_4 tetrahedra, and is a distorted derivative of beta-quartz.

References

Kostov (1968) 386.
Povarennykh (1972) 291.
Wyckoff (1963) Vol. 1, 312-314.
Zoltai + Stout (1984) 298.

COESITE C 2/c

$^{3}_{\infty}\left[Si^{t}\ O_{2}\right]^{(2,IV^{11}_{by})}s$

a = 7.17 Å
b = 12.38 Å
c = 7.17 Å
β = 120°
Z = 16

Si_I	(8f)	x = 0.1403, y = 0.0735, z = 0.1084
Si_{II}	(8f)	x = 0.5063, y = 0.5388, z = 0.1576
O_I	(4a)	x = 0, y = 0, z = 0
O_{II}	(4e)	z = 0.3834
O_{III}	(8f)	x = 0.2094, y = 0.9405, z = 0.1256
O_{IV}	(8f)	x = 0.3080, y = 0.3293, z = 0.1030
O_V	(8f)	x = 0.123, y = 0.4726, z = 0.2122

(Str. deter. based on B2/b)

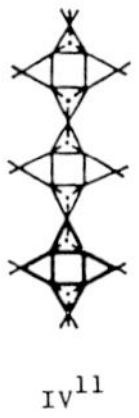

Fig. 1

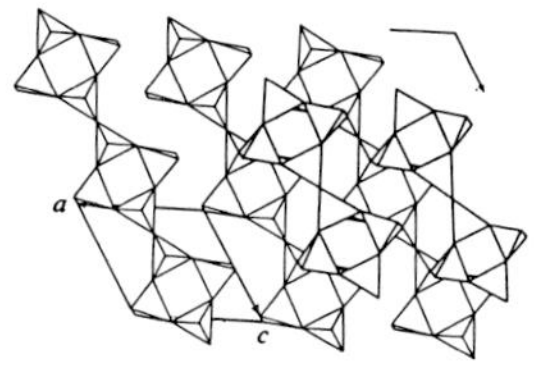

Fig. 2

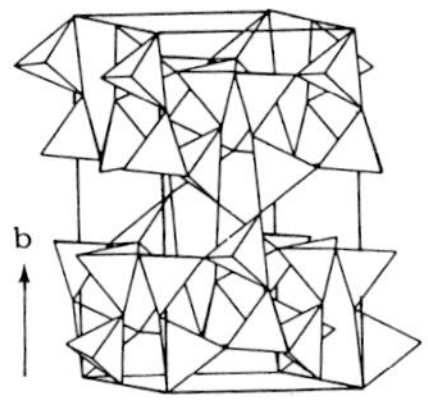

Fig. 3

Properties

Habit	Cleav.	Fract.	Twin.	Hardn.	Dens.	Colour	Transp.
tabular		conchoidal	(100) (021)	7-8	2.93	colourless	transparent

Refr. index/Reflect.	Birefr.	Luster	Streak	Melt.p.	CPI
n_α = 1.59 n_β = 1.60 n_γ = 1.60	(+) 2V = 64°	vitreous	white		(SPI) 52

Figures

Fig. 1. Tetrahedral silicate infinite chain of the feldspar type which constitutes the connected unit of the coesite structure.

Fig. 2. Polyhedral description of the way the silicate chains are linked together in the coesite structure (after Zoltai+Stout, 1984).

Fig. 3. Polyhedral drawing in perspective, of the coesite structure (after Zoltai + Stout, 1984).

Description

The coesite structure is a framework of $Si^{t}O_4$ tetrahedra formed by interlinked infinite chains of the feldspar type.

References

Kostov (1968) 390.
Povarennykh (1972) 293.
Wyckoff (1963) Vol. 1, 320, 321.
Zoltai + Stout (1984) 299, 300, 303

SANIDINE C 2/m

$K^{[9]}\ {}^{3}_{\infty}\left[Si^{t}_{3}\ Al^{t}\ O_{8}\right](2IV^{11}_{by})_{s}$

a = 8.5642 Å
b = 13.0300 Å
c = 7.1749 Å
β = 115.994°
Z = 4

Atom	Position	x	y	z
K	(4i)	0.2829	0	0.1361
$(Si,Al)_I$	(8j)	0.0089	0.1842	0.2237
$(Si,Al)_{II}$	(8j)	0.7073	0.1178	0.3437
O_I	(4g)	0	0.1455	0
O_{II}	(4i)	0.6303	0	0.2840
O_{III}	(8j)	0.8273	0.1447	0.2271

...

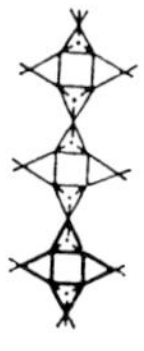

Fig. 1

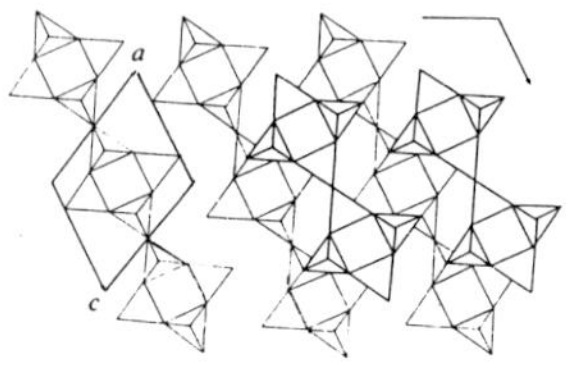

Fig. 2

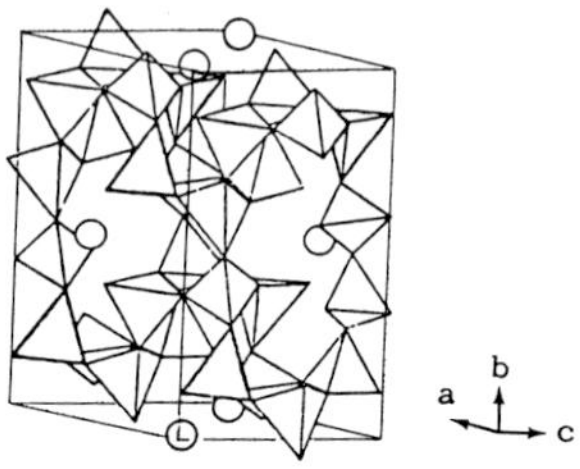

Fig. 3

Properties

Habit	Cleav.	Fract.	Twin.	Hardn.	Dens.	Colour	Transp.
prismatic	perfect (001)	uneven	Carlbad [001] comp. plane (010)	6	2.56	white, variable	transparent to translucent

Refr. index /Reflect.	Birefr.	Luster	Streak	Melt.p.	CPI
n_α = 1.521 n_β = 1.525 n_γ = 1.528	(-) 2V = 80°-85°	vitreous	white		(SPI) 42

Figures

Fig. 1. Tetrahedral silicate infinite chain which constitutes the connected unit of the feldspar structure, in this case of sanidine. It is the same connected unit as that of coesite, which is one of the forms of silica.

Fig. 2. Polyhedral description of the way the silicate chains are linked together in a feldspar structure (after Zoltai+Stout, 1984).

Fig. 3. Polyhedral drawing, in perspective, of the sanidine structure (after Zoltai, 1977).

Description

Sanidine is a framework of $(Si,Al)^{t}O_4$ tetrahedra with potassium atoms in the voids. Its connected unit is the infinite feldspar chain.

Crystallographic data (continued)

Atom	Position	x	y	z
O_{IV}	(8j)	0.0335	0.3094	0.2631
O_V	(8j)	0.1804	0.1256	0.4051

References

Kostov (1968) 392-397.

Povarennykh (1072) 346, 347.

Wyckoff (1968) Vol. 4, 449, 450.

Zoltai + Stout (1984) 197, 303, 304.

Zoltai (1977) 6-88.

ORTHOCLASE

C 2/m

$K^{[9]} \, {}^{3}_{\infty}\left[Si^{t}_{3}\, Al^{t}\, O_{8}\right](2IV^{11}_{by})_{s}$

a = 8.5616 Å
b = 12.9962 Å
c = 7.1934 Å
β = 116.015°
Z = 4

Atom	Position	x	y	z
K	(4i)	0.2843	0	0.1361
$(Si,Al)_I$ (Al 0.30, Si 0.70)	(8j)	0.0101	0.1848	0.2237
$(Si,Al)_{II}$ (Al 0.19, Si 0.81)	(8j)	0.7084	0.1175	0.3437
O_I	(4g)	0	0.1461	0
O_{II}	(4i)	0.6335	0	0.2840
O_{III}	(8j)	0.8276	0.1465	0.2271
...				

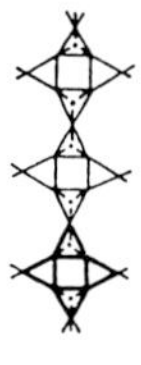

Fig. 1

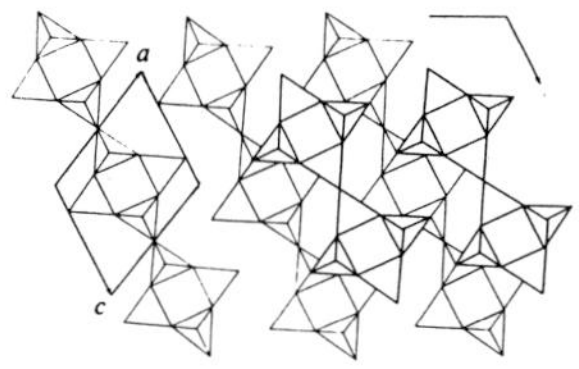

Fig. 2

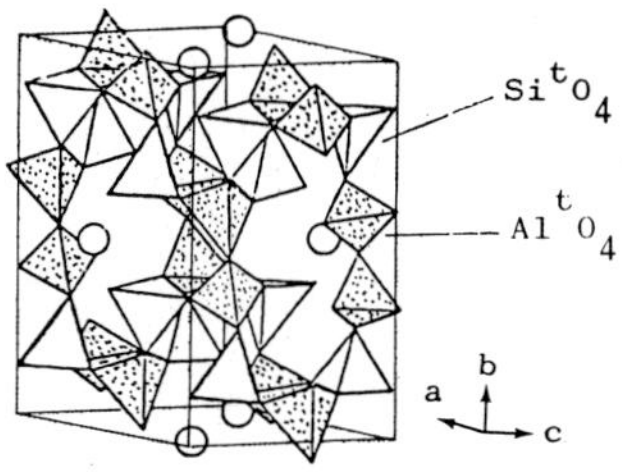

Fig. 3

Properties

Habit	Cleav.	Fract.	Twin.	Hardn.	Dens.	Colour	Transp.
prismatic	perfect (001)	uneven	Carlsbad 001, Baveno 021	6	2.56	white turbid	translucent

Refr. index/Reflect.	Birefr.	Luster	Streak	Melt.p.	CPI
$n\alpha$ = 1.521, $n\beta$ = 1.525, $n\gamma$ = 1.528	(-) 2V = 60°-65°	pearly, vitreous	white	1117°C	(SPI) 42

Population

Hyalophane

$(K,Ba)^{[9]} \, {}^{3}_{\infty}\left[Si^{t}_{2}\,(Al,Si)^{t}_{2}\, O_{8}\right](2VI^{11}_{by})_{s}$

Description

The orthoclase structure is an ordered sanidine structure.

Figures

Fig. 1. Tetrahedral silicate infinite chain of the feldspar type which constitutes the connected unit of the orthoclase structure.

Fig. 2. Polyhedral description of the way the feldspar chains are linked together (after Zoltai + Stout, 1984).

Fig. 3. Polyhedral representation, in perspective, of the orthoclase structure (after Zoltai, 1977).

Crystallographic data (continued)

Atom	Position	x	y	z
O_{IV}	(8j)	0.0347	0.3104	0.2631
O_V	(8j)	0.1801	0.1256	0.4051

References

Kostov (1968) 391-401.
Wyckoff (1968) Vol. 4, 449-451.
Zoltai + Stout (1984) 303, 304, 306.
Jones + Taylor (1961) 453.
Ingerson (1955) 352.
Zoltai (1977) 6-88.

MICROCLINE $P\bar{1}$

$K^{[10]}\ {}^{3}_{\infty}\left[Si^{t}_{3}\ Al^{t}\ O_{8}\right](2IV^{11}_{by})_{s}$

a = 8.643 Å	K (4i)	x = 0.2867, y = 0.9985, z = 0.1406
b = 12.929 Å	Si_I (4i)	x = 0.7105, y = 0.1184, z = 0.3413
c = 7.190 Å	$(Si,Al)_I$ (4i)	x = 0.0095, y = 0.1844, z = 0.2225
α = 90.1°	Si_{II}(4i)	x = 0.7095, y = 0.8826, z = 0.3433
β = 116.2°	$(Si,Al)_{II}$(4i)	x = 0.0091, y = 0.8168, z = 0.2249
γ = 89.6°	O_I (4i)	x = 0.0005, y = 0.1440, z = 0.9977
Z = 4	...	

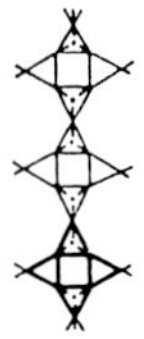

IV^{11}

Fig. 1

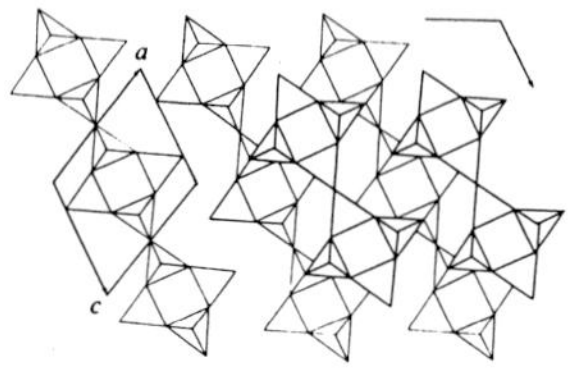

Fig. 2

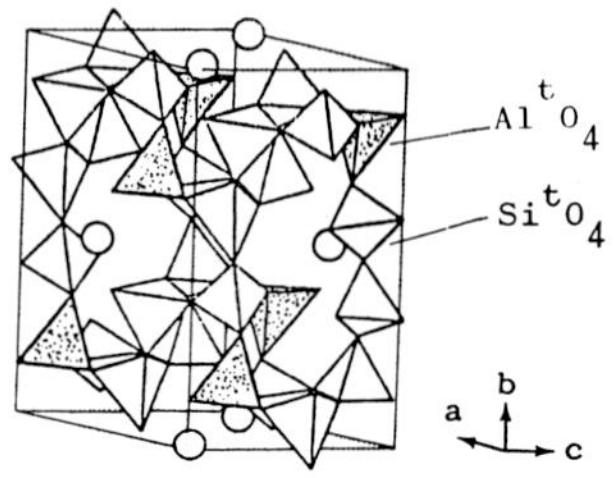

Fig. 3

Properties

Habit	Cleav.	Fract.	Twin.	Hardn.	Dens.	Colour	Transp.
prismatic	perfect		Albite [010] Pericline [010]	6	2.56	white, green	translucent

Refr. index/Reflect.	Birefr.	Luster	Streak	Melt.p.	CPI
nα = 1.518	(-)	pearly,	white		(SPI)
nβ = 1.524	2V = 77°-84°	vitreous			43
nγ = 1.528					

Figures

Fig.1. Tetrahedral silicate infinite chain of the feldspar type, which constitutes the connected unit of the microcline structure.

Fig. 2. Polyhedral description of the way the felspar chains are linked together (after Zoltai+Stout, 1984).

Fig. 3. Polyhedral representation, in perspective, of the microcline structure (after Zoltai, 1977).

Description

The microcline structure is an ordered sanidine structure, but different from orthoclase.

Crystallographic data (continued)

O_{II}	(4i)	x = 0.6382, y = 0.0011, z = 0.2809
O_{VI}	(4i)	x = 0.0378, y = 0.6903, z = 0.2602
O_{III}	(4i)	x = 0.8251, y = 0.1473, z = 0.2228
O_{VII}	(4i)	x = 0.1815, y = 0.1240, z = 0.4075
O_{IV}	(4i)	x = 0.8278, y = 0.8534, z = 0.2274
O_{VIII}	(4i)	x = 0.1790, y = 0.8749, z = 0.4085
O_V	(4i)	x = 0.0367, y = 0.3116, z = 0.2580

References

Kostov (1968) 395.

Povarennykh (1972) 346, 347.

Wyckoff (1968) Vol. 4, 443-447.

Zoltai + Stout (1984) 303-306.

Ribbe (1979) 402-408.

Zoltai (1977) 6-89.

ALBITE

C 1

(low-temperature form)

$(Na,Ca)^{[8]} \; {}^{3}_{\infty}\left[Si_2^t\ (Si,Al)^t\ Al^t\ O_8\right] \ (2IV^{11}_{by})_s$

a = 8.152 Å	Na (4i)	x = 0.2683	y = 0.9890	z = 0.1463
b = 12.784 Å	T_1(o)(4i)	x = 0.0089	y = 0.1682	z = 0.2080
c = 7.165 Å	T_1(m)(4i)	x = 0.037	y = 0.8205	z = 0.2374
α = 94.28°	...			
β = 116.67°				
γ = 87.74°				
Z = 4				

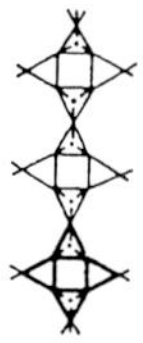

Fig. 1

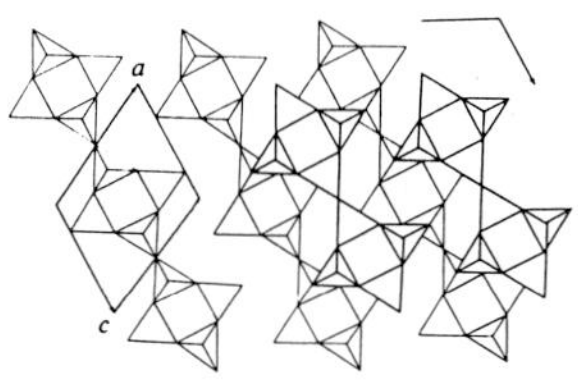

Fig. 2

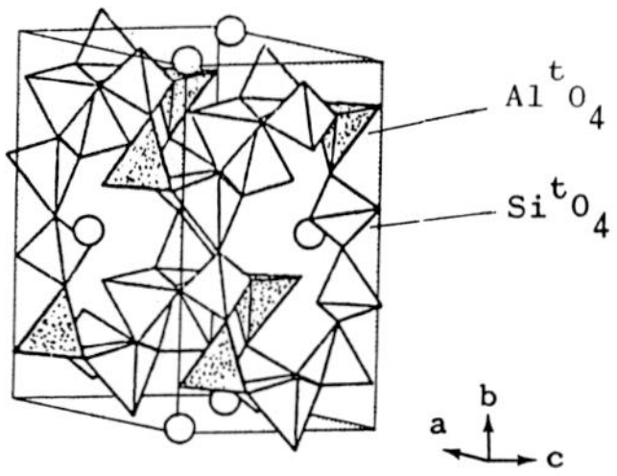

Fig. 3

Properties

Habit	Cleav.	Fract.	Twin.	Hardn.	Dens.	Colour	Transp.
prismatic, tabular	perfect (001)	uneven	Albite [010] Pericline [010]	6-6.5	2.62	white, grey, green	translucent

Refr. index/Reflect.	Birefr.	Luster	Streak	Melt.p.	CPI
nα = 1.527 nβ = 1.531 nγ = 1.538	(+) 2V = 77°	pearly, vitreous	white	1118°C	(SPI) 45

Population

Plagioclases:

Albite-Anorthite series

$Na^{[8]}\left[Si_3\,Al\,O_8\right]$ — $Ca^{[8]}\left[Si_2\,Al_2\,O_8\right]$

Albite	0	10% An.
Oligoclase	10	30% An.
Andesine	30	50% An.
Labradorite	50	70% An.
Bytownite	70	90% An.
Anorthite	90	100% An.

Description

The albite structure has the same atomic arrangement as sanidine, but the Al and Si tetrahedra have a certain order, however different from orthoclase.

Figures

Fig. 1. Tetrahedral silicate infinite chain of the feldspar type.

Fig. 2. Polyhedral description of the way the feldspar chains are linked together (after Zoltai+Stout, 1984). 303).

Fig.3. Packing drawing of the low albite structure (after Zoltai, 1977).

Crystallographic data (continued)

Atom	x	y	z
T_2(o)(4i)	0.6917	0.1102	0.3147
T_2(m)(4i)	0.6815	0.8818	0.3607
O_{A1} (4i)	0.0051	0.1312	0.9662
O_C(o)(4i)	0.0128	0.3012	0.2708
O_C(m)(4i)	0.0239	0.6939	0.2291
O_{A2} (4i)	0.5922	0.9973	0.2809
O_B(o)(4i)	0.8125	0.1099	0.1902
O_B(m)(4i)	0.8196	0.8512	0.2583
O_D(o)(4i)	0.2073	0.1089	0.3891
O_D(m)(4i)	0.1840	0.8681	0.4368

References

Kostov (1968) 393-395.

Povarennykh (1972) 344.

Zoltai + Stout (1984) 305.

Winter et al. (1977) 921-931.

Ingerson (1955) 352.

Ford (1932) 541, 542.

Zoltai (1977) 6-89.

ANORTHITE $P\bar{1}$

$(Ca,Na)^{[8]}\ {}^{3}_{\infty}\left[Si^{t}_{2}\ (Al,Si)^{t}\ Al^{t}\ O_{8}\right]^{(2,IV^{11}_{by})_{s}}$

a =	8.1768 Å
b =	12.8768 Å
c =	14.1690 Å
α =	93° 10'
β =	115° 51'
γ =	91° 13'
Z =	8

Atom	Site	x	y	z
Ca_{I}	(2i)	0.2647	0.9844	0.0873
Ca_{II}	(2i)	0.2684	0.0312	0.5438
Ca_{III}	(2i)	0.7732	0.5354	0.5422
Ca_{IV}	(2i)	0.7636	0.5067	0.0740
Si_{I}	(2i)	0.0099	0.1584	0.1043
Si_{II}	(2i)	0.5061	0.6567	0.6033
Si_{III}	(2i)	0.0059	0.8154	0.6129
Si_{IV}	(2i)	0.5034	0.3207	0.1091
Si_{V}	(2i)	0.6818	0.1029	0.6644
Si_{VI}	(2i)	0.1714	0.6062	0.1504
Si_{VII}	(2i)	0.6749	0.8828	0.1881
Si_{VIII}	(2i)	0.1759	0.3802	0.6744
Al_{I}	(2i)	0.0069	0.1609	0.6125
Al_{II}	(2i)	0.4987	0.6675	0.1125
Al_{III}	(2i)	0.9928	0.8146	0.1190
Al_{IV}	(2i)	0.5078	0.3154	0.6212
Al_{V}	(2i)	0.6841	0.1134	0.1512
Al_{VI}	(2i)	0.1906	0.6123	0.6681
Al_{VII}	(2i)	0.6799	0.8717	0.6715
Al_{VIII}	(2i)	0.1865	0.3790	0.1815
O_{AI}	(2i)	0.0276	0.1234	0.9957
O_{AII}	(2i)	0.9814	0.1237	0.4808
O_{AIII}	(2i)	0.4873	0.6256	0.4865
O_{AIV}	(2i)	0.5129	0.6256	0.9970
O_{AV}	(2i)	0.5724	0.9909	0.1451
O_{AVI}	(2i)	0.5732	0.9901	0.6398
O_{AVII}	(2i)	0.0732	0.4876	0.6330
O_{AVIII}	(2i)	0.0755	0.4925	0.1363
O_{BI}	(2i)	0.8114	0.1027	0.0792
O_{BII}	(2i)	0.8071	0.0996	0.6046
O_{BIII}	(2i)	0.3363	0.5938	0.6045
O_{BIV}	(2i)	0.2912	0.6036	0.0819
O_{BV}	(2i)	0.8148	0.8516	0.1454
O_{BVI}	(2i)	0.8090	0.8510	0.6018
O_{BVII}	(2i)	0.2983	0.3575	0.6113
O_{BVIII}	(2i)	0.3382	0.3606	0.1309
O_{CI}	(2i)	0.0120	0.2780	0.1351
O_{CII}	(2i)	0.0177	0.2900	0.6486
O_{CIII}	(2i)	0.5082	0.7778	0.6311
O_{CIV}	(2i)	0.5102	0.7969	0.1500
O_{CV}	(2i)	0.0004	0.6802	0.1063
O_{CVI}	(2i)	0.0083	0.6898	0.6017
O_{CVII}	(2i)	0.5150	0.1794	0.6090
O_{CVIII}	(2i)	0.5067	0.1947	0.0974
O_{DI}	(2i)	0,1795	0.1072	0.1919

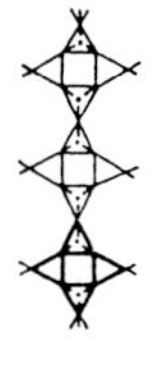

Fig. 1

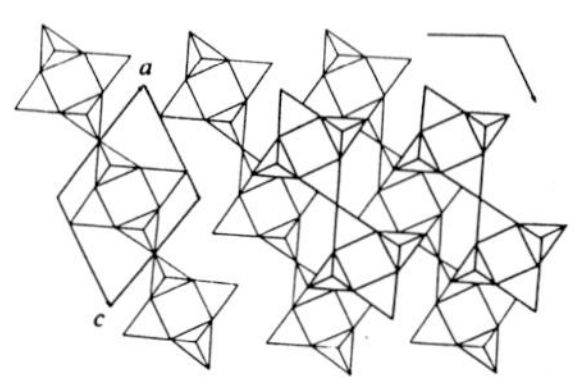

Fig. 2

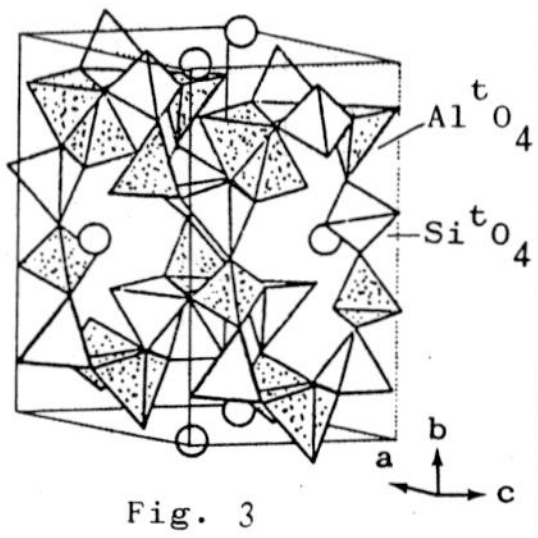

Fig. 3

Properties

Habit	Cleav.	Fract.	Twin.	Hardn.	Dens.	Colour	Transp.
prismatic, tabular	perfect (001)	uneven	Albite [010] Pericline [010]	6-6.5	2.76	white, grey	transparent

Refr. index /Reflect.	Birefr.	Luster	Streak	Melt.p.	CPI
n_{α} = 1.577 n_{β} = 1.585 n_{γ} = 1.590	(-) 2V = 78°	pearly, vitreous	white	1550°C	(SPI) 48

Figures

Fig. 1. Tetrahedral silicate infinite chain of the feldspar type.

Fig. 2. Polyhedral description of the way the feldspar chains are linked together (after Zoltai + Stout, 1984).

Fig. 3. Polyhedral representation, in perspective, of the anorthite structure. Only one half of the unit cell is shown along the c axis (after Zoltai, 1977).

Description

The anorthite structure has the same atomic arrangement as sanidine, but the Al and Si tetrahedra have a certain order, however different from the other feldspars.

Crystallographic data (continued)

Atom	Site	x	y	z
O_{DII}	(2i)	0.2153	0.1057	0.6862
O_{DIII}	(2i)	0.6992	0.6031	0.6779
O_{DIV}	(2i)	0.6921	0.6014	0.2000
O_{DV}	(2i)	0.2027	0.8723	0.2106
O_{DVI}	(2i)	0.1754	0.8557	0.7170
O_{DVII}	(2i)	0.6861	0.3637	0.7335
O_{DVIII}	(2i)	0.6999	0.3691	0.1993

References

Kostov (1968) 393.

Povarennykh (1972) 344.

Wyckoff (1968) Vol. 4, 447-449.

Zoltai + Stout (1984) 303, 305.

Ingerson (1955) 352.

Zoltai (1977) 6-89.

KEATITE

$P\,4_1 2_1 2$

$\overset{3}{\infty}\left[Si^{t}\ O_2\right]^{III^{1}_{Qd}}$

a = 7.456 Å
c = 8.604 Å
Z = 12

Si_I	(8b)	x = 0.326, y = 0.120, z = 0.248
Si_{II}	(4a)	x = 0.410, y = 0.410, z = 0
O_I	(8b)	x = 0.445, y = 0.132, z = 0.400
O_{II}	(8b)	x = 0.117, y = 0.123, z = 0.296
O_{III}	(8b)	x = 0.344, y = 0.297, z = 0.143

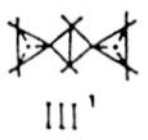

Fig. 1

Fig. 2

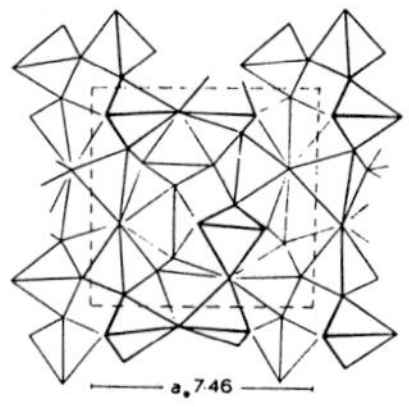

Fig. 3

Properties

Habit	Cleav.	Fract.	Twin.	Hardn.	Dens.	Colour	Transp.
					2.50		

Refr. index/Reflect.	Birefr.	Luster	Streak	Melt.p.	CPI
n_ω = 1.522 n_ϵ = 1.513	(-)				

Figures

Fig. 1. Group of three silicon tetrahedra which constitutes the connected unit of the keatite structure.

Fig. 2. Polyhedral description of the keatite structure (after Liebau, 1985).

Fig. 3. Polyhedral drawing of the keatite structure projected along the <u>c</u> axis (after Kostov, 1968).

Description

The keatite structure is similar to that of quartz but instead of threefold spirals, it has four-fold spirals of silicon tetrahedra along the tetragonal axis. Morphological data is lacking because it is only known as microcrystalline synthetic product.

References

Kostov (1968) 390, 391.
Wyckoff (1963) Vol. 1, 321-324.
Liebau (1985) 127.
Frondel (1982) 307.

CUPRITE

Pn3m

a = 4.2696 Å

Z = 2

Cu(4b)

O (2a)

$\overset{3}{\infty}\left[Cu_2^{[2]} O^t\right]$

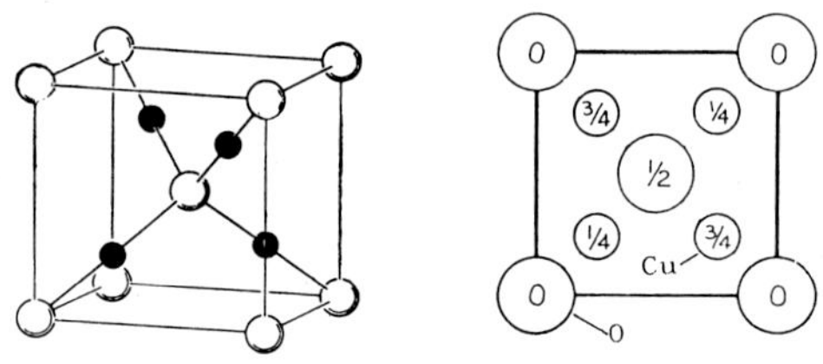

(a) Fig. 1 (b)

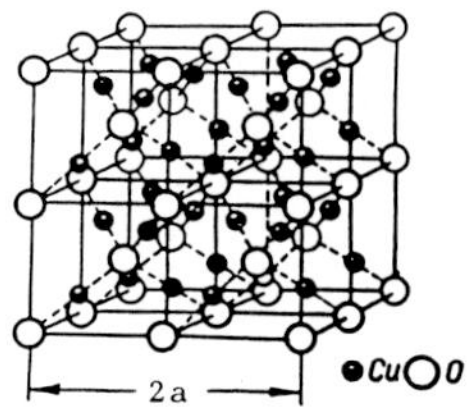

Fig. 2

Fig. 3

Properties

Habit	Cleav.	Fract.	Twin.	Hardn.	Dens.	Colour	Transp.
cubic, octahedral	poor (111)	conchoidal		3.5-4	5.9-6.1	red	opaque

Refr. index	/Reflect.	Birefr.	Luster	Streak	Melt.p.	CPI
n = 2.849	23%		submetallic	brown red	1235°C	(SPI) 33

Distortion derivatives

Argentite

$\overset{3}{\infty}\left[Ag_2^{[2]} S^t\right]$ Im3m

Naumannite

$\overset{3}{\infty}\left[Ag_2^{[2]} Se^t\right]$ Im3m

Figures

Fig. 1. (a) Ball and spoke drawing of the cuprite structure, and (b) corresponding unit cell projection along an a axis.

Fig. 2. Ball and spoke description of the cuprite structure (after Povarennykh, 1972).

Fig. 3. Polyhedral drawing of the cuprite structure (after Povarennykh, 1972).

Description

The cuprite structure is a framework of O^tCu_4 tetrahedra.

References

Kostov (1968) 258.
Povarennykh (1972) 294.
Wyckoff (1963) Vol. 1, 331, 332.
Zoltai + Stout (1984) 419.
Ingerson (1955) 351.

MILLERITE

R3m

$\overset{3}{\infty}\left[Ni^{[5]} S\right]$

$a_R = 5.655Å$
$\alpha = 116^\circ 36'$
$Z_R = 3$
$a_H = 9.612Å$
$c = 3.259Å$
$Z_H = 9$

For hex. cell:

Ni(9b) u = -0.088
v = 0.088

S (9b) u = 0.114
v = 0.596

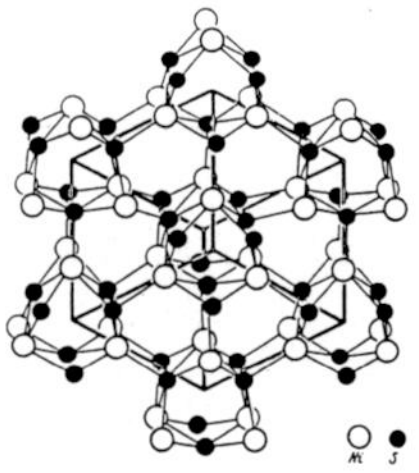

Fig. 1

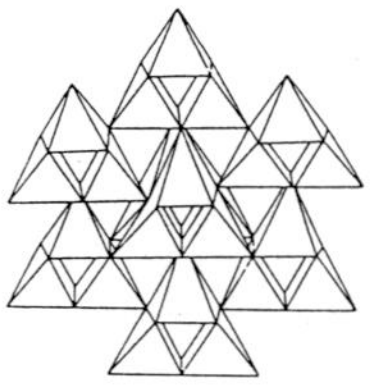

Fig. 2

Properties

Habit	Cleav.	Fract.	Twin.	Hardn.	Dens.	Colour	Transp.
fibrous	perfect (10$\bar{1}$1) (01$\bar{1}$2)	uneven		3-3.5	5.41	brass yellow	

Refr. index /Reflect.	Birefr.	Luster	Streak	Melt.p.	CPI
54%		metallic	greenish black	990°C	

Figures

Fig. 1. Ball and spoke representation of the millerite structure (after Strukturbericht, 1931, Vol. 1).

Fig. 2. Polyhedral representation of the millerite structure (after Kostov, 1968),

Description

The millerite structure is a framework of interlinked NiS_5 tetragonal pyramides.

References

Kostov (1968) 121-123.

Wyckoff (1963) Vol. 1, 122.

Zoltai + Stout (1984) 390.

Roberts et al. (1974) 406.

Strukturbericht (1931) Vol. 1, 739.

Ingerson (1955) 350.

BERYL	P 6/m c c	a = 9.206 Å c = 9.205 Å Z = 2	Al+0.3R (4c)	x = 1/3 y = 2/3 z = 1/4	O_I (12l)	x = 0.294 y = 0.242 z = 0
$Al_2^o \, {}^3_\infty\left[Be_3^t\, Si_6^t\, O_{18}\right] \; IX_s^{201}$			5.68Be+ 0.32Al (6f)	x = 1/2 y = 0 z = 1/4	O_{II} (24m)	x = 0.499 y = 0.143 z = 0.138
			Si+0.02Al (12l)	x = 0.382 y = 0.118 z = 0		

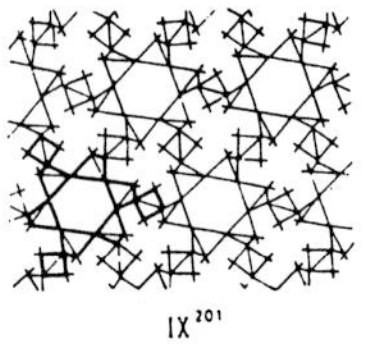

Fig.1

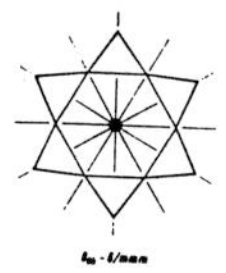

Fig. 2

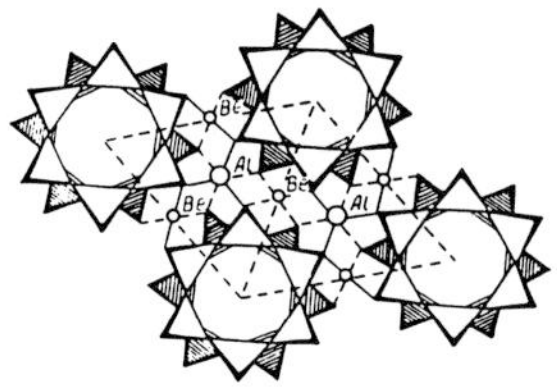

(a)

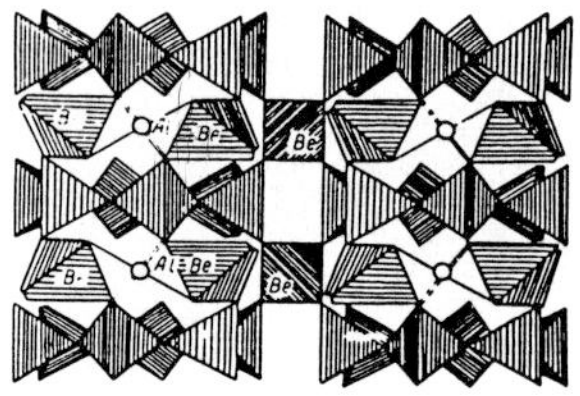

(b)

Fig. 3

Properties

Habit	Cleav.	Fract.	Twin.	Hardn.	Dens.	Colour	Transp.
prismatic	poor (001)	even	(311), (110)	7.5-8	2.7-2.9	colourless, variable	transparent to translucent

Refr. index/Reflect.	Birefr.	Luster	Streak	Melt.p.	CPI
n_ω = 1.568 n_ϵ = 1.562	(-)	vitreous	white	1420°C	(SPI) 50

Figures

Fig. 1. Layer of Be O_4 and Si O_4 tetrahedra which constitutes the connected unit of the beryl structure.

Fig. 2. Beryllium oxygen ring (after Ramdohr + Strunz, 1980).

Fig. 3. (a) Polyhedral description of the beryl structure projected on the c axis, and (b) viewed along the c axis (after Povarennykh, 1972).

Description

The beryl structure is formed by six-member rings of SiO_4 tetrahedra linked by BeO_4 tetrahedra giving rise, in the whole, to a framework. The aluminium atoms are located in the voids with octahedral coordination.

References

Kostov (1968) 277, 278.
Povarennykh (1972) 376.
Wyckoff (1968), Vol. 4, 277-279.
Zoltai + Stout (1984) 314.
Ramdohr + Strunz (1980) 151, 706.
Ingerson (1955) 352.

CORDIERITE

Cccm

$Mg^{o}_{2} \; {}^{3}_{\infty}\left[Al^{t}_{4} Si^{t}_{5} O_{18}\right] \; IX^{201}_{s}$

a =	17.083 Å
b =	9.738 Å
c =	9.335 Å
Z =	4

Atom	Site	x	y	z
Mg	(8g)	0.3374	0	1/4
$(Al,Si)_{I}$	(8k)	1/4	1/4	0.2500
$(Al,Si)_{II}$	(4b)	0	1/2	1/4

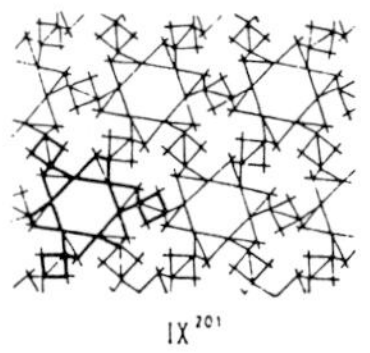

Fig. 1

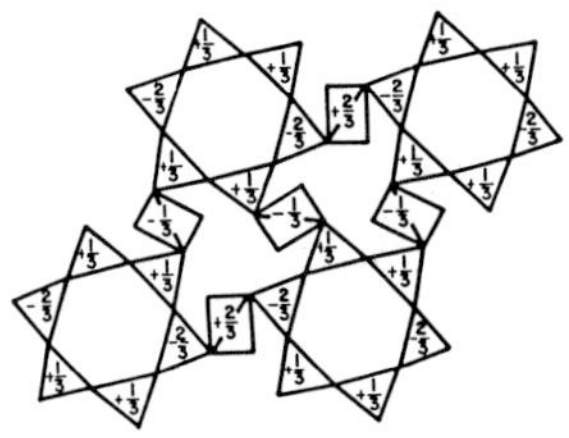

Fig. 2

Fig. 3

Properties

Habit	Cleav.	Fract.	Twin.	Hardn.	Dens.	Colour	Transp.
prismatic	good (010)	even	(110), (130)	7	2.5-2.8	grey blue	transparent to trans-lucent

Refr. index/Reflect.	Birefr.	Luster	Streak	Melt.p.	CPI
n_{α} = 1.54 n_{β} = 1.55 n_{γ} = 1.56	(+), (-) 2V = 65°-105°	vitreous	white	1470° C	(SPI) 48

Figures

Fig. 1. Layer of $(Al,Si)^{t}O_{4}$ which constitutes the connected unit of the cordierite structure, and that is the same of the beryl structure.

Fig. 2. Polyhedral description of the cordierite structure, showing only the tetrahedra, and projected on the c axis (after Putnis + Angel, 1985).

Fig. 3. Polyhedral representation of the cordierite structure showing the structural relationship between the six member rings of $(Si,Al)^{t}O_{4}$ tetrahedra (after Armbruster, 1986).

Description

The cordierite structure is a distortion substitution derivative of the beryl structure.

Crystallographic data (continued)

Atom	Site	x	y	z
$(Al,Si)_{III}$	(81)	0.1923	0.0781	0
$(Al,Si)_{IV}$	(81)	0.1351	-0.2372	0
$(Al,Si)_{V}$	(81)	0.0506	0.3084	0
O_{I}	(16m)	0.2466	-0.1040	0.3591
O_{II}	(16m)	0.0616	-0.4167	0.3494
O_{III}	(16m)	-0.1730	-0.3091	0.3583
O_{IV}	(81)	0.0434	-0.2453	0
O_{V}	(81)	0.1224	0.1848	0
O_{VI}	(81)	0.1639	-0.0788	0
channel atoms (H_2O)	(4a)	0	0	1/4

References

Kostov (1968) 280.

Povarennykh (1972) 376.

Wyckoff (1968) Vol. 4, 437-439.

Zoltai + Stout (1984) 314.

Ingerson (1955) 352.

Putnis + Angel (1985) 220.

Armbruster (1986) 747.

SPHENE (Titanite)	C 2/c	a = 6.55 Å	Ca(4e) u = 0.082	O_{II} (8f) x = 0.028, y = 0.322, z = 0.100
$^3_\infty[Ca^p Ti^o O Si^t O_4]$		b = 8.70 Å	Ti(4c)	O_{III}(4e) u = -0.187
		c = 7.43 Å	Si(4e) u = 0.319	
		β = 119° 43'	O_I(8f) x = 0.25, y = 0.039, z = 0.100	
		Z = 4		

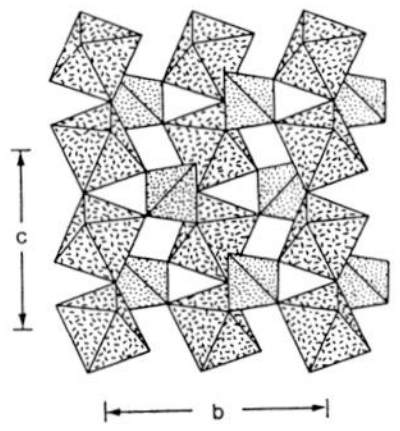

Fig. 1

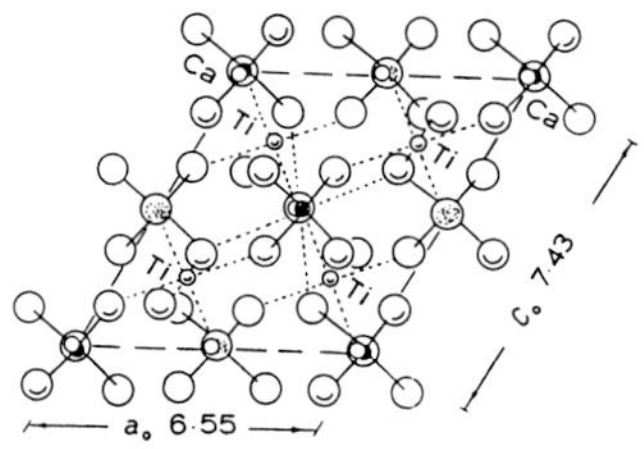

Fig. 2

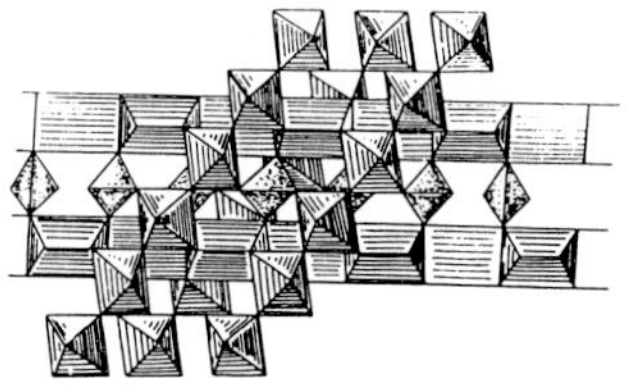

Fig. 3

Properties

Habit	Cleav.	Fract.	Twin.	Hardn.	Dens.	Colour	Transp.
tabular	perfect (110)	uneven	(100), (221) lamellar	5	3.50	grey, variable	transparent to trans-lucent

Refr. index /Reflect.	Birefr.	Luster	Streak	Melt.p.	CPI
n_α = 1.86, n_β = 1.93, n_γ = 2.10	(+) 2V = 23°-50°	adaman-tine	white		(SPI) 55

Population

Tilasite $^3_\infty[Ca^p Mg^o F As^t O_4]$

Durangite $^3_\infty[Na^p Al^o F As^t O_4]$

Figures

Fig. 1. Partial representation of the structure of sphene (after Groat et al., 1990).

Fig. 2. Ball and spoke representation of the structure of sphene projected along the b axis (after Kostov, 1968).

Fig. 3. Polyhedral description of the structure of sphene projected on the b axis (after Povarennykh, 1972).

Description

The structure of sphene is a framework. It consists of infinite chains of Ti^oO_6 octahedra linked by their corners; between these chains and parallel to the c axis run columns of distorted calcium prisms. SiO_4 tetrahedra link the Ti chains in the (010) plane.

References

Kostov (1968) 298, 299.

Povarennykh (1972) 388, 389, 392.

Wyckoff (1965) Vol. 3, 181, (1968) Vol. 4, 190.

Zoltai + Stout (1984) 366.

Groat + al. (1990) 992-1008.

KIESERITE C2/c

$^{3}_{\infty}\left[Mg^{o}\ (H_2O)\ S^{t}\ O_4\right]$

a = 6.886 Å
b = 7.610 Å
c = 7.630 Å
β = 117° 43'
Z = 4

Mg (4b)
S (4e) y = 0.158
H_2O(4e) y = 0.640

O_I (8f) x = 0.113, y = 0.273, z = 0.167
O_{II} (8f) x = 0.167, y = 0.043, z = 0.413

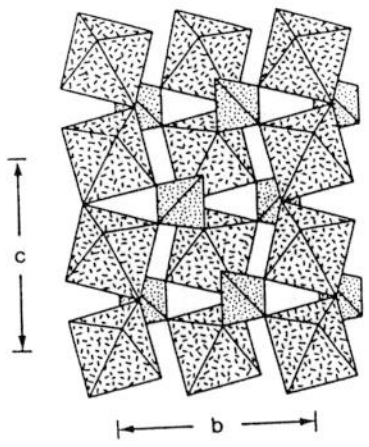

Fig. 1

Properties

Habit	Cleav.	Fract.	Twin.	Hardn.	Dens.	Colour	Transp.
massive, granular	perfect (101) (111)			3.5	2.571	colourless, white, greyish, yellowish	transparent to translucent

Refr. index /Reflect.	Birefr.	Luster	Streak	Melt.p.	CPI
n_α = 1.520 n_β = 1.533 n_γ = 1.584	(+) 2V = 55°	vitreous			

Figures

Fig. 1. Tetrahedral-octahedral network of the kieserite structure (after Groat et al., 1990).

Description

The kieserite structure consists of individual SO_4 tetrahedra linked by Mg atoms in octahedra coordination.

References

Kostov (1968) 494.
Povarennykh (1972) 590.
Wyckoff (1965) Vol. 3, 554.
Groat et al. (1990) 1000.
Structure Reports (1964) Vol. 21, 361, 362.
Roberts et al. (1974) 324.

AMBLYGONITE $P\bar{1}$

$^{3}_{\infty}\left[Li^{[6]}Al^{o}(F,OH)P^{t}O_4\right]$

a = 5.184 Å	Li (2i) x = 0.615, y = 0.305, z = 0.560	(F,OH)(2i) x = 0.959, y = 0.225, z = 0.360
b = 7.155 Å	Al_I (1a) x = 0, y = 0, z = 0	P (2i) x = 0.358, y = 0.267, z = 0.941
c = 5.040 Å	Al_{II} (1g) x = 0, y = 1/2, z = 1/2	O_I (2i) x = 0.236, y = 0.425, z = 0.784
α = 112° 7'		...
β = 97° 48'		
γ = 67° 53'		
Z = 2		

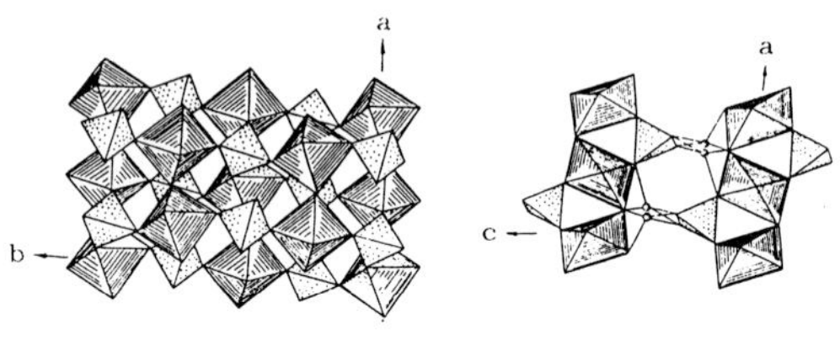

(a) (b)

Fig. 1

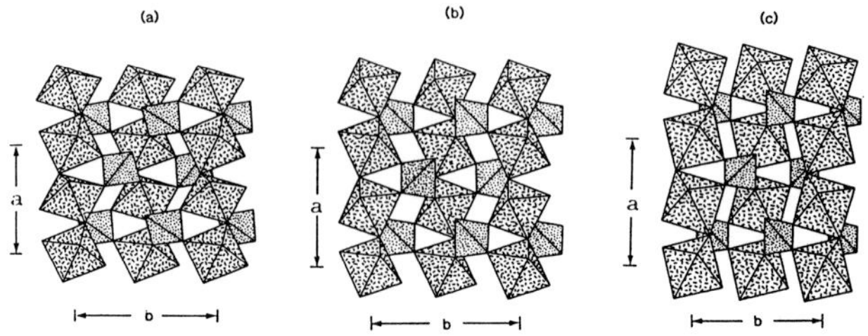

Fig. 2

Properties

Habit	Cleav.	Fract.	Twin.	Hardn.	Dens.	Colour	Transp.
equant	perfect (100)	subconchoidal	(111)	6	3.0	white, green	transparent to translucent

Refr. index/Reflect.	Birefr.	Luster	Streak	Melt.p.	CPI
n_α = 1.59 n_β = 1.60 n_γ = 1.62	(-) 2V = 52°-90°	vitreous greasy	white		(SPI) 52

Figures

Fig. 1. Polyhedral partial representation of the amblygonite structure: (a) projected on the c axis, and (b) projected on the b axis (after Povarennykh, 1972).

Fig. 2. Polyhedral partial representation of: (a) amblygonite, (b) titanite, and (c) kieserite structures, in order to realise their relationship (after Groat et al., 1990).

Description

The structure of amblygonite is a framework. It consists of infinite chains of $Al^{o}O_6$ octahedra linked by their corners by PO_4 tetrahedra. The chains are parallel to the a axis. Li atoms are located in large voids and have 5 coordination.

References

Kostov (1968) 450.
Povarennykh (1972) 530, 531.
Wyckoff (1965) Vol. 3, 192, 193.
Zoltai + Stout (1984) 450.
Groat et al. (1990) 1000.

Crystallographic data (continued)

O_{II} (2i) x = 0.325, y = 0.391, z = 0.269

O_{III} (2i) x = 0.213, y = 0.103, z = 0.842

O_{IV} (2i) x = 0.670, y = 0.160, z = 0.856

LEUCITE

$I\,4_1/a$

$K^{[6+6]}\ {}^{3}_{\infty}\left[Si^{t}_{2}Al^{t}O_{6}\right]$

a = 13.09 Å
c = 13.75 Å
Z = 16

Atom	x	y	z
K (16f)	0.3660	0.3645	0.1147
$(Si,Al)_I$ (16f)	0.0579	0.3964	0.1666
$(Si,Al)_{II}$ (16f)	0.1676	0.6115	0.1283
$(Si,Al)_{III}$ (16f)	0.3924	0.6418	0.0860
O_I (16f)	0.1318	0.3131	0.1100
O_{II} (16f)	0.0921	0.5107	0.1303
O_{III} (16f)	0.1453	0.6798	0.2275
O_{IV} (16f)	0.1333	0.6841	0.0354
O_V (16f)	0.2900	0.5772	0.1205
O_{VI} (16f)	0.4826	0.6174	0.1667

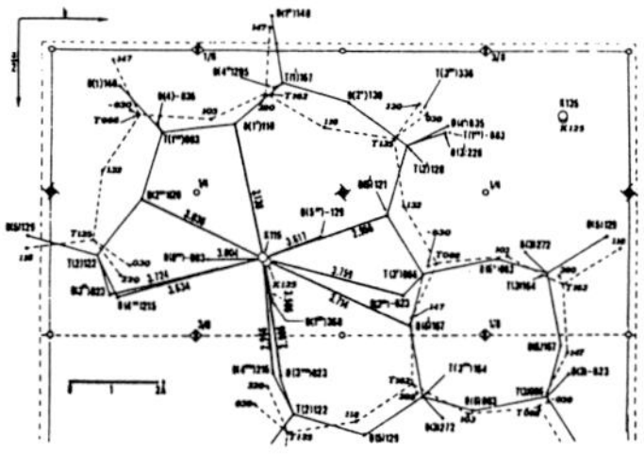

Properties

Habit	Cleav.	Fract.	Twin.	Hardn.	Dens.	Colour	Transp.
trapezohedral	poor (100)	conchoidal	(100) (112)	5.5-6	2.48	white, grey	transparent to translucent

Refr. index/Reflect.	Birefr.	Luster	Streak	Melt.p.	CPI
n_ω = 1.508 n_ϵ = 1.509	(+)	vitreous	white	1686° C	(SPI) 39

Figures

Fig. 1. Structure of tetragonal (full lines) and cubic (dashed lines) forms of leucite (after Structure Reports, 1978, Vol. 42A).

Description

The leucite structure is a framework of $(Si,Al)O_4$ tetrahedra, with potassium atoms located within the voids.

References

Kostov (1968) 403.
Wyckoff (1968) Vol. 4, 400,401.
Zoltai + Stout (1984) 310.
Klein + Hurlbut (1985) 456.
Mazzi et al. (1976) 108-115.
Papike + Cameron (1976) 74.
Ingerson (1958) 352.
Structure Reports (1978) 42A, 408.

BORACITE

Pc2a

$\overset{[6]}{Mg_3} Cl \overset{3}{\infty} \left[\overset{t}{B_3} \overset{tr}{B_4} O_{13} \right]$

a = 8.54 Å
b = 8.54 Å
c = 12.07 Å
Z = 4

Atom	Position	x	y	z
Mg_I	(4a)	0.500	0	0.981
Mg_{II}	(4a)	0.231	-0.269	0.250
Mg_{III}	(4a)	0.231	0.269	0.250
B_I	(4a)	0.250	-0.250	0
B_{II}	(4a)	0	0	0.250
B_{III}	(4a)	-0.250	0.250	0
B_{IV}	(4a)	0.500	0.330	0.415
B_V	(4a)	0.500	-0.330	0.415
B_{VI}	(4a)	0.170	0	0.085

...

Fig. 1

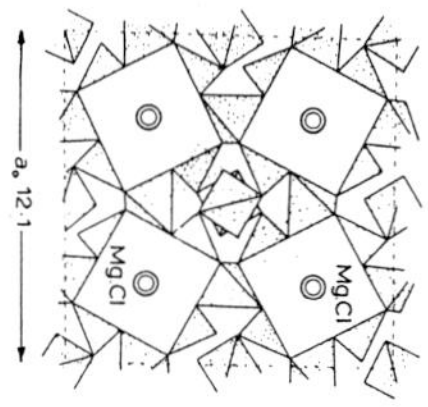

Fig. 2

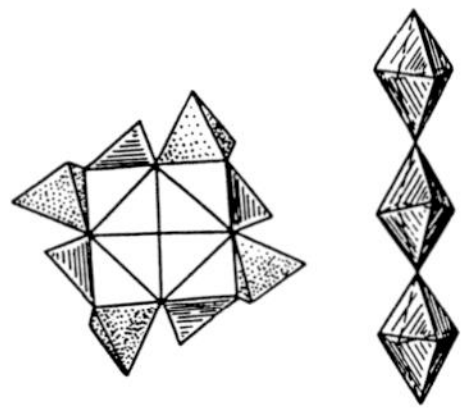

Fig. 3

Properties

Habit	Cleav.	Fract.	Twin.	Hardn.	Dens.	Colour	Transp.
pseudo-cubic	poor (111)	conchoidal		7	2.95	white, yellow	transparent to translucent

Refr. index/Reflect.	Birefr.	Luster	Streak	Melt.p.	CPI
n_α = 1.662 n_β = 1.647 n_γ = 1.673	(+) 2V = 82°	vitreous	white		(SPI) 61

Figures

Fig. 1. Polyhedral partial description of the boracite structure (after Povarennykh, 1972).

Fig. 2. Polyhedral description of the high temperature form of boracite which is cubic, $F\bar{4}3c$, a = 12.1 (after Kostov, 1968).

Fig. 3. Polyhedral representation of the Mg octahedron position in a square section of the framework and chain of distorted Mg octahedra (after Povarennykh, 1972).

Description

The boracite structure is a framework consisting of regular B^tO_4 tetrahedra and $B^{tr}O_3$ triangles, with large square holes where are located Mg and Cl atoms forming infinite octahedral chains.

References

Kostov (1968) 432, 433.
Povarennykh (1972) 462, 463.
Wyckoff (1968) Vol. 4, 60-64.
Zoltai + Stout (1984) 435.

Crystallographic data (continued)

Atom	Position	x	y	z
B_{VII}	(4a)	0.170	0	-0.415
O_I	(4a)	0	0	0
O_{II}	(4a)	0.082	-0.278	0.479
O_{III}	(4a)	0.159	0.201	0.402
O_{IV}	(4a)	0.077	-0.119	0.320
O_V	(4a)	0.418	0.222	0.479
O_{VI}	(4a)	0.341	-0.299	0.402
O_{VII}	(4a)	0.423	0.381	0.320
O_{VIII}	(4a)	0.119	0.076	0.180
O_{IX}	(4a)	0.201	-0.159	0.098
O_X	(4a)	0.222	-0.418	0.021
O_{XI}	(4a)	0.381	-0.423	0.180
O_{XII}	(4a)	0.299	0.341	0.098
O_{XIII}	(4a)	0.278	0.082	0.021
Cl	(4a)	0.524	0	0.262

TURQUOISE

$P\bar{1}$

$Cu^{[6]}(H_2O)_4 \; {}^3_\infty \left[Al^{o}_6 (OH)_8 (P^{t}O_4)_4 \right]$

a = 7.424 Å
b = 7.629 Å
c = 9.910 Å
α = 68°61'
β = 69°71'
γ = 65°08'
Z = 1

Atom	Site	x	y	z
Cu	(1a)	0	0	0
Al_I	(2i)	0.2843	0.1766	0.7521
Al_{II}	(2i)	0.7520	0.1862	0.2736
Al_{III}	(2i)	0.2448	0.5023	0.2438
P_I	(2i)	0.3504	0.3867	0.9429
P_{II}	(2i)	0.8423	0.3866	0.4570

...

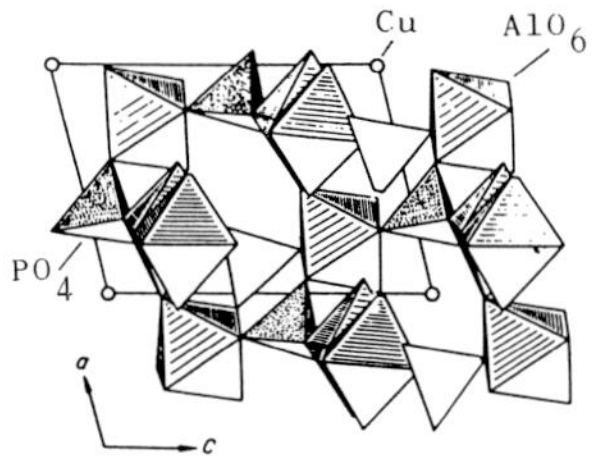

Fig. 1

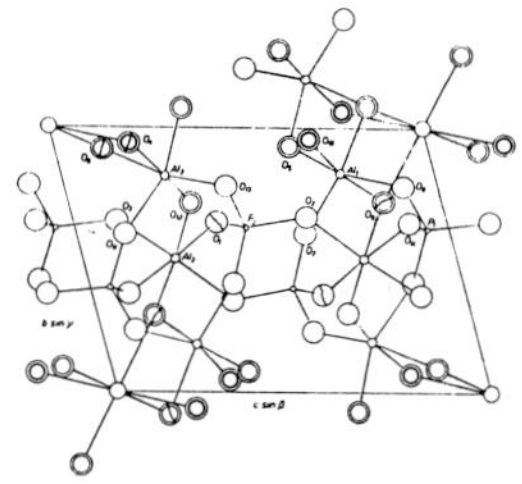

Fig. 2

Properties

Habit	Cleav.	Fract.	Twin.	Hardn.	Dens.	Colour	Transp.
reniform	perfect (001)	subconchoidal, britle		6	2.7	blue green	translucent

Refr. index /Reflect.	Birefr.	Luster	Streak	Melt.p.	CPI
n_α = 1.61 n_β = 1.62 n_γ = 1.65	(+) 2V = 40°	resinous, waxy	white, green		(SPI) 35

Figures

Fig. 1. Polyhedral representation of the structure of turquoise projected along the b axis (after Povarennykh, 1972).

Fig. 2. Ball and spoke representation of the structure of turquoise projected along the a axis (after Cid-Dresdner, 1965).

Description

The turquoise structure is a framework of P^tO_4 tetrahedra and two types of Al octahedra, $AlO_2(OH)_3(H_2O)$ and $AlO_4(OH)_2$, with Cu atoms and water molecules located within the large holes.

Atom	Site	x	y	z
O_I	(2i)	0.0675	0.3633	0.3841
O_{II}	(2i)	0.8058	0.3435	0.6262
O_{III}	(2i)	0.2757	0.3554	0.1129
O_{IV}	(2i)	0.0663	0.0639	0.1973
O_V	(2i)	0.2375	0.0739	0.6287
O_{VI}	(2i)	0.7334	0.0857	0.1243
O_{VII}	(2i)	0.2978	0.4016	0.6060
O_{VIII}	(2i)	0.3249	0.2227	0.9049
O_{IX}	(2i)	0.9857	0.2807	0.8471
O_X	(2i)	0.5756	0.0467	0.6855
O_{XI}	(2i)	0.7866	0.4067	0.1319
O_{XII}	(2i)	0.4630	0.2950	0.3277
O_{XIII}	(2i)	0.7864	0.2281	0.4323
O_{XIV}	(2i)	0.5779	0.3660	0.8987

References

Kostov (1968) 473.
Povarennykh (1972) 534, 535.
Zoltai + Stout (1984) 452.
Cid-Dresdner(1965) 87-113.

VARISCITE

P cab

$(H_2O)_2 \overset{3}{\infty} \left[(Al,Fe)^{o} P^{t} O_4 \right]$

a = 9.822 Å
b = 8.561 Å
c = 9.630 Å
Z = 8

Atom	x	y	z
Al(8c)	0.13389	0.15500	0.16841
P (8c)	0.14779	0.46844	0.35284
O_I(8c)	0.11180	0.29870	0.31525
O_{II} (8c)	0.04030	0.58186	0.29453
O_{III}(8c)	0.28545	0.51245	0.29006
O_{IV} (8c)	0.14997	0.47916	0.51224

...

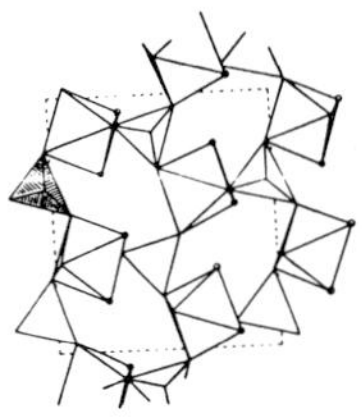

Fig. 1

Fig. 2

Properties

Habit	Cleav.	Fract.	Twin.	Hardn.	Dens.	Colour	Transp.
prismatic	good (001)	subconchoidal, splintery		4	2.5	green, yellow	transparent to translucent

Refr. index/Reflect.	Birefr.	Luster	Streak	Melt.p.	CPI
n_α = 1.55-1.56 n_β = 1.57-1.58 n_γ = 1.58-1.59	(-) 2V = 48°-54°	resinous	white		(SPI) 39

Population

Scorodite

$(H_2O)_2 \overset{3}{\infty} \left[Fe^{o} As^{t} O_4 \right]$

Figures

Fig. 1. Polyhedral representation of the structure of variscite (after Kniep et al., 1977).

Fig. 2. Polyhedral representation of the structure of metavariscite, which is a monoclinic modification of variscite (after Kniep + Mootz, 1973).

Description

The variscite a framework structure consisting of interlinked $Al^{o}O_4(H_2O)_2$ octahedra and $P^{t}O_4$ tetrahedra. Metavariscite is the monoclinic modification of variscite.

Crystallographic data (continued)

Atom	x	y	z
O_V (8c)	0.06041	0.32564	0.05469
O_{VI}(8c)	0.30726	0.23597	0.11499
H_I (8c)	0.063	0.310	-0.028
H_{II}(8c)	-0.012	0.317	0.068
H_{III}(8c)	0.362	0.201	0.168
H_{IV} (8c)	0.314	0.334	0.085

References

Kostov (1968) 448.
Povarennykh (1972) 531.
Zoltai + Stout (1984) 452.
Kniep et al. (1977) 263-265.
Kniep + Mootz (1973) 2292-2294.

TETRAHEDRITE $I\bar{4}3m$

$$^{3}_{\infty}\left[Cu^{t}_{6}\,Cu^{tr}_{6}\,Sb^{[3n]}_{4}\,S_{13}\right]$$

a = 10.323 Å
Z = 2

Cu_I (12d)
Cu_{II} (12e) x = 0.2180
Sb (8c) x = 0.2683
S_I (24g) x = 0.1144, z = 0.3635
S_{II} (2a)

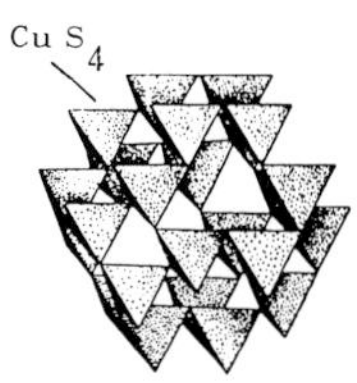

Fig. 1

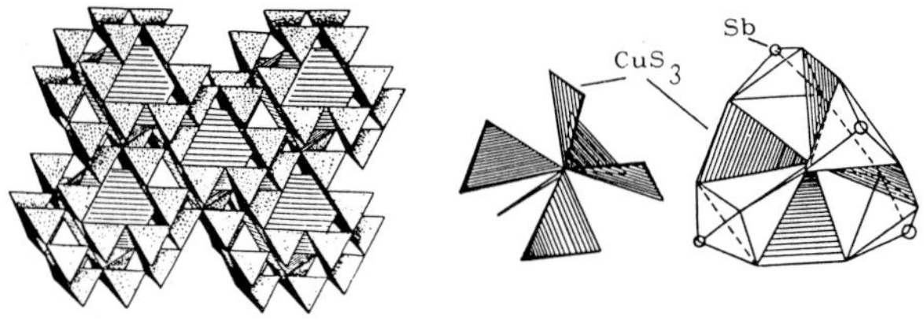

Fig. 2

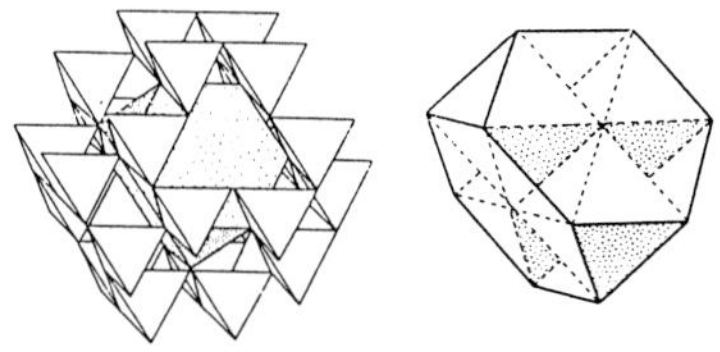

Fig. 3

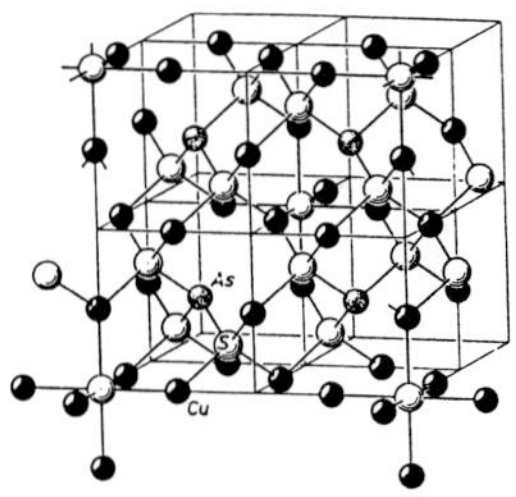

Fig. 4

Properties

Habit	Cleav.	Fract.	Twin.	Hardn.	Dens.	Colour	Transp.
tetrahedral, massive, compact		uneven, subcon-choidal	111	3-4.5	4.6-5.1	steel grey, iron black	opaque

Refr. index	Reflect.	Birefr.	Luster	Streak	Melt.p.	CPI
n = 2.72	24%		metallic	brown, black		

Population

Tennantite

$$^{3}_{\infty}\left[Cu^{t}_{6}\,Cu^{tr}_{6}\,As^{[3n]}_{4}\,S_{13}\right]$$

Description

The tetrahedrite structure is a framework of Cu^tS_4 tetrahedra with Sb and Cu atoms in the large holes forming Friauf-Laves polyhedra. It is related to chalcopyrite, and sphalerite.

References

Kostov (1968) 169, 170.
Povarennykh (1972) 122, 123, 233.
Wyckoff (1964) Vol. 2, 512.
Zoltai + Stout (1984) 397.
Johnson et al. (1988) 393.
Roberts et al. (1974) 611.
Ramdohr + Strunz (1980) 434.

Figures

Fig. 1. Polyhedral description of the skeleton of the tetrahedrite structure, formed only by Cu^tS_4 tetrahedra (after Povarennykh, 1972).

Fig. 2. Complete polyhedral description of the tetrahedrite structure, including the Friauf-Laves polyhedra which contains Sb, S and Cu atoms (after Povarennykh, 1972).

Fig. 3. Tetrahedrite structure showing the position of the Friauf-Laves polyhedron which contains Sb and Cu atoms (after Kostov, 1968).

Fig. 4. Ball and spoke description of the tennantite structure, which is isotypic with tetrahedrite (after Ramdohr + Strunz, 1980).

MARIALITE — I 4/m

(Scapolite)

$Na_4^{[9]} Cl \, {}^{3}_{\infty}\left[Si_3^{t} Al^{t} O_8\right]_3$

a = 12.060 Å	Cl (2a) x = 0, y = 0, z = 0	O_I (8h) x = 0.4587, y = 0.3483, z = 0
c = 7.572 Å	Si (8h) x = 0.3388, y = 0.4104, z = 0	O_{II} (8h) x = 0.3066, y = 0.1206, z = 0
Z = 2		
Na(8h) x = 0.1340, y = 0.2113, z = 0	(Si,Al)(16i) x = 0.3374, y = 0.0851, z = 0.2060	O_{III}(16i) x = 0.0517, y = 0.3500, z = 0.2148

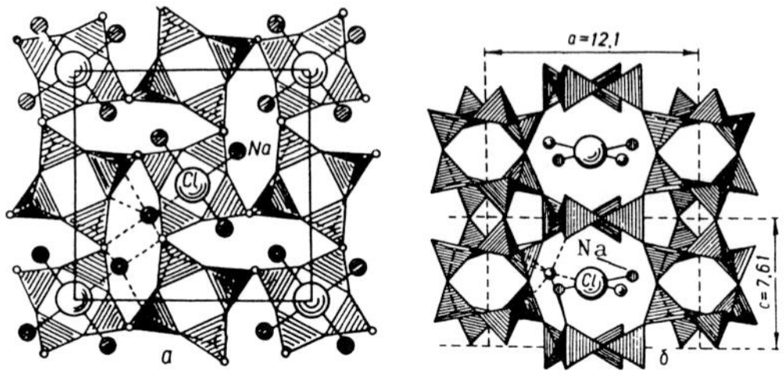

(a) Fig. 1 (b)

Properties

Habit	Cleav.	Fract.	Twin.	Hardn.	Dens.	Colour	Transp.
prismatic	good (100)	conchoidal		5-6	2.55	colourless, variable	transparent to translucent

Refr. index/Reflect.	Birefr.	Luster	Streak	Melt.p.	CPI
n_ω = 1.540 n_ϵ = 1.536	(-)	vitreous	white		(SPI) 45

Figures

Fig. 1. Polyhedral representation of the marialite structure: (a) viewed along the c axis, and (b) viewed along an a axis (after Povarennykh, 1972).

Description

Its structure consists of a framework of Si^tO_4 and $(Si,Al)^tO_4$ tetrahedra, with sodium and chlorine atoms in the holes.

Crystallographic data (continued)

O_{IV}(16i) x = 0.2293, y = 0.1289, z = 0.3281

References

Kostov (1968) 404, 405.
Povarennykh (1972) 348.
Wyckoff (1968) Vol. 4, 434-437.
Zoltai + Stout (1984) 313.

SODALITE

$P\bar{4}3n$

$|Na_4Cl| \, {}^3_\infty[Si^t_3Al^t_3O_{12}]$

a = 8.91 Å

Z = 2

Cl(2a)

Al(6c)

Si(6d)

Na(8e) u = 0.175

O(24i) x = 0.150, y = 0.135, z = 0.440

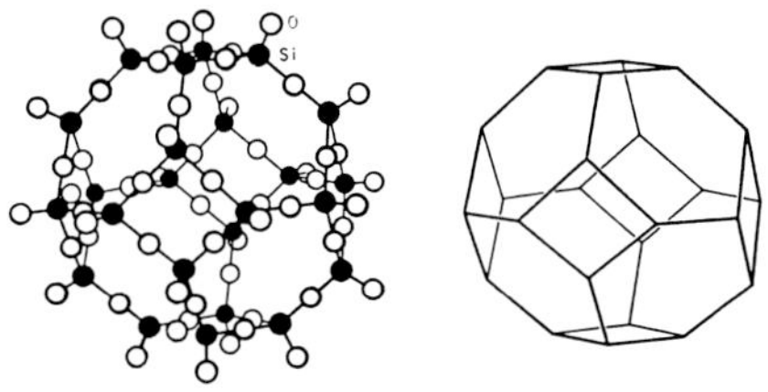

(a) Fig. 1 (b)

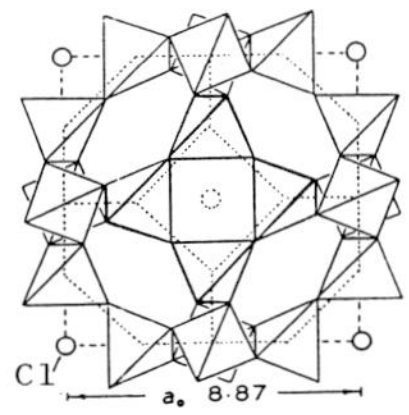

Fig. 2

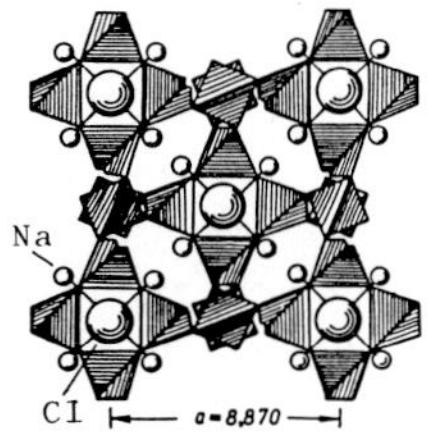

Fig. 3

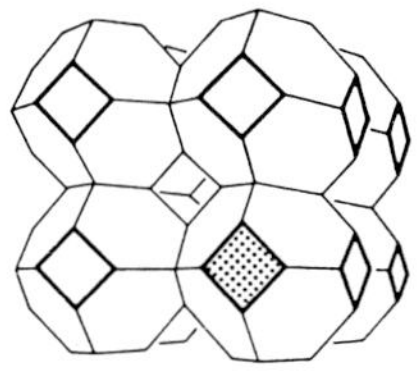

Fig. 4

Properties

Habit	Cleav.	Fract.	Twin.	Hardn.	Dens.	Colour	Transp.
dodeca-hedral	good (110)	conchoi-dal	(111)	5.5-6	2.3	blue, white	transparent to translu-cent

Refr. index/Reflect.	Birefr.	Luster	Streak	Melt.p.	CPI
n = 1.485		vitreous	white		(SPI) 40

Population

Helvine (or helvite)

$|(Mn,Fe)_4S| \, {}^3_\infty[Si^t_3Be^t_3O_{12}]$

Genthelvite

$|Zn_4S| \, {}^3_\infty[Si^t_3Be^t_3O_{12}]$

Danalite

$|Fe_4S| \, {}^3_\infty[Si^t_3Be^t_3O_{12}]$

Description

Sodalite is a zeolite. Its structure is an opened framework of SiO_4 and AlO_4 tetrahedra, with Na_4Cl radicals located in large holes.

Figures

Fig.1. (a) Ball and spoke representation of the polyhedron of silicate tetrahedra, and (b) the corresponding truncated cuboctahedron (after Mumpton, 1981).

Fig. 2. Polyhedral representation of the sodalit framework (after Kostov, 1968).

Fig. 3. Polyhedral representation of the sodalite structure (after Povarennykh, 1972).

Fig. 4. Representation of the sosalite structure showing the way the silicate polyhedra are linked together (after Liebau, 1985).

References

Kostov (1968) 406-408, 418.
Povarennykh (1972) 349,350.
Wyckoff (1968) Vol. 4, 430-432.
Zoltai + Stout (1984) 316.
Liebau (1985) 146.

GMELINITE

$P6_3/mmc$

$(Na,Ca)_2\ (H_2O)_{\sim 6}\ {}^{3}_{\infty}\left[Si^{t}_{4}\ Al^{t}_{2}\ O_{12}\right]$

a = 13.756 Å
c = 10.048 Å
Z = 4

Atom	Position	x	y	z
(Al,Si)	(24l)	0.4419	0.1058	0.0944
Na_I	(4f)	1/3	2/3	0.0743
0.33 Na_{II}	(12k)	0.2246	0.1123	-0.0617
O_I	(12k)	0.4158	0.2079	0.0605
...				

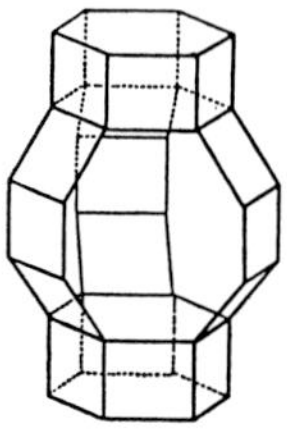

Fig. 1

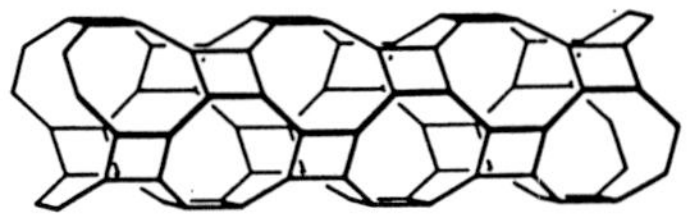

Fig. 2

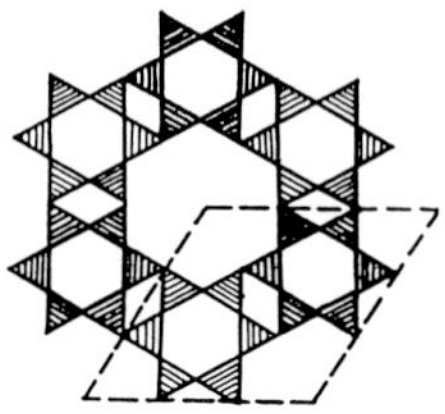

Fig. 3

Properties

Habit	Cleav.	Fract.	Twin.	Hardn.	Dens.	Colour	Transp.
pyramidal	distinct (10$\bar{1}$0)	uneven		4.5	2.04--2.17	colourless, white, yellowish	transparent to trans-lucent

Refr. index /Reflect.	Birefr.	Luster	Streak	Melt.p.	CPI
n_ω = 1.476 – 1.494 n_ϵ = 1.474 – 1.480	(-)	vitreous			

Figures

Fig. 1. Structural scheme of the silica-oxygen framework of the gmelinite structure (after Barrer + Kerr, 1959, quoted by Bragg + Claringbull, 1965).

Fig. 2. Tubular building unit of the silicate framework of the gmelinite structure (after Liebau, 1985).

Fig. 3. Projection on the basal plane of the gmelinite structure (after Povarennykh 1972).

Description

Gmelinite is a zeolite mineral. Its structure consists of a framework of $(Si,Al)O_4$ tetrahedra, with (Na,Ca) atoms and water molecules located in large holes.

Crystallographic data (continued)

Atom	Position	x	y	z
O_{II}	(12k)	0.8524	0.4262	0.0630
O_{III}	(12j)	0.4126	0.0664	1/4
O_{IV}	(12i)	0.3559	0	0
0.50 H_2O_I	(12j)	0.1997	0.5430	1/4
H_2O_{II}	(6h)	0.3356	0.1678	-1/4
0.66 H_2O_{III}	(12k)	0.1544	0.0772	0.1271

References

Kostov (1968) 414.
Povarennykh (1972) 352.
Liebau (1985) 150.
Roberts et al. (1974) 240.
Structure Reports (1989) Vol. 49A, 337.
Bragg + Claringbull (1965) 351.

CHABAZITE

$R\bar{3}m$

$Ca_2 \square_{10} (H_2O)_{13} \ {}^3_{\infty} [Si^t_8 Al^t_4 O_{24}]$

a_R = 9.42 Å a_H = 13.78 Å
α = 94° 28' c = 14.97 Å
Z_R = 1 Z_H = 3

(rhomboh. descript.)
Ca 1/6 (12i) x = 0.354, y = 0.498, z = 0.575

$(H_2O)_I$ (6h) u = 0.244, v = 0.573
$(H_2O)_{II}$ (3e)
$(H_2O)_{III}$ (2c) u = 0.194
$(H_2O)_{IV}$ (2c) u = 0.416
...

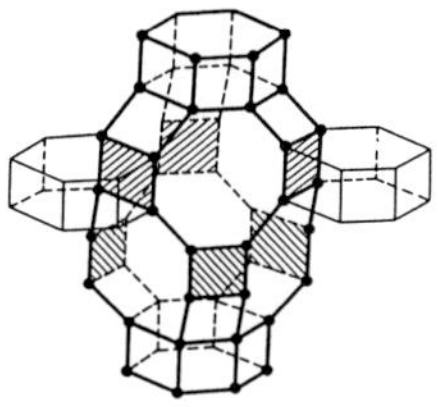
Fig. 1

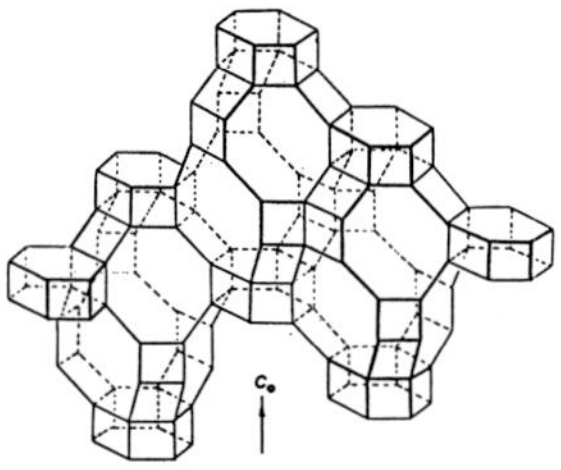

Fig. 2

Fig. 3

Properties

Habit	Cleav.	Fract.	Twin.	Hardn.	Dens.	Colour	Transp.
rhombo-hedral	poor (10$\bar{1}$1)	uneven	(0001)	4-5	2.1	colourless, red	transparent to rans-lucent

Refr. index/Reflect.	Birefr.	Luster	Streak	Melt.p.	CPI
n_ω = 1.484, n_ϵ = 1.481	(-)	vitreous	white		(SPI) 30

Figures

Fig. 1. Silicon-oxygen framework in the chabazite structure. Combination of two-level hexagonal rings with octagonal ones (after Belov, 1963).

Fig. 2. Structural scheme of chabazite (after Kostov, 1968).

Fig. 3. Polyhedral representation of the chabazite structure projected along the c axis (after Povarennykh 1972).

Description

Chabazite is a zeolite mineral. Its structure consists of a framework of $(Si,Al)O_4$ tetrahedra with calcium atoms and water molecules located in large holes.

Crystallographic data (continued)

(Si,Al)(12i) x = 0.1044, y = 0.3330, z = 0.8756

O_I (6f) u = 0.264

O_{II} (6g) u = 0.153

O_{III} (6h) u = 0.253, v = 0.896

O_{IV} (6h) u = 0.026, v = 0.324

References

Kostov (1968) 414, 417.
Povarennykh (1972) 351-352.
Wyckoff (1968) Vol. 4, 408-411.
Zoltai + Stout (1984) 316.
Strunz (1982) 492.
Roberts et al. (1974) 117.
Belov (1963) 29.

LEVYNE R$\bar{3}$m

$(Ca,Na)(H_2O)_6 \,{}^{3}_{\infty}\left[Si^{t}_{4}Al^{t}_{2}O_{12}\right]$

a_R = 10.87 Å
α = 75°42'
Z_R = 3
a_H = 13.338Å
c = 23.014Å
Z_H = 9

Atom	x	y	z
$(Si,Al)_I$ (36i)	0.001	0.2322	0.0697
$(Si,Al)_{II}$ (18g)	0.2396	0	1/2
O_I (36i)	0.0339	0.3495	0.1079
O_{II} (18h)	0.0920	-0.0920	0.0827
O_{III} (18h)	0.1275	-0.1275	-0.0910
O_{IV} (18h)	0.2643	0	0

...

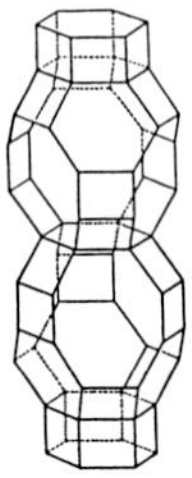

Fig. 1

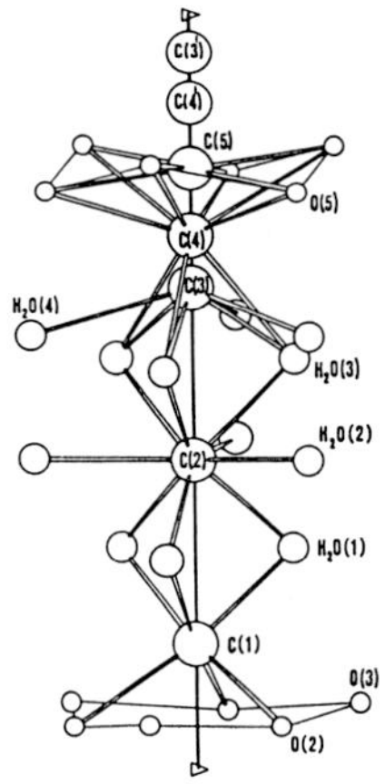

Fig. 2

Properties

Habit	Cleav.	Fract.	Twin.	Hardn.	Dens.	Colour	Transp.
tabular (0001)		uneven		4-4.5	2.09--2.16	colourless, white, greyish	

Refr. index/Reflect.	Birefr.	Luster	Streak	Melt.p.	CPI
n_ω = 1.496-1.505 n_ϵ = 1.491-1.500	(-)	vitreous			

Figures

Fig. 1. Structural scheme of the silica-oxygen framework of the levyne structure (after Barrer + Kerr, 1959, quoted by Bragg + Claringbull, 1965).

Fig. 2. Coordination of the cations (Ca,Na), The vertical line is the ternary axis (after Merlino et al., 1975).

Description

Levyne is a zeolite mineral. Its structure consists in a framework of layers of single and double x-membered rings of $(Si,Al)O_4$ tetrahedra. The cations are distributed over five sites on the threefold axis. One site is fully occupied by Ca, with the other four sites partially occupied. Three of the four water sites are also only partially occupied.

Crystallographic data (continued)

Atom	x	y	z
O_V (18h)	0.2219	-0.2219	0.1793
Ca_I (6c)	0	0	0.1389
$0.28(Ca,Na)_{II}$ (6c)	0	0	0.2782
$0.27(Ca,Na)_{III}$ (6c)	0	0	0.4095
$0.15(Ca,Na)_{IV}$ (6c)	0	0	0.4498
$0.25(Ca,Na)_V$ (3b)	0	0	1/2
$(H_2O)_I$ (18h)	0.2567	-0.2567	-0.1241
$0.73(H_2O)_{II}$ (18h)	0.1222	-0.1222	0.2852
$0.54(H_2O)_{III}$ (18h)	0.2547	-0.2547	0.0187
$0.27(H_2O)_{IV}$ (18h)	0.5429	-0.5429	-0.0445

References

Kostov (1968) 414.
Bragg + Claringbull (1965) 351.
Merlino et al. (1975) 117-129.

NATROLITE

Fd2d

$Na^{o}_{2}(H_{2}O)_{2}\ {}^{3}_{\infty}\left[Si^{t}_{3}\ Al^{t}_{2}\ O_{10}\right]$

a = 18.30 Å
b = 18.63 Å
c = 6.60 Å
Z = 8

Atom	x	y	z
Si_{I} (8a)	0	0	0.000
Si_{II} (16b)	0.1532	0.2112	0.6181
Al (16b)	0.0376	0.0936	0.6119
Na (16b)	0.2208	0.0305	0.6120
O_{I} (16b)	0.0227	0.0683	0.8594
O_{II} (16b)	0.0704	0.1824	0.6011

...

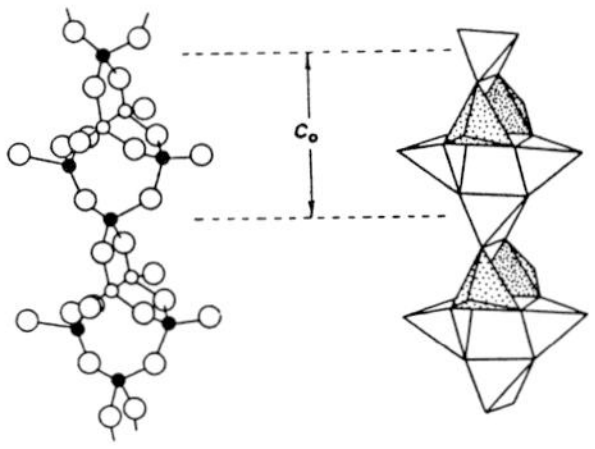

Fig. 1

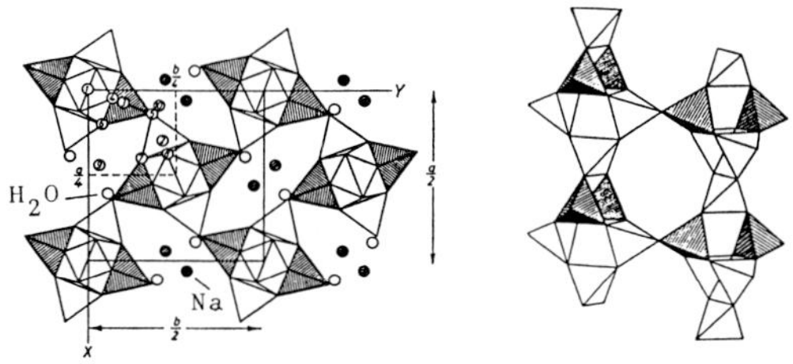

(a) Fig. 2 (b)

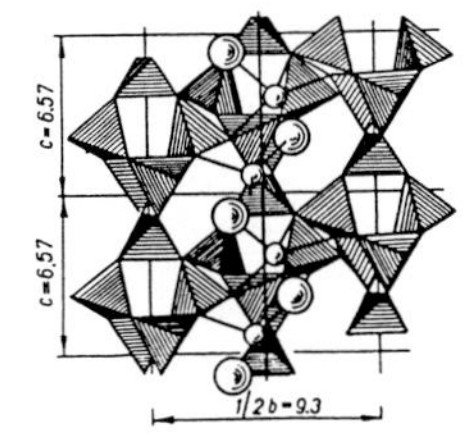

Fig. 3

Properties

Habit	Cleav.	Fract.	Twin.	Hardn.	Dens.	Colour	Transp.
acicular, fibrous	perfect (110)	uneven		5-5.5	2.23	colourless, grey	transparent to translucent

Refr. index/Reflect.	Birefr.	Luster	Streak	Melt.p.	CPI
n_α = 1.48 n_β = 1.48 n_γ = 1.49	(+) 2V = 38°-62°	vitreous	white		(SPI) 34

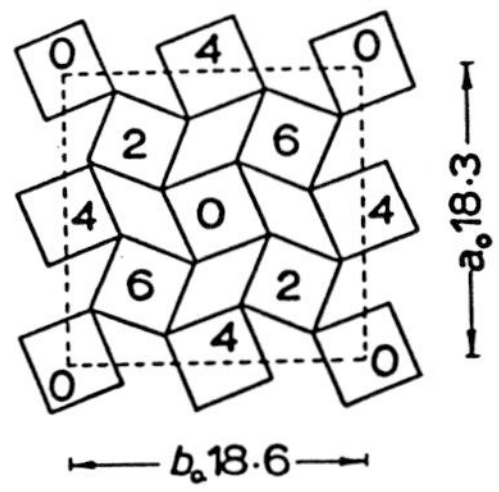

Fig. 4

Figures

Fig. 1. Typical chain of $(Si,Al)O_4$ tetrahedra in the natrolite structure (Kostov, 1968).

Fig. 2. (a) Polyhedral representation of the natrolite structure viewed along the c axis, and (b) parts of two tetrahedra chains showing the way they are linked together (after Bragg + Claringbull, 1965).

Fig. 3. Polyhedral description of the natrolite structure viewed along the a axis (after Povarennykh, 1972).

Fig. 4. Structural scheme of the natrolite structure. Squares represent $(Si,Al)^{t}O_4$ tetrahedral chains viewed along their elongation (after Kostov, 1968).

Description

Natrolite is a zeolite mineral. Its structure consists of a framework of $Si^{t}O_4$ and $Al^{t}O_4$ tetrahedra with sodium and water molecules located in the holes.

Crystallographic data (continued)

Atom	x	y	z
O_{III} (16b)	0.0986	0.0346	0.4997
O_{IV} (16b)	0.2063	0.1526	0.7166
O_{V} (16b)	0.1799	0.2272	0.3860
(H_2O) (16b)	0.0564	0.1893	0.1085

References

Kostov (1968) 418, 419.
Povarennykh (1972) 356.
Wyckoff (1968) Vol. 4, 411-413.
Zoltai + Stout (1984) 317.
Bragg + Claringbull (1965) 362.

MESOLITE

Fdd2

$Na_2^{[6]}Ca_2^{[7]}(H_2O)_8\ {}^3_\infty\left[Si_3^t\ Al_2^t\ O_{10}\right]_3$

a = 18.4049 Å
b = 56,655 Å
c = 6.5443 Å
Z = 8

Atom	Site	x	y	z
Na	(16b)	0.96736	0.76089	0.63172
Ca	(16b)	0.22807	0.82737	0.88068
Si_I	(16b)	0.75000	0.75000	0.00000
Si_{II}	(16b)	0.15499	0.76432	0.87210
Si_{III}	(16b)	0.90188	0.81979	0.65040
Si_{IV}	(16b)	0.08794	0.84688	0.65213

...

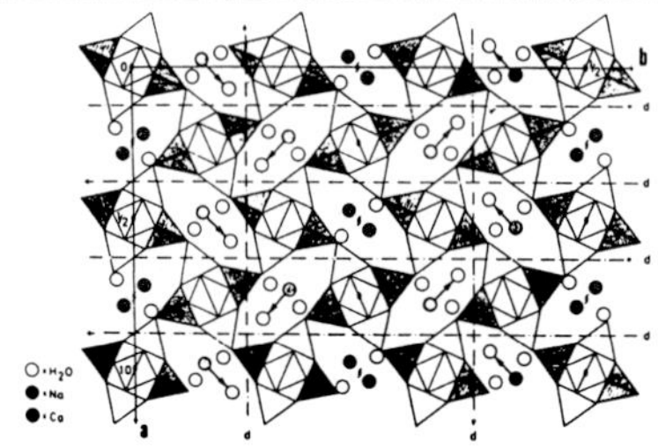

Fig. 1

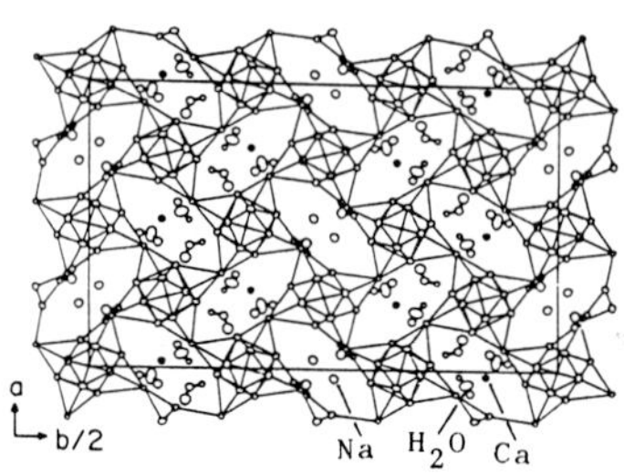

Fig. 2

Properties

Habit	Cleav.	Fract.	Twin.	Hardn.	Dens.	Colour	Transp.
acicular, tufts of prismatic needles	perfect (110) (1$\bar{1}$0)		(010)	5	2.257	colourless, white	transparent

Refr. index	/Reflect.	Birefr.	Luster	Streak	Melt.p.	CPI
n_α = 1.505 n_β = 1.506 n_γ = 1.507		(+) 2V = 84°	vitreous, silky			

Figures

Fig. 1. The structure of mesolite projected along the c axis, symmetry Fdd2, with tripled b (after Gottardi + Galli, 1985).

Fig. 2. Mesolite structure projected down the c axis (after Artioli et al., 1986).

Description

The mesolite is a zeolite of the natrolite group. It has the same framework topology, and complete Si/Al order. The Na atoms coordinate four oxygen atoms and two water molecules in a distorted trigonal prism, and the Ca atoms coordinate four framework oxygen atoms and three water oxygen atoms in a distorted pentagonal bipyramid.

References

Kostov (1968) 418.
Artioli et al. (1986) 937-942.
Gottardi + Galli (1985) 35, 39.
Roberts et al. (1974) 394.

Crystallographic data (continued)

Atom	Site	x	y	z
Si_V	(16b)	0.00435	0.83180	0.27740
Al_I	(16b)	0.78689	0.78025	0.61310
Al_{II}	(16b)	0.03630	0.80211	0.89287
Al_{III}	(16b)	0.95155	0.86233	0.89294
O_I	(16b)	0.76818	0.77289	0.86482
O_{II}	(16b)	0.02298	0.80800	0.15228
O_{III}	(16b)	0.98498	0.85319	0.12809
O_{IV}	(16b)	0.82365	0.80837	0.59682
O_V	(16b)	0.07262	0.77385	0.86425
O_{VI}	(16b)	0.16644	0.85935	0.67096
O_{VII}	(16b)	0.84335	0.75866	0.51529
O_{VIII}	(16b)	0.09631	0.82339	0.79447
O_{IX}	(16b)	0.88933	0.84183	0.80185
H_{IV}	(16b)	0.8085	0.8232	0.004
H_V	(16b)	0.2274	0.8560	0.247
H_{VI}	(16b)	0.1685	0.8438	0.260
O_X	(16b)	0.95568	0.80062	0.75319
O_{XI}	(16b)	0.20947	0.78462	0.95950
O_{XII}	(16b)	0.02193	0.86358	0.71805
O_{XIII}	(16b)	0.93735	0.82799	0.43383
O_{XIV}	(16b)	0.18020	0.75695	0.64166
O_{XV}	(16b)	0.07638	0.83892	0.41273
OW_I	(16b)	0.05347	0.76839	0.37684
OW_{II}	(16b)	0.78167	0.81598	0.07685
OW_{III}	(16b)	0.20069	0.84820	0.18338
OW_{IV}	(16b)	0.81793	0.85687	0.36556
H_I	(16b)	0.0450	0.7813	0.322
H_{II}	(16b)	0.0881	0.7679	0.420
H_{III}	(16b)	0.7773	0.8037	0.018
H_{VII}	(16b)	0.8072	0.8702	0.387
H_{VIII}	(16b)	0.8615	0.8525	0.359

SCOLECITE   C c

$Ca^{[7]}(H_2O)_3\ {}^{3}_{\infty}\left[Si^t_3\ Al^t_2\ O_{10}\right]$

a = 18.508 Å
b = 18.981 Å
c = 6.527 Å
β = 90.64°
Z = 8

(Described in the pseudo-orthorhombic setting of Fd)

Atom	Coordinates	Atom	Coordinates
Ca	x = 0.22378 y = 0.01824 z = 0.6141	Si(2)	x = 0.14964 y = 0.20681 z = 0.6197
Si(1)	x = 0.00 y = 0.00438 z = 0.00	Si(20)	x = -0.16558 y = -0.20765 z = 0.6250

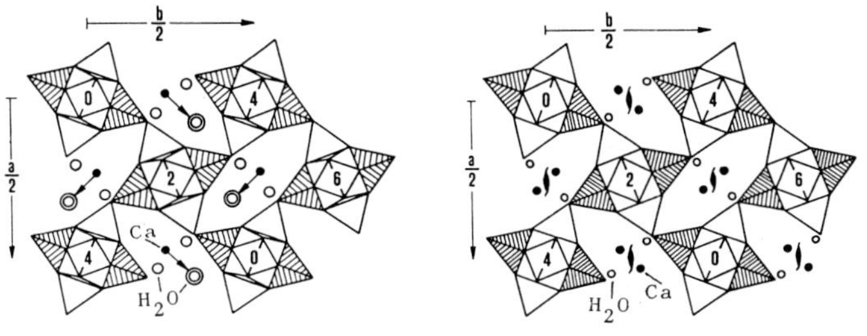

(a) Fig. 1 (b)

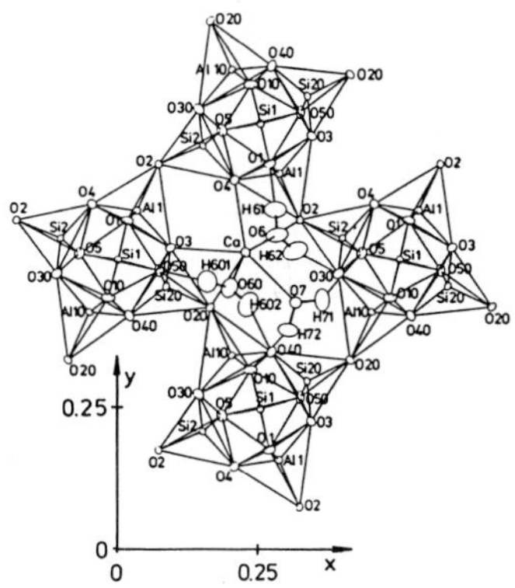

Fig. 2

Properties

Habit	Cleav.	Fract.	Twin.	Hardn.	Dens.	Colour	Transp.
prismatic, massive	perfect (110)	uneven	(100)	5	2.27	colourless, white	transparent to translucent

Refr. index/Reflect.	Birefr.	Luster	Streak	Melt.p.	CPI
n_α = 1.507-1.513 n_β = 1.516-1.520 n_γ = 1.517-1.521	(-) 2V = 36°-56°	vitreous, silky			

Figures

Fig. 1. (a) Polyhedral representation of the scolecite structure, projected along the c axis on the Fldl symmetry description, and (b) corresponding representation of the natrolite structure to enable the comparison (after Gottardi+ Galli, 1985).

Fig. 2. Projection of the scolecite structure along the fibre and channel direction (after Joswig et al., 1984).

Description

Scolecite is a zeolite of the natrolite group. Its aluminosilicate framework is iso typical with the framework of natrolite, except for a small monoclinic distortion. The Ca ion coordinates three water molecules and four oxygen atoms of the framework in a distorted pentagonal bipyramid.

Crystallographic data (continued)

Atom	Coordinates	Atom	Coordinates
Al(1)	x = 0.03279 y = 0.09168 z = 0.6107	O(50)	x = -0.17841 y = -0.23515 z = 0.3886
Al(10)	x = -0.04985 y = -0.08722 z = 0.6154	O(6)	x = 0.02959 y = +0.20108 z = 0.0616
O(1)	x = 0.01586 y = 0.07440 z = 0.8682	O(60)	x = -0.05416 y = -0.20522 z = 0.1633
O(10)	x = -0.01706 y = -0.06189 z = 0.8557	O(7)	x = 0.06289 y = 0.32057 z = 0.3577
O(2)	x = 0.07071 y = 0.17454 z = 0.5749	H(61)	x = 0.02608 y = 0.15595 z = -0.0059
O(20)	x = -0.08738 y = -0.17097 z = 0.6395	H(62)	x = 0.06191 y = 0.22954 z = -0.0167
O(3)	x = 0.09207 y = 0.02618 z = 0.5237	H(601)	x = -0.09260 y = -0.21643 z = 0.2548
O(30)	x = -0.10754 y = -0.02053 z = 0.5303	H(602)	x = -0.02131 y = -0.17597 z = 0.2368
O(4)	x = 0.20445 y = 0.14628 z = 0.7085	H(71)	x = 0.11235 y = 0.31614 z = 0.3353
O(40)	x = -0.23041 y = -0.15649 z = 0.6869	H(72)	x = 0.04938 y = 0.36853 z = 0.3779
O(5)	x = 0.18274 y = +0.23351 z = 0.4024		

References

Kostov (1968) 418.
Joswig et al. (1984) 219-223.
Gottardi + Galli (1985) 38.
Roberts et al. (1974) 548.

THOMSONITE — Pbmn

$Na\,Ca_2\,(H_2O)_6\ {}^{3}_{\infty}\left[Si^t_5\,Al^t_5\,O_{20}\right]$

a = 13.07 Å
b = 13.09 Å
c = 13.25 Å
Z = 2

Site	x	y	z
Ca(2c)	1/2	0	1/2
(Ca,Na)(4h)	0.069	0	0.25
$(H_2O)_I$ (4e)	0	0.139	0
$(H_2O)_{II}$(4h)	0.403	0	0.75
$(H_2O)_{III}$(4h)	0.111	0	0.75
$(Si,Al)_I$(4g)	1/4	1/4	0.875
$(Si,Al)_{II}$(8i)	0.125	0.194	0.500

...

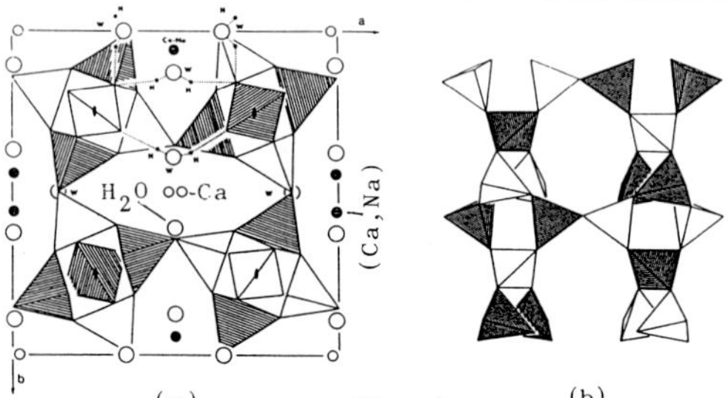

(a) Fig. 1 (b)

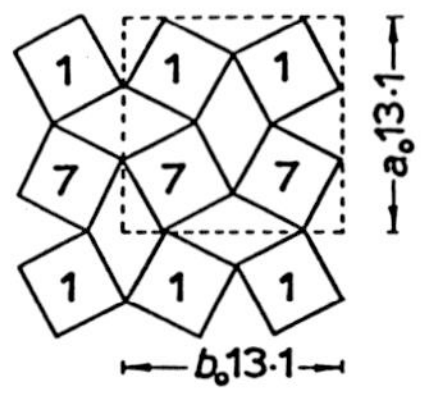

Fig. 2

Properties

Habit	Cleav.	Fract.	Twin.	Hardn.	Dens.	Colour	Transp.
prismatic, acicular, compact	perfect (010)	uneven, subconchoidal	(110)	5-5.5	2.25--2.40	colourless, white, yellowish	transparent to translucent

Refr. index /Reflect.	Birefr.	Luster	Streak	Melt.p.	CPI
n_α = 1.497-1.530 n_β = 1.513-1.533 n_γ = 1.518-1.544	(+) 2V = 42°-75°	vitreous, pearly	uncoloured		

Figures

Fig. 1. Polyhedral representation of the thomsonite structure (a) viewed along the c axis, and (b) two thomsonite chains at the same height with their (Si,Al) order. Striped tetrahedra are occupied by Al, white tetrahedra by Si (after Gottardi + Galli, 1985).

Fig. 2. Structural scheme of the thomsonite structure. Squares represent $(Si,Al)^tO_4$ tetrahedral chains viewed along their elongation (after Kostov, 1968).

Description

Thomsonite is a zeolite of the natrolite group. Its framework is similar to that of natrolite.

Crystallographic data (continued)

Site	x	y	z
$(Si,Al)_{III}$(8i)	0.305	0.125	0.250
O_I (4f)	1/2	0.361	1/2
O_{II}(4h)	0.402	0	0.25
O_{III}(8i)	0.167	0.194	0.75
O_{IV} (8i)	0.180	0.119	0.375
O_V (8i)	0.305	0.139	0.000
O_{VI} (8i)	0.375	0.194	0.375

(structure determination based on half cell, with c' = 6.61Å, Z = 2)

References

Kostov (1968) 418, 419.
Povarennykh (1972) 355.
Wyckoff (1968) Vol. 4, 414-415.
Gottardi + Galli (1985) 58, 59.

ANALCIME $I\ 4_1/a\ \bar{3}\ 2/d$

$Na\ (H_2O)\ {}^{3}_{\infty}\left[Si^{t}_{2}\ Al^{t}\ O_6\right]$

a = 13.43 Å
Z = 16
1/2 Na (24c)
$(H_2O)(16b)$

(Si,Al)(48g) u = 0.66239

O(96h) x = 0.10506, y = 0.13490, z = 0.71976

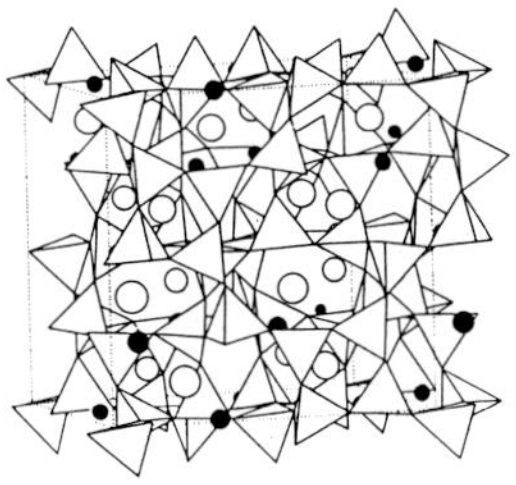

Fig. 1

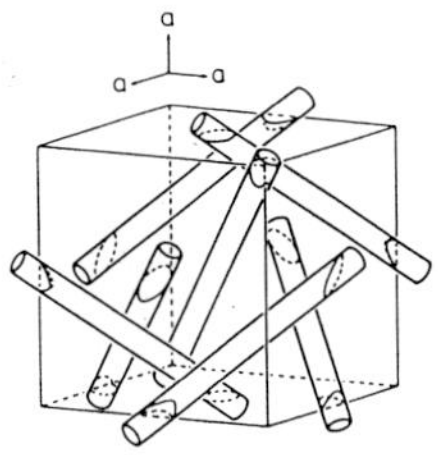

Fig, 2

Properties

Habit	Cleav.	Fract.	Twin.	Hardn.	Dens.	Colour	Transp.
trapezo-hedral	poor (100)	uneven	(100) (110)	5.5	2.26	white, grey, pink	transparent to trans-cent

Refr. index/Reflect.	Birefr.	Luster	Streak	Melt.p.	CPI
n = 1.482		vitreous	white		(SPI) 38

Figures

Fig 1. Polyhedral representation of the structure of analcime (after Zoltai + Stout, 1984).

Fig. 2. System of non intersecting one-dimensional channels in the analcime structure, running parallel to the [111] zone axis (after Liebau, 1985).

Description

Analcime is a zeolite mineral. Its structure consists of a framework of $(Si,Al)^{t}O_4$ tetrahedra, with sodium and water molecules in the holes.

References

Kostov (1968) 416.
Povarennykh (1972) 351.
Wyckoff (1968) Vol. 4, 399-401.
Zoltai + Stout (1984) 309, 310.
Liebau (1985) 158.

LAUMONTITE

C 2

$Ca(H_2O)_4 \; {}^{3}_{\infty}\left[Si^{t}_{4}\, Al^{t}_{2}\, O_{12}\right]$

a = 14.90 Å
b = 13.17 Å
c = 7.55 Å
β = 111° 30'
Z = 4

Atom	Site	x	y	z
Ca	(4c)	0.273	0.258	0.254
Si_I	(4c)	0.238	0.147	0.662
Si_{II}	(4c)	0.076	0.148	0.823
Si_{III}	(4c)	0.088	0.382	0.832
Si_{IV}	(4c)	0.239	0.383	0.652
Al_I	(4c)	0.134	0.077	0.238

...

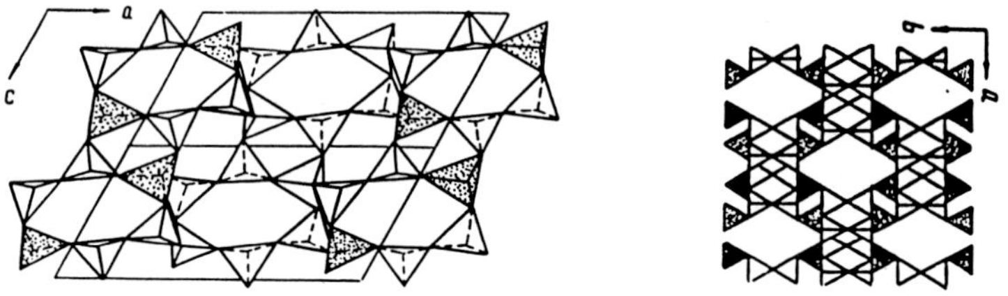

(a) Fig. 1 (b)

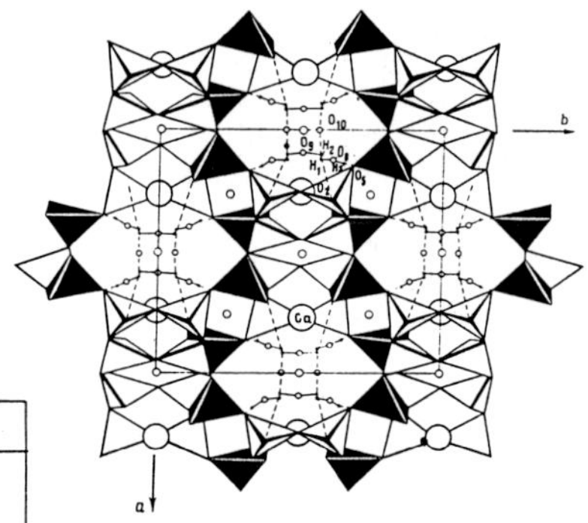

Fig. 2

Properties

Habit	Cleav.	Fract.	Twin.	Hardn.	Dens.	Colour	Transp.
prismatic	perfect (010) (110)	uneven	(100)	3-4	2.3	white	transparent to translucent

Refr. index/Reflect.	Birefr.	Luster	Streak	Melt.p.	CPI
n_α = 1.51 n_β = 1.52 n_γ = 1.52	(-) 2V = 25°-45°	vitreous	white		(SPI) 33

Figures

Fig. 1. Structure of laumontite: (a) projected along the b axis, and (b) projection along the c axis (after Povarennykh, 1972).

Fig. 2. Structure of laumontite projected along the c axis (after Yakubovich + Simonov, 1985).

Description

Laumontite is a zeolite mineral, having a framework of $(Si,Al)O_4$ tetrahedra with four -, six - and eight - - sided rings.

Crystallographic data (continued)

Atom	Site	x	y	z
Al_{II}	(4c)	0.122	0.455	0.226
O_I	(4c)	0.262	0.270	0.743
O_{II}	(4c)	0.141	0.153	0.051
O_{III}	(4c)	0.150	0.387	0.058
O_{IV}	(4c)	0.222	0.141	0.436
O_V	(4c)	0.201	0.388	0.422
O_{VI}	(4c)	0.344	0.079	0.779
O_{VII}	(4c)	0.330	0.450	0.761
O_{VIII}	(4c)	0.161	0.108	0.741
O_{IX}	(4c)	0.142	0.426	0.689
O_X	(4c)	0.051	0.267	0.760
O_{XI}	(4c)	0.008	0.460	0.218
O_{XII}	(4c)	0.010	0.085	0.210
$(H_2O)_I$	(4c)	0.383	0.137	0.254
$(H_2O)_{II}$	(4c)	0.377	0.375	0.150

References

Kostov (1968) 422.
Povarennykh (1972) 357.
Zoltai + Stout (1984) 316.
Yakubovich + Simonov (1985) 624-626.
Amirov et al. (1967) 121-124.
Structure Reports (1975) Vol. 32A, 482,483.

GISMONDINE

(Gismondite)

$Ca\,(H_2O)_4\ {}^3_\infty\left[Si^t_2\ Al^t_2\ O_8\right]$

$P\,2_1/c$

a = 10.02 Å
b = 10.62 Å
c = 9.84 Å
β = 92° 25'
Z = 4

Atom	Site	x	y	z
Ca	(4e)	0.720	0.077	0.354
$(H_2O)_I$	(4e)	0.257	0.107	0.505
$(H_2O)_{II}$	(4e)	0.590	0.127	0.539
$(H_2O)_{III}$	(4e)	0.911	0.119	0.501
$1/2(H_2O)$	(4e)	0.77	0.21	0.17
$1/2(H_2O)$	(4e)	0.74	0.18	0.89
...				

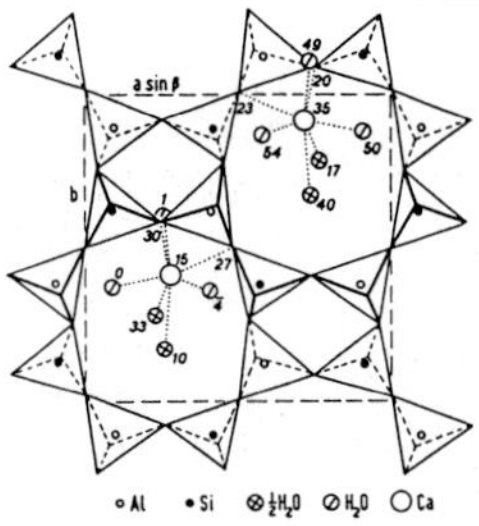

Fig. 1

Properties

Habit	Cleav.	Fract.	Twin.	Hardn.	Dens.	Colour	Transp.
bipyramidal	distinct (101)		(110) (001)	4.5	2.27	colourless, white	transparent to translucent

Refr. index/Reflect.	Birefr.	Luster	Streak	Melt.p.	CPI
n_α = 1.520-1.521 n_γ = 1.522-1.525	(-) 2V ≃ 86°	vitreous			

Figures

Fig. 1. Polyhedral representation of the gismondine structure projected along the c axis and showing the coordination of calcium. Only the lower half of the cell is represented (after Bragg + Claringbull, 1965).

Description

The structure of gismondine, which is a zeolite, consists of a framework of SiO_4 and AlO_4 tetrahedra, which form three types of four-membered rings. Eight-membered rings define aperture channels parallel to [100] and [001], within which Ca and water molecules are located.

Crystallographic data (continued)

Atom	Site	x	y	z
Si_I	(4e)	0.415	0.113	0.182
Si_{II}	(4e)	0.908	0.870	0.160
Al_I	(4e)	0.097	0.113	0.170
Al_{II}	(4e)	0.590	0.867	0.149
O_I	(4e)	0.078	0.154	-0.001
O_{II}	(4e)	0.262	0.075	0.212
O_{III}	(4e)	0.438	0.145	0.026
O_{IV}	(4e)	0.246	0.407	0.303
O_V	(4e)	0.000	-0.017	0.215
O_{VI}	(4e)	0.044	0.242	0.261
O_{VII}	(4e)	0.463	0.224	0.276
O_{VIII}	(4e)	0.511	-0.005	0.226

References

Kostov (1968) 408.
Wyckoff (1968) Vol. 4, 427, 428.
Bragg + Claringbull (1965) 367.
Roberts et al. (1974) 237.

PHILLIPSITE

$P2_1m$

$(K,Ca)_5(H_2O)_{10}\ {}^3_\infty\left[Si^t_{11}\ Al^t_5\ O_{32}\right]$

a = 9.865 Å
b = 14.300 Å
c = 8.668 Å
β = 124.20°
Z = 1

Atom	Site	Coordinates	Atom	Site	Coordinates
K	(2e)	x = 0.8480 y = 1/4 z = 0.2076	$(Si,Al)_{II}$	(4f)	x = 0.4206 y = 0.1409 z = 0.0019
Ca 1.65	(4f)	x = 0.6080 y = 0.6262 z = 0.4401	$(Si,Al)_{III}$	(4f)	x = 0.0604 y = 0.0078 z = 0.2844
$(Si,Al)_I$	(4f)	x = 0.7362 y = 0.0248 z = 0.2805	$(Si,Al)_{IV}$	(4f)	x = 0.1204 y = 0.1396 z = 0.0421
...					

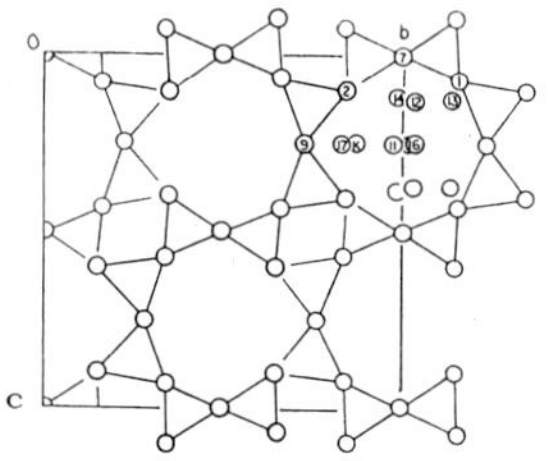

Fig. 1

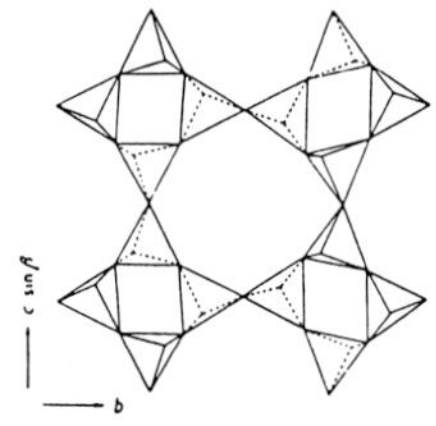

Fig. 2

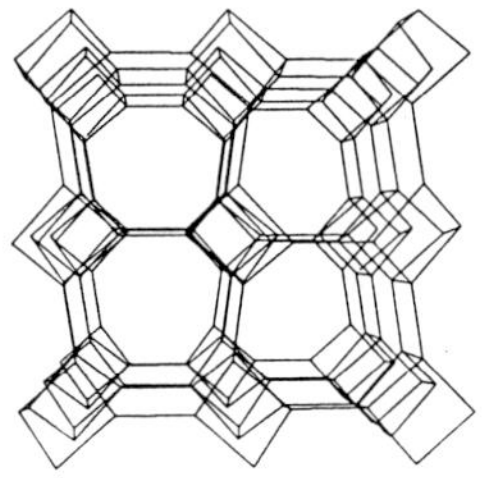

Fig. 3

(a)

(b)

Fig. 4

Properties

Habit	Cleav.	Fract.	Twin.	Hardn.	Dens.	Colour	Transp.
spherulites	distinct (010)	uneven	(001) (021)	4-4.5	2.20	colourless, white, redish	transparent to translucent

Refr. index/Reflect.	Birefr.	Luster	Streak	Melt.p.	CPI
n_α = 1.483-1.504 n_β = 1.484-1.509 n_γ = 1.486-1.514	(+) 2V = 60°-80°	vitreous			

Population

Harmotome

$Ba_2Ca_{0.6}(H_2O)_{12}\ {}^3_\infty\left[(Si,Al)^t_{16}O_{32}\right]$

Figures

Fig. 1. Ball and spoke representation of the phillipsite structure (pseudo-symmetry, B2mb), where the occupied interstitial sites by water and large cations are shown in one channel only (after Steinfink, 1962, adapted from Structure Reports, 1971, Vol. 27).

Fig. 2. Projection of the harmotome structure, which is isotypic with phillipsite, along the a axis (after Liebau, 1985).

Fig. 3. Phillipsite framework (ORTEP drawing) looking down the a direction (after Rinaldi et al., 1974).

Fig. 4. Polyhedral representation of the structure of phillipsite (a) looking along the a axis (B2mb description) and (b) looking along the b axis (same description) (after Steinfink, 1962).

Description

Phillipsite is a zeolite mineral. Its structure consists of a framework of SiO_4 and AlO_4 linked tetrahedra, with large channels parallel to the a axis.

References

Kostov (1968) 423.
Wyckoff (1988) Vol. 4, 422, 423.
Rinaldi et al. (1974) 2426-2433.
Steinfink (1962) 644-651.
Roberts et al. (1974) 475, 476.

Crystallographic data (continued)

Atom	Site	Coordinates	Atom	Site	Coordinates
O_I	(4f)	x = 0.1335 y = 0.0976 z = 0.2289	O_{VIII}	(2)	x = 0.5814 y = 3/4 z = 0.0616
O_{II}	(4f)	x = 0.6445 y = 0.5766 z = 0.1878	O_{IX}	(2e)	x = 0.0665 y = 1/4 z = 0.0196
O_{III}	(4f)	x = 0.6100 y = 0.1130 z = 0.1728	$(H_2O)_I$	(2)	x = 0.7551 y = 3/4 z = 0.4733
O_{IV}	(4f)	x = 0.0254 y = 0.9154 z = 0.1494	$(H_2O)_{II}$	(2)	x = 0.1552 y = 3/4 z = 0.4382
O_V	(4f)	x = 0.8957 y = 0.0440 z = 0.2713	$(H_2O)_{III}$	(4f)	x = 0.3208 y = 0.8525 z = 0.1740
O_{VI}	(4f)	x = 0.3022 y = 0.3738 z = 0.0783	$(H_2O)_{IV}$	(2)	x = 0.5085 y = 1/4 z = 0.4384
O_{VII}	(4f)	x = 0.7872 y = 0.4795 z = 0.4982	$(H_2O)_V$	(2)	x = 1/2 y = 1/2 z = 1/2

MORDENITE

Cmcm

$Na(H_2O)_3\ {}^3_\infty\left[\underbrace{Si^t_5\ Al^t}\ O_{12}\right]$

a = 18.13 Å
b = 20.49 Å
c = 7.52 Å
Z = 8

Na(4b)

(4 Na not determined)

$(Si,Al)_I$	(16h)	x = 0.196, y = 0.427, z = 0.536
$(Si,Al)_{II}$	(16h)	x = 0.196, y = 0.191, z = 0.540
$(Si,Al)_{III}$	(8g)	x = 0.089, y = 0.384, z = 1/4
...		

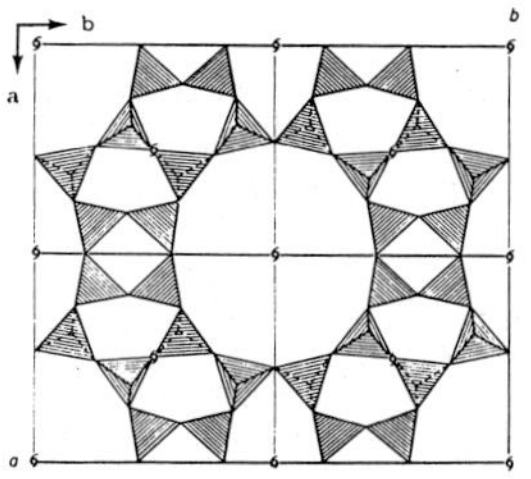

Fig. 1

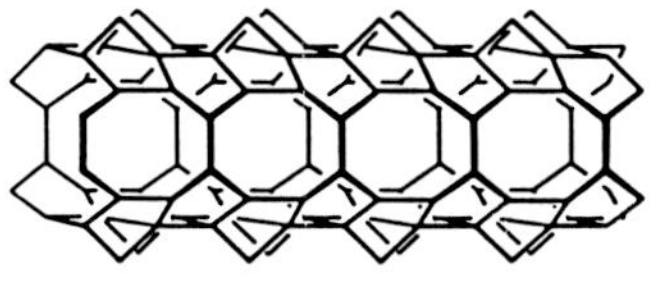

Fig. 2

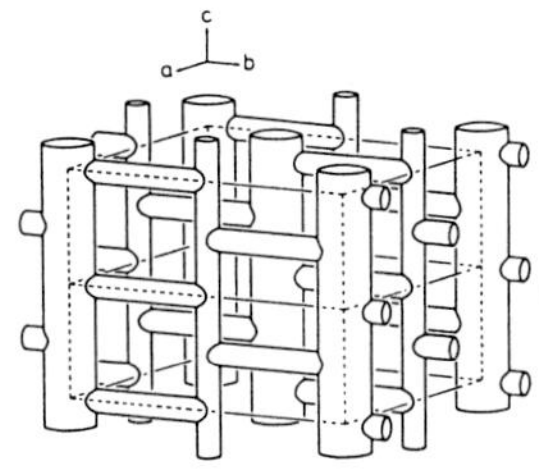

Fig. 3

Properties

Habit	Cleav.	Fract.	Twin.	Hardn.	Dens.	Colour	Transp.
prismatic, acicular, compact	perfect (100)			4-5	2.12--2.15	colourless, white, yellowish	transparent to translucent

Refr. index/Reflect.	Birefr.	Luster	Streak	Melt.p.	CPI
n_α = 1.472-1.483 n_β = 1.475-1.485 n_γ = 1.477-1.487	(+) or (-) 2V = 76°-104°	vitreous, silky			

Figures

Fig. 1 Alumino-silicate framework in the mordenite structure (after Gottardi + Meier, 1963).

Fig. 2. Tabular building units of the mordenite structure (after Liebau, 1985).

Fig. 3. System of nonintersection one-dimensional channels in the mordenite structure (after Liebau, 1985).

1985).

Description

Mordenite is a zeolite mineral. Its structure consists of a framework of $(Si,Al)^tO_4$ tetrahedra with calcium atoms and water molecules in the holes.

Crystallographic data (continued)

$(Si,Al)_{IV}$	(8g)	x = 0.087, y = 0.225, z = 1/4	O_V	(8g)	x = -0.184, y = -0.199, z = 1/4
O_I	(16h)	x = 0.124, y = 0.418, z = 0.416	O_{VI}	(8g)	x = -0.186, y = -0.423, z = 1/4
O_{II}	(16h)	x = 0.128, y = 0.197, z = 0.416	O_{VII}	(8e)	u = -0.269
O_{III}	(16h)	x = 0.263, y = 0.373, z = 0.504	O_{VIII}	(8d)	
			O_{IX}	(4c)	u = 0.398
O_{IV}	(8g)	x = 0.083, y = 0.305, z = 1/4	O_X	(4c)	u = 0.210

References

Kostov (1968) 410,411,

Wyckoff (1968) Vol. 4, 415-418.

Liebau (1985) 150, 158

HEULANDITE

C 2/m

$$\mathrm{Ca\,(H_2O)_6}\ \ {}^{3}_{\infty}\left[\mathrm{Si_7^t\ Al_2^t\ O_{18}}\right]$$

a = 17.77 Å
b = 17.95 Å
c = 7.435 Å
β = 116.46°
Z = 1

For $Ba_{0.1},Ca_{2.2},Na_{1.4},Sr_{0.1},Al_{7.9},Si_{28.5},O_{72} \cdot 24.5\ H_2O$

(see below the positional parameters)

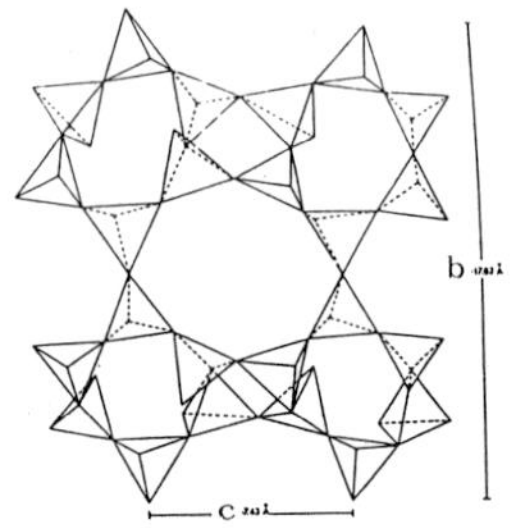

Fig. 1

(a)

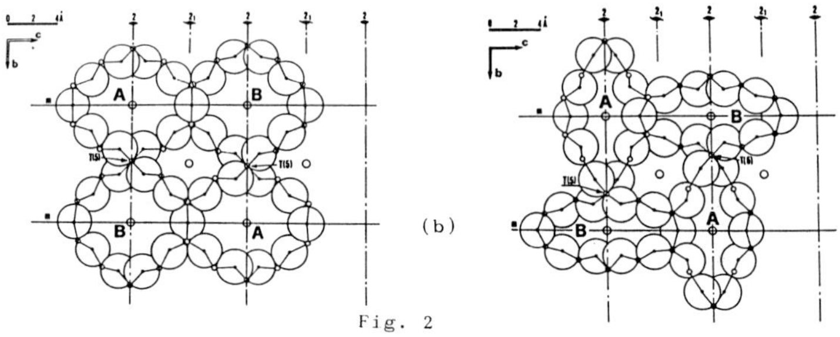

(b)

Fig. 2

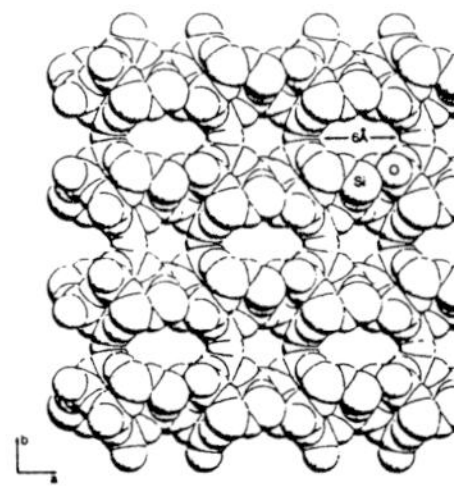

Fig. 3

Properties

Habit	Cleav.	Fract.	Twin.	Hardn.	Dens.	Colour	Transp.
platy	perfect (010)	subcon-choidal		3.5-4	2.15	white, variable	transparent to trans-lucent

Refr. index/Reflect.	Birefr.	Luster	Streak	Melt.p.	CPI
n_α = 1.49 n_β = 1.50 n_γ = 1.50	(+) 2V = 35°	vitreous	white		(SPI) 32

Figures

Fig. 1. Schematic view of part of the framework of the heulandite structure (after Merkle + Slaughter, 1968).

Fig. 2. Structure of heulandite: (a) hydrated where the channels are blown up, and (b) dehydrated where the channels are flattened (after Gottardi + Galli, 1985). quoted by Gottardi + Galli, 1985, 271).

Fig. 3. The structure of natural heulandite viewed along the c axis (after Hambley + Taylor, 1984).

Description

Heulandite is a zeolite mineral. Its structure consists of a framework of $(Si,Al)O_4$ tetrahedra, with (Ca,Na) atoms and water molecules located in relatively open channels.

Crystallographic data (continued)

Atom	Coordinates	Atom	Coordinates
Si(1)	x = 0.1791 y = 0.1696 z = 0.0950	0.6 O(12)	x = 0.0786 y = 0 z = 0.8901
Si(2)	x = 0.2114 y = 0.4098 z = 0.5002	0.6 H(2)	x = 0.0799 y = 0.0425 z = 0.9327
Si(3)	x = 0.2086 y = 0.1911 z = 0.7169	0.4 O(12')	x = 0.0955 y = 0 z = 0.8091
Si(4)	x = 0.0650 y = 0.2991 z = 0.4117	0.4 H(2')	x = 0.0341 y = 0 z = 0.8180
Si(5)	x = 0 y = 0.2129 z = 0	0.4 H(2")	x = 0.1234 y = 0 z = 0.8163
O(1)	x = 0.1968 y = 1/2 z = 0.4551	0.9 O(13)	x = 0.0758 y = 0.4179 z = 0.9678
O(2)	x = 0.2315 y = 0.1202 z = 0.6126	0.4 H(3)	x = 0.0826 y = 0.3857 z = 0.9808
O(3)	x = 0.1828 y = 0.1544 z = 0.8835	0.4 H(4)	x = 0.0861 y = 0.3985 z = 0.0949
O(4)	x = 0.2382 y = 0.1067 z = 0.2539	0.5 H(3')	x = 0.0646 y = 0.3612 z = 0.9329
O(5)	x = 0 y = 0.3254 z = 1/2	0.5 H(4')	x = 0.1307 y = 0.4421 z = 0.9702
O(6)	x = 0.0821 y = 0.1591 z = 0.0615	O(14)	x = 0 y = 1/2 z = 1/2
O(7)	x = 0.1282 y = 0.2347 z = 0.5506	H(5)	x = 0 y = 0.4548 z = 1/2
O(8)	x = 0.0098 y = 0.2669 z = 0.1858	0.3 O(15)	x = 0.0285 y = 0.0951 z = 0.4970
O(9) ...	x = 0.2102 y = 0.2542 z = 0.1761	0.3 H(6)	x = 0.0344 y = 0.1369 z = 0.5004

Crystallographic data (continued)

Atom	Coordinates	Atom	Coordinates
O(10)	x = 0.1154 y = 0.3734 z = 0.4000	0.3 H(7)	x = 0.0096 y = 0.0921 z = 0.6250
O(11)	x = 0.2214 y = 1/2 z =-0.0174	0.72 O(16)	x = 0.0984 y = 0 z = 0.2798
H(1)	x = 0.2312 y = 0.4627 z =-0.0925	0.72 H(8)	x = 0.0991 y = 0.0471 z = 0.2342
0.6 CS(1)	x = 0.1567 y = 0 z = 0.6701	0.14 O(16')	x = 0.0805 y = 0.0385 z = 0.3766
0.52 CS(2)	x = 0.0418 y = 1/2 z = 0.2069	0.28 H(8')	x = 0.0297 y = 0 z = 0.2939

References

Kostov (1968) 414.
Zoltai + Stout (1984) 317.
Gottardi + Galli (1985) 271.
Hambley + Taylor (1984) 1-9.
Merkle + Slaughter (1968) 1131.

STILBITE

C 2/m

(Desmine)

$(Ca,Na)(H_2O)_7 \; {}^3_\infty\left[Si^t_7 \, Al^t_2 \, O_{18}\right]$

a = 13.64 Å
b = 18.24 Å
c = 11.27 Å
β = 128°.0
Z = 4

(for $Ca_{4.18}Na_{4.18}Mg_{.18}\ 34.07\ H_2O$ $Si_{25.71}Al_{10.30}O_{72}$

Atom	x	y	z
Si_I (8j)	0.4830	0.3042	0.2420
Si_{III}(8j)	0.1892	0.0893	0.4846
Si_{II}(8j)	0.2653	0.3097	0.2619
Si_{IV} (8j)	0.1124	0.3166	0.5013

...

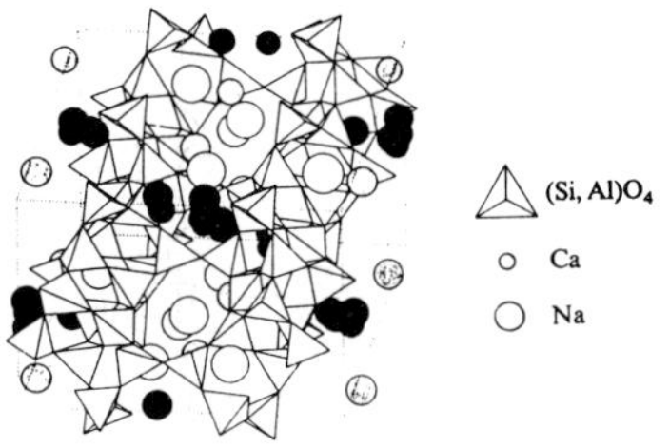

Fig. 1

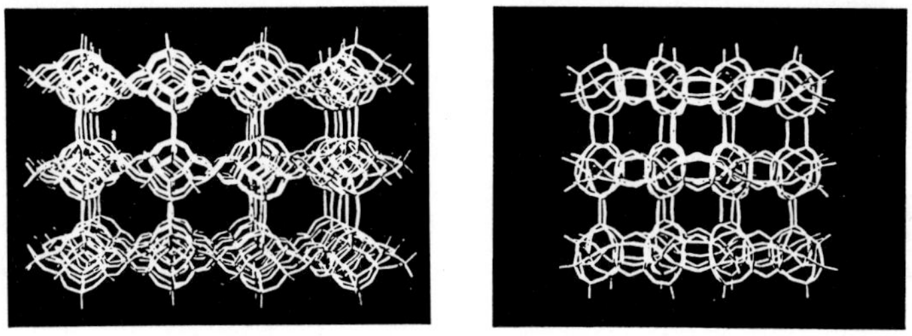

(a) Fig. 2 (b)

Properties

Habit	Cleav.	Fract.	Twin.	Hardn.	Dens.	Colour	Transp.
prismatic, striated	perfect (010)	subcon-choidal		3.5-4	2.15	grey	transparent to trans-lucent

Refr. index /Reflect.	Birefr.	Luster	Streak	Melt.p.	CPI
n_α = 1.49, n_β = 1.50, n_γ = 1.50	(-) 2V = 30°–50°	pearly	grey		(SPI) 31

Figures

Fig. 1. Polyhedral representation, in perspective, of the stilbite structure (after Zoltai + Stout, 1984).

Fig. 2. Representations of the stilbite framework: (a) viewed along the a axis, and (b) along the [102] axis (after Galli, 1971).

Description

Stilbite is a zeolite mineral. Its structure consists of a framework of $(Si,Al)O_4$ tetrahedra, with Ca and Na atoms and water molecules located in relatively open channels.

References

Kostov (1968) 412.
Zoltai + Stout (1984) 317, 319.
Galli (1971) 833-841.

Crystallographic data (continued)

Atom	x	y	z
Si_V (4g)	0	0.2610	0
O_X (4g)	0	0.3509	0
O_I (8j)	0.4645	0.2911	0.0842
Ca (4i)	0.2805	0	0.0949
O_{II} (8j)	0.1164	0.3162	0.1189
0.22Na(8j)	0.5055	0.0659	0.0392
O_{III} (8j)	0.0512	0.2661	0.3516
$0.84H_2O_I$(8j)	0.1347	0.0776	0.1067
O_{IV} (8j)	0.0638	0.1199	0.3226
$0.91H_2O_{II}$(8j)	0.3306	0.1258	0.1013
O_V (8j)	0.2935	0.2308	0.3453
$0.85H_2O_{III}$(4i)	0.3691	0	0.3549
O_{VI} (8j)	0.3001	0.3796	0.3714
H_2O_{IV}(4i)	0.5124	0	0.2416
O_{VII} (8j)	0.3406	0.3162	0.1914
H_2O_V (4)	0.4028	0.5000	0.1632
O_{VIII}(8j)	0.3157	0.1119	0.5024
H_2O_6 (4)	0.1712	0.5000	0.1040
O_{IX} (4i)	0.1863	0	0.4876

CANCRINITE

$P6_3$

$Na_6 Ca\ CO_3 (H_2O)_2\ {}^3_\infty\left[Si^t_6 Al^t_6 O_{24}\right]$

a = 12.75 Å
c = 5.14 Å
Z = 1

Atom	x	y	z
$(Na,Ca)_I$(2b)	0.6667	0.3333	0.1440
$(Na,Ca)_{II}$(6c)	0.1232	0.2486	0.2920
C (2a)	0.0	0.0	0.6835
Si(6c)	0.0750	0.4124	0.7500

...

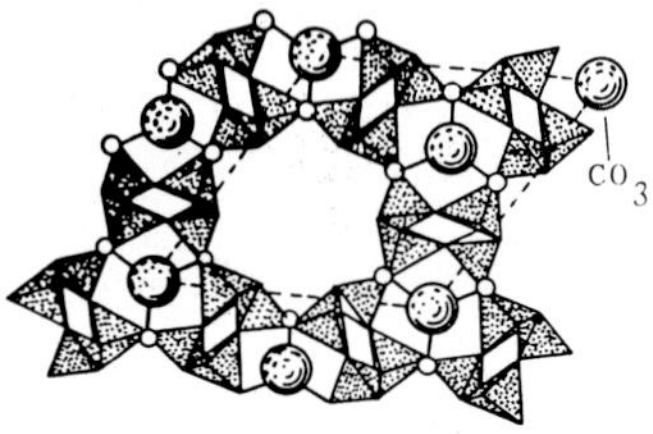

Fig. 1

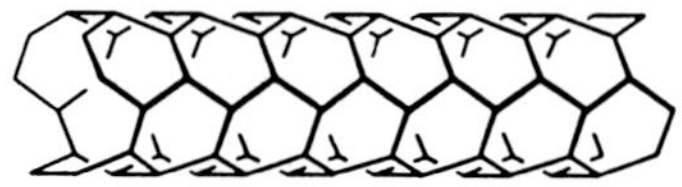

Fig. 2

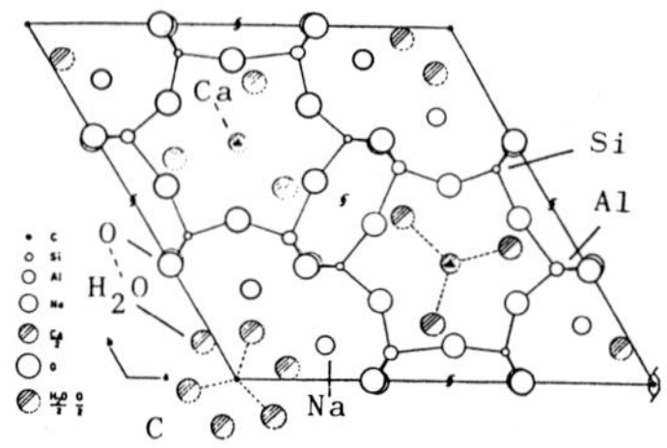

Fig. 3

Properties

Habit	Cleav.	Fract.	Twin.	Hardn.	Dens.	Colour	Transp.
prismatic, massive	perfect (1010)	uneven		5-6	2.42- -2.51	colourless, white, yellow	transparent to translucent

Refr. index/Reflect.	Birefr.	Luster	Streak	Melt.p.	CPI
n_ω = 1.507-1.528 n_ϵ = 1.495-1.503	(-)	vitreous, pearly	colourless	1190° C	

Figures

Fig. 1. Polyhedral representation of the cancrinite structure projected on the basal plane (after Povarennykh, 1972).

Fig. 2. Tubular building unit of the silicate framework of the cancrinite structure (after Liebau, 1985).

Fig. 3. The structure of cancrinite projected along the c axis (after Nithollon, 1955, quoted in Structure Reports, 1963, Vol. 19).

Description

Cancrinite is a zeolite mineral. Its structure consists of framework of $(Si,Al)O_4$ tetrahedra, with (Na,Ca) atoms, CO_3 groups and water molecules located in the channels.

Crystallographic data (continued)

Atom	x	y	z
Al (6c)	0.3272	0.4101	0.7508
O_I (6c)	0.2039	0.4045	0.6589
O_{II}(6c)	0.1141	0.5623	0.7229
O_{III} (6c)	0.0269	0.3485	0.0619
O_{IV} (6c)	0.3150	0.3580	0.0407
ca.3 O_V(6c)	0.0584	0.1200	0.6857
ca.2 H_2O(6c)	0.6150	0.3130	0.6492

References

Kostov (1968) 406.

Povarennykh (1972) 349.

Wyckoff (1968) Vol. 4, 433.

Liebau (1985) 148-150.

Roberts et al. (1974) 105.

Ingerson (1955) 352.

Structure Reports (1963) Vol. 19, 481.

Jarchow (1965) 407-422.

8.4.6. Structures not classified

HESSITE $Ag_2^t Te$

$P 2_1/c$

a = 8.09 Å
b = 4.48 Å
c = 8.96 Å
β = 123° 20'
Z = 4

Atom	Site	Coordinates
Ag_I	(4e)	x=0.018 y=0.152 z=0.371
Ag_{II}	(4e)	x=0.332 y=0.837 z=0.995
Te	(4e)	x=0.272 y=0.159 z=0.243

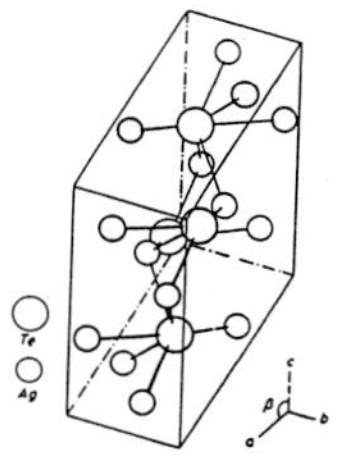

(a)

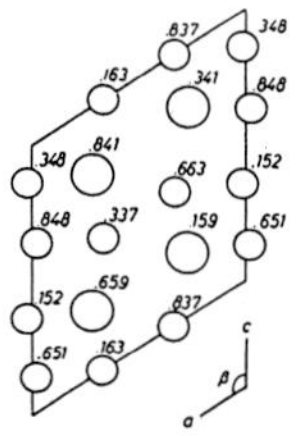

(b)

Fig. 1

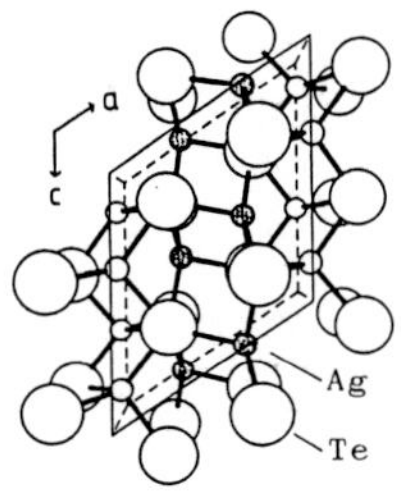

Fig. 2

Properties

Habit	Cleav.	Fract.	Twin.	Hardn.	Dens.	Colour	Transp.
pseudo-cubic, massive, compact	indistinct (100)	even		2.3	8.24-8.45	lead grey, steel grey	opaque

Refr. index	Reflect.	Birefr.	Luster	Streak	Melt.p.	CPI
			metallic		955°-959°C	

Figures

Fig. 1. (a) Perspective view of the hessite structure, and (b) projection on (010) (after Frueh, 1959).

Fig. 2. Perspective projection along the b axis of the structure of hessite (after Lee + + Boer, 1993).

Description

In the hessite structure the Ag atoms are approximately tetrahedrally coordinated by the Te atoms. Hessite can be regarded as a strongly distorted antifluorite structure (Lee+Boer, 1993).

References

Kostov (1968) 158.
Povarennykh (1972) 211.
Wyckoff (1963) Vol. 1, 340.
Frueh, Jr. (1959) 44-52.
Lee+Boer (1993) 1444-1446.
Palache et al. (1944) Vol. 1, 184.
Roberts et al. (1974) 270.

CHALCOCITE

(Chalcosine)

$P2_1/c$

$Cu^{[2]}Cu^{[3]}S$

a = 15.246 Å
b = 11.884 Å
c = 13.494 Å
β = 116°35'
Z = 48

Atom	Position	x	y	z
S_I	(4e)	0.9575	0.0829	C.8422
S_{II}	(4e)	0.9413	0.0768	0.3462
S_{III}	(4e)	0.7940	0.0824	0.5068
S_{IV}	(4e)	0.7917	0.0817	0.0060
S_V	(4e)	0.4491	0.0883	0.6133
S_{VI}	(4e)	0.4444	0.0726	0.0957

...

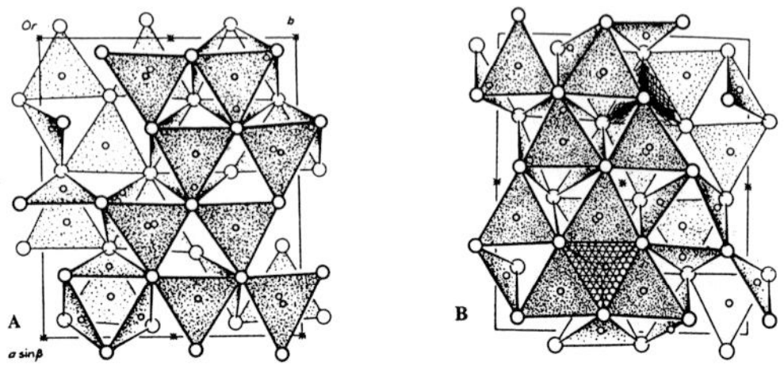

Fig. 1

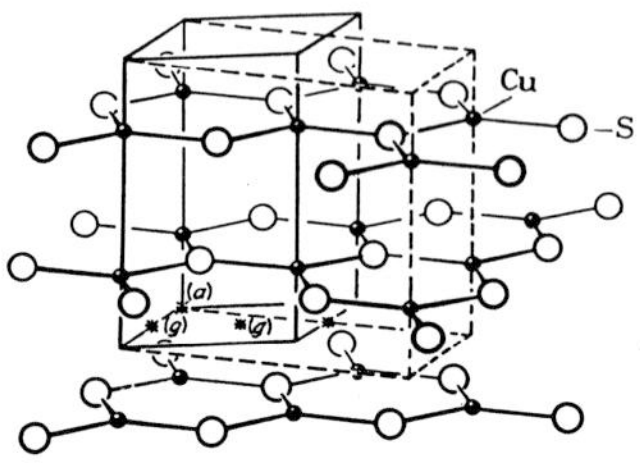

Fig. 2

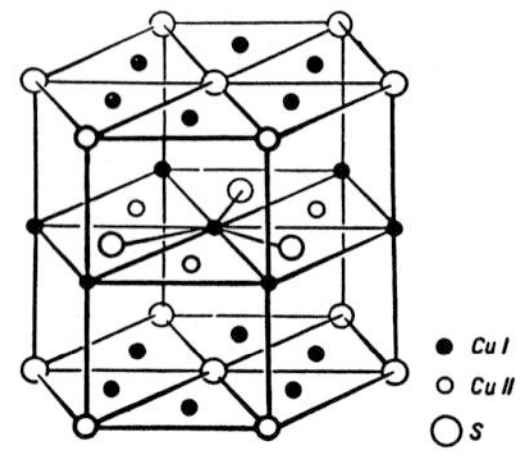

Fig. 3

Properties

Habit	Cleav.	Fract.	Twin.	Hardn.	Dens.	Colour	Transp.
short prismatic, thick tabular		conchoi-dal		2.5	5.5--5.8	blackish grey, black	opaque

Refr. index/Reflect.	Birefr.	Luster	Streak	Melt.p.	CPI
		metallic	blackish	1130°C	

Crystallographic data (continued)

Atom	Position	x	y	z
S_{VII}	(4e)	0.2999	0.0781	0.7868
S_{VIII}	(4e)	0.2843	0.0832	0.2869
S_{IX}	(4e)	0.6960	0.2481	0.7220
S_X	(4e)	0.5479	0.2237	0.4167
S_{XI}	(4e)	0.1970	0.2384	0.4766
S_{XII}	(4e)	0.0483	0.2324	0.1332
Cu_I	(4e)	0.8645	0.2496	0.2927
Cu_{II}	(4e)	0.6171	0.0740	0.6767
Cu_{III}	(4e)	0.6102	0.0916	0.1677
Cu_{IV}	(4e)	0.3628	0.2400	0.0731
Cu_V	(4e)	0.1276	0.0849	0.9451
Cu_{VI}	(4e)	0.1065	0.0783	0.4429
Cu_{VII}	(4e)	0.9345	0.1233	0.9923
Cu_{VIII}	(4e)	0.9414	0.1412	0.5099
Cu_{IX}	(4e)	0.7615	0.2504	0.4109
Cu_X	(4e)	0.4429	0.1477	0.9348
Cu_{XI}	(4e)	0.4254	0.1229	0.4388
Cu_{XII}	(4e)	0.2578	0.2357	0.8507
Cu_{XIII}	(4e)	0.8209	0.0358	0.6830
Cu_{XIV}	(4e)	0.7830	0.0624	0.1671
Cu_{XV}	(4e)	0.0261	0.2045	0.7722
Cu_{XVI}	(4e)	0.5026	0.0795	0.2834
Cu_{XVII}	(4e)	0.3022	0.0434	0.6230
Cu_{XVIII}	(4e)	0.3050	0.0431	0.1339
Cu_{XIX}	(4e)	0.5243	0.2082	0.7543
Cu_{XX}	(4e)	0.9992	0.0856	0.2166
Cu_{XXI}	(4e)	0.6227	0.1032	0.9531
Cu_{XXII}	(4e)	0.7037	0.1944	0.5659
Cu_{XXIII}	(4e)	0.2028	0.2069	0.1398
Cu_{XXIV}	(4e)	0.1308	0.0966	0.6791

Figures

Fig. 1. Two different slabs, A and B, of the chalcocite structure (low form) viewed along the c axis. CuS_3 are shown as shaded triangles and distorted CuS_4 as line-hatched tetrahedra (after Evans Jr., 1979).

Fig. 2. The partial structure of high chalcocite (> 103°C) showing the characteristic CuS laylers. Dashed lines indicates ortho-hexagonal unit cell (after Evans Jr., 1979).

Fig. 3. Structure of the high-temperature form of chalcocite, $P6_3/mmc$ (after Povarennykh, 1972, 262),

Description

The structure of chalcocite is possibly based on a hexagonal closest packing of sulphur atoms, with copper atoms occupying mainly triangular interstices, and some tetrahedral voids. It is monoclinic but a pseudo-orthorhombic cell has been described Abm2 a = 11.92 Å, b = 27.84 Å, c = 13.44 Å and Z = 96 (according to Strunz, 1982,108). The high temperature form (> 103°C) of chalcocite is hexagonal $P6_3/mmc$, a = 3.96 Å, c = 6.72 Å, Z = 2, which corresponds to an hexagonal closest packing of sulphur atoms with copper in triangular interstices, but with an entirely different arrangement (Evans Jr., SR, 45-A, 64)

References

Kostov, 1968, 151
Povarennykh, 1972, 241-242, 262
Wyckoff, Vol. 1, 1963, 332-333
Ramdohr + Strunz, 1980, 417
Palache et al., Vol. 1, 1955, 187-188
Ingerson, 1955, table II
Evans Jr., 1979, Zeit. Krist., 150, 299-320

GLASERITE $P\bar{3}m1$

$K^{[12]} K_2^{[10]} Na^{o} S_2^{t} O_8$

a = 5.6801 Å
c = 7.309 Å
Z = 1

$K^{[12]}$ (1a)	x = 0, y = 0, z = 0	S (2d)	x = 2/3, y = 1/3, z = 0.2338
$K^{[10]}$ (2d)	x = 2/3, y = 1/3, z = 0.6831	O_I (2d)	x = 2/3, y = 1/3, z = 0.0316
Na (1b)	x = 0, y = 0, z = 1/2	O_{II} (6i)	x = 0.8073, y = -0.8073, z = 0.3018

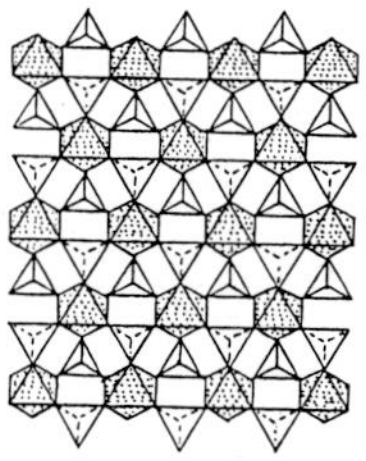

Fig. 1

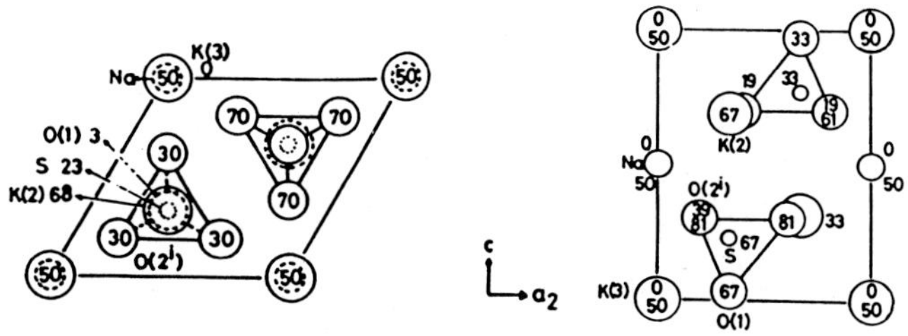

(a) Fig. 2 (b)

Properties

Habit	Cleav.	Fract.	Twin.	Hardn.	Dens.	Colour	Transp.
tabular, massive	fair (10$\bar{1}$0)	conchoidal, uneven	(0001) (11$\bar{2}$0)	3	2.71	colourless, white, grey	transparent to translucent

Refr. index/Reflect.	Birefr.	Luster	Streak	Melt.p.	CPI
n_ω = 1.487-1.491 n_ϵ = 1.492-1.499	(+)	vitreous, resinous			

Figures

Fig. 1. Tetrahedral-octahedral layer of the glaserite (after Zoltai, 1975).

Fig. 2. Representation of the glaserite structure: (a) viewed along the c axis, and (b) viewed along an a axis (after Okada + Ossaka, 1980).

Description

The glaserite structure consists of alternate tetrahedral-octahedral layers where all the tetrahedra, or all the octahedra are replaced by a larger Na or K cation.

Glaserite is possibly a distorted close-packed structure.

References

Kostov (1968) 502.
Okada + Ossaka (1980) 919-921.
Zoltai (1975) VI-5.
Roberts et al. (1974) 29-30.

BISCHOFITE

C 2/m

$Mg^{o}\ Cl_2\ (H_2O)_6$

Cell		Atom	Position	Coordinates
a = 9.90 Å		Mg	(2a)	
b = 7.15 Å		Cl	(4i)	x = 0.318, z = 0.615
c = 6.10 Å				
β = 93° 20'		$(H_2O)_I$	(4i)	x = 0.20, z = 0.11
Z = 2		$(H_2O)_{II}$	(8g)	x = 0.96, y = 0.20, z = 0.225

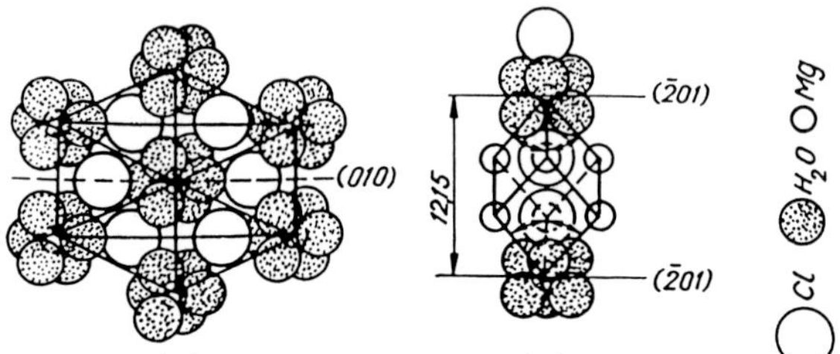

(a) (b)

Fig. 1

Properties

Habit	Cleav.	Fract.	Twin.	Hardn.	Dens.	Colour	Transp.
prismatic, granular, fibrous	none	conchoidal to uneven		1-2	1.56	colourless, white	transparent to translucent

Refr. index/Reflect.	Birefr.	Luster	Streak	Melt.p.	CPI
n_α = 1.495 n_β = 1.507 n_γ = 1.528	(+) 2V = 79°24'	vitreous, dull			

Figures

Fig. 1. Structure of bischofite in projection: (a) on (201), and (b) on a plane perpendicular to the previous one (after Povarennykh, 1972).

Description

The structure of bischofite has distinct $Mg(H_2O)_6Cl_2$ groups formed by $Mg(H_2O)_6$ octahedra, with Cl located at the octahedron faces above and below. Six Cl atoms not in this group form a ring in a plane perpendicular to the axis of the molecule.

This structure is possibly based on a defect close packing of Cl and water molecules.

References

Kostov (1968) 195.

Povarennykh (1972) 639.

Andress + Gundermann (1934) 345-369.

Roberts et al. (1974) 72.

Strukturbericht (1937) Vol. 3, 124, 125, 489-491.

VIVIANITE   C 2/ m

$Fe^{o}_{3}\ P^{t}_{2}\ O_{8}\ (H_2O)_8$

a = 10.08 Å
b = 13.43 Å
c = 4.70 Å
β = 104° 30'
Z = 2

Atom	Position	Coordinates
Fe_I	(2a)	x = 0, y = 0, z = 0
Fe_{II}	(4g)	x = 0, y = 0.390, z = 0
P	(4i)	x = 0.315, y = 0, z = 0.410
O_I	(4i)	x = 0.155, y = 0, z = 0.350
O_{II}	(4i)	x = 0.400, y = 0, z = 0.750
O_{III}	(8j)	x = 0.365, y = 0.100, z = 0.245
...		

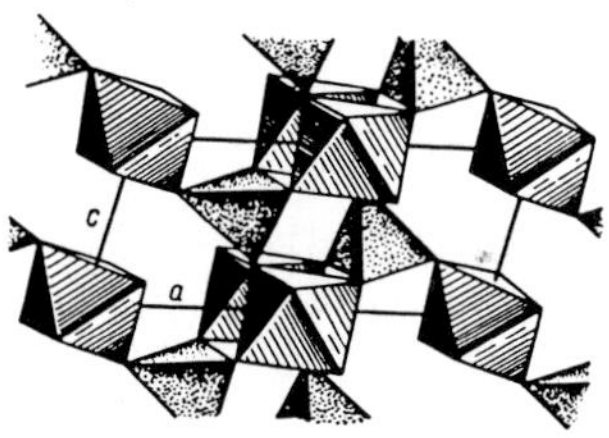

Fig. 1

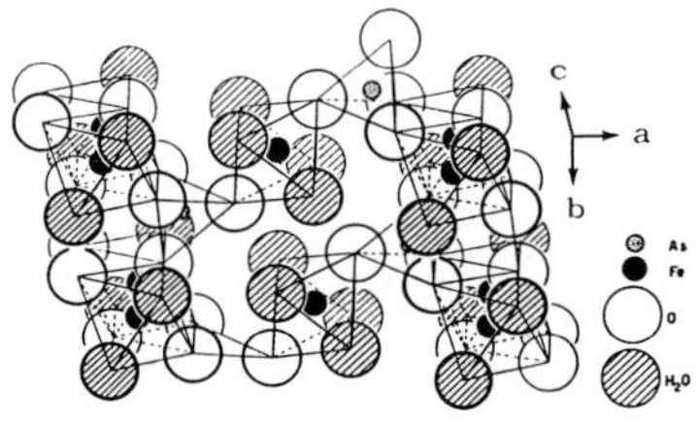

Fig. 2

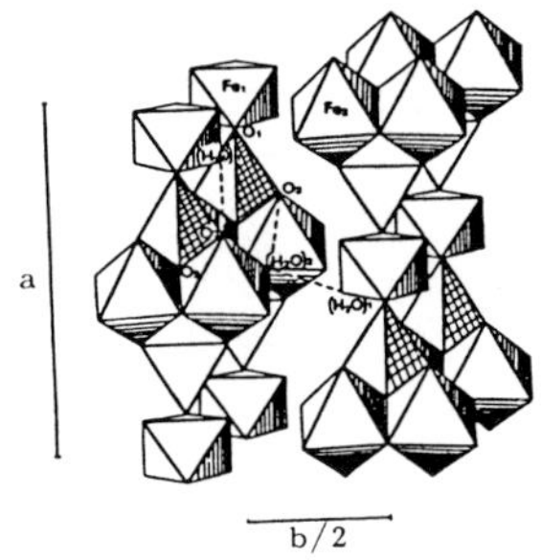

Fig. 3

Properties

Habit	Cleav.	Fract.	Twin.	Hardn.	Dens.	Colour	Transp.
reniform, fibrous	perfect (010)	sectile		2	2.58	colourless, green	transparent to translucent

Refr. index/Reflect.	Birefr.	Luster	Streak	Melt.p.	CPI
n_α = 1.579 n_β = 1.603 n_γ = 1.633	(+) 2V = 83°	pearly, vitreous	white		(SPI) 38

Population

Erythrite	$Co^{o}_{3}\,As^{t}_{2}\,O_8\,(H_2O)_8$
Köttigite	$Zn^{o}_{3}\,As^{t}_{2}\,O_8\,(H_2O)_8$
Parasymplesite	$Fe^{o}_{3}\,As^{t}_{2}\,O_8\,(H_2O)_8$
Hoernesite	$Mg^{o}_{3}\,As^{t}_{2}\,O_8\,(H_2O)_8$

Description

The vivianite structure is probably based on a distorted closest packing of oxygen and water molecules, with iron atoms in octahedral voids, and phosphorus atoms in tetrahedral voids.

Figures

Fig. 1. Structure of vivianite, showing the $FeO_2(OH)_4$ octahedra and the PO_4 tetrahedra forming layers parallel to (010) (after Povarennykh, 1972).

Fig. 2. Layer of the clino-symplesite structure (after Mori + Ito, 1950, quoted in Structure Reports, 1954, 13).

Fig. 3. Polyhedral representation of the vivianite structure, in a section parallel to (001) (after Fedji et al., 1980, quoted in Structure Reports, 1982, 46A).

Crystallographic data (continued)

Atom	Position	Coordinates
$(OH)_I$	(8j)	x = 0.085, y = 0.110, z = 0.820
$(OH)_{II}$	(8j)	x = 0.400, y = 0.220, z = 0.750

References

Kostov (1968) 453.
Povarennykh (1972) 523, 558.
Wyckoff (1965) Vol. 3, 852-854.
Zoltai + Stout (1984) 451.
Structure Reports (1954) vol. 13, 308.
Structure Reports (1982) Vol. 46A, 327.

MERWINITE

$P2_1/a$

$Mg^{o}Si^{t}_{2}Ca_{3}O_{8}$

a = 13.254 Å
b = 5.293 Å
c = 9.328 Å
β = 91.90°
Z = 4

Atom	Pos.	x	y	z
Ca_I	(4e)	0.2563	0.1789	0.2234
Ca_{II}	(4e)	0.0811	0.2271	-0.0753
Ca_{III}	(4e)	0.0978	0.7333	0.4254
Mg	(4e)	0.0043	0.2566	0.2535
Si_I	(4e)	0.1326	0.2293	0.6008
Si_{II}	(4e)	0.1412	0.7280	0.0931

...

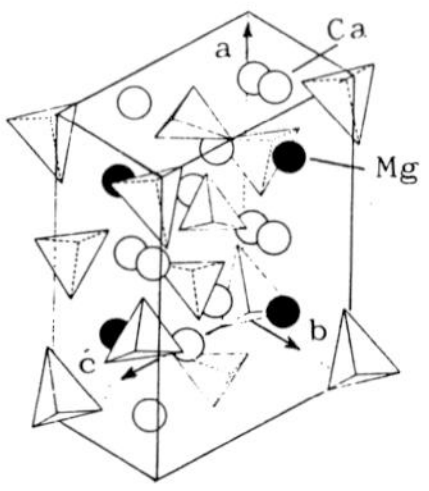

Fig. 1

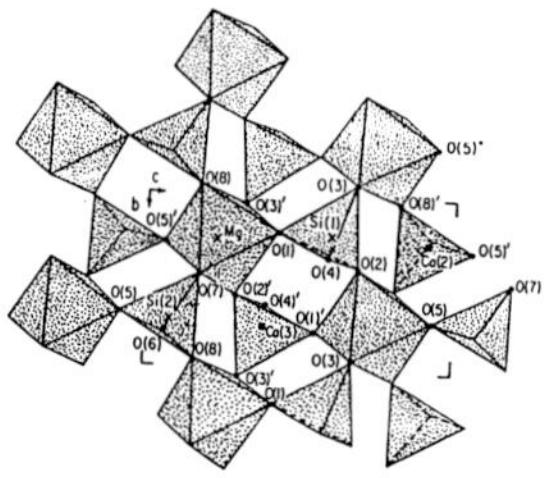

Fig. 2

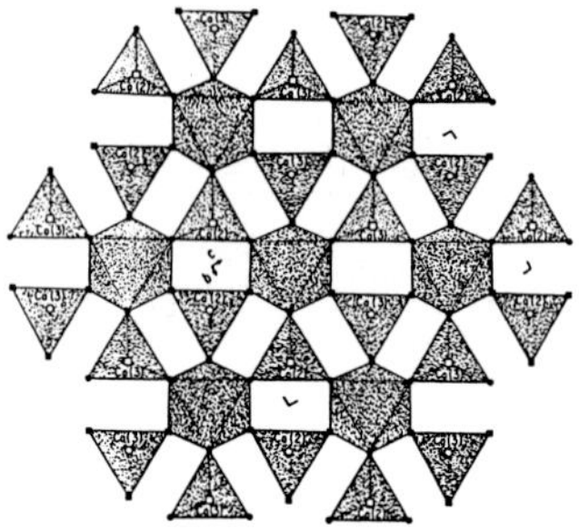

Fig. 3

Properties

Habit	Cleav.	Fract.	Twin.	Hardn.	Dens.	Colour	Transp.
granular, massive, compact	perfect (100)			6	3.15	colourless, pale greenish	transparent to translucent

Refr. index/Reflect.	Birefr.	Luster	Streak	Melt.p.	CPI
n_α = 1.706 n_β = 1.711-1.712 n_γ = 1.724	(+) 2V = 71°	vitreous, greasy			

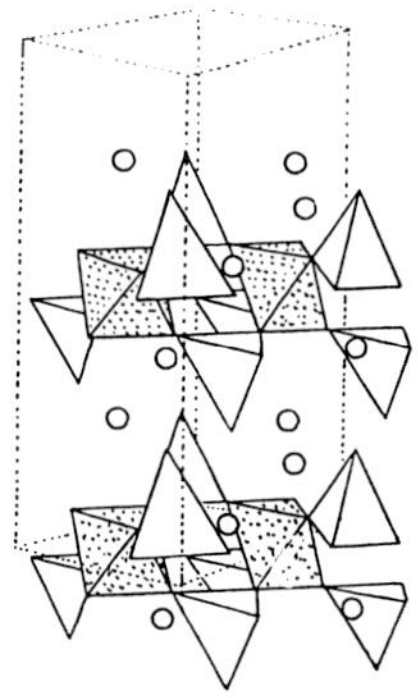

Fig. 4

Figures

Fig. 1. The structure of merwinite, in perspective (after Yamaguchi + Suzuki, 1967, quoted in Structure Reports 1975, 32A).

Fig. 2. Polyhedral diagram of a slab of MgO_6 octahedra and SiO_4 tetrahedra down the X^*-axis (after Moore + Araki, 1972).

Fig. 3. Geometrical idealized arrangement of the MgO_6 octahedra and SiO_4 tetrahedra in the merwinite structure. The black squares are the anions and the open squares the cations in the A-layer, and black circles are the anions and open circles the cations in the B-layer (after Moore + Araki, 1972).

Fig. 4. Polyhedral drawing of the merwinite structure (after Zoltai, 1975).

Description

The structure of merwinite consists of individual SiO_4 tetrahedra bound together by Ca and Mg atoms.

The Ca and O atoms probably form a distorted close packing.

It is related to glaserite.

Crystallographic data (continued)

Atom	Pos.	x	y	z
O_I	(4e)	0.0740	0.2123	0.4450
O_{II}	(4e)	0.0632	0.4193	0.6937
O_{III}	(4e)	0.1253	-0.0475	0.6773
O_{IV}	(4e)	0.2414	0.3618	0.5940
O_V	(4e)	0.0768	0.7064	-0.0591
O_{VI}	(4e)	0.2548	0.8130	0.0683
O_{VII}	(4e)	0.1266	0.4726	0.1853
O_{VIII}	(4e)	0.0832	-0.0424	0.1751

References

Kostov (1968) 303.

Yamaguchi + Suzuki (1967) 220-229.

Moore + Araki (1972) 1355-1374.

Roberts et al. (1974) 394.

Structure Reports (1975) Vol. 32A, 433.

Zoltai (1975) V-2.

CARNALLITE

Pnna

$K^o Mg^o Cl_3 (H_2O)_6$

a = 16.119 Å
b = 22.472 Å
c = 9.551 Å
Z = 12

Atom	Position	x	y	z
K_I	(4c)	1/4	0	0.2486
K_{II}	(8e)	0.08860	0.15668	0.74995
Mg_I	(4d)	0.25657	1/4	1/4
Mg_{II}	(8e)	0.42094	0.08981	0.74770
Cl_I	(4d)	0.23939	1/4	3/4
Cl_{II}	(8e)	0.16583	0.07525	0.98526
...				

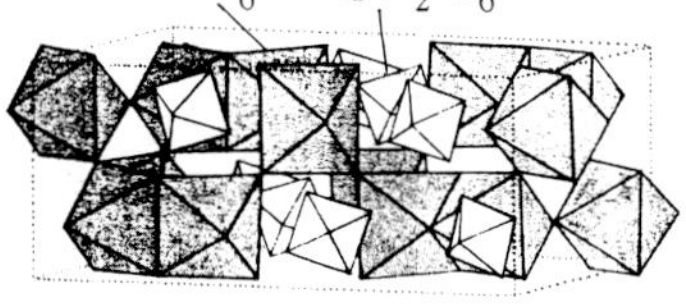

Fig. 1

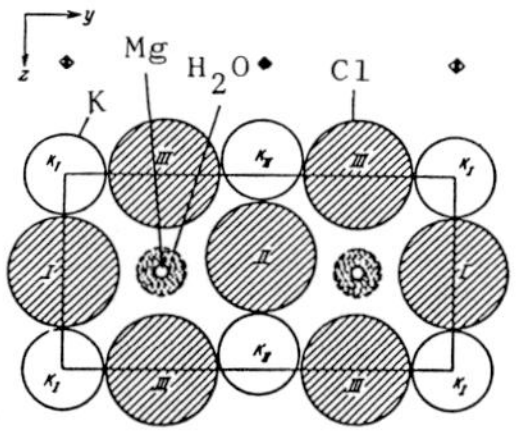

Fig. 2

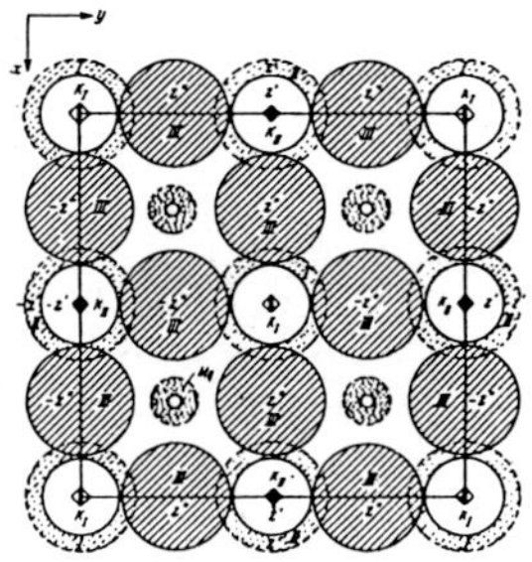

Fig. 3

Properties

Habit	Cleav.	Fract.	Twin.	Hardn.	Dens.	Colour	Transp.
pyramidal, tabular, massive, granular	none	conchoidal		2.5	1.602	colourless, white, reddish	transparent to translucent

Refr. index /Reflect.	Birefr.	Luster	Streak	Melt.p.	CPI
n_α = 1.4665 n_β = 1.4753 n_γ = 1.4937	(+) 2V = 70° 3'	shining, greasy			

Figures

Fig. 1. Polyhedral drawing of the carnallite structure formed by KCl_6 octahedra and $Mg(H_2O)_6$ octahedra (hydrogen not shown) (after Schlemper et al., 1985).

Fig. 2. Unit cell content of the brom-carnallite structure projected along the an a axis (old and incorrect structure determination by Andress + Saffe, 1939, quoted in Strukturbericht, 1943, 7).

Fig. 3. Unit cell content of the brom-carnallite structure projected along the c axis (old and incorrect structure determination, by Andress + Saffe, 1939, quoted in Strukturbericht, 1943, Vol. 7).

Description

The structure of carnallite consists of a network of face-sharing KCl_6 octahedra linked to $Mg(H_2O)_6$ octahedra.

Crystallographic data (continued)

Atom	Position	x	y	z
Cl_{III}	(8e)	0.16928	0.08113	0.48834
Cl_{IV}	(8e)	0.41462	0.08176	0.25235
Cl_V	(8e)	0.01987	0.24957	0.97686
O_I	(4d)	0.1308	1/4	1/4
O_{II}	(4d)	0.3834	1/4	1/4
O_{III}	(8e)	0.2554	0.20704	0.4384
O_{VIII}	(8e)	0.3956	0.16912	0.8444
O_{IV}	(8e)	0.2673	0.16936	0.1506
O_V	(8e)	0.4466	0.01041	0.6500
O_{VI}	(8e)	0.2967	0.07117	0.7349
O_{VII}	(8e)	0.4259	0.04662	0.9354
O_{IX}	(8e)	0.5448	0.10769	0.7631
O_X	(8e)	0.4179	0.13155	0.5580

+ 18 hydrogen positions, each in (8e)

References

Kostov (1968) 200.

Schlemper et al. (1985) 1309-1313.

Strukturbericht (1943) Vol. 7, 19-21.

Roberts et al. (1974) 108.

THENARDITE

$Na^{o}_{2}\ S^{t}\ O_{4}$

F d d d

a = 5.863 Å
b = 12.304 Å
c = 9.821 Å
Z = 8

Na(16g) u = 0.436

S (8a)

O (32h) x = -0.022, y = 0.056, z = 0.214

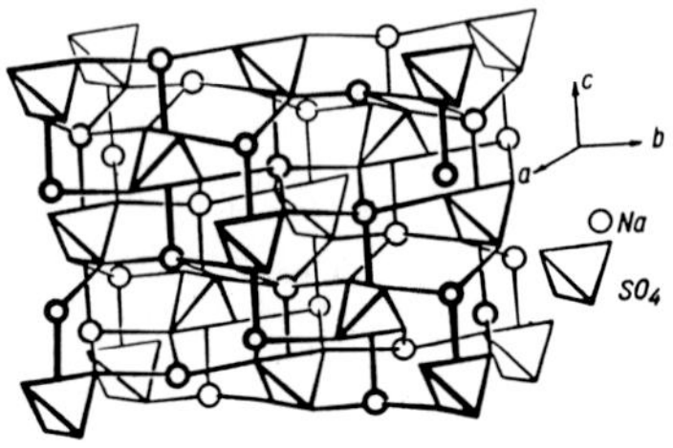

Fig. 1

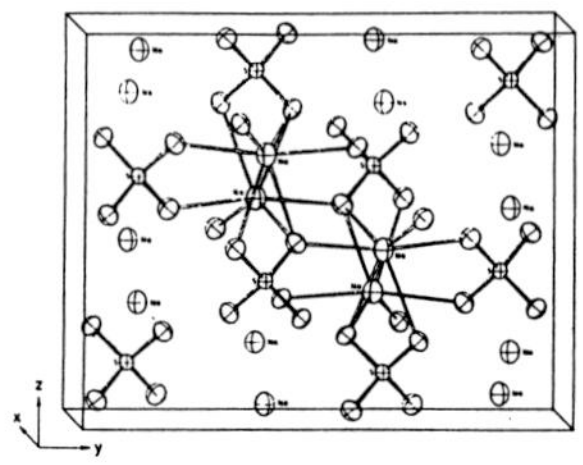

Fig. 2

Properties

Habit	Cleav.	Fract.	Twin.	Hardn.	Dens.	Colour	Transp.
dipyramidal, tabular	perfect (010)	uneven to hackly	(011)	2.5-3	2.664	colourless, greyish white	transparent to translucent

Refr. index /Reflect.	Birefr.	Luster	Streak	Melt.p.	CPI
n_α = 1.464-1.471 n_β = 1.473-1.477 n_γ = 1.4812-1.485	(+) 2V = 82°35'	vitreous, resinous		885°C	

Figures

Fig. 1. Polyhedral representation of the thernadite structure (after Povarennykh, 1972).

Fig. 2. Ball and spoke representation of the thernadite structure (after Nord, 1973, quoted in Structure Reports, 1975,Vol.39A).

Description

The thenardite structure consists of isolated SO_4 tetrahedra linked by Na atoms, which have distorted octahedral coordination.

References

Kostov (1968) 502, 503.
Povarennykh (1972) 582.
Wyckoff (1965) Vol. 3, 109, 110.
Roberts et al. (1974) 612, 613.
Ingerson (1955) 351.
Structure Reports (1975) Vol. 39A, 306.

ALUNITE

R 3 m

$Al^{o}_{3}\ S^{t}_{2}\ K\ O_8\ (OH)_6$

$a_R = 7.05$ Å
$\alpha = 59°2'$
$Z_R = 1$
$a_H = 6.96$ Å
$c = 17.35$ Å
$Z_H = 3$

Atom	Position	x	y	z
K	(3a)	0	0	0
Al	(9b)	0.167	-0.167	0.167
S_I	(3a)	0	0	0.305
S_{II}	(3a)	0	0	-0.305
O_I	(3a)	0	0	0.39
O_{II}	(3a)	0	0	-0.39

...

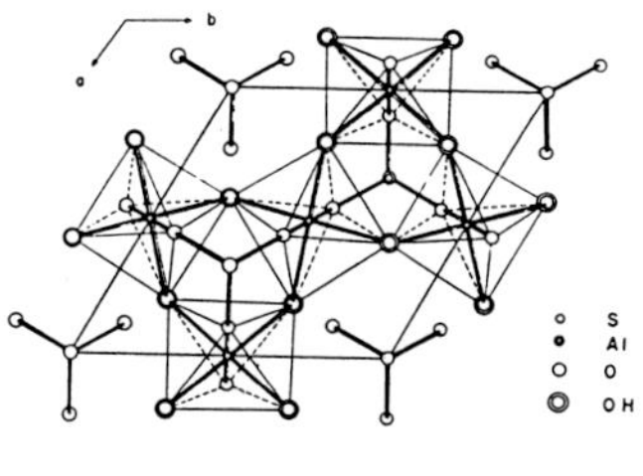

Fig. 1

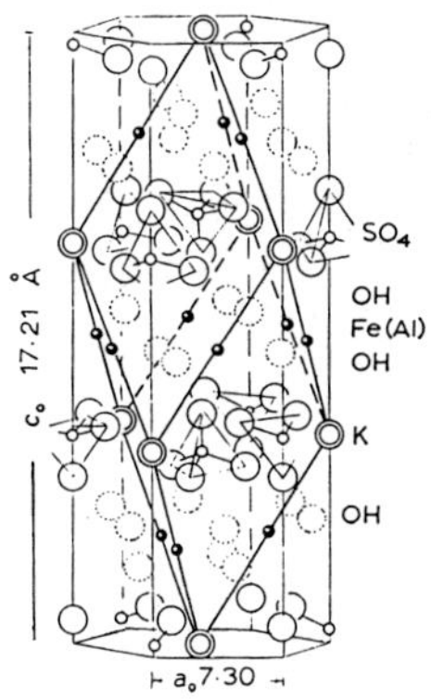

Fig. 2

Properties

Habit	Cleav.	Fract.	Twin.	Hardn.	Dens.	Colour	Transp.
rhombo-hedral, fibrous	good (0001)	conchoi-dal		3.5-4	2.6--2.9	white, grey	transparent to trans-lucent

Refr. index/Reflect.	Birefr.	Luster	Streak	Melt.p.	CPI
$n_\omega = 1.572$ $n_\epsilon = 1.592$	(+)	vitreous			(SPI) 71

Population

Jarosite	$Fe^{o}_{3}\ S^{t}_{2}\ K\ O_8(OH)_6$
Plumbo-jarosite	$Fe^{o}_{3}\ S^{t}_{2}\ Pb_{0.5}\ O_8(OH)_6$
Hamlinite	$Al^{o}_{3}\ P^{t}_{2}\ Sr\ O_7(OH)(OH)_6$
Hidalgoite	$Al^{o}_{3}\ As^{t}S^{t}\ Pb\ O_8(OH)_6$
Crandalite	$Al^{o}_{3}\ P^{t}_{2}\ Ca\ O_7(OH)(OH)_6$

Description

The structure of alunite is probably based on a distorted closest packing of O, OH and K, with Al atoms in octahedral voids and S atoms in tetrahedral interstices.

Figures

Fig. 1. Polyhedral representation of the alunite structure projected along the c axis (after Wang et al., 1965).

Fig. 2. Structure of jarosite, which is isotypic with alunite (after Kostov, 1968).

Crystallographic data (continued)

Atom	Position	x	y	z
O_{III}	(9b)	-0.215	0.215	0.058
O_{IV}	(9b)	0.215	-0.215	0.058
$(OH)_I$	(9b)	-0.150	0.150	0.126
$(OH)_{II}$	(9b)	0.150	-0.150	-0.126

References

Kostov (1968) 493, 501.
Wyckoff (1965) Vol. 3, 210-213.
Zoltai + Stout (1984) 441.
Wang et al. (1965) 251.

BARYTE

$Ba^{[12]} S^{t} O_4$

P n m a

a = 8.8701 Å
b = 5.4534 Å
c = 7.1507 Å
Z = 4

Atom	x	y	z
Ba(4c)	0.1844	1/4	0.1587
S (4c)	0.064	1/4	0.690
O_I(4c)	-0.092	1/4	0.612
O_{II} (4c)	0.187	1/4	0.543
O_{III}(8d)	0.079	0.034	0.813

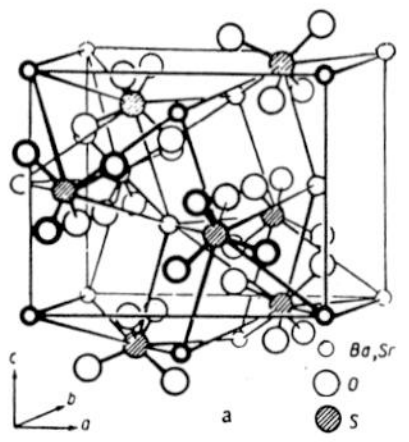

Fig. 1

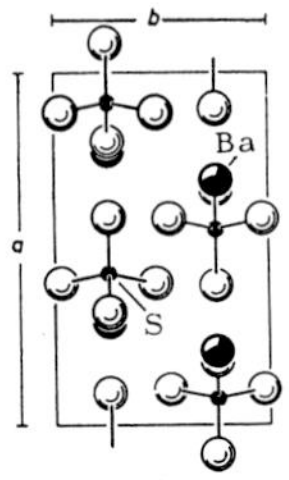

Fig. 2

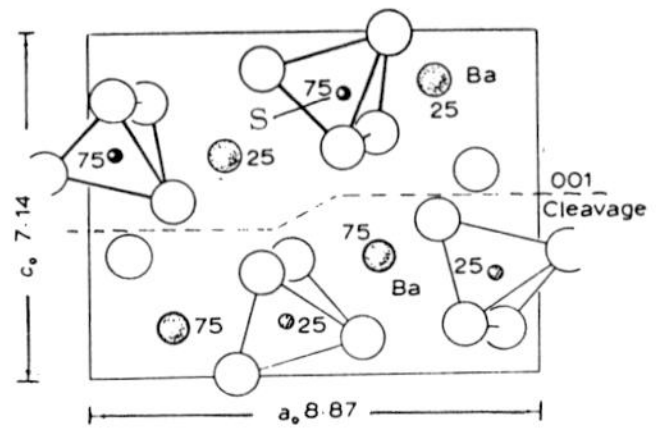

Fig. 3

Properties

Habit	Cleav.	Fract.	Twin.	Hardn.	Dens.	Colour	Transp.
tabular, prismatic, equant	perfect (001)	uneven		3-3.5	4.50	colourless, white, greyish	transparent to subtranslucent

Refr. index/Reflect.	Birefr.	Luster	Streak	Melt.p.	CPI
n_α = 1.6362 n_β = 1.6373 n_γ = 1.6482	(+) 2V = 37°	vitreous, reginous, pearly		158°C	(SPI) 60

Population

Anglesite $Pb^{[12]} S^{t} O_4$

Celestite $Sr^{[12]} S^{t} O_4$

Description

The baryte structure consists of isolated SO_4 tetrahedra linked by Ba atoms with coordination [12] in relation to oxygen.

Figures

Fig. 1. Ball and spoke representation of the baryte structure (after Povarennykh, 1972).

Fig. 2. Baryte structure projected along the c axis (after Ramdohr + Strunz, 1980).

Fig. 3. Polyhedral representation of the baryte structure showing the cleavage direction (after Kostov, 1968).

References

Kostov (1968) 508.
Povarennykh (1972) 581, 582.
Wyckoff (1965) Vol. 3, 45-47.
Ramdohr + Strunz (1980) 147.
Zoltai + Stout (1984) 439.
Roberts et al. (1974) 409, 410.
Ingerson (1955) 351.

EPSOMITE $P\,2_1 2_1 2_1$

$Mg^{o}\,(H_2O)_7\,S^{t}\,O_4$

a = 11.868 Å
b = 11.996 Å
c = 6.857 Å
Z = 4

Atom		x	y	z
Mg	(4a)	0.4229	0.1063	0.0340
S	(4a)	0.7266	0.1839	0.4905
O_I	(4a)	0.6853	0.0750	0.4279
O_{II}	(4a)	0.8506	0.1868	0.4821
O_{III}	(4a)	0.6884	0.2063	0.6904
O_{IV}	(4a)	0.6808	0.2716	0.3605

...

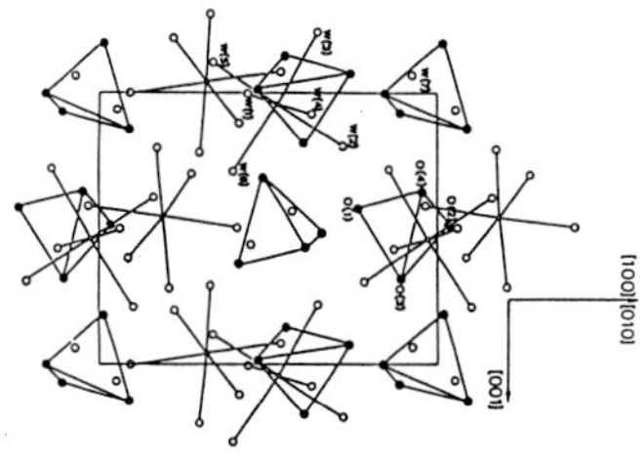

Fig. 1

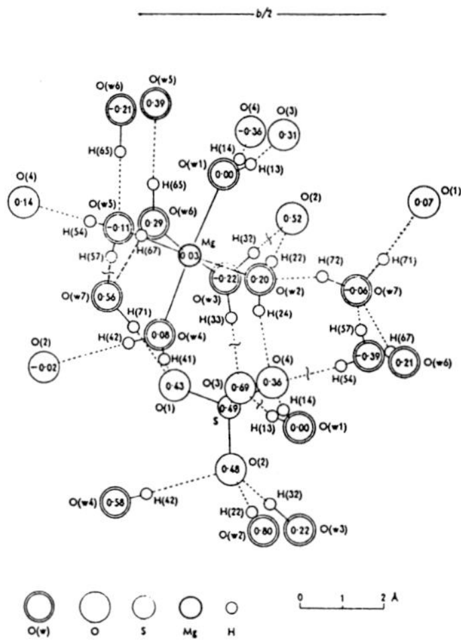

Fig. 2

Properties

Habit	Cleav.	Fract.	Twin.	Hardn.	Dens.	Colour	Transp.
short prismatic, fibrous	perfect (010)	conchoidal		2-2.5	1.677	colourless, white	transparent to translucent

Refr. index/Reflect.	Birefr.	Luster	Streak	Melt.p.	CPI
n_α = 1.433 n_β = 1.455 n_γ = 1.461	(-) 2V = 52°	vitreous, silky	white		(SPI) 30

Figures

Fig. 1. Representation of the epsomite structure (a) projected along $1\bar{1}0$ (after Calleri et al., 1984).

Fig. 2. Schematic representation of the epsomite structure projected along the c axis, showing the hydrogen bonding system which are indicated by dashed lines (after Baur, 1964).

Crystallographic data (continued)

Atom		x	y	z
Ow_I	(4a)	0.2655	0.1746	0.0038
Ow_{II}	(4a)	0.4716	0.2478	0.1992
Ow_{III}	(4a)	0.4691	0.1758	0.7796
Ow_{IV}	(4a)	0.5828	0.0463	0.0773
Ow_V	(4a)	0.3761	0.9620	0.8889
Ow_{VI}	(4a)	0.3622	0.0342	0.2907
Ow_{VII}	(4a)	0.4916	0.4377	0.9384
H_I	(4a)	0.244	0.229	0.105
H_{II}	(4a)	0.234	0.206	0.888
H_{III}	(4a)	0.435	0.270	0.330
H_{IV}	(4a)	0.533	0.246	0.268
H_V	(4a)	0.419	0.234	0.664
H_{VI}	(4a)	0.549	0.188	0.755
H_{VII}	(4a)	0.632	0.055	0.212
H_{VIII}	(4a)	0.600	0.983	0.025
H_{IX}	(4a)	0.356	0.906	0.961
H_X	(4a)	0.428	0.946	0.768
H_{XI}	(4a)	0.281	0.033	0.320
H_{XII}	(4a)	0.385	0.008	0.414
H_{XIII}	(4a)	0.431	0.492	0.975
H_{XIV}	(4a)	0.465	0.370	0.008

Description

The structure of epsomite consists essentially of isolated SO_4 tetrahedra and $Mg(H_2O)_6$ octahedra.

References

Kostov (1968) 494.
Povarennykh (1972) 592.
Wyckoff (1965) Vol. 3, 837-839.
Zoltai + Stout (1984) 442.
Baur (1964) 1361-1369.
Calleri et al. (1984) 221.
Roberts et al. (1974) 194.

MELANTERITE

$Fe^{o} (H_2O)_7 S^{t}O_4$

P 2_1/c

a = 14.07 Å
b = 6.51 Å
c = 11.04 Å
β = 105°36'
Z = 4

Atom	Position	x	y	z
Fe_I	(2a)	0	0	0
Fe_{II}	(2d)	1/2	1/2	0
S	(4e)	0.227	0.471	0.177
O_I	(4e)	0.205	0.472	0.038
O_{II}	(4e)	0.139	0.533	0.215
O_{III}	(4e)	0.307	0.618	0.225

...

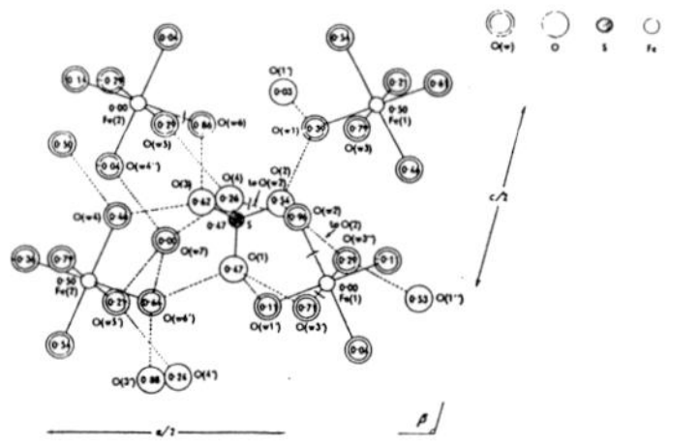

Fig. 1

Properties

Habit	Cleav.	Fract.	Twin.	Hardn.	Dens.	Colour	Transp.
equant, short prismatic	perfect (001)	conchoidal		2	1.90	green, greenish blue	translucent

Refr. index/Reflect.	Birefr.	Luster	Streak	Melt.p.	CPI
n_α = 1.4913 n_β = 1.478 n_γ = 1.4856	(+) 2V = 85°27'	vitreous, silky	colourless		

Figures

Fig. 1. Schematic representation of the melanterite structure projected parallel to 010 showing the hydrogen bonding system, which are indivated by dashed lines (after Baur, 1964).

Description

The structure of melanterite consists of $Fe(H_2O)_6$ octahedra, SO_4 tetrahedra, and a seventh water molecule which is not coordinated to Fe. They are all linked by hydrogen bonds.

References

Kostov (1968) 498.

Wyckoff (1965) Vol. 3, 839-841.

Baur (1964) 1167-1174.

Roberts et al. (1974) 390.

Crystallographic data (continued)

Atom	Position	x	y	z
O_{IV}	(4e)	0.255	0.258	0.227
$(H_2O)_I$	(4e)	0.115	0.384	0.433
$(H_2O)_{II}$	(4e)	0.102	0.957	0.182
$(H_2O)_{III}$	(4e)	0.030	0.795	0.433
$(H_2O)_{II}$	(4e)	0.478	0.456	0.181
$(H_2O)_V$	(4e)	0.429	0.285	0.441
$(H_2O)_{VI}$	(4e)	0.354	0.860	0.440
$(H_2O)_{VII}$	(4e)	0.363	0.004	0.114

GLAUBERITE $Na_2^7 Ca^8 S_2^t O_8$

C 2/c

a = 10.129 Å
b = 8.306 Å
c = 8.533 Å
β = 112.19°
Z = 4

Atom	x	y	z
Ca(4e)	1/2	0.4356	1/4
Na(8f)	0.1371	0.4445	0.4394
S (8f)	0.1856	0.2143	0.1888
O_I (8f)	0.1252	0.0894	0.0583
O_{II} (8f)	0.1616	0.1633	0.3399
O_{III} (8f)	0.3400	0.2294	0.2303

...

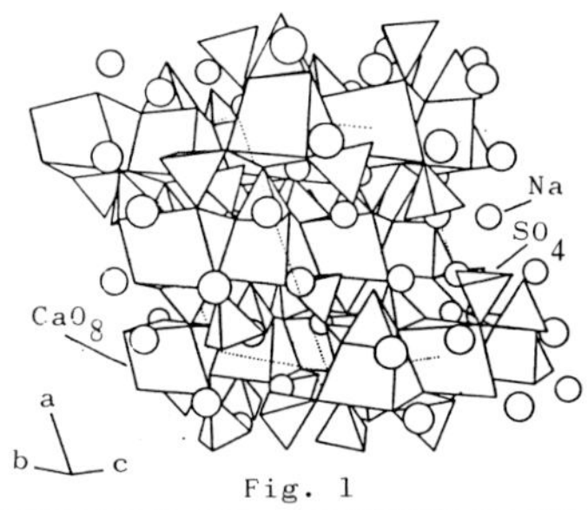

Fig. 1

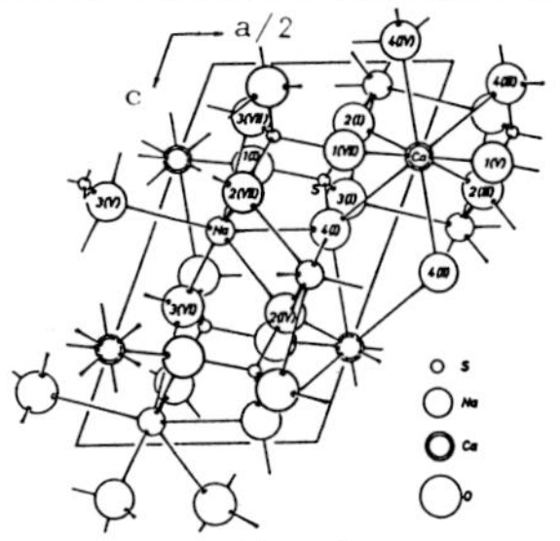

Fig. 2

Properties

Habit	Cleav.	Fract.	Twin.	Hardn.	Dens.	Colour	Transp.
tabular, prismatic, dipyramidal	perfect (001)	conchoi-dal		2.5-3	2.80	colourless, grey, yellowish	transparent to trans-lucent

Refr. index /Reflect.	Birefr.	Luster	Streak	Melt.p.	CPI
n_α = 1.515 n_β = 1.535 n_γ = 1.536	(-) 2V = 7°	vitreous, pearly	white		

Figures

Fig. 1. Polyhedral description of the glauberite structure (after Araki + Zoltai, 1967).

Fig. 2. Ball and spoke representation of the glauberite structure projected along the b axis (after Cocco et al., 1965).

Description

The glauberite structure consists of regular SO_4 tetrahedra, with 8 coordinated Ca atoms, at the corners of an irregular antiprism. Na is coordinated by 7 oxygens forming a distorted mono-capped octahedron.

Crystallographic data (continued)

Atom	x	y	z
O_{IV}(8f)	0.1134	0.3697	0.1325

References

Kostov (1968) 504.
Araki + Zoltai (1967) 1272-1277.
Cocco et al. (1965) 182.
Roberts et al. (1974) 237, 238.

LINARITE

$P2_1/m$

$Pb^3\ Cu^6\ (OH)_2\ S^t\ O_4$

a = 9.81 Å
b = 5.65 Å
c = 4.70 Å
β = 104° 42'
Z = 2

Atom	Position	x	y	z
Pb	(2e)	0.3432	1/4	0.015
Cu	(2a)	0	0	0
S	(2e)	0.666	1/4	0.550
$(OH)_I$	(2e)	0.966	1/4	0.253
$(OH)_{II}$	(2e)	0.094	1/4	0.822
O_I	(2e)	0.522	1/4	0.590

...

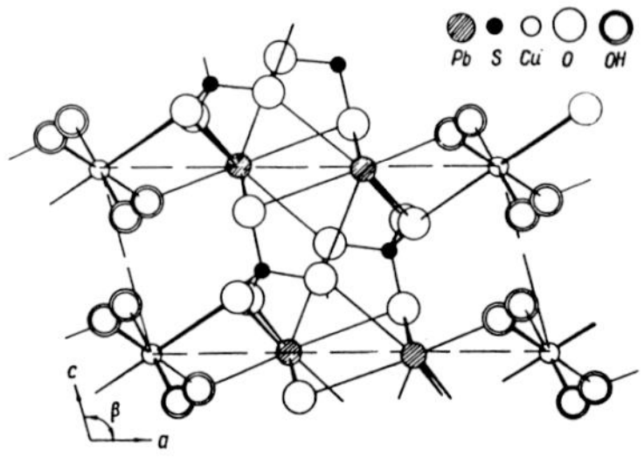

Fig. 1

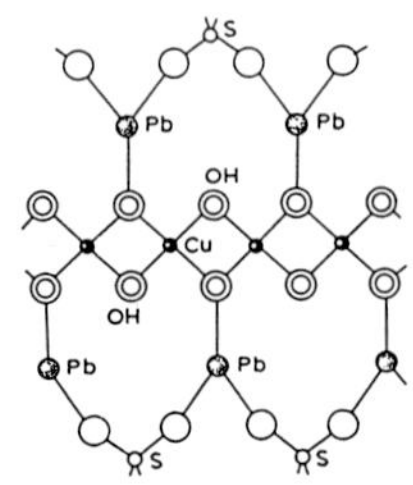

Fig. 2

Properties

Habit	Cleav.	Fract.	Twin.	Hardn.	Dens.	Colour	Transp.
tabular, prismatic, aggregates	perfect (100)	conchoidal	(100)	2.5	5.30	dark azure blue	transparent to translucent

Refr. index/Reflect.	Birefr.	Luster	Streak	Melt.p.	CPI
n_α = 1.809 n_β = 1.838 n_γ = 1.859	(-) 2V = 80°	vitreous, subadamantine	pale blue		

Figures

Fig 1. Ball and spoke representation of the linarite structure projected along the b axis (after Povarennykh, 1972).

Fig. 2. Partial representation of the linarite structure (after Kostov, 1968).

Description

The structure of linarite consists of $Cu(OH)_2$ infinite chains of $Cu(OH)_4$ squares sharing edges, SO_4 isolated tetrahedra linked together by PbO_3 pyramids.

Crystallographic data (continued)

Atom	Position	x	y	z
O_{II}	(2e)	0.658	1/4	0.235
O_{III}	(4f)	0.252	0.540	0.305

References

Kostov (1968) 515.

Povarennykh (1972) 587.

Wyckoff (1965) Vol. 3, 196-199.

Structure Reports (1969) Vol. 26, 452.

Bachmann + Zemann (1961) 747-753.

Roberts et al. (1974) 358.

CHALCANTHITE

$Cu^{o}\ (H_2O)_5\ S^{t}\ O_4$

P $\bar{1}$

a = 6.141	Cu_I (1a) x = 0, y = 0, z = 0	O_I (2i) x = 0.907, y = 0.152, z = 0.675
b = 10.736	Cu_{II} (1e) x = 1/2, y = 1/2, z = 0	O_{II} (2i) x = 0.243, y = 0.318, z = 0.797
c = 5.986	S (2i) x = 0.011, y = 0.287, z = 0.624	O_{III} (2i) x = 0.859, y = 0.373, z = 0.635
α = 82°16'		...
β = 107°26'		
γ = 102°40'		
Z = 2		

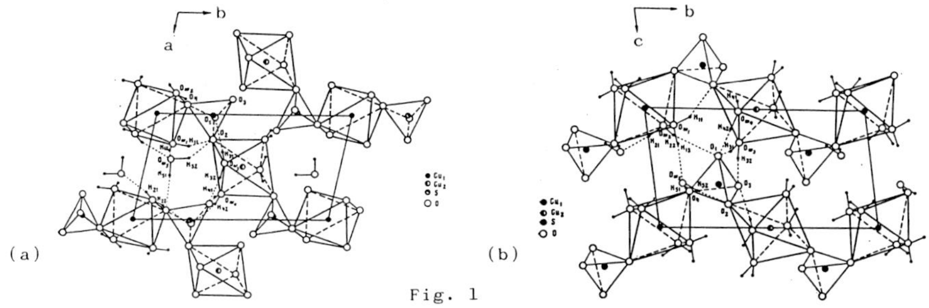

Fig. 1

Properties

Habit	Cleav.	Fract.	Twin.	Hardn.	Dens.	Colour	Transp.
short prismatic, tabular, massive	imperfect (1$\bar{1}$0)	conchoidal		2.5	2.286	pale blue, greenish	transparent to translucent

Refr. index/Reflect.	Birefr.	Luster	Streak	Melt.p.	CPI
n_α = 1.516	(-)	vitreous,	colourless		(SPI)
n_β = 1.539	2V = 56°	resinous			35
n_γ = 1.546					

Population

Pentahydrite $Mg^{o}(H_2O)_5\ S^{t}\ O_4$

Siderotil $Fe^{o}\ (H_2O)_5\ S^{t}\ O_4$

Figures

Fig. 1. Structure of chalcanthite (a) projected on the (001) plane, and (b) projected on the (100) plane (after Iskhakova et al., 1983).

Description

The structure of chalcanthite consists of isolated SO_4 tetrahedra linked to $Cu(H_2O)_6$ octahedra forming zig-zag infinite chains with composition $Cu(H_2O)_4SO_4$.

The structure determination of Iskhakova et al., 1983, gives very similar results to those of Bacon + Curry, 1962 .

Crystallographic data (continued)

Atom	x	y	z	Atom	x	y	z
O_{IV} (2i)	0.045	0.300	0.384	H_{III} (2i)	0.300	0.202	0.067
O_V (2i)	0.817	0.074	0.154	H_{IV} (2i)	0.335	0.128	0.322
O_{VI} (2i)	0.290	0.118	0.149	H_V (2i)	0.320	0.378	0.342
O_{VII} (2i)	0.465	0.406	0.299	H_{VI} (2i)	0.603	0.394	0.425
O_{VIII} (2i)	0.756	0.416	0.019	H_{VII} (2i)	0.805	0.400	0.888
O_{IX} (2i)	0.434	0.125	0.630	H_{VIII} (2i)	0.853	0.384	0.159
H_I (2i)	0.897	0.141	0.250	H_{IX} (2i)	0.601	0.132	0.670
H_{II} (2i)	0.719	0.011	0.229	H_X (2i)	0.410	0.196	0.697

References

Kostov (1968) 512.

Povarennykh (1972) 591.

Zoltai + Stout (1984) 442.

Bacon + Curry (1962) 95-108.

Structure Reports (1971) Vol. 27, 614-616.

Iskhakova et al. (1983) 385, 386.

Roberts et al. (1974) 489.

MIRABILITE

$Na_2^o(H_2O)_{10}S^tO_4$

P $2_1/c$

a = 11.48 Å
b = 10.35 Å
c = 12.82 Å
β = 107° 40'
Z = 4

Atom		Coordinates
Na_I	(4e)	x = 0.249, y = 0.613, z = 0.250
Na_{II}	(4e)	x = 0.251, y = 0.755, z = 0.492
S	(4e)	x = 0.251, y = 0.139, z = 0.260
O_I	(4e)	x = 0.130, y = 0.464, z = 0.115
O_{II}	(4e)	x = 0.406, y = 0.613, z = 0.130
O_{III}	(4e)	x = 0.140, y = 0.780, z = 0.133
...		

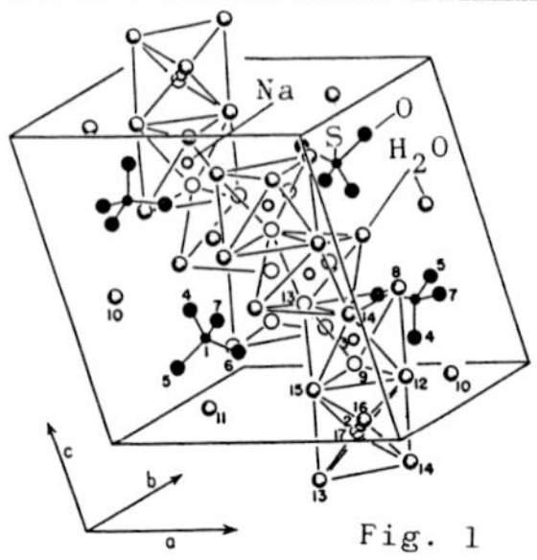

Fig. 1

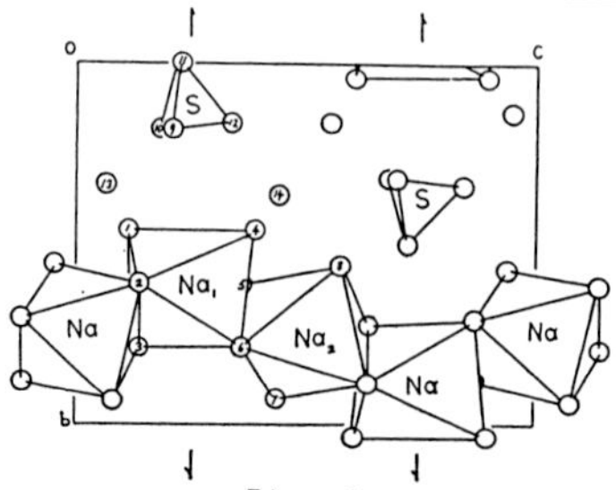

Fig. 2

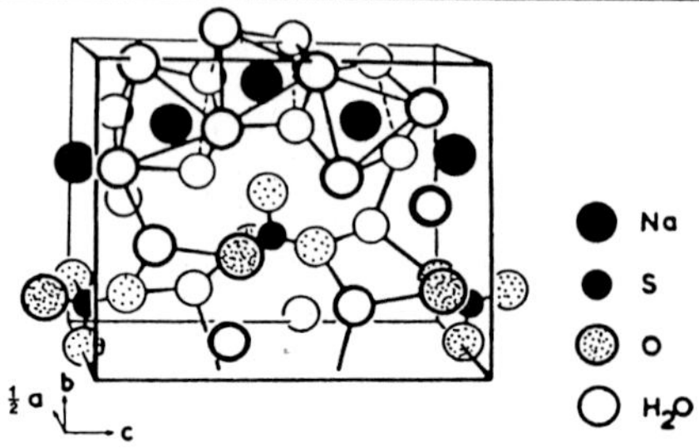

Fig. 3

Properties

Habit	Cleav.	Fract.	Twin.	Hardn.	Dens.	Colour	Transp.
prismatic, acicular, fibrous	perfect (001) (100)	conchoidal		1.5-2	1.464	colourless, white	transparent to opaque

Refr. index /Reflect.	Birefr.	Luster	Streak	Melt.p.	CPI
n_α = 1.396 n_β = 1.4103 n_γ = 1.419	(-) 2V = 75°56'	vitreous			

Figures

Fig. 1 Crystal structure of $Na_2SO_4(H_2O)_{10}$ (after Ruben et al., 1961, quoted in Structure Reports, 1969, Vol. 26).

Fig. 2 Section in the structure of mirabilite centred on x = 1/4 The other half of the structure is generated by the screw axes which are at x = 0 and x = 1/2 (after Cocco, 1962, quoted in Structure Reports, 1971, Vol.27).

Fig. 3. Representation of the structure of $Na_2SO_4(H_2O)_{10}$ (half of a cell) (after Meulendijk, 1956, quoted in Structure Reports, 1963, Vol. 20).

Description

The mirabilite structure consists of isolated SO_4 tetrahedra and infinite chains of $Na(H_2O)_6$ octahedra which share their edges. Two of the water molecules, in formula unit, are interstitial.

Crystallographic data (continued)

Atom		Coordinates
O_{IV}	(4e)	x = 0.367, y = 0.466, z = 0.400
O_V	(4e)	x = 0.111, y = 0.618, z = 0.367
O_{VI}	(4e)	x = 0.356, y = 0.780, z = 0.367
O_{VII}	(4e)	x = 0.120, y = 0.932, z = 0.438
O_{VIII}	(4e)	x = 0.367, y = 0.558, z = 0.583
O_{IX}	(4e)	x = 0.355, y = 0.186, z = 0.209
O_X	(4e)	x = 0.144, y = 0.186, z = 0.192
O_{XI}	(4e)	x = 0.240, y = 0.000, z = 0.233
O_{XII}	(4e)	x = 0.250, y = 0.167, z = 0.341
O_{XIII}	(4e)	x = 0.408, y = 0.342, z = 0.058
O_{XIV}	(4e)	x = 0.105, y = 0.364, z = 0.442

References

Kostov (1968) 503.
Cocco (1962) 690-698.
Ruben et al. (1961) 820-824.
Meulendijk (1956) 493-495.
Structure Reports (1971) Vol. 27, 620, 621.
Structure Reports (1969) Vol. 26, 433, 434.
Structure Reports (1963) Vol. 20, 345, 346.

KAINITE

C 2/m

$K Mg^{o} (H_2O)_3 Cl S^{t} O_4$

a = 19.72 Å
b = 16.23 Å
c = 9.53 Å
β = 94°55'
Z = 16

Atom	Position	x	y	z
K_{I}	(4i)	0.19158	1/2	0.4203
K_{II}	(4i)	0.19318	0	0.9398
K_{III}	(8j)	0.19487	0.3064	0.8483
S_{I}	(8j)	0.09695	0.3308	0.5175
S_{II}	(8j)	0.09805	0.1672	0.0136
Mg_{I} ...	(2d)	0	1/2	1/2

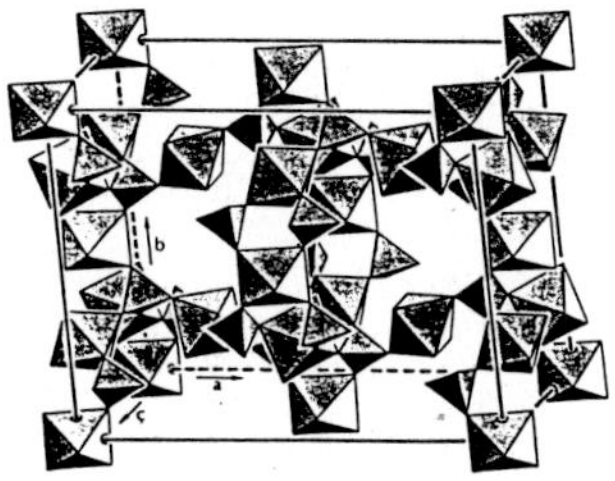

Fig. 1

Properties

Habit	Cleav.	Fract.	Twin.	Hardn.	Dens.	Colour	Transp.
tabular, equant, massive	perfect (001)	smooth to splintery		2.5-3	2.15	colourless, grey, blue	transparent to translucent

Refr. index/Reflect.	Birefr.	Luster	Streak	Melt.p.	CPI
n_α = 1.494 n_β = 1.505 n_γ = 1.516	(-) 2V ≃ 90°	vitreous			

Figures

Fig. 1. Polyhedral representation of the kainite structure (K and Cl atoms omitted) (after Robinson et al., 1972).

Description

The kainite structure consists of polyhedral layers of corner-sharing Mg-octahedra and S-tetrahedra, with K, Cl atoms and water molecules between the layers. The geometry of the Mg octahedra and S tetrahedra is close to ideal.

References

Kostov (1968) 496.
Robinson et al. (1972) 1325-1332.
Roberts et al. (1974) 315, 316.

Crystallographic data (continued)

Atom	Position	x	y	z
Mg_{II}	(2a)	0	0	0
Mg_{III}	(4f)	1/4	1/4	1/2
Mg_{IV}	(8j)	0.00026	0.2484	0.2481
Cl_{I}	(4i)	0.21470	0	0.2730
Cl_{II}	(4i)	0.13985	0	0.6231
Cl_{III}	(8j)	0.17890	0.3892	0.1357
O_{I}	(8j)	0.0710	0.2317	0.9152
O_{II}	(8j)	0.0707	0.2694	0.4142
O_{III}	(8j)	0.0749	0.3107	0.6569
O_{IV}	(8j)	0.0734	0.4133	0.4744
O_{V}	(8j)	0.0715	0.0859	0.9646
O_{VI}	(8j)	0.0760	0.1844	0.1552
O_{VII}	(8j)	0.1721	0.3316	0.5261
O_{VIII}	(8j)	0.1722	0.1668	0.0183
OW_{I}	(4i)	0.0399	0	0.2197
OW_{II}	(4i)	0.0193	1/2	0.7192
OW_{III}	(8j)	0.2133	0.1727	0.6510
OW_{IV}	(8j)	0.0259	0.3588	0.1468
OW_{V}	(8j)	0.0200	0.1354	0.6489
OW_{VI}	(8j)	0.1963	0.1849	0.3438
OW_{VII}	(4i)	0.1454	1/2	0.8430

POLYHALITE

$P\bar{1}$

$K_2^{[11]} Ca_2^{[8]} Mg^{o} (H_2O)_2 S_4^{t} O_{16}$

a = 11.69 Å
b = 16.33 Å
c = 7.60 Å
α = 91°6'
β = 90°0'
γ = 91°9'
Z = 4

Atom	Site	x	y	z
K^+	(8i)	0.13189	0.28615	0.63004
Ca^{2+}	(8i)	0.11914	0.37120	0.12619
Mg^{2+}	(4a)	0	0	0
Si_I	(8i)	0.12582	0.7054	0.63225
Si_{II}	(8i)	0.12061	0.17896	0.11415
O_I	(8i)	0.02782	0.12014	0.59472
O_{II}	(8i)	0.22760	0.12428	0.66306
O_{III}	(8i)	0.10304	0.01950	0.78843
O_{IV}	(8i)	0.14787	0.1101	0.48555
O_V	(8i)	0.13167	0.22955	0.27763
O_{VI}	(8i)	0.09617	0.23414	0.96862
O_{VII}	(8i)	0.02245	0.11814	0.13130
O_{VIII}	(8i)	0.22550	0.13457	0.08081
H_2O	(8i)	0.13515	0.45748	0.63492

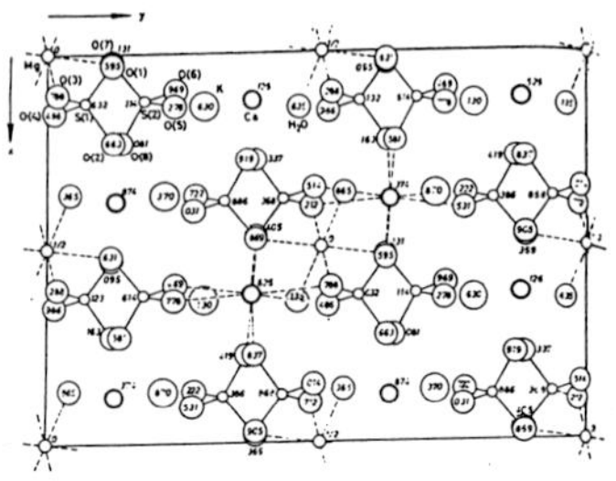

Fig. 1.

Properties

Habit	Cleav.	Fract.	Twin.	Hardn.	Dens.	Colour	Transp.
fibrous, foliated, tabular				3.5	2.78	colourless, white, grey	transparent to translucent

Refr. index /Reflect.	Birefr.	Luster	Streak	Melt.p.	CPI
n_α = 1.547 n_β = 1.560 n_γ = 1.567	(-) 2V = 62°	vitreous to resinous			

Figures

Fig. 1. Projection of the structure of polyhalite parallel to (001) (after Schlatti et al., 1970).

Description

The polyhalite structure consists of isolated SO_4 tetrahedra linked by Mg atoms octahedrally coordinated, Ca atoms (8 coordinated) and K atoms (11-coordinated). Each water molecule is coordinated by one Mg and one K.

References

Kostov (1968) 504.
Schlatti et al. (1970) 75-86.
Roberts et al. (1974) 489.

MONAZITE

$P\,2_1/n$

$(Ce, La, Y, Th)^{[9]} P^{t} O_4$

a = 6.76 Å
b = 7.00 Å
c = 6.44 Å
β = 103° 38'
Z = 4

Atom		
Ce(4e)	x = 0.286	y = 0.158	z = 0.094
P (4e)	x = 0.325	y = 0.167	z = 0.611
O_I(4e)	x = 0.306	y = -0.008	z = 0.472
O_{II} (4e)	x = 0.375	y = 0.342	z = 0.500
O_{III}(4e)	x = 0.500	y = 0.133	z = 0.806
O_{IV} (4e)	x = 0.128	y = 0.200	z = 0.681

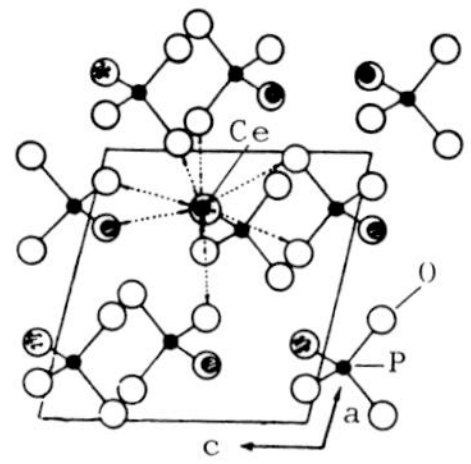

Fig. 1

Fig. 2

Properties

Habit	Cleav.	Fract.	Twin.	Hardn.	Dens.	Colour	Transp.
tabular, equant, prismatic	distinct (100)	conchoidal to uneven	(100)	5-5.5	4.6-5.4	reddish brown, brown	transparent to subtranslucent

Refr. index/Reflect.	Birefr.	Luster	Streak	Melt.p.	CPI
n_α = 1.785 n_β = 1.787 n_γ = 1.840	(+) 2V = 12°	resinous, waxy, vitreous	white, slightly coloured		(SPI) 58

Population

Huttonite $Th^{[9]} Si^{t} O_4$

Crocoite $Pb^{[9]} Cr^{t} O_4$

Figures

Fig. 1. Ball and spoke representation of the monazite structure projected along the b axis (after Povarennykh, 1972).

Description

The monazite structure consists of isolated PO_4 tetrahedra linked by Ce atoms, which have coordination nine (they link 6 tetrahedra). Another structure determination has been worked out by Ghouse, 1968, which gives very similar results.

References

Kostov (1968) 461.
Povarennykh (1972) 536.
Wyckoff (1965) Vol. 3, 33-35.
Zoltai + Stout (1984) 448.
Roberts et al. (1974) 413, 414.
Ghouse (1968) 265-268.

LAZULITE $Mg^{o}\ Al_2^{o}\ (OH)_2\ P_2^{t}\ O_8$

$P\,2_1/c$

a = 7.16 Å
b = 7.26 Å
c = 7.24 Å
β = 120° 40'
Z = 2

Atom	x	y	z
Mg(2a)	0	0	0
Al(4e)	0.732	0.266	0.004
P (4e)	0.250	0.113	0.749
(OH)(4e)	0.754	0.135	0.749
O_I (4e)	0.222	0.000	0.901
O_{II}(4e)	0.286	0.000	0.589
...			

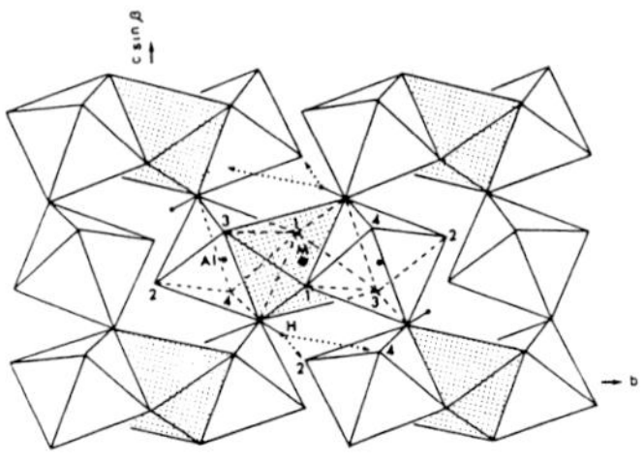

Fig. 1

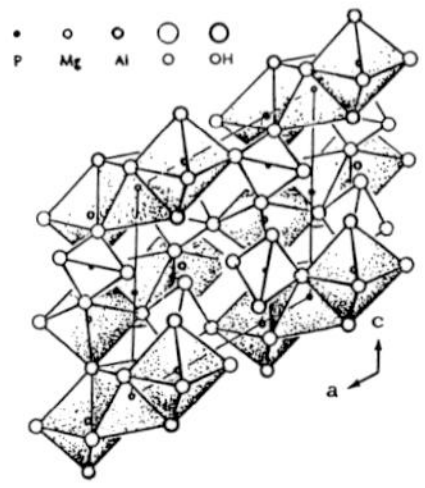

Fig. 2

Properties

Habit	Cleav.	Fract.	Twin.	Hardn.	Dens.	Colour	Transp.
pyramidal, tabular, massive		uneven, splintery		5.5-6	3.08	azure blue, light blue	translucent to opaque

Refr. index /Reflect.	Birefr.	Luster	Streak	Melt.p.	CPI
n_α = 1.604-1.635 n_β = 1.633-1.663 n_γ = 1.642-1.673	(-) 2V = 68.9°	vitreous, dull			

Population

Scorzalite $Fe^{o}Al_2^{o}(OH)_2(P^{t}O_4)_2$

Barbosalite $Fe^{o}Fe_2^{o}(OH)_2(P^{t}O_4)_2$

Figures

Fig. 1. Polyhedral representation of the lazulite structure projected along the a axis (after Giuseppetti + Tadini, 1983, quoted in Structure Reports (1988) Vol. 50A).

Fig. 2. Polyhedral representation of the lazulite structure viewed along the b axis (after Lindberg + Christ, 1959, quoted in Structure Reports, 1965, Vol. 23).

Description

The structure of lazulite consists of the packing of $(Mg,Fe)O_6$ octahedra and PO_4 tetrahedra. It contains chains of AlO_6 corner-sharing octahedra, along the c axis, linked by face-sharing $(Mg,Fe)O_6$ octahedra. The resulting layers are linked by PO_4 tetrahedra.

Crystallographic data (continued)

Atom	x	y	z
O_{III}(4e)	0.048	0.253	0.130
O_{IV} (4e)	0.443	0.244	0.377

References

Kostov (1968) 45.

Wyckoff (1965) Vol. 3, 197-199.

Lindberg + Christ (1959) 695-697.

Giuseppetti + Tadini (1983) 410-416.

Structure Reports (1965) Vol. 23. 434, 435.

Structure Reports (1988) Vol. 50A, 264.

WAVELLITE

P c m n

$Al^{o}_{3} (H_2O)_5 (OH)_3 P^{t}_{2} O_8$

a = 9.621 Å
b = 17.363 Å
c = 6.994 Å
Z = 4

Atom	Position	x	y	z
Al_I	(4c)	0.22384	1/4	0.12326
Al_{II}	(8d)	0.75605	0.01638	0.14186
P	(8d)	0.06061	0.09221	0.10399
O_I	(8d)	0.90525	0.08349	0.06432
O_{II}	(8d)	0.08916	0.17642	0.15555
O_{III}	(8d)	0.10095	0.04183	0.27355

...

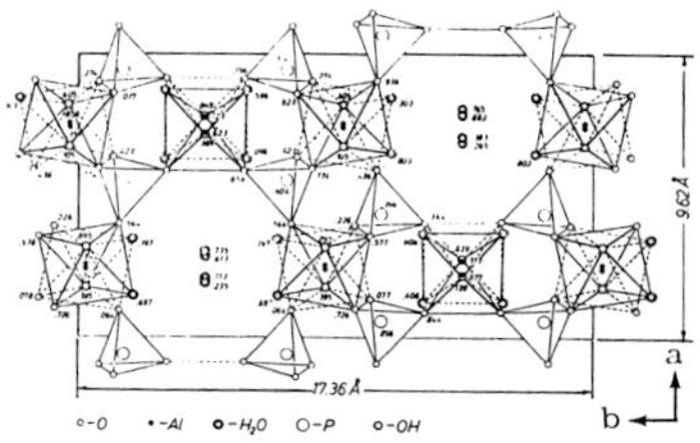

Fig. 1

Properties

Habit	Cleav.	Fract.	Twin.	Hardn.	Dens.	Colour	Transp.
prismatic, acicular aggregates	perfect (110)	subconchoidal uneven		3.25	2.36	white, greenish, yellowish	transparent to translucent

Refr. index/Reflect.	Birefr.	Luster	Streak	Melt.p.	CPI
n_α = 1.520-1.535	(+)	vitreous,	white		(SPI)
n_β = 1.526-1.543	2V ≃ 71°	pearly,			35
n_γ = 1.545-1.561		resinous			

Figures

Fig. 1. The structure of wavellite projected along the c axis (after Araki + Zoltai, 1968).

Description

The wavellite structure is formed by two kinds of AlO_6 octahedra along the c axis, linked together by PO_4 tetrahedra. This linkage creates a dense layer of octahedra and tetrahedra in the plane (010), which are in turn connected by alternating AlO_4 octahedra and water molecules.

Possibly a framework of AlO_6 octahedral chains linked by PO_4 tetrahedra.

Crystallographic data (continued)

Atom	Position	x	y	z
O_{IV}	(8d)	0.36037	0.07246	0.42258
OH_I	(4c)	0.27997	1/4	0.36925
OH_{II}	(8d)	0.82173	0.01851	0.39490
H_2O_I	(8d)	0.37060	0.17055	0.09551
H_2O_{II}	(8d)	0.64979	0.11144	0.19740
$0.5H_2O_{III}$	(4c)	0.81314	1/4	0.23453
$0.5H_2O_{IV}$	(4c)	0.78308	1/4	0.11326

References

Kostov (1968) 445.

Povarennykh (1972) 549.

Zoltai + Stout (1984) 451.

Structure Reports (1975) Vol. 33A, 404, 405.

Araki + Zoltai (1968) 21-33.

Roberts et al. (1974) 663.

DUMORTIERITE

P m c n

$(Al,Fe)^{o}_{7}\ (Si^{t}O_4)_3O_3\ B^{tr}O_3$

a = 11.79 Å
b = 20.209 Å
c = 4.701 Å
Z = 4

Al_I (4c)	x = 1/4, y = 0.249, z = 0.392	Al_{IV}(8d)	x = 0.641, y = 0.211, z = 0.057
Al_{II}(8d)	x = 0.390, y = 0.028, z = 0.560	Si_I (4c)	x = 1/4, y = 0.094, z = 0.087
Al_{III} (8d)	x = 0.508, y = 0.069, z = 0.057	Si_{II}(8d) ...	x = 0.474, y = 0.172, z = 0.588

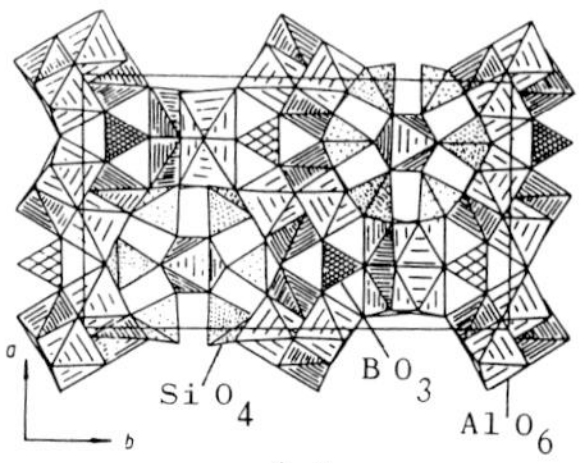

(a)

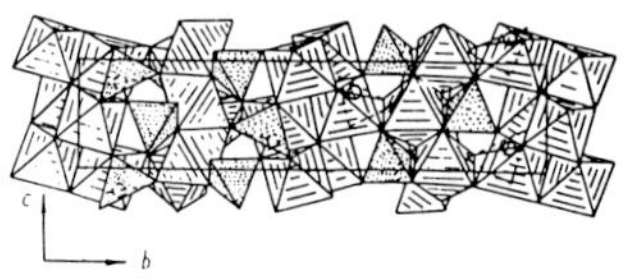

(b)

Fig. 1

Properties

Habit	Cleav.	Fract.	Twin.	Hardn.	Dens.	Colour	Transp.
prismatic, massive, fibrous	good (100)			8.5	3.41	blue, violet, pinkish	transparent to trans-lucent

Refr. index/Reflect.	Birefr.	Luster	Streak	Melt.p.	CPI
n_α = 1.6860 n_β = 1.722 n_γ = 1.7229	(-) 2V = 13°	vitreous, dull	white		

Figures

Fig. 1. Polyhedral representation of the dumortierite structure: (a) projected along the c axis, showing different types of columns of AlO_6 octahedra, as well as disposition of SiO_4 tetrahedra and BO_3 triangles (double hatching), and (b) projected along the a axis, showing the columns of AlO_6 octahedra along the c axis (after Povarennykh, 1972).

Description

The structure of dumortierite contains three types of chains parallel to the c axis; partly disordered AlO_3 face-sharing chains, and two types of Al_4O_{12} chains, one of which forms Al_4O_{11} sheets.

Two recent structure determinations have confirmed the presented structure: Moore + Araki, 1978, and Alexander et al., 1986.

Crystallographic data (continued)

B (4c)	x = 1/4, y = -0.085, z = -0.225	O_{VI} (4c)	x = 1/4, y = 0.047, z = 0.388
O_I (4c)	x = 1/4, y = 0.173, z = 0.122	O_{VII} (8d)	x = 0.544, y = 0.150, z = 0.875
O_{II}(8d)	x = 0.360, y = 0.213, z = 0.650	O_{VIII}(4c)	x = 1/4, y = -0.225, z = 0.268
O_{III}(8d)	x = 0.361, y = 0.075, z = 0.895	O_{IX} (4c)	x = 1/4, y = -0.150, z = -0.165
O_{IV} (8d)	x = 0.449, y = 0.106, z = 0.390	O_X (8d)	x = 0.650, y = 0.052, z = 0.255
O_V (8d)	x = 0.570, y = 0.216, z = 0.400	O_{XI}(8d)	x = 0.537, y = 0.013, z = 0.748

References

Kostov (1968) 282.

Povarennykh (1972) 392, 393.

Wyckoff (1968) Vol. 4, 213-215.

Moore + Araki (1978) 231-241.

Alexander et al. (1986) 786-794.

Roberts et al. (1974) 181.

DESCLOIZITE   Pmcn

[8ap] [6by]
Pb (Zn,Cu) (OH) $V^t O_4$

a = 6.074 Å
b = 9.446 Å
c = 7.607 Å
Z = 4

Atom	Coordinates	Atom	Coordinates
Pb (4c)	x = 1/4, y = 0.663, z = 0.118	(OH)(4a)	x = 0, y = 0, z = 0
(Zn,Cu)(4c)	x = 1/4, y = 0.0000, z = 0.125	O_I (4c)	x = 1/4, y = 0.125, z = 0.375
V (4c)	x = 1/4, y = 0.250, z = 0.250	O_{II}(4c) ...	x = 1/4, y = 0.375, z = 0.375

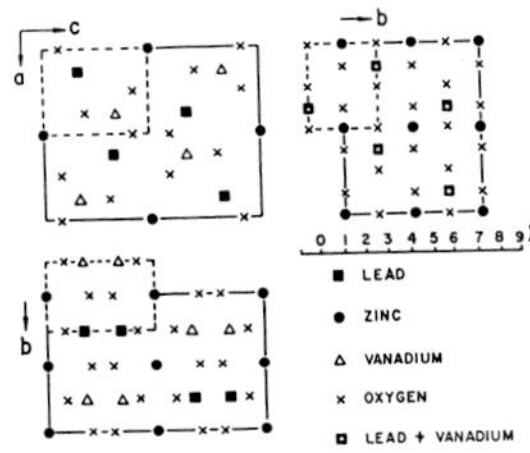

Properties

Habit	Cleav.	Fract.	Twin.	Hardn.	Dens.	Colour	Transp.
pyramidal, prismatic, tabular	none	uneven, conchoidal		3-3.5	6.24--6.26	orange red, reddish brown	transparent to opaque

Refr. index/Reflect.	Birefr.	Luster	Streak	Melt.p.	CPI
n_α = 2.185 n_β = 2.265 n_γ = 2.35	(-) 2V ≃ 90°	vitreous, greasy			

Figures

Fig. 1. Approximate structure of descloizite deduced from Patterson maps (after Structure Reports, 1963, Vol. 17).

Description

Descloizite is built of VO_4 tetrahedra, $ZnO_4(OH)_2$ tetragonal bipyramids, and square $PbO_7(OH)$ antiprisms.

Crystallographic data (continued)

O_{III}(8d) x = 0.450, y = 0.250, z = 0.125

References

Kostov (1968) 471.

Wyckoff (1965) Vol. 3, 183, 184.

Roberts et al. (1974) 169.

Structure Reports (1963) Vol.17, 502.

CHLORITOID

C 2/c

$\overset{o}{(Fe,Mg)}\overset{o}{(Al,Fe)}_2(H_2O)\ Si^tO_6$

a = 9.52 Å
b = 5.47 Å
c = 18.19 Å
β = 101° 39'
Z = 8

(Fe,Mg)	(8f)	x = 0.086 y = 0.750 z = 0.001
$(Al,Fe)_I$	(4c)	
$(Al,Fe)_{II}$	(4e)	u = 0.4036
$(Al,Fe)_{III}$	(8f)	x = 0.250 y = 0.342 z = -0.250
...		

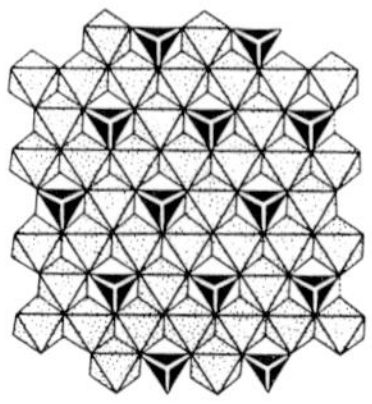

Fig. 1

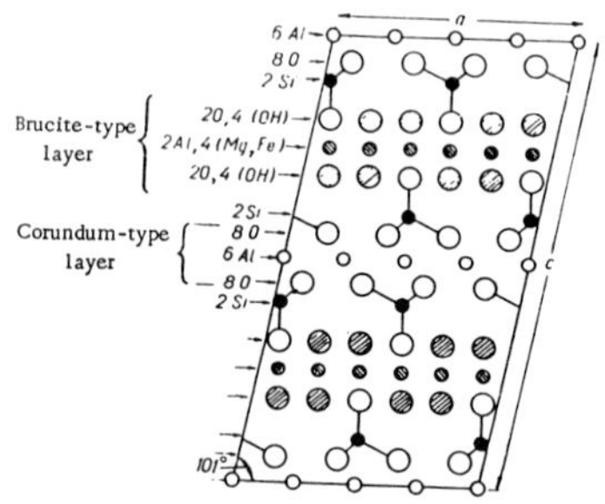

Fig. 2

Properties

Habit	Cleav.	Fract.	Twin.	Hardn.	Dens.	Colour	Transp.
platy, foliated	good (001)	uneven	(001) (100)	6.5	3.50	dark green	transparent

Refr. index/Reflect.	Birefr.	Luster	Streak	Melt.p.	CPI
n_α = 1.715 n_β = 1.720 n_γ = 1.725	(+) 2V = 45°-65°	pearly	grey		(SPI) 62

Figures

Fig. 1. The octahedral and tetrahedral sheet of the chloritoid structure (after Kostov, 1968).

Fig. 2. Chloritoid structure projected along the b axis (after Povarennykh, 1972).

Description

The chloritoid structure is probably a close-packed structure, and not a sheet structure. This structure is formed by alternating octahedral layers one of the brucite type and another of the corundum type, both parallel to (001), and linked by Si^tO_4 tetrahedra.

Crystallographic data (continued)

O_I	(8f)	x = 0.069 y = 0.085 z = -0.066	O_V	(8f)	x = -0.143 y = 0.598 z = -0.197
O_{II}	(8f)	x = 0.105 y = 0.347 z = -0.186	$0.85(OH)_I$	(8f)	x = -0.105 y = 0.585 z = -0.050
O_{III}	(8f)	x = 0.105 y = 0.847 z = -0.186	$0.85(OH)_{II}$	(8f)	x = 0.233 y = 0.585 z = -0.052
O_{IV}	(8f)	x = -0.143 y = 0.098 z = -0.197	Si	(8f)	x = 0.034 y = 0.090 z = -0.156

References

Kostov (1968) 365.

Povarennykh (1972) 428, 429.

Wyckoff (1968) Vol. 4, 363,364.

Zoltai + Stout (1984) 366.

References

Adams, F.S. (1954) *The birth and development of the geological sciences,* Dover Publications, New York.

Agricola, G. (Georg Bauer) (1546) De natura fossilium. Froben, Basel.

Ahrens, L. H. (1952) "The use of ionization potentials". *Geochim. Cosmochim. Acta, 2,* 155–169.

Akao, M. & Iawai, S. (1977) "The hydrogen bonding of hydromagnesite". *Acta Cryst., B33,* 1273–1275.

Akao, M., Marumo, F. & Iawai, S. (1974) "The crystal structure of hydromagnesite". *Acta Cryst., B30,* 2670–2672.

Alexander, V. D., Griffen, D. T. & Martin, T. J. (1986) "Crystal chemistry of some Fe- and Ti- poor dumortierites". *Amer. Miner., 71,* 786–794.

American Society for Testing and Materials (1957) "What can be done to improve alloy phase nomenclature?" A report of Nomenclature Subcommittee E-4 on Metallography. *ASTM Bulletin,* 226, 23–30.

Amirov, S. T., Iljukhin, V. V. & Belov, N. V. (1967) "The crystal structure of leonhardite (laumontite) $CaAl_2Si_4O_{12} \bullet nH_2O$ ($2 < n < 4$)". *Dokl. Akad. Nauk SSSR, 174,* 667–670. English translation *Dokl. Acad. Sci. URSS, Earth Sci. Section, 174,* 121–124.

Andersson, S. & O'Keeffe, M. (1977) "Body-centred cylinder packing and garnet structure". *Nature, 267,* 605–606.

Andress, K. R. & Gundermann, K. R. (1934) "Kristallhydrate. I. Die Struktur von Magnesiumchlorid- und Magnesium-bromidhexahydrat". *Zeit. Krist.,* (A), *87,* 345–369.

Andress, K. R. & Saffe, O. (1939) "Rontgenographische Untersuchung der Mischkristallreihe Karnallit-Bromkarnallit". *Zeit. Krist.,* (A), *101,* 451–469.

Appleman, D. E. & Evans, H. T. Jr. (1965) "The crystal structure of synthetic anhydrous carnotite, $K_2(UO_2)_2V_2O_8$ and its cesium analogue, $Cs_2(UO_2)_2V_2O_8$". *Amer. Miner., 50,* 825–842.

Araki, T. & Zoltai, T. (1967) "Refinement of the crystal structure of a glauberite". *Amer. Miner., 52,* 1272–1277.

Araki, T. & Zoltai, T. (1968) "The crystal structure of wavellite". *Zeit. Krist., 127,* 21–38.

Armbruster, T. (1986) "Role of Na in the structure of low-cordierite. A single-crystal X-ray study". *Amer. Miner., 71,* 746–757.

Artioli, G., Smith, J. V. & Pluth, J. J. (1986) "X-ray structure refinement of mesolite". *Acta Cryst., C42,* 937–942.

Avicenna (Ibn Siná) (980–1037) Quoted by A. S. Povarennykh (1972), pp. 4 and 12.

Bachmann, H. G. & Zemann, J. (1961) "Die Kristallstruktur von Linarit, $PbCuSO_4(OH)_2$". *Acta Cryst., 14,* 747–753.

Bacon, G. E. & Curry, N. A. (1956) "A neutron diffraction study of sodium sesquicarbonate". *Acta Cryst., 9,* 82–85.

Bacon, G. E. & Curry, N. A. (1962) "The water molecules in $CuSO_4.5H_2O$". *Proc. Royal Soc. Lond., A 266,* 95–108.

Ball, M. C. & Taylor, H. F. W. (1961) "The dehydration of brucite". *Miner. Mag., 32,* 754–766.

Barlow, W. (1883) "Probable nature of the internal symmetry of crystals". *Nature* (London) *29,* 186–188, 205–207.

Barrer, R. M. & Kerr, I. S. (1959) *Trans. Faraday Soc., 55,* 1915. Quoted by W. L. Bragg & G. F. Claringbull (1965), p. 351.

Baur, W. H. (1964a) "On the crystal chemistry of salt hydrates. III. The determination of the crystal structure of $FeSO_4 \bullet 7H_2O$ (melanterite)". *Acta Cryst., 17,* 1167–1174.

Baur, W. H. (1964b) "On the crystal chemistry of salt hydrates. IV. The refinement of the crystal structure of $MgSO_4 \bullet 7H_2O$ (epsomite)". *Acta Cryst., 17,* 1361–1369.

Belov, N. V. (1939) "Classification of closest and close packings". *Comptes Rendus (Doklady) de l'Académie des Sciences de l'URSS,* Vol. XXIII, N⁰ 2, 170–174.

Belov, N. V. (1947) *Structure of ionic crystals and metallic phases* (in Russian), Izd. Akad. Nauk SSSR, Moscow. Quoted by I. Kostov (1981), p. 707.

Belov, N. V. (1951) "Studies in structural mineralogy. II" (in Russian). *Mineralogitscheski Sbornik Lvovskoho Geologitscheskoho Obschestva,* N° 5, 13–36.

Belov, N. V. (1956) "Essays of structural mineralogy" (in Russian) *Mineral. Sb. (Lvov),* (10), 10–32.

Belov, N. V. (1963) *Crystal chemistry of large-cation silicates.* Consultants Bureau, New York. English translation of the Russian original (1961). Akademy of Sciences Press, Moscow.

Berry, L. G. & Mason, B. (1959) *Mineralogy. Concepts, descriptions, determinations.* W. H. Freeman and Company, San Francisco.

Berzelius, J. J. (1819) *Nouveau système de Mineralogie.* Méquignon-Marvis, Paris.

Bloss, F. D. (1971) *Crystallography and chemistry. An introduction.* Holt, Rinehart and Wilson, New York.

Bokii, G. B. (1954/1960) *Introduction to crystal chemistry.* United States Joint Publ. Research Service, New York (1960). English translation of the Russian original (1954). Izdatel'stvo Moskovskogo Universiteta.

Born, L. & Hellner, E. (1960) "A structural proposal for boulangerite". *Amer. Miner., 45,* 1266–1271.

Born, M. (1919) "A thermo-chemical application of the lattice theory". *Verh. Dtsch. Phys. Ges., 21,* 13–24.

Born, M. & Landé, A. (1918) "The calculation of the compressibility of cubic crystals from the space lattice theory". *Verh. Dtsch. Phys. Ges.*, *20*, 210–216.

Bragg, W. L. (1913) "The structure of some crystals as indicated by their diffraction of X-rays." *Proc. R. Soc. London Ser. A*, *89*, 248–277.

Bragg, W. L. (1929) "Atomic arrangement in the silicates". In *Crystal structure and chemical constitution. A general Faraday discussion held by the Faraday Society* (pp. 291–314).

Bragg, W. L. (1930) "The structure of the silicates". *Zeit. Krist.*, *74*, 237–305.

Bragg, W. L. (1964) "Minerals". *Proceedings of the Royal Institution of Great Britain*, *40*, Part I, N° 182, 64–81.

Bragg, W. L. & Claringbull, G. F. (1965) *Crystal structures of minerals*. G. Bell and Sons, London.

Brindley, G. W. (1961) "Crystal structure and thermal behaviour of some ceramic materials" (in Japanese). *J. Jap. Ass.*, *69*, C 189–194. Quoted by G. W. Brindley (1963), p. 8.

Brindley, G. W. (1963) "Crystallographic aspects of some decomposition and recrystallization reactions". In *Progress in ceramic science*, *Vol. 3* edited by J.E. Burke. Pergamon Press, Oxford (pp. 1–55).

Brown, C. J., Peiser, H. S. & Turner-Jones, A. (1949) "The crystal structure of sodium sesquicarbonate", *Acta Cryst.*, *2*, 167. Quoted in *Structure Reports* (1952), *12*, 238–240.

Brown, I. D. (1981) "The bond-valence method: an empirical approach to chemical structure and bonding". In *Structure and bonding in crystals II*, edited by M. O'Keeffe and A. Navrotsky. Academic Press, New York (pp. 1–30).

Buerger, M. J. (1947) "Derivative crystal structures". *J. Chem. Physics*, *15*, 1–16.

Buerger, M. J. (1967) (Private communication).

Buerger, M. J. (1971) *Introduction to crystal geometry*. McGraw-Hill, New York, p.5.

Buerger, M. J., Burnham, C. W. & Peacor, D. R. (1962) "Assessment of the several structures proposed for tourmaline". *Acta Cryst.*, *15*, 583–590.

Bunn, C. (1964) *Crystals. Their role in nature and in science*. Academic Press, New York.

Burke, J. H. (1966) *Origins of the science of crystals*. University of California Press, Berkeley.

Burns, R. G. & Burns, V. M. (1979) "Manganese oxides". In *Marine minerals*. Reviews in Mineralogy, Vol. 6, Mineralogical Society of America (pp. 1–46).

Bystrom, A. & Mason, B. (1943) "The crystal structure of braunite $3Mn_2O_3 \bullet MnSiO_3$". *Ark. Kemi, Min. Geol.*, *16*, N° 15, 1–8.

Caillère, S. & Hénin, S. (1961) "Palygorskite". In *The X-ray identification and crystal structures of clay minerals*, edited by G. Brown. Mineralogical Society, London (pp. 343–353).

Calleri, M., Gavetti, A., Ivaldi, G. & Rubbo, M. (1984) "Synthetic epsomite, $MgSO_4 \bullet 7H_2O$: Absolute configuration and surface features of the complementary {111} forms". *Acta Cryst.*, *B40*, 218–222.

Chaskolskaia, M. (1959) *Crystals*. Editions en Langues Etrangères, Moscow.

Christ, C. L., Clark, J. R. & Evans, H. T. Jr. (1958) "Borate minerals (III). The crystal structure of colemanite, $CaB_3O_4(OH)_3 \bullet H_2O$". *Acta Cryst.*, *11*, 761–770.

Christian, J. W. (1951) "A theory of the transformation in pure cobalt". *Proc. Roy. Soc. A*, *206*, 51–64.

Cid-Dresdner, H. (1965) "Determination and refinement of the crystal structure of turquoise, $CuAl_6(PO_4)_4(OH)_8 \bullet 4H_2O$". *Zeit. Krist.*, *121*, 87–113.

Cocco, G. (1962) "La struttura della mirabilite". *Rend. Accad. Naz. dei Lincei Ser. 8*, *32*, 690–698.

Cocco, G., Corazza, E. & Sabelli, C. (1965) "The crystal structure of glaubite, $CaNa_2(SO_4)_2$". *Zeit. Krist.*, *122*, 175–184.

Cooper, W. F., Larsen, F. K. & Coppens, P. (1973) "Electron population analysis of accurate diffraction data. V. Structure and one-center charge refinement of light-atom mineral kernite $Na_2B_4O_6(OH)_2 \bullet 3H_2O$". *Amer. Miner.*, *58*, 21–31.

Corbridge, D. E. C. (1971) "The structural chemistry of phosphates". *Bull. Soc. Miner. Cristallographie*, *94*, 271–299.

Cronstedt, F. (1758) *An essay towards a system of mineralogy* (in Swedish), Stockholm.

Dachs, H. von (1963) "Neutronen- und Röntgenuntersuchungen an Manganit, MnOOH". *Zeit. Krist.*, *118*, 303–326.

Dana, J. D. (1850) *The system of mineralogy*. Third edition John Wiley, New York.

Dana, J. D. (1854) *The system of mineralogy*. Fourth edition John Wiley, New York.

Deer, W. A., Howie, R. A. & Zussman, J. (1962) *Rock-forming minerals, Vol. 1. Ortho and ring silicates*. Longmans, London.

Deer, W. A., Howie, R. A. & Zussman, J. (1962) *Rock-forming minerals, Vol. 3. Sheet silicates*. Longmans, London.

Deer, W. A., Howie, R. A. & Zussman, J. (1962) *Rock-forming minerals, Vol. 5. Non-silicates*. Longmans, London.

De Jong, W. F. (1959) *General Crystallography. A brief compendium*. Freeman, San Francisco.

Dent Glasser, L. S., Glasser, F. P. & Taylor, H. F. (1962) "Topotatic reactions in inorganic oxy-compounds". *Quarterly Reviews. Chem. Soc. London*, *16*, 343–360.

Dollase, W. A. (1967) "The crystal structure at 220 °C of orthorhombic high tridymite from the Steinbach meteorite". *Acta Cryst.*, *23*, 617–623.

Dollase, W. A. (1968) "Refinement and comparison of the structures of zoisite and clinozoisite". *Amer. Miner.*, *53*, 1882–1898.

Dollase, W. A. (1971) "Refinement of the crystal structure of epidote, allanite and hancockite". *Amer. Min.*, *56*, 447–464.

Dollase, W. A. & Baur, W. H. (1976) "The superstructure of meteoritic low tridymite solved by computer simulation". *Amer. Miner.*, *61*, 971–978.

Donnay, J. D. H., Donnay, G., Cox, E. G., Kennard, O. & King, M. V. (1963) "Crystal data. Determinative tables". Second edition. *American Crystallographic Association*.

Donnay, J. D. H., Hellner, E. & Niggli, A. (1964) "Coordination polyhedra". *Zeit. Krist.*, *120*, 364–374.

Donnay, J. D. H., Hellner, E. & Niggli, A. (1966) "Symbolism for lattice complexes, revised by a Kiel Symposium". *Zeit. Krist.*, *123*, 255–262.

Eitel, W. (1964) *Silicate science, Vol. 1. Silicate structures*. Academic Press, New York.

Ernst, W. G. (1968) *Amphiboles*. Springer-Verlag, Berlin.

Ervin, G. (1952) "Structural interpretation of diaspore-corundum and boehmite-γ-Al_2O_3 transitions". *Acta Cryst.*, *5*, 103–108.

Evans, H. T. Jr. (1979) "The crystal structures of low chalcocite and djurleite". *Zeit. Krist.*, *150*, 299–320.

Evans, H. T. Jr. & Konnert, J. A. (1976) "A crystal structure refinement of covellite". *Amer. Miner.*, *61*, 996–1000.

Evans, R. C. (1939) *An introduction to crystal chemistry*. First edition. Cambridge University Press, Cambridge.

Evans, R. C. (1964) *An introduction to crystal chemistry*. Second edition. Cambridge University Press, Cambridge.

Ewald, P. & Hermman, C., Editors (1931) *Strukturbericht*, Vol. 1 (for 1913–1928). Akademische Verlagsgesellschaft MBH, Leipzig.

Ewing, F. J. (1935) *J. Chem. Phys.*, *3*, 203. Quoted by A. F. Wells (1962), p. 556.

Fedorov, E. S. (1901) *A course of crystallography* (in Russian). Rikker, St. Petersburgh. Quoted by B.K. Vainshtein. (1981) p. 28.

Fedorov, E. S. (1913) "Mineralogy". *Encyclopedia Assoc. Granat.*, *28*, 681. Quoted by Povarennyk (1972), p. 15.

Fejdi, P., Poullen, J. F. & Gasperin, M. (1980) *Bull. Miner.*, *103*, 135–138. Quoted in *Structure Reports* (1982), *46A*, 327.

Fersman, A. E. (1935) "The geochemistry of alkaline magmas" (in Russian). *Bull. Acad. Sci. URSS*, 1419–1424.

Figueiredo, M. O. (1973) "Tesselations and plane symmetry groups as applied to the derivation of close-packed binary layers". *Acta Cryst.*, *A29*, 234–243.

Figueiredo, M. O. (1976a) "Discriminatory characters for structure-type identification isopoint crystal structures". *III European Crystallographic Meeting*, Zurich (Abstract I-54-M).

Figueiredo, M. O. (1976b) (Private communication on condensed models).

Figueiredo, M. O. (1977) "Estabilidade e estrutura cristalina: o caso tipo das fases da sílica". *Comunicação dos Serviços Geológicos de Portugal, Tomo LXII*, 19–34.

Figueiredo, M. O. (1979a) *Características de empilhamento e modelos condensados das micas e filossilicatos afins*. Junta de Inv. Cient. do Ultramar. Estudos, Ensaios e Documentos, N° 131, Lisboa.

Figueiredo, M. O. (1979b) (Private communication on condensed models).

Figueiredo, M. O. (1981) "A structural model for feldspars and related compounds" (in Portuguese). Presented at the *VII Congresso Iberoamericano de Cristalografia*, Coimbra (Abstract).

Figueiredo, M. O. (1985) "An interplay of chemical and structural data in the classification of minerals". Presented at the *IX European Crystallographic Meeting*, Torino (Abstract).

Figueiredo, M. O. (1986) "A unified geometrical scheme for polytypism in phylosilicates". *Bull. Minéralogique*, *109*, 31–44.

Figueiredo, M. O. & Lima-de-Faria, J. (1977) "Theoretical approach to the derivation of condensed models of crystal structures based on square-type layers". *Acta Cryst.*, *A33*, 395–398.

Figueiredo, M. O. & Lima-de-Faria, J. (1978) "Condensed models of structures based on loose packings". *Zeit. Krist.*, *148*, 7–19.

Figueiredo, M. O. & Lima-de-Faria, J. (1983) "Condensed models of pyroxenes and amphiboles". *Garcia de Orta, Sér. Geologia*, *6*, 101–108.

Figueiredo, M. O. & Lima-de-Faria, J. (1991a) "Standard sheets for condensed models of crystal structures. III – $CuAl_2$-like intermetallic phases and affine structures". *Garcia de Orta, Sér. Geologia*, *14*, 29–35.

Figueiredo, M. O. & Lima-de-Faria, J. (1991b) "Standard sheets for condensed models of crystal structures. IV – Sillén phases and related structures". *Garcia de Orta, Sér. Geologia*, *14*, 37–43.

Finger, L. W. & Hazen, R. M. (1991) "Crystal chemistry of six-coordinated silicon: a key to understanding the earth's deep interior". *Acta Cryst.*, *B47*, 561–580.

Ford, W. E. (1932) *Dana's textbook of mineralogy*. Fourth edition. John Wiley & Sons, London.

Francombe, M. H. & Rooksby, H. P. (1959) "Structure transformations effected by the dehydration of diaspore, goethite and delta ferric oxide". *Clay Miner. Bull.*, *4*, 1–14.

Frank, F. C. & Kasper, J. S. (1958) "Complex alloy structures regarded as sphere packings. I. Definitions and basic principles". *Acta Cryst.*, *11*, 184–190.

Frondel, C. (1982) *The system of mineralogy, Vol III. Silica minerals*. John Wiley, New York.

Frondel, C. (1983) "An overview of crystallography in North America". In *Crystallography in North America*, edited by Dan McLachlan Jr. and Jenny P. Glusker. Amer. Cryst. Association, New York (pp. 1–24).

Frueh, A. J. Jr. (1959) "The structure of hessite, Ag_2Te". *Zeit. Krist.*, *112*, 44–52.

Galli, E. (1971) "Refinement of the crystal structure of stilbite". *Acta Cryst.*, *B27*, 833–841.

Garrido, J. & Orlando, J. (1946) *Los rayos-X y la estructura fina de los cristales*. Dossat, Madrid.

Geber (Jabir Ibn Hayan) (721–c.803) Quoted by J. W. Spronsen (1969), p. 25.

Ghose, S. (1964) "The crystal structure of hydrozincite, $Zn_5(OH)_6(CO_3)_2$". *Acta Cryst.*, *17*, 1051–1057.

Ghouse, K. M. (1968) "Refinement of the crystal structure of heat-treated monazite crystal". *Indian J. Pure and Applied Physics*, *6*, 265–268.

Giacovazzo, C., Monaco, H. L., Viterbo, D., Scordari, F., Gilli, G., Zanotti, G. & Catti, M. (1992) *Fundamentals of crystallography*. Published by Oxford University Press for the International Union of Crystallography.

Giuseppetti, G. & Tadini, C. (1983). *Neues Jb. Miner., Mh.*, 410–416. Quoted in *Structure Reports*, *50A*, p. 264.

Goble, R. J. (1985) "The relationship between crystal structure, bonding and cell dimensions in the copper sulfides". *Canad. Miner.*, *23*, 61–76.

Goldschmidt, V. M. (1926) "Geochemische Verteilungsgesetze der Elemente". *Skryfter det Norske Videnskaps Akad. Oslo*. I. Mat. Natur, Kl.

Goldschmidt, V. M. (1929) "Crystal structure and chemical constitution". In *Crystal structure and chemical constitution. A general Faraday discussion held by the Faraday Society* (pp. 253–283).

Gottardi, G. & Galli, E. (1985) *Natural zeolites*. Springer-Verlag, Berlin.

Gottardi, G. & Meier, W.M. (1963) "The cyrstal structure of dachiardite". *Zeit. Krist.*, *119*, 53–64.

Grigoriev, D.P. (1964) *Fundamentals of the constitution of minerals*. Israel Program for Scientific Translations, Jerusalem.

Groat, L. A., Randseff, M., Hawthorne, F. C., Ercitt, T. S., Sherriff, B. L. & Hartman, S. T. (1990) "The ambligonitemontebrasite series: Characterization by single-crystal structure refinement, infrared spectroscopy, and multinuclear MAS-NMR spectroscopy", *Amer. Miner.*, *75*, 992–1008.

Groth, P. H. R. von (1874) *Tabellarische Ubersicht der Mineralien nach ihren Kristallographysch-chemischen Beziehungen*. Braunschweig.

Haber, F. (1919) "Betrachtungen zur Theórie der Wärmetönung". *Verh. Dtsch. Phys. Ges.*, *21*, 750–768.

Hambley, T. W. & Taylor, J. C. (1984) "Neutron diffraction studies on natural heulandite and partially dehydrated heulandite". *J. Solid State Chem.*, *54*, 1–9.

Haüy, R.J. (1801), *Traité de Mineralogie*. 5 vols. Delance, Paris.

Hawthorne, F. C. (1983) "Graphical enumeration of polyhedral clusters". *Acta Cryst.*, *A39*, 724–736.

Hawthorne, F. C. (1984) "Towards a structural classification of minerals". Presented at the *XIII International Congress of Crystallography*, Hamburg (Abstract).

Hawthorne, F. C. (1985) "Towards a structural classification of minerals: The $^{VI}M^{IV}T_2|_n$ minerals". *Amer. Miner.*, *70*, 455–473.

Hellner, E. (1958) "The structural scheme for sulfide minerals". *The Journal of Geology*, *66*, 503–525.

Hellner, E. (1965) "Descriptive symbols for crystal-structure types and homeotypes based on lattice complexes". *Acta Cryst.*, *19*, 703–712.

Hellner, E. (1984) "Frameworks and a classification scheme for inorganic and intermetallic structure types". Presented at the *XIII International Congress of Crystallography*, Hamburg. *Acta Crystal.* (Abstract).

Hellner, E. (1986) "Einfürung in eine anorganische Strukturchemie". *Zeit. Krist.*, *175*, 227–248.

Hellner, E. & Leineweber, G. (1956) "Uber complex zusammengesetz sulfidische Erze. I. Sruktur des Bourmonits $CuPbSbS_3$ and Seligmannits, $CuPbAsS_3$". *Zeit. Krist.*, *107*, 150–154.

Hermann, C., Editor (1935) *Internationale Tabellen zur Bestimmung von Kristallstrukturen*. Gebrüder Bornträger, Berlin.

Hermann, C. (1960) "Zur Nomenklatur der Gitterkomplexe". *Zeit. Krist.*, *113*, 142–154.

Hermann, C., *et al.* (1937) See *Strukturbericht* (1937), Vol. II, p. 126.

Hey, M. H. (1955) *An index of mineral species and varieties arranged chemically*. British Museum, London.

Hooke, R. (1665) *Micrographia*. Jo. Martin and Js. Allestry for the Royal Society, London. Quoted by J. H. Burke, (1966), p. 38.

Iida, S. (1957) "Layer structures of magnetic oxides". *Journ. Physic, Soc. Japan*. *12*, 222–233.

Iitaka, Y. & Nowacki, W. (1962) "A redetermination of the crystal structure of galenobismutite, $PbBi_2S_4$". *Acta Cryst.*, *5*, 691–698.

Infeld, L. (1950) *Albert Einstein. His work and its influence on our world*. Scribner Publishers, New York.

Ingerson, E. (1955) "Methods and problems of geologic thermometry". *Economic Geology*, Fiftieth Anniversary Volume, Part I (1905–1955) pp. 341–410.

International Tables of Crystallography, Vol. A, Space-group symmetry, edited by T. Hahn. Published for the IUCr by Reidel Publishing Co., Dordrecht, Holland, 1983, and 1987 (Second edition).

Iskhakova, L. D., Trunov, V. K., Schegoleva, T. M., Ilyukhin, V. V. & Vedernikov, A. A. (1983) "Crystal structure of chalcantite $CuSO_4 \bullet 5H_2O$ grown under microgravity" (in Russian). *Kristallografiya*, *28*, 651–657. English translation *Soviet Physics-Crystallography* (1983), *28*, 383–386.

Ito, T. (1950) "The crystal structure of antigorite". In *X-ray studies on polymorphism*. Maruzen, Tokyo (pp. 160–167).

IUPAC (1990) *Nomenclature of inorganic chemistry*. Blackwell, Oxford.

Jarchow, O. (1965) "Atomanordnung und Strukturverfeinerung von Cancrinite". *Zeit. Krist.*, *122*, 407–422.

Johnson, N. E., Craig, J. R. & Rimstidt, J. D. (1988) "Crystal chemistry of tetrahedrite". *Amer. Miner.*, *73*, 389–397.

Jones, J. B. & Taylor, W. H. (1961) "The structure of orthoclase". *Acta Cryst.*, *14*, 443–456.

Joswig, W., Bartl, H. & Fuess, H. (1984) "Structure refinement of scolecite by neutron diffraction". *Zeit. Krist.*, *166*, 219–223.

Kapustinskii, A. F. (1933) "General formula for the lattice of crystals of arbitrary structure". *Z. Phys. Chem. Abt.*, *22*, 257–260.

Kepler, J. (1611) *Strena Seu de Nive Sexangula*. Gottfried Tampach, Frankfurt.

Kern, R. & Gindt, R. (1958) "Contribution à l'étude des accolements réguliers des feldspaths potassiques et des plagioclases". *Bull. Soc. Franc. Min. Crist.*, *81*, 263–266.

Kerrick, D. M. (1990) *The Al_2SiO_5 polymorphs*. Reviews in Mineralogy, Vol. 22. Miner. Soc. America.

Kitaigorodskii, A. I. (1955) *Organic chemical crystallography* (in Russian). Press of the Academy of Sciences of the URSS, Moscow. English translation and revision (1961), Consultants Bureau Enterprises, New York.

Klein, C. & Hurlbut, C. S. Jr. (1985) *Manual of Mineralogy*. John Wiley, New York.

Kniep, R. & Mootz, D. (1973) "Metavariscite. A redetermination of its crystal structure". *Acta Cryst.*, *B29*, 2292–2294.

Kniep, R., Mootz, D. & Vegas, A. (1977) "Variscite". *Acta Cryst.*, *B33*, 263–265.

Kondrasev, J. M. D. & Fedorova, N. N. (1954) "The crystal structure of $CoHO_2$" (in Russian). *Dokl. Akad. Nauk.*, *94*, 2, 229–231. Quoted in *Structure Reports* (1961), *18*, 515–516.

Koptsik, V. A. & Belov, N. V. (1977) "Towards a theory of translationally centered Fedorov groups" (in Russian). *Kristallografiya*, *22*, 1140–1146. English translation *Soviet Physics-crystallography*, (1977), *22*, 650–653.

Kostov, I. (1954) "A note on a more rational classification of minerals" (in Russian). *Zap. Vses. Min. obshch.* Part 83, N° 4, 328–347.

Kostov, I. (1968) *Mineralogy*. Oliver and Boyd, Edinburgh.

Kostov, I. (1981) "Structural patterns and structural genetic trends in minerals". *Kristallografiya*, *26*, 1244–1247. English translation *Soviet Phys. Crystallogr.*, (1981), *26*, 707–708.

Krebs, H. (1968) *Fundamentals of inorganic chemistry*. English translation, McGraw Hill, London.

Kripyakevich, P. I. (1963) "A systematic classification of types of intermetallic structures" (In Russian). *Zhurnal Strukturnoi Khimii*, *4*, N° 1, p. 117 and N° 2 p. 282. English translation, Consultants Bureau Enterprises, New York (1964).

Kripyakevich, P. I. (1973) "Loose packings of dense layers". *Kristallografiya*, *18*, 730–736. English translation *Sov. Phys.*

Crystallography, (1974), *18*, 460–463.
Landé, A. (1920) "Bemerkungen über die Grosse der Atome". *Z. Physik 1*, 87–89.
Langlet, G. A. (1975) "Extension of the FIGATOM program to the automatic plotting of layers in close-packed structures". *J. Appl. Cryst.*, *8*, 515–519.
Langlet, G. A. (1976) "Extension of the FIGATOM program to the automatic plotting of layers in close-packed structures: erratum". *J. Appl. Cryst.*, *9*, 320.
Langlet, G. A., Figueiredo, M. O. & Lima-de-Faria, J. (1977) "Determination of spherical voids in layered structures (Void program)". *J. Applied Crystallography*, *10*, 21–23.
Lapparent, A. (1884) *Mineralogie*, Paris.
Laves, F. (1930) "XVI. Die Bau-Zusammenhänge innerhalb der Kristallstrukturen. I. Teil". *Zeit. Krist.*, *73*, 202–265.
Laves, F. (1956) "Crystal structure and atomic size". In *Theory of alloy phases*, edited by the American Society for Metals. Cleveland, Ohio (pp. 124–198).
Laves, F. (1963) "Factors governing the structure of intermetallic phases". In *Advances in X-ray analysis*, Vol. 6, edited by W. M. Mueller and M. Fay. Plenum Press, New York (pp. 43–61).
Lee, A. van der & Boer, J. L. de (1993) "Redetermination of the structure of hessite, Ag_2Te". III. *Acta Cryst.*, *C49*, 1444–1446.
Liebau, F. (1956) "Bemerkungen zur Systematik der Kristallstrukturen von Silikaten mit hochkondensierten Anionen". *Physik. Chem.*, *206*, 73–92.
Liebau, F. (1966) "Die Kristallchemie der Phosphate". *Fortschr. Miner.*, *42*, 266–302.
Liebau, F. (1982) "Classification of silicates". *Rev. Mineral.*, *5*, 1–24.
Liebau, F. (1985) *Structural chemistry of silicates. Structure bonding, and classification*. Springer-Verlag, Berlin.
Lima-de-Faria, J. (1963) "Dehydration of goethite and diaspore". *Zeit. Krist.*, *119*, 176–203.
Lima-de-Faria, J. (1965a) "A condensed way of representing inorganic close-packed structures". *Zeit. Krist.*, *122*, 346–358.
Lima-de-Faria, J. (1965b) "Systematic derivation of inorganic close-packed structures: AX and AX_2 compounds, sequence of equal layers". *Zeit. Krist.*, *122*, 359–374.
Lima-de-Faria, J. (1966) "Space group representation in condensed models of inorganic close-packed structures". *Nature*, *211*, 281.
Lima-de-Faria, J. (1967) "Anomalous orientations of cubic close packing in the dehydration of goethite in an inert atmosphere". *Acta Cryst.*, *23*, 733–736.
Lima-de-Faria, J. (1978a) "Rules governing the layer organization of inorganic crystal structures". *Zeit. Krist.*, *148*, 1–5.
Lima-de-Faria, J. (1978b) "Improved notation for inorganic group structures derived from atomic structures by coalescence of atoms". *R. Iberoam. Crist. Miner. Metalogen.*, *1*, 47–52.
Lima-de-Faria, J. (1983) "A proposal for a structural classification of minerals". *Garcia de Orta, Sér. Geologia*, *6*, 1–14.
Lima-de-Faria, J. (1986) "The need of a structural classification of minerals". *Rendiconti Soc. Ital. Miner. Petrolog.*, *41*, 157–179.
Lima-de-Faria, J. (1988a) "The two kinds of ideal analogues of mineral structures and the Laves principles of stability". *Garcia de Orta, Sér. Geologia*, *11*, 33–42.
Lima-de-Faria, J. (1988b) "The hierarchy of symmetry". Presented at the *XI European Crystallographic Meeting* (Vienna) (Abstract). Published in *Zeit. Krist.* (1988), *185*, 286.
Lima-de-Faria, J. (1990) "The packing analogues of some carbonates and borates". *Acta Cryst.*, *A46*. C267 Supplement (Abstract).
Lima-de-Faria, J. (1991) "On the problem of the measure of the symmetry of crystal structures". *Garcia de Orta, Sér. Geologia*, *14*, 45–50.
Lima-de-Faria, J., Buerger, M. J., Glusker, J. P., Megaw, H. D., Moore, P. B., Senechal, M. & Wooster, W. A. (1990) *Historical Atlas of Crystallography*, edited by J. Lima-de-Faria. Published by Kluwer Academic Publishers, Dordrecht. Holland, for the International Union of Crystallography.
Lima-de-Faria, J. & Figueiredo, M. O. (1969) "Systematic derivation of inorganic basic structure types: X_mY_n and A_mX_n compounds, X and Y in cubic or hexagonal close packing, A in octahedral voids". *Zeit. Krist.*, *130*, 54–67.
Lima-de-Faria, J. & Figueiredo, M. O. (1976) "Classification, notation, and ordering on a table of inorganic structure types". *J. Solid State Chem.*, *16*, 7–20.
Lima-de-Faria, J. & Figueiredo, M. O. (1978) "General chart of inorganic structural units and building units". *Garcia de Orta. Sér. Geologia*, *2*, 69–76.
Lima-de-Faria, J. & Figueiredo, M. O. (1990 a) "Standard sheets for condensed models of crystal structures. I. Structures based on close packings". *Garcia de Orta, Sér. Geologia*, *13*, 43–58.
Lima-de-Faria, J. & Figueiredo, M. O. (1990 b) "Standard sheets for condensed models of crystal structures. II. Pyroxenes and amphiboles". *Garcia de Orta, Sér. Geologia*, *13*, 59–72.
Lima-de-Faria, J., Hellner, E., Liebau, F., Makovicky, E. & Parthé, E. (1990) "Nomenclature of inorganic structure types". Report of the International Union of Crystallography Comission on Crystallographic Nomenclature Sub-Committee on the Nomenclature of inorganic structure types. *Acta Cryst.*, *A46*, 1–11.
Lindberg, M. L. & Christ, C. L. (1959) "Crystal structures of the isostructural minerals lazulite, scorzalite and barbosalite". *Acta Cryst.*, *12*, 695–697.
Linnaeus, C. von (1735) *Systema naturae, sive regna tria naturae systematice proposita per classes, ordinis, genera et species*. Theodorum Haak, Leiden.
Loeb, A. L. (1970) "A systematic survey of cubic crystal structures". *J. Solid State Chem.*, *1*, 237–267.
Lonsdale, K. (1966) "A new kind of twinning". *Acta Cryst.*, *21*, 5–7.
Machatschki, F. (1928) "Zur Frage der Struktur und Konstitution der Feldspate". Cbl. Min. Geol. Paläont., *Abteilung A. Miner. Petr.*, 97–104.
Machatschki, F. (1947) "Konstitutionsformeln für den festen Zustand". *Monatsch. Chem.*, *77*, 333–342.
Machatschki, F. (1953) *Spezielle Mineralogie auf geochemischer Grundlage mit einem Anhang ein kristallchemisches Mineralsystem*. Springer-Verlag, Wien.
Madelung, E. (1918) "Das elektrische Feld in Systemen von

regelmässig angeordeneten Punktladungen". *Phys. Z.*, *19*, 524–533.

Makovicky, E. (1983) "Archetypes structure building principles and homologous series for sulphosalts of As, Sb and Bi". *VIII European Crystallography Congress*, Liege (Abstract).

Makovicky, E. (1985) "Cyclically twinned sulphosalt structures and their approximate analogues". *Zeit. Krist.*, *173*, 1–23.

Matkovich, V. I., Giese, R. F. Jr. & Economy, J. (1965) "Packing of B_{12} groups in boron and boride structures". *Zeit. Kris.*, *122*, 116–130.

Mazzi, F., Galli, E. & Gottardi, G. (1976) "The crystal structure of tetragonal leucite". *Amer. Miner.*, *61*, 108–115.

McDonald, W. S. & Cruischank, D. W. J. (1967) "Refinement of the structure of hemimorphite". *Zeit. Krist.*, *124*, 180–191.

Megaw, H. D. (1973) *Crystal structures. A working approach.* W. B. Saunders Company, Philadelphia.

Meier, W. M. & Olson, D. H. (1978) "Atlas of zeolite structure types". Structure Commission of the International Zeolite Association. Quoted by J. V. Smith (1982), 165–169.

Merkle, A. B. & Slaughter, M. (1968) "Determination and refinement of the structure of heulandite". *Amer. Miner.*, *53*, 1120–1138.

Merlino, S., Galli, E. & Alberti, A. (1975) "The crystal structure of levyne". *Tschermaks Min. Petrogr. Mitt.*, *22*, 117–129.

Meulendijk, P. N. (1956) "An X-ray study of glaubersalt". *Proc. K. Ned. Akad. Wet.*, *B59*, 493–495. Quoted in *Structure Reports* (1963), *20*, 345–346.

Moore, P. B. (1973) "Pegmatite phosphates. Descriptive mineralogy and crystal chemistry". *Mineral. Rec.*, *4*, 103–130.

Moore, P. B. (1992) "Betapakdalite unmasked, and a comment on bond valences". *Aust. J. Chem.*, *45*, 1335–1354.

Moore, P. B. (1993) (Private communication).

Moore, P. B. & Araki, T. (1972) "Atomic arrangement of merwinite, $Ca_3Mg[SiO_4]_2$, an unusual dense-packed structure of geophysical interest". *Amer. Miner.*, *57*, 1355–1374.

Moore, P. B. & Araki, T. (1976) "Braunite: its structure and relationship to bixbyite, and some insights on the geneology of fluorite derivative structures". *Amer. Miner.*, *61*, 1226–1240.

Moore, P. B. & Araki, T. (1977) "Gerstmannite, a new zinc silicate mineral and a novel cubic close-packed oxide structure". *Amer. Miner.*, *62*, 51–59.

Moore, P. B. & Araki, T. (1978) *Neus Jb. Min. Ab.*, *132*, 231–241.

Mori, H. & Ito, T. (1950) "The structure of vivianite and symplesite". *Acta Cryst.*, *3*, 1–6.

Mumpton, F. A. (1981) "Natural zeolites". In *Mineralogy and Geology of natural zeolites*. Reviews in Mineralogy, Vol. 4. Amer. Soc. America (pp. 1–17).

Náray-Szabó, St. v. (1930) "Ein auf der Kristallstruktur basierendes Sylicatsystem". *Zeit. Phys., Abteilung B*, *2*, 356–377.

Náray-Szabó, St. v. (1943) "Der Strukturtyp des Perovskits ($CaTiO_3$)". *Naturwiss.*, *31*, 202.

Nickel, E. H. & Nichols M. C. (1991) *Mineral reference manual.* Van Nostrand Reinhold, New York.

Niggli, P. (1919) *Geometrische Kristallographie des Diskontinuums*. Gebr. Bornträger, Leipzig.

Niggli, P. (1945) *Grundlagen der Stereochemie*. Verlag Birkhäuser, Basel. French translation "Les bases de la Stéréochimie". Dunod, Paris (1952).

Nithollon, P. (1955) *Structure cristalline de la cancrinite*. Publications Scientifiques et Techniques du Ministère de l'Air, France, N. T. 53.

Nord, A. G. (1973). *Acta Chem. Scand.*, *27*, 814–822. Quoted in *Structure Reports* (1975), *39A*, 306.

Okada, K. & Ossaka, J. (1980) "Structures of potassium sodium sulphate and tripotassium disulphate". *Acta Cryst.*, *B36*, 919–921.

Pabst, A. (1950) "A structural classification of fluoaluminates". *Amer. Miner.*, *35*, 149–165.

Palache, C., Berman, H. & Frondel, C. (1944) *The system of mineralogy of James Dawght Dana and Edward Salisbury Dana. Vol. 1. Elements, sulfides, sulfosalts, oxides*. Seventh edition. John Wiley, London.

Palache, C., Berman, H. & Frondel, C. (1951) *The system of mineralogy of James Dawght Dana and Edward Salisbury Dana. Vol. 2. Halides, nitrates, borates, carbonates, sulfates, phosphates, arsenates, tungstates, molybdates, etc.* Seventh edition. John Wiley, London.

Papike, J. J. & Cameron, M. (1976) *Review of Geophysics and Space Physics*, *14*, 74. Quoted by C. Klein & C. S. Hurlbut Jr. (1985), 456.

Papike, J. J., Prewitt, C. T., Sueno, S. & Cameron, M. (1973) "Pyroxenes: comparisons of real and ideal structural topologies". *Zeit. Krist.*, *138*, 254–273.

Papike, P. P. (1987) "Chemistry of the rock-forming silicates: ortho, ring and single-chain structures". *Review of Geology*, *25*, 1483–1526.

Papike, P. P. & Zoltai, T. (1967) "Ordering of tetrahedral aluminium in phrenite $Ca_2(Al,Fe^{+3})Si_3AlO_{10}(OH_2)$". *Amer. Miner.*, *52*, 974–984.

Parthé, E. (1964) *Crystal chemistry of tetrahedral structures.* Gordon and Breach, New York.

Parthé, E. (1980) "Crystal-chemical formulae for simple inorganic crystal structures". *Acta Cryst.*, *B36*, 1–7.

Parthé, E. (1990) *Elements of inorganic structural chemistry. A course on selected topics.* K. Sutter Parthé, Petit-Lanay, Switzerland.

Pauling, L. (1927) "The sizes of ions and the structure of ionic crystals". *J. Amer. Chem. Soc.*, *49*, 765–790.

Pauling, L. (1928) *Proc. Nat. Acad. Sci., USA*, *14*, 603–606. Quoted in N. V. Belov (1936) "Classification of closest and close packings". *Comptes Rendus (Doklady) de l'Académie des Sciences de l'URSS*, *XXIII*, No. 2, 170–174.

Pauling, L. (1929) "The principles determining the structure of complex ionic crystals". *J. Amer. Chem. Soc.*, *51*, 1010–1026.

Pauling, L. (1960) *The nature of the chemical bond.* Cornell University Press, London.

Pearson, W. B. (1967) *Handbook of lattice spacings and structures of metals and alloys*, Vol. 2. Pergamon Press, Oxford.

Pearson, W. B. (1972) *The crystal chemistry and physics of metals and alloys.* Wiley-Interscience, New York.

Petrova, I. V., Kaplunnik, L. N., Bortnikov, N. S., Pobedimskaya, E. A. & Belov, N. V. (1978) "The crystal structure of synthetic robinsonite" (in Russian). *Dokl. Acad. Nauk SSSR*, *241*, 88–90.

Plinius (the older) Gajus Secundus (77 A. D.) *Natural history* (in Latin). Quoted by Povarennykh (1972), p. 12.

Povarennykh, A. S. (1972) *Crystal chemical classification of minerals*. English translation by Plenum Press, New York

(2 Vols.) of the Russian edition (1966), Naukova Dunka, Kiev.

Putnis, A. & Angel, R. J. (1985) "Al, Si ordering in cordierite using «magic angle spinning» NMR. II. Models of Al, Si order from NMR data". *Phys. Chem. Minerals*, *12*, 217–222.

Prewitt, C. T. (1967) "Refinement of the structure of pectolite". *Zeit. Krist.*, *125*, 298–316.

Prewitt, C. T. & Buerger, M. J. (1963) "Comparison of the crystal structure of wollastonite and pectolite". *Miner. Soc. Amer. Sp. Paper 1*, 293–302.

Radoslovich, E. W. (1963) "The cell dimensions and symmetry of layer-lattice silicates. IV Interatomic forces". *Amer. Miner.*, *48*, 76–99.

Ramdohr, P. & Strunz, H. (1978/1980) *Klockmanns Lehrbuch der Mineralogie*. Sixteenth edition. Ferdinand Enke Verlag, Stuttgart.

Ribbe, P. H. (1979) "The structure of a strained intermediate microcline in cryptoperthitic association with twinned plagioclase". *Amer. Miner.*, *64*, 402–408.

Ribbe, P. H., Gibbs, G. V. & Norris, W. J. (1968) "Cation and anion substitutions in the humite minerals". *Miner. Mag.*, *36*, 966–975.

Rinaldi, R., Pluth, J. J. & Smith, J. V. (1974) "Zeolites of the phillipsite family. Refinement of the crystal structures of phillipsite and harmotome". *Acta Cryst.*, *B30*, 2426–2433.

Roberts, W. L., Rapp, G. R. Jr. & Weber, J. (1974) *Encyclopedia of minerals*. Van Nostrand Reinhold & Company, New York.

Robinson, P. D., Fang, J. H. & Ohya, Y. (1972) "The crystal structure of kainite". *Amer. Miner.*, *57*, 1325–1332.

Ruben, H. W., Templeton, D. H., Rosenstein, R. D. & Olovsson, I. (1961) "Crystal structure and entropy of sodium sulphate decahydrate". *Amer. Chem. Soc.*, *83*, 820–824.

Sanderson, R. T. (1960) *Chemical periodicity*. Reinhold Publishing Corporation, New York.

Schlatti, M., Sahl, K., Zemann, A. & Zemann, J. (1970) "Die Kristallstruktur des Polyhalits, $K_2Ca_2Mg[SO_4]_2 \bullet 2H_2O$". *Tschermaks Min. u. Petrogr. Mitt.*, *14*, 75–86.

Schlemper, E. O., Gupta, P. K. S. & Zoltai, T. (1985) "Refinement of the structure of carnallite, $Mg(H_2O)_6KCl_3$". *Amer. Miner.*, *70*, 1309–1313.

Schubert, K. (1964) *Kristallstrukturen zweikomponentiger Phasen*. Springer-Verlag, Berlin.

Shannon, R. D. & Prewitt, C. T. (1969) "Effective ionic radii in oxides and fluorides". *Acta Cryst.*, *B25*, 925–946.

Shubnikov, A. V. & Koptsik, V. A. (1972) *Symmetry in science and art* (in Russian). Naüka Press, Moscow. English translation Plenum Press, New York (1974).

Smirnova, N. L. (1956) "Structure types with atomic close packings. Possible structure types for the composition AB_3" (in Russian). *Kristallographiya, Acad. Sc. URSS. 1*, 165–170. English translation *Soviet Phys.-Crystallography*, (1956), *1*, 128–131.

Smith, D. K., Gruner, J. W. & Lipscomb, W. N. (1957) "The crystal structure of uranophane $[Ca(H_3O)_2](UO_2)_2(SiO_4)_3 \bullet 3H_2O$". *Amer. Miner.*, *42*, 594–618.

Smith, J. V. (1982) *Geometrical and structural crystallography*. John Wiley, New York.

Sololeva, L. P. & Belov, N. V. (1964) "Precise determination of the crystal structure of bertrandite $Be_4[SiO_7](OH)_2$" (in Russian). *Kristallogr.*, *9*, 551–553. English translation *Soviet Phys., Crystallography* (1965), *9*, 458–460.

Spice, J. E. (1964) *Chemical binding and structure*. Pergamon Press, Oxford.

Spronsen, J. W. van (1969) *The periodic system of the chemical elements. An history of the first hundred years*. Elsevier, New York (p. 1).

Srikrishnan, T. & Novacki, W. (1974) "A redetermination of the crystal structure of cosalite, $Pb_2Bi_2S_5$". *Zeit. Krist.*, *140*, 114–136.

Steele, I. M. & Pluth, J. J. (1990) "Crystal structure of synthetic joshiokaite, a stuffed derivative of the tridymite structure". *Amer. Miner.*, *75*, 1186–1191.

Steinfink, H. (1962) "The crystal structure of the zeolite phillipsite". *Acta Crystal.*, *15*, 644–651.

Structure Reports (1951) Vol. 11, (1952) Vol. 12, (1954) Vol. 13, (1955) Vol. 9, (1959) Vol. 16, (1961) Vol. 18, (1963) Vol. 17, Vol. 20, (1964) Vol. 21, (1965) Vol. 23, (1969) Vol. 26, (1971) Vol. 27, (1975) Vol. 32A, (1975) Vol. 33A, (1975) Vol.39A, (1978) Vol. 42A, (1982) Vol. 46A, (1988) Vol. 50A, (1989) Vol. 49A. International Union of Crystallography, Utrecht-Dordrecht.

Strukturbericht (1931) Vol. 1, (1937) Vol. 2, (1937) Vol. 3, (1943) Vol. 7. Akademische Verlagsgesellschaft, Leipzig.

Strunz, H. (1982) *Mineralogische Tabellen*. Eighth edition. Akademische Verlagsgesellschaft, Leipzig.

Taga, T. (1969) "Crystal structure of $Na_2CO_3 \bullet 10H_2O$". *Acta Cryst.*, *B25*, 2656–2658.

Takéuchi, Y., Ozawa, T., Ito, T., Araki, T., Zoltai, T. & Finney, J. J. (1974) "The $B_2Si_8O_{30}$ groups of tetrahedra in axinite and comments on the deformation of Si tetrahedra in silicates". *Zeit. Krist.*, *140*, 289–312.

Tendeloo, G., van Wenk, H. R. & Gronsky, R. (1985) "Modulated structures in calcium dolomite. A study by electron microscopy". *Phys. Chem. Miner.*, *12*, 333–341.

Theophrastus (372–287 B.C.) Quoted by A. S. Povarennykh (1972), pp. 4 and 12.

Thompson, J. B. Jr. (1970) "Geometrical possibilities for amphiboles structures: model biopyriboles". *Amer. Miner.*, *55*, 292–293.

Uklonskii, A. S. (1940) *Mineralogy* (in Russian). Gostoptekhizdat, Moscow.

Vainshtein, (1981) *Modern Crystallography. I. Symmetry of crystals. Methods of structural crystallography*. Springer-Verlag, Berlin.

Wadsley, A. D. (1952) "The structure of lithiophorite". *Acta Cryst.*, *5*, 676–680.

Wang, R., Bradley, W. F. & Steinfink, H. (1965) "The crystal structure of alunite". *Acta Cryst.*, *18*, 249–252.

Wasastjerna, J. A. (1923) *Soc. Sci. Ferm. Comm. Phys. Math.*, *38*, 1 Quoted by L. Pauling (1960) p. 517.

Wells, A. F. (1962) *Structural inorganic chemistry*. Third edition. Clarendon Press, Oxford.

Werner, A. G. (1774) *Von den ausserlichen Kennzeichen der Fossilien*. Siegfried Lebrecht Crusius, Leipzig.

Werner, A. G. (1817) *Abraham Gottlob Werner's letztes mineral System*, Freiberg. Quoted by A. D. Adams (1954) p. 204.

Whewell, W. (1837) *History of the inductive sciences*. First edition. London, *3*, 227. Quoted by C. Frondel, (1983), p. 12.

Whittaker, E. J. W. (1956) "The structure of chrysotile. II.

Clino-chrysotile ". *Acta Cryst.*, *9*, 855–862.

Winter, J. K. & Ghose, S. (1979) "Thermal expansion and high-temperature crystal chemistry of the Al_2SiO_5 polymorphs". *Amer. Miner.*, *64*, 573–586.

Winter, J. K., Ghose, S. & Okamura, F. P. (1977) "A high-temperature study of the thermal expansion and the anisotropy of the sodium atom in low albite". *Amer. Miner.*, *62*, 921–931.

Wuensch, B. J. (1974) "Sulfide crystal chemistry". In *Sulfide mineralogy, short course notes*. Vol. 1 Edited by P. H. Ribbe. Miner. Soc. America, W-21–W-44.

Wyckoff, R. W. G. (1963) *Crystal structures*, Vol. 1. Second edition. John Wiley Interscience Publishers, New York.

Wyckoff, R. W. G. (1964) *Crystal structures, Vol. 2. Inorganic compounds* RX_n, R_nMX_2, R_nMX_3. Second edition. John Wiley Interscience Publishers, New York.

Wyckoff, R. W. G. (1965) *Crystal structures, Vol. 3. Inorganic compounds* $R_x(MX_4)_y$, $R_x(M_nX_p)_y$, *hydrates and ammoniates*. Second edition. John Wiley Interscience Publishers, New York.

Wyckoff, R. W. G. (1968) *Crystal structures, Vol. 4. Miscellaneous inorganic compounds, silicates, and basic structural information*. Second edition. John Wiley Interscience Publishers, New York.

Yakubovich, O. V. & Simonov, M. A. (1985) "Refined crystal structure of zeolite laumontite $Ca(H_2O)_2 \bullet {}_8[Al_2Si_4O_{12}]O \bullet 5H_2O$" (in Russian). *Kristallografiya*, *30*, 1072–1076. English translation *Soviet Physics, Crystallography* (1985), *30*, 624–626.

Yamaguchi, G. & Suzuki, K. (1967) "Structural analyses of merwinite". *J. Ceram. Assoc. Japan*, *75*, 220–229. Quoted in *Structure Reports* (1975), *32A*, 431–433.

Zemann, J. (1969) "Part I. Crystallography". In *Introduction to mineralogy, crystallography and petrology*, edited by C. W. Correns. Springer-Verlag, Berlin (pp.3–178).

Zoltai, T. (1960) "Classification of silicates and other minerals with tetrahedral structures". *Amer. Miner.*, *45*, 960–973.

Zoltai, T. (1974) *Systematics of simple sulfide structures*. University of Minnesota, Minneapolis (Private communication).

Zoltai, T. (1975) *Notes on the systematics of close-packed silicate structures*. University of Minnesota, Minneapolis (Private communication).

Zoltai, T. (1977) *Crystal structures*. University of Minnesota, Minneapolis (Private communication).

Zoltai, T. & Stout, J. H. (1984) *Mineralogy. Concepts and principles*. Burgess Publishing Company, Minneapolis, U.S.A.

Mineral index

(= means isotypic, ≅ distortion derivative)

Subject index

Author index